Stress

Stress

From Synapse to Syndrome

edited by

S. Clare Stanford

and

Peter Salmon

Consultant Editor
Jeffrey A. Gray

ACADEMIC PRESS
Harcourt Brace & Company, Publishers
London   San Diego   New York   Boston
Sydney   Tokyo   Toronto

ACADEMIC PRESS LIMITED
24–28 Oval Road
London NW1 7DX

US edition published by
ACADEMIC PRESS INC
San Diego, CA 92101

A catalogue record for this book is available from the British Library

ISBN 0–12–663370–3

Typeset by Phoenix Photosetting, Chatham, Kent
Printed and bound in Great Britain by
the University Printing House, Cambridge

This book is dedicated to our senior colleagues,
Marianne Fillenz and Jeffrey Gray,
who inspired a line of thought

Contents

Part One
General introduction

1 The stress response
Holger Ursin and Miranda Olff

2 The role of life events in the aetiology of depressive and anxiety disorders
George W. Brown

Part Two
Clinical effects of stress

3 Stress and psychiatric disorder: reconciling social and biological approaches
Paul Glue, David Nutt and Nick Coupland

4 Stress and the cardiovascular system: central and peripheral physiological
  mechanisms
Béla Bohus and Jaap M. Koolhaas

Part Three
Animal models of stress coping and resistance

6 Animal models of stress: an overview
Paul Willner

7 Coping with stress
Robert Dantzer

8 Stress and behavioural inhibition
Neil McNaughton

9 Learned helplessness: relationships with fear and anxiety
Steven F. Maier

Part Four
Neurochemistry of stress: Introduction to techniques

10 Neurochemistry of stress: introduction to techniques
Marianne Fillenz

11 Monoamines in response and adaptation to stress
S. Clare Stanford

12 The role of GABA in the regulation of the stress response
Win Sutanto and E. Ron de Kloet

13 Role(s) of neuropeptides in responding and adaptation to stress: A focus on corticotropin-releasing factor and opioid peptides
Diane M. Hayden-Hixson and Charles B. Nemeroff

Part Five
Future prospects

14 Emotional effects of physical exercise
Peter Salmon

15 Posttraumatic stress disorder
David Sturgeon

Contributors

Béla Bohus Department of Animal Physiology, University of Groningen, Haren, The Netherlands

George W. Brown Department of Social Policy and Social Science, Royal Holloway and Bedford New College, University of London, Egham Hill, UK

Nick Coupland Psychopharmacology Unit, School of Medical Sciences, University of Bristol, Bristol, BS8 1TD, UK

Robert Dantzer Unité de Recherches de Neurobiologie Integrative, Institut National de la Santé et de la Recherche Medicale, Institut de la Recherche Agronomique, Bordeaux Cedex, France

Marianne Fillenz University Laboratory of Physiology, University of Oxford, Oxford, UK

Paul Glue Schering-Plough Research Institute, Kenilworth, New Jersey, USA

Diane M. Hayden-Hixson Integrated Toxicology Program & Department of Pharmacology, Duke University Medical Center, Durham, North Carolina, USA

E. Ron de Kloet Division of Medical Pharmacology, Leiden-Amsterdam Center for Drug Research, University of Leiden, 2300 RA Leiden, The Netherlands

Jaap M. Koolhaas Department of Animal Physiology, University of Groningen, Haren, The Netherlands

Steven F. Maier Department of Psychology, University of Colorado, Boulder, Colorado, USA

Neil McNaughton Department of Psychology and Centre for Neuroscience, University of Otago, Dunedin, New Zealand

Charles B. Nemeroff Department of Psychiatry, Emory University School of Medicine, Atlanta, Georgia, USA

David Nutt Psychopharmacology Unit, School of Medical Sciences, University of Bristol, Bristol, BS8 1TD, UK

Miranda Olff Department of Clinical Psychology and Health, University of Utrecht, Utrecht, The Netherlands

Peter Salmon Department of Clinical Psychology, University of Liverpool, Whelan Building, PO Box 147, Liverpool, L69 3BX, UK

S. Clare Stanford Department of Pharmacology, University College London, London, UK

Andrew Steptoe Department of Psychology, St. George's Hospital Medical School, University of London, London, UK

David Sturgeon Department of Psychological Medicine, University College Hospital, Gower Street, London, UK

Win Sutanto Division of Medical Pharmacology, Leiden-Amsterdam Center for Drug Research, University of Leiden, 2300 RA Leiden, The Netherlands

Holger Ursin, Institute of Biological and Medical Psychology, University of Bergen, Bergen, Norway

Paul Willner Department of Psychology, University College of Swansea, Swansea, UK

Abbreviations

ACTH	adrenocorticotropic hormone (adrenocorticotropin, corticotropin)
AMPA	α-amino-3-hydroxy-5-methyl-isoxazole
ATP	adenosine triphosphate
BDZ	benzodiazepine
BNST	bed nucleus stria terminalis
cAMP	adenosine 3′,5′-cyclic monophosphate
β-CCE	β-carboline-3-carboxylic acid ethyl ester
CEA	central amygdala (amygdaloid) nucleus
cDNA	complementary deoxyribonucleic acid
CER	conditioned emotional response
CNS	central nervous system
COMT	catechol-O-methyl transferase
CR	conditioned response
CRF	corticotropin releasing factor
CRH	corticotropin releasing hormone
CSF	cerebrospinal fluid
DBH	dopamine-β-hydroxylase
DBI	diazepam binding inhibitor
5,7-DHT	5,7-dihydroxytryptamine
DMH	dorsomedial hypothalamus
DMI	defence mechanism inventory
DNA	deoxyribonucleic acid
DMVN	dorsal motor vagal nucleus
DOC	deoxycorticosterone
l-DOPA	l-dihydroxyphenylalanine
DOPAC	3,4-dihydroxyphenylacetic acid
DOPEG	3,4-dihydroxyphenylglycol
DPAG	dorsal periaqueductal gray
DRN	dorsal raphe nucleus
DSM-III-R	diagnostic and statistical manual, 3rd edition, revised
DYN A	dynorphin A
DYN B	dynorphin B
ECG	electrocardiogram; electrocardiographic
EEA	excitatory amino acid
EEG	electroencephalogram
FSL	Flinders sensitive line (rat)
GABA	γ-aminobutyric acid
GAD	glutamate decarboxylase
GHB	γ-hydroxybutyric acid

GHQ	general health questionnaire
GR	glucocorticoid receptor
5-HIAA	5-hydroxyindoleacetic acid
HPA	hypothalamus-pituitary-adrenal (axis)
HPG	hypothalamus-pituitary-gonadal (axis)
HPLC	high pressure (performance) liquid chromatography
HPLC-ECD	high pressure liquid chromatography with electrochemical detection
HPT	hypothalamus-pituitary-thyroid
5-HT	5-hydroxytryptamine
HVA	homovanillic acid
ICD-10	international classification of diseases, 10th edition
ICSS	intracranial self-stimulation
ICV	intracerebroventricular
IES	impact of events scales
IML	intermedio-lateral column
LEDS	life events and difficulties schedule
LH	lateral hypothalamus
MAO	monoamine oxidase
MBH	mediobasal hypothalamus
mCPP	1-(m-chlorophenyl)piperazine
ME	median eminence
5-MeODMT	5-methoxy-N,N-dimethyltryptamine
MHPG	3-methoxy-4-hydroxyphenyl glycol
MIF	melanocyte-stimulating hormone inhibitory factor
MPOA	medial preoptic area
MR	mineralocorticoid receptor
mRNA	messenger ribonucleic acid
3-MT	3-methoxytyramine
NMDA	N-methyl-D-aspartate
NPY	neuropeptide Y
NTS	nucleus tractus solitarius
PAG	periaqueductal gray
PAR	population attributable risk
pCPA	*para*chlorophenylalanine
PET	positron emission tomography
POA	preoptic area
POMC	pro-opiomelanocortin
pro-DYN	pro-dynorphin
pro-ENK	pro-enkephalin
PSE	present state examination
PTSD	posttraumatic stress disorder
PVG	periventricular gray
PVN	periventricular nucleus
pvPVN	parvocellular division of the hypothalamic periventricular nucleus
RHA	Roman high avoidance (rat)
RIA	radioimmuno assay

RLA	Roman low avoidance (rat)
SESS	self-evaluation and social support
SHR	spontaneously hypertensive rat
SIA	stress-induced analgesia
SIP	schedule-induced polydipsia
SPET	single photon emission tomography
SP-SHR	stroke prone spontaneously hypertensive rat
SCL	symptom check list
TRF	thyrotropin releasing factor
TRH	thyrotropin releasing hormone
UCR	unconditioned response
UCS	unconditioned stimulus
VMH	ventromedial hypothalamus
VO2max	maximal oxygen uptake
VPAG	ventral periaqueductal gray
Wky	Wistar-Kyoto (rat)

Preface

When Selye outlined his concept of stress and its long-term consequences, he integrated contemporary knowledge of the physiological and psychological processes affected by stress in a coherent scheme relevant to clinical problems. For many years, the study of stress was generally synonymous with the study of Selye's model, but it has now fallen victim to the divisive effect of specialization. Often, the only factor linking different studies is an interest in the noxious effects of environmental demands. Another difficulty is that specialized concepts and terminology in the literature from one field can make it impenetrable to researchers in others. This means that although psychology, physiology and the clinical sciences each has much to gain from each other, interdisciplinary communication is usually arduous, sometimes extremely difficult, and occasionally unsuccessful. It is clear that expanding our knowledge is at the cost of limiting individuals' ability to integrate findings from different disciplines. In this book, we hope to help to reverse this process.

In recent years, stress research has progressed along three, traditionally distinct, lines. First, clinical investigations have explored the links between stress and disorders such as heart disease and psychopathology. The second approach has been to investigate the psychological processes which mediate the effects of stress and the psychological qualities which protect against them. The third approach has been to investigate the physiological processes involved in response and adaptation to stress. Inevitably, many of the psychological and physiological studies have involved the study of animals as well as humans. This book attempts to link, on the one hand, clinical work with that done in animals and, on the other, neurochemical changes with psychological findings. To this end, we have included aspects of research from all the major disciplines, and have included discussions of common experimental procedures and the rationale behind them. At the same time, by recruiting contributions from experts in a range of aspects of stress research, we have aimed to provide up-to-date information for those already familiar with the field.

The objectives of the book have dictated its format. In the general introduction, Holger Ursin and Mirander Olff discuss historical aspects of research of the stress response and trace its development, while at the same time linking neuroendocrine and psychological theory. George Brown introduces clinical aspects, providing a critical appraisal of the impact of the environment ('life events') on psychopathology.

Part two deals with the physiological basis of clinical disorders caused by, or aggravated by stress. Paul Glue, Nick Coupland and David Nutt discuss neurochemical changes induced by stress and their implications for explaining causation and as markers for psychopathology. Bela Bohus and Jaap Koolhaas discuss the current status of research of the effects of stress on the cardiovascular system, both in health and disease, and Andrew Steptoe shows how an understanding of the cardiovascular response to psychological stress helps to explain the incidence of cardiovascular disease.

Part three concentrates on the use of animal models to gain insight into the impact of stress in man. Paul Willner highlights limitations of animal models and stresses the difficulties in deciding precisely what is being modelled. Robert Dantzer focuses on a topic which, although of central importance in human research, is often neglected in animals: whether neurochemical and behavioural changes reflect adverse effects of stress, or compensatory (coping) mechanisms. Peter McNaughton outlines the type of models used to study anxiety, and shows how attention to endocrine, as well as behavioural effects, raises some challenging questions for those trying to explain the therapeutic effects of antianxiety drugs. Finally, Steven Maier reviews the current status of research into one specific model: learned helplessness. In appraising experimental evidence he suggests that learned helplessness has a place in the modelling of anxiety.

Part four outlines progress in research of the role of some major neurotransmitters in stress. Developments in experimental techniques are reviewed by Marianne Fillenz, who also discusses their limitations and pitfalls. This chapter provides a background for those which follow. Monoamines and γ-aminobutyric acid are covered by Clare Stanford, Win Sutanto and Ron de Kloet. Charles Nemeroff and Diane Haydon-Hixson detail the enormous progress in our understanding of the role of peptides in stress, with special emphasis on corticotropin releasing factor. Although discovered less than fifteen years ago, the pivotal role of this peptide in stress is unassailable.

The final part deals with future prospects for stress research. Peter Salmon argues that the use of physical exercise to study stress adaptation might provide a way of drawing together clinical phenomena with experimental stress research. By contrast, in the final chapter, David Sturgeon describes the recent, increased public awareness and clinical profile of a phenomenon which is not yet amenable to interpretation in terms of the processes discussed in earlier sections: posttraumatic stress disorder. In this respect, the book ends with a major challenge. The preceding chapters make it clear that, despite a vast bank of knowledge, we still have difficulties in explaining psychological changes which take place over a period of seconds to days. We are even more bemused by posttraumatic stress disorder which often has a latency of many months.

It is inevitable that we have been unable to cover all aspects of this enormous topic, and readers will doubtless have strong views about others which should have been included. For instance, one field which could inform, and be informed by, approaches presented here is the effect of stress on prognosis in cancer and infectious disease. Although we were defeated in our attempts to secure (even promised) contributions in this field, our objective is to highlight the scope of the field, rather than to review all its facets. In so doing, we hope to kindle further interest in stress research and to counter criticisms of its scientific value. If we have succeeded at any level in this objective, it will be due largely to the enormous support of those who have helped us with this task. Recognizing that we are also constrained by the very barriers this book is aimed to demolish, we must thank the countless scientific colleagues to whom we have turned for help. In particular, we are indebted to Academic Press for their constant support and encouragement.

S. Clare Stanford
Peter Salmon
March 1993

Part One
General introduction

1

The stress response

HOLGER URSIN[1] AND MIRANDA OLFF[2]

[1]*INSTITUTE OF BIOLOGICAL AND MEDICAL PSYCHOLOGY*
UNIVERSITY OF BERGEN
AARSTADVEIEN 21
5009 BERGEN, NORWAY
[2]*DEPARTMENT OF CLINICAL PSYCHOLOGY AND HEALTH*
UNIVERSITY OF UTRECHT
UTRECHT
THE NETHERLANDS

Humpty-Dumpty sat on the Wall
Humpty-Dumpty had a great Fall
All the King's Horses
and all the King's Men
cannot keep Humpty-Dumpty from
getting together again–
(*From Mother Goose, slightly modified*)

ISBN 0–12–663370–3

1.1. INTRODUCTION

Our slightly modified Mother Goose nursery rhyme is meant to signify the unbelievable robustness of the term 'stress'. The term does not seem to die even if a number of authors, including several contributors to this volume, have attempted to destroy it. This indicates that we must learn to live with this concept.

The term stress has high face validity. All human subjects we have tested seem to have their own understanding of the word: even children use it. There seems to be no language barrier, at least within the Western world. North American, Dutch and Scandinavian subjects all have personal experience of using the term in their native tongue. We have also tested Europeans using English as their second language, but have never encountered any difficulties in communicating with research subjects, job applicants, students or colleagues from other continents and cultures about the area of human experience we are addressing. Difficulties arise only when we try to provide a concise and stringent definition with which all stress researchers will agree. It is even difficult to find definitions that the *majority* of researchers will accept.

A major problem with stress is that we are faced with a composite, multidimensional concept. The three main components that can be identified are (1) the input (stress stimuli), (2) the processing systems, including the subjective experience of stress, and (3) stress responses. One may also add that the experience and attributions of the sensory input from the stress response is a fourth possible way of using the stress concept. An additional difficulty is that these separate components of the stress concept interact with each other.

This chapter deals with the stress response, and concentrates on its physiological aspects. However, stress responses also comprise verbal reports of and complaints about stress, behaviours indicative of stress and interactions with performance. We will not deal with stress stimuli here, but we cannot exclude some of the processing systems since it is believed that specific processing systems or strategies might be linked to particular types of response. We will also address the fact that the *experience* of the stress response is often referred to as 'stress', and may be an important part of the subjective experience of the situation.

The essential picture that we want to convey is one of a complex system with feedback and control loops, no less and no more complicated than any other of the body's self-regulated systems. This system affects many organic processes and may function as a common alarm and drive system whenever there is a real or apparent challenge to the self-regulating systems of the organism. There is no *a priori* reason why this physiological response should lead to pathology; we shall briefly discuss where this belief came from, and outline circumstances in which it may be true.

1.2 SELYE'S CONCEPT OF STRESS

The essential element in Selye's definition of stress and, indeed, in medical thinking before him, is that there is a non-specific effect of any demand upon the body. This may influence recuperation from disease, reduce the resistance to disease, or even produce disease directly. Selye addressed this problem as a medical student; he was interested in studying why ill people looked and appeared ill. His teachers thought that he should start with something easier, but Selye returned to the same problem, *via* his initial work on the suprarenal (adrenal) cortex. In his first paper on stress (a letter to *Nature* published on July 4 1936) he described a syndrome produced by 'diverse nocuous agents'. In his very first sentence he makes it clear that he describes a non-specific response to non-specific agents:

"Experiments on rats show that if the organism is severely damaged by acute non-specific nocuous agents such as exposure to cold . . ., a typical syndrome appears, the symptoms of which are independent of the nature of the damaging agent . . .' (Selye, 1936).

In this initial description he included cold, surgical injury, spinal shock, muscular exercise and intoxications. The syndrome developed in three stages. The first appeared 6–48 h after initial injury. This involved a rapid decrease in the size of thymus, spleen, lymph glands and liver, disappearance of fat tissue, oedema formation, pleural and peritoneal transudate, loss of muscle tone, fall in body temperature, erosions in the digestive tract and loss of lipoid and chromaffin substance from the adrenal cortex. The second stage was said to begin 48 h after the injury, and involved enlarged adrenals, regaining the lipoid granulates, disappearance of the oedema, increase of the thyroid size and atrophy of the gonads. This was interpreted as reflecting a loss of anterior pituitary production of growth hormone, gonadotropic hormones and prolactin and an increase in 'thyrotropic and adrenotropic principles'. Continuation of relatively small amounts of the insult, or slight injuries, would lead to resistance being built up in this second stage and the organs would return to normal. However, if the insult continued at the same high level, the animals would lose their resistance after a couple of months and succumb with symptoms similar to those seen in the first stage, entering a phase of exhaustion. This was regarded as the third stage of the syndrome.

Selye considered the first stage to be a 'general alarm reaction' and he called the whole syndrome 'the general adaptation syndrome'. The alarm reaction seemed close to a histamine toxicosis but, even so, Selye suggested that more or less pronounced forms of this three-stage reaction represented the usual response of the organism to stimuli such as temperature changes, drugs and exercise. He also stressed that habituation could occur.

Selye worked on this concept for almost 50 years after this first publication. The concept went through many changes and shifts in emphasis; in particular, it was difficult to reconcile his original notions of non-specific costs with the apparent harmless, or even beneficial, effects of stressors. The literature refers to 'anti-stressors', characterizing events that function as 'uplifts' (the opposite of daily hassles; Lazarus and Launier, 1978), and gains from stress (Haan, 1978), or 'eustress' to use the term coined by Selye himself (Selye, 1974).

It has been a source of confusion that Selye used the term 'stress' differently from its original use in Hooke's equation (1635–1703) describing the relationship between stress and strain in an elastic body. Selye did not use the term strain, and interpreted 'stress' as being produced by stimuli he referred to as 'stressors', a terminology that has been adopted by many of the writers in this field. Likewise, we will deal with a general response and refer to this as the 'stress response' without implying that this involves 'strain'.

In a recent review of the stress literature, Levine and Ursin (1991) found that emotional 'loads' were the most frequently reported stress stimuli, and are given as the most common reason for stress responses. They are also a major component of commonly used stimuli in animal research, and must be taken into account even when the experimenters assume they are dealing with something analogous to a physical load.

Since the emotional aspect of stress is so important, and as this must depend upon psychological aspects of individual perceivers, there is no reason to expect to find a linear relationship between stimuli imposed and responses evoked. This would presume that the brain functioned as an automaton, with inborn and stereotyped reflex pathways linking stimulus to response. However, the brain is a complex network of self-regulating systems, and we have no *a priori* reason to expect to be able to predict the effects of a given stimulus without knowledge of how the brain of an individual organism is going to process a specific stimulus. This depends upon information about the particular stimulus that is stored in each individual's brain. In other words, the response depends on previous experience, or learning.

A key concept in Selye's original formulation is that stressors may summate and that this summation may be important for the course of healing after serious damage or disease. That is, the total input or load on the organism must be taken into account. This follows from the non-specificity of Selye's stress concept. There are many situations where the total sum of inputs is important for the outcome, such as when we deal with a sick patient or try to analyse a work environment. This was an essential element in the original medical interest in stress. However, the principle was not invented by Selye: it has been the rationale for therapy and nursing since Hippocrates, at least. After all, animals seek isolation from their usual world of challenges when sick or wounded. Multicausality has therefore been a part of medicine long before the term stress was invented. We believe the long-lasting, tonic activation to be the pathogenic part of the stress response.

1.3 THE BRAIN STEM ACTIVATION SYSTEM

Although Selye's work should not and cannot be ignored, it concerns only one aspect of the total organismic response to a challenge. The most important contribution to our understanding of the total 'stress response' was, in our opinion, the discovery of a general activation or alarm system in the brain (Moruzzi and Magoun, 1949). Selye never worked this into his model since he dealt mainly with the adreno-cortical axis. Nevertheless, there is no particular reason for concentrating on that axis, since the general alarm response seems to affect all somatic processes, either directly or indirectly.

In this chapter, the stress response will be regarded as a general arousal, or activation, as it is known in neurophysiology (Moruzzi and Magoun, 1949; Lindsley, 1951; Vanderwolf and Robinson, 1981). Activation may be defined as the process in the central nervous system (CNS) which raises the activity in the CNS from one level to a higher level and maintains this higher level. The strong correlations between the subjective feeling of wakefulness *versus* drowsiness and the corresponding changes in the electroencephalogram (EEG) (Jones, 1981) were the basis for the original classification of the EEG by Berger (1930). This was also the foundation for the neurophysiological concept of activation.

The normal activation response to a stressor consists of several phases. The initial phase immediately follows exposure to the stressor. This seems to be similar to Selye's alarm response (Selye, 1936), except that it is definitely not a 'histamine toxicosis', and does not reach the intensity he initially described. It consists of many responses and affects almost all regulatory systems and organs. All endocrine systems, the autonomic system and immune systems have been shown to be sensitive to environmental changes *via* these mechanisms. This has also been shown for biochemical activity in the brain (Modigh, 1974; Anisman, 1978; Coover *et al.*, 1983). The changes produced in these subsystems are to be regarded as parts of the total activation or arousal response. The various subsystems are contributory parts, but none of them is an essential element. Surgical or chemical manipulation of one or several of these systems may *dampen* activation, but such procedures do not seem to *eliminate* it.

However, there is no consensus on the extent to which there really is a general activation system dependent upon the reticular formation. It is generally agreed that specific ascending systems from the brain stem are involved, but these specific subsystems have not been written into a larger systematic theory to replace the general activation theory. We will review what we consider to be the most important reasons for assuming that there are subsystems within general activation, and briefly discuss suggestions for more specific pathways in the brain mediating the effects of specific stimuli.

1.4. PSYCHOENDOCRINE AND PSYCHOIMMUNOLOGICAL STRESS RESPONSES

Over recent decades, man has gained a new insight which seems to us to represent a paradigm shift in Western thought and beliefs. We have learnt that the brain not only controls the so-called 'autonomic' nervous system, but there are also central nervous influences on endocrine and immunological responses. This is seen most clearly in the 'stress' response, simply because this response is general, affecting almost everything influenced by the brain.

The immediate stress response affects the rate of discharge of sympathetic neurones and the hormonal secretion of catecholamines into the blood. The sympathetic response has many physiological and psychological concomitants or consequences. These include accelerated heart rate, increased blood pressure and respiration rate, dilation of pupils, perspiration and pallor. At the peak of the response, physiological resources are mobilized. This part of the activation response is easy to observe and monitor. It may be experienced as pleasant (thrill, 'trip'), or unpleasant ('stress'), depending upon the context of the situation and the attributions of the individual.

Later stages of the response are characterized by release of slower reacting hormones such as cortisol. This 'tail' of the acute reaction may function as a suppressive mechanism to dampen the acute response and re-establish physiological balance (Munck *et al.*, 1984; Levine and Ursin, 1991). Possible pathological consequences appear to be associated with prolonged, sustained states where the homeostatic mechanisms are taxed (Ursin, 1980).

Activation also affects the immune system: catecholamines and cortisol have all been shown to affect several immune parameters. However, findings are often contradictory. For instance, sympathetic activation is not necessarily related to the number of immune cells. In general, catecholamines have a dual effect on the immune system. Chronically high levels of catecholamine in the circulation may dampen the immune response either through downregulation of adrenoceptors on lymphocytes or effects on leucocyte function. Short exposure to high levels of catecholamines may result in activation or suppression of the immune system, depending upon the initial metabolic status of the cell (Kavelaars *et al.*, 1990). Cortisol in low physiological doses increases immune responses, whereas it usually suppresses the immune response in high (stress-induced or pharmacological) doses. Catecholamines as well as cortisol can influence homing patterns of leucocytes in various organs. In addition, these hormones may interfere with the adherence of leucocytes to the vessel wall (the composition of the marginal pool).

The final pattern of leucocyte distribution in the peripheral blood is determined by a myriad of hormonal factors and neurotransmitters. Therefore, it is difficult to relate, causally and precisely, changes in the level of a given hormone, or sympathetic nerve activity, to changes in the total distribution of leucocytes. The speed of change and the relative independence of hormones has led many researchers to suggest direct effects by sympathetic nerves. Although such a specific aspect should be kept in mind, the effect seems to be part of the general, non-specific activation.

1.5 STRESS: THE ALARM SYSTEM IN A HOMEOSTATIC ORGANISM

The general response to novelty, danger signals and real or potential threats to homeostasis should be regarded as an alarm system which forms part of the homeostatic system of the organism. Levine and Ursin (1991) suggested that it may be easier to regard the stress response as a response to something missing, rather than to the presence of a 'stressor'. This is an interpretation of the stress response which may be more in line with contemporary thinking about the homeostasis of the body, in terms of information or control theory.

This view of stress clearly differs from that fostered by Selye, and comprises a vast number of studies in which severely traumatic stimuli were used to elicit the stress response. The perspective adopted by Levine and Ursin (1991) suggests a very different approach. The stress response occurs as a response to diverse stimuli such as maternal loss, frustration, electric shock and surgical trauma. According to the Levine and Ursin hypothesis, the element common to all these inputs is that they represent the absence of critical features in the environment. By this it is meant that there is a lack of information (uncertainty or unpredictability) on the ability to attain positive outcomes, or avoid negative outcomes. Thus stress stimuli, or signals representing such stimuli, indicate that something, which is highly relevant and desirable to the organism, is missing or about to disappear. Therefore stress is presumed to be the state that is created whenever the processor (brain) registers this informational discrepancy.

In both animals and man there has been a consistent finding that uncertainty, lack of information, or the absence or loss of control produce alarm states. Conversely, the presence of information in the form of clear and salient safety signals and behaviours, that lead to positive outcomes or result in control, reduce or eliminate the alarm state with a concomitant reduction of physiological responses.

There is, therefore, reasonable agreement that the alarm system is activated whenever there is a discrepancy between what the organism is set for, regarding an important variable (set value), and what really exists (actual value of the same variable). This is a reformulation of homeostatic theory.

1.6 THE STRESS RESPONSE IS NOT AN ATAVISTIC OR UNDESIRABLE RESPONSE

The stress response, as we have defined it, is an integral part of an adaptive biological system. Both behavioural and physiological responses to stress are required for humans and other animals to function within the confines of a dynamic and frequently challenging environment. It should not be viewed as an atavistic mechanism no longer suited for civilized man, as was suggested in earlier literature (Charvat *et al.*, 1964). Some of the perceived sensations from the stress response may be described as uncomfortable, but this is not a signal that the process is necessarily pathogenic.

Since activation may be regarded as the driving force behind solution of problems, stress responses are not to be accepted as something which should be avoided at all costs. Activation may be regarded not only as an alarm system, but also as the driving force that makes an animal or a human act to reduce needs. Activation, in this context, therefore, is an essential element of the total adaptive system of the organism (Ursin, 1988). Feelings of stress or related states, such as anxiety, are not necessarily evils to be dampened, even if they appear to be unpleasant. They may be appropriate responses to stimuli requiring full attention and integrated action which would entail a reduction of the source of the stress response. The purpose of the response is to eliminate not only itself, but also the situation which gave rise to it. Thus, when there is a discrepancy between the set value and the actual value for a particular variable, the alarm system or driving force will remain activated until there is an agreement between them, or until the brain gives lower priority to the set value. It is of particular interest that activation also seems to depend on the probability or expectancy of a solution to a particular discrepancy (Coover *et al.*, 1984).

The initial phase of the response may function as a positive feedback mechanism. Increased muscle tension, together with the vegetative components of the response, give rise to sensations that are essential elements of the experience of stress; this tends to facilitate further development of the state of stress. The process is described in the original James-Lange theory of emotions which emphasized the experience of the peripheral components of emotion as essential for the emotional experience.

At the peak of the response, physiological resources are mobilized and result in enhanced function; also, from a psychological point of view, there is often improved performance (Frankenhaeuser, 1980). In complex tasks, the limit for performance may be surpassed. In such cases, high stress levels may be incompatible with peak performance (Hockey, 1986; see also Hamilton *et al.*, 1987). However, the general effect of the response is to improve the chances of dealing with the stressful situation and to eliminate the source of stress. The uncomfortable aspect of the sensation is not pathogenic: it should drive the subject to appropriate behaviours. It is only when the situation becomes chronic that the stress response may be inadequate and pathological (Ursin, 1980).

1.7 TIME CHARACTERISTICS OF THE ACTIVATION RESPONSE

The activation response may be regarded as a cascade of events. If we define as time zero the start of a change in activity from one stage to another, higher, stage (our definition of activation), the events could be defined partly on the basis of their latency.

Within *milliseconds* of time zero the brain state is changing, with frequency shifts in the EEG and changes in the slow components of evoked cortical potentials. There may also be changes in the activity of sympathetic nerves, affecting the classical indicators of activation used by psychophysiologists, such as sweating and increased heart rate. There are also short latency shifts in muscle tone and immune function.

Within *seconds* the anterior pituitary is affected by signal peptides from the hypothalamus, such as corticotropin releasing factors. The full response from the anterior pituitary may not reach peak levels until 20–30 s after time zero. At this time the full impact of the sympathetic arousal feeds back to the brain. As stated above, this feedback constitutes an important part of the experience of the change in state.

For cortisol, at least, the full endocrine effect resulting from activation of the anterior pituitary does not reach its peak until some 10 *minutes* after the time zero. The effects on other stress indicators are even slower. There is reason to believe that some of the immunoglobulins will not respond until *days* or *weeks* after the challenge.

In real life, no subject is exposed to only one stressor. In animal research we may obtain reasonably stable levels of exposure by establishing strict routines in the animal quarters. Any minor change will change basal levels of hormones; expectancy of change also produces endocrine changes before the change actually occurs (Coover *et al.*, 1984). In human research, these expectancies are even stronger; this makes it difficult to control the experimental situation. Expectancies associated with the experimental situation, such as an expectancy of pain or fear connected with blood sampling, will influence the hormone measurements. It is very difficult to obtain a blood sample which represents a basal level without implanting catheters; even this does not represent a normal life situation.

1.8 DAMPENING THE ALARM: DEFENCE AND COPING

The importance of the evaluation of a stimulus, rather than the stimulus itself, is a central theme in most contemporary writing on stress. There seems to be a consensus that the main reason for the lack of linearity between the stimulus input and what occurs within the organism is that the stimuli are being evaluated, or filtered, before they gain access to any response system. We believe that the most important filters are the evaluation of the potential threat ('stimulus expectancy', according to Ursin; 'primary appraisal', according to Lazarus) and the evaluation of the efficiency of available responses ('response outcome expectancy' or 'secondary appraisal'). We will deal with the best described stress-related 'filters' within this framework of expectancies: one of these filters is 'defence', the other is 'coping'. Both reduce the impact of the potential stressor on the level of activation. Lack of dampening the response may have health consequences. The ways in which these may arise are discussed in subsequent chapters of this book. Figure 1.1 illustrates the overall scheme.

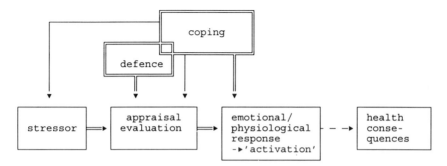

Figure 1.1 Model of the role of defence and coping in the stressor-health chain

1.8.1 DEFENCE: DISTORTED STIMULUS EXPECTANCIES

Stimulus expectancy (Bolles, 1972) refers to the storing of the information that one stimulus precedes another. i.e. classical conditioning. When an aversive event is expected (i.e. signalled by other stimuli), the stress response occurs. When this relationship is misperceived by the individual through mechanisms distorting the true relationship, the individual is said to use cognitive defence. Denial, for instance, one of the most primitive defence mechanisms, is defined as the distortion of true relationships between stimuli by the refusal to accept a threat as real.

This filtering mechanism has only been demonstrated in humans, and many researchers would claim that it is specific to this species, since it seems to depend on complex cognitive functions. When humans perceive a threat signal they may distort or deny the true or probable consequences of that signal. Since Freud's time, this has been referred to as 'defence'. In its most primitive forms, defence may prevent threat signals from producing stress responses. The price for this stress attenuation may be a failure to behave appropriately in dangerous situations (Kragh, 1960; Værnes, 1982). We do not know whether other animals have similar mechanisms. There is no *a priori* reason why they could not have mechanisms which distort stimulus expectancies although none have yet been identified.

1.8.2 COPING: RESPONSE-OUTCOME EXPECTANCIES

The second type of expectancy results from learning about the consequences of acts. This type of expectancy is referred to as response-outcome expectancy (Bolles, 1972), and is related to instrumental conditioning. There are, in principle, three classes of such expectancies that are important in determining the internal, physiological state of the individual, i.e. whether the individual will or will not show stress responses. These three types of expectancies have been referred to as coping, helplessness and hopelessness.

In the literature, the term 'coping' is used in many different ways. The meaning which will be used in this chapter refers to established positive response-outcome expectancies. Coping, therefore, is the result of a learning process. An individual expects a positive, attractive outcome with a high probability. This definition of coping predicts reduced stress in the individual as a result of successful coping, whereas the inability to cope results in high levels of stress (Ursin and Murison, 1983). This has been demonstrated both in animals (Coover et al., 1973) and humans (Ursin et al., 1978).

The development of a positive response-outcome expectancy is dependent upon whether the individual can control, or perceive that they can control, a specific situation or experimental paradigm. This depends on the capacity to make active responses in the presence of aversive stimuli and to register the results of such acts. If the responses result in a positive outcome, either by escaping or avoiding a noxious event, or by being able to obtain positive reinforcement, the brain will then store these relationships as positive outcome expectancies. The condition for this type of learning to occur is not only that such positive events occur, but that the brain must also be able to record these results. This is often referred to as feedback, and rapid and clear feedback is a condition for control and coping to occur, as it is for all learning.

Stress responses may also be reduced when an individual is exposed to signals that a particular need is to be satisfied. The empirical background for this comes from a long series of experiments in man and other animals. Presence of positive reinforcers, or signals representing positive reinforcers, reduce stress responses (Goldman et al., 1973; Levine and Coover, 1976). It is perhaps more surprising that absence of positive reinforcers, and clear signals that such reinforcement will not occur, also reduces the stress response. This latter phenomenon is less well known, but seems to be a very important mechanism for survival. A hungry rat does not run around after food in an environment where there is obviously no food present. It is only when signals occur that may indicate food that we observe increased activity (Coover et al., 1984).

1.9 DEFENCE *VERSUS* COPING

So far we have treated defence and coping as theoretically two different ways of dealing with potentially stressful situations. However, coping and defence are not that easy to discriminate in practice and some authors do not even attempt to discriminate between them (Lazarus and Folkman, 1984). One possible way of evaluating the psychometric structure of defence and coping in humans is to subject the results of tests to factor analysis. This identifies groups of tests which are statistically inter-related and which may, therefore, reflect a common underlying process. This could show whether these skills represent one, two or several independent factors. Olff and collaborators have done this for scales currently used in this research in the Netherlands (Olff *et al.*, 1991). Factor analysis was performed on the subscales of the following questionnaires: the Defence Mechanism Inventory (DMI) (Gleser and Ihilevich, 1969), the Life Style Index (LSI) (Plutchik *et al.*, 1979), the Internal, Powerful Others and Chance Locus of Control scales (IPC) (Levenson, 1974), and the Utrecht Coping List (UCL) (Schreurs *et al.*, 1988). The Locus of Control scales were included to evaluate the response-outcome expectancies essential for the Levine–Ursin definition of coping.

Instead of two factors (defence and coping), the factor analysis yielded four clusters: two clusters consisted mainly of coping subscales (from the UCL and IPC), while the other two mostly contained defence subscales (from the DMI and LSI). The two 'coping factors' were:

1. *Instrumental mastery-oriented coping*: an instrumental, active, goal oriented coping style where the individual has a sense of personal control.
2. *Emotion-focused coping*: which basically involves showing an emotional, expressive reaction towards problems.

The two 'defence factors' were:

1. *Cognitive defence*: a mainly cognitive defensive style, reflecting internalization of negative emotions aroused by conflict situations, such as intellectualization, denial or having comforting cognitions.
2. *Defensive hostility*: an outward directed defensive style, containing elements of anger, hostility and acting out, including projection.

Thus, as we have theorized, defensive strategies were separable from coping strategies, although within each category further clusters could be differentiated. Thus, defence and coping apparently constitute fundamentally different approaches to stressful situations.

1.10 SPECIFIC ACTIVATION SYSTEMS: PHASIC AND TONIC ACTIVATION

The first differentiation of stress responses to be made is between the types of response seen in coping *versus* non-coping subjects. The early, general activation in an individual faced with a challenge is reduced when a positive response-outcome has been established, i.e. when the individual is coping with the task. However, when no such response-outcome expectancy is established, the general activation has been postulated to persist ('tonic' activation). This has been assumed to generate psychosomatic complaints (Ursin, 1980). However, when coping has been established, there is still a shortlasting activation which has been referred to as 'phasic' activation (Ursin, 1980). The phasic activation in the coping individual is characterized by release of adrenaline (rather than noradrenaline; Hansen *et al.*,

1978), increased pulse rate (Strømme *et al.*, 1978) and a modest but significant rise in plasma levels of testosterone (Davidson *et al.*, 1978).

The second differentiation of stress responses is more complex. Pribram and McGuiness (1975) identified three separate, but interacting, neural systems for the control of attention. The immediate behavioural and physiological responses to a novel stimulus, the orienting response, were referred to as 'arousal'. Tonic readiness necessary for solving a problem was called 'activation', and the coordination between these two was referred to as 'effort'. They suggested that the effort mechanism was dependent upon the amygdala, and the arousal mechanism on the hippocampus.

Henry and Meehan (1981) have suggested that there is endocrine specificity to these activation mechanisms. They believe the amygdala effort mechanism may be tied to catecholamines, and the hippocampus arousal mechanism to cortisol.

We find it hard to accept the neuropsychological specificity of this hypothesis, but the notion of two separate effector systems receives increasing support from different parts of the literature. Bohus *et al.* (1987) have reported extensive experimental animal data to support these notions. In humans, factor analyses of endocrine responses have consistently turned up at least three reasonably stable factors: a catecholamine, a cortisol and a testosterone factor (Rose *et al.*, 1967; Ellertsen *et al.*, 1978; Ursin, 1980). Further statistical analyses have revealed stable relationships between these factors and psychological traits. It is generally accepted that the tendency to react with sympathetic activation and catecholamine secretion is reliably related to the behavioural pattern commonly identified as Type A behaviour. This is believed to be coronary-prone, at least when it occurs with hostility and lack of coping (Glass, 1977; Ursin, 1980; Jenkins, 1982). The cortisol factor is less well accepted. In our own opinion, there seem to be consistent data on relations between the tendency to react with cortisol and psychological defence mechanisms (Vaernes *et al.*, 1982; Endresen and Ursin, 1991). There also seems to be a consistent relationship between the cortisol factor and psychological defence, on the one hand, and plasma levels of immunoglobulins on the other (Ursin *et al.*, 1984; Endresen and Ursin, 1991). Similar relationships have been found independently by the Utrecht group, including a link with leucocytes (Olff, 1991).

Levine and Ursin (1991) have pointed out that pituitary-adrenal and sympathetic activation are highly related from a physiological point of view but, statistically, they show different relationships to personality. This may be related to the differences in time course, discussed above. Sympathetic activation may indicate a 'fight-flight' response or 'effort to control' as opposed to the pituitary-adrenal axis which may be related to the 'conservation-withdrawal response' or 'distress'. The sympathetic activation may be related to the 'on' aspects of the stress response, the pituitary-adrenal activation to the 'off'- or 'keep on'-aspects of the response, since cortisol may be regarded as suppressive, dampening the acute stress response (Munck *et al.*, 1984; Levine and Ursin, 1991).

1.11 ONTOGENETIC AND GENETIC FACTORS IN THE STRESS RESPONSE

There is a great deal of evidence that stress responses are modified in animals by ontogenetic factors. It is assumed, but not demonstrated, that these factors are also relevant in human development. In general, there are critical periods in development when exposure to either handling or mild electric shocks produces an overall diminution of stress responses in later life (Levine, 1969). Long-standing changes in reactivity can also be produced by experiences in adulthood. Exposure to inescapable and unpredictable shock may result in a general hyperreactivity to input stimuli (Overmier and Murison, 1991; Levine *et al.*, 1973). The exaggerated responses do not appear to be task dependent and seem to generalize to numerous input stimuli.

Although the evidence presented above represents the best documented demonstration of experientially-induced individual differences, there is evidence that other ontogenetic variables also permanently alter the processing systems. Amongst these we can mention: prenatal stress, differential mother-infant interactions, nutritional factors and altered hormonal states during development (Levine, 1983). It seems reasonable to assume that animal data in these cases are relevant for human preventive medicine.

Finally, although the evidence is not yet compelling, there is a growing body of information suggesting a strong genetic component as a determinant of individual differences in sub-human species, and probably also in humans. In humans, there are also social factors that produce individual differences. It seems well established that the presence of social support increases tolerance to stressful life events and has a general beneficial effect on health (Karasek and Theorell, 1990). Similar stress-reducing effects of the social structure may be seen also in animals, both in the mother-infant relationship (Levine, 1983) and in more complex social structures of primates (Stanton *et al.*, 1985) and birds (Myhre *et al.*, 1981).

1.12 DEFENCE AND COPING IN RELATION TO BIOLOGICAL PARAMETERS

No apparent psychobiological explanation has been offered for the correlations found between psychological defence and plasma levels of cortisol and immunglobulins (Endresen and Ursin, 1991). Recent empirical data may shed light on these issues. There seem to be relationships between the specific way people handle stressful situations and the immune system (Olff, 1991).

In a laboratory stress situation, Olff *et al.* (see Olff, 1991) found significant correlations of defence and coping with several changes in immune parameters, in particular with B-cells. Changes in endocrine parameters appeared to be even more sensitive to defence and coping. Instrumental Mastery-Oriented Coping and Emotion-Focused Coping were negatively related to changes in most of the stress hormones. Perceptual defence, measured by the tachistoscopic Defence Mechanism Test, was negatively related to changes in most of the stress hormones, but positively related to changes in noradrenaline.

These findings are complex. They support the hypothesis that it is not the laboratory stressor which determines the endocrine and immune responses, but the defence or coping style used by the individual. The findings also support the idea that defensive and coping subjects both show low stress responses or 'activation' and that both mechanisms are 'stress dampening'. Effective *copers* (Instrumental

Mastery-oriented Coping) did not show any signs of increased activation to the stressor. The later responses in hormones related to the pituitary-adrenal axis (adrenocorticotropic hormone (ACTH), cortisol and endorphins) also showed low levels: 'efficient copers' do not need to mobilize the pituitary-adrenal axis. This coping style involves the perception of controllability of stress situations and the expectancy of positive outcomes. As expected these subjects showed low levels of health complaints and self-reported distress.

The *defenders* also show low activation, but we believe this to be a different psychological mechanism. An interesting possibility is that these mechanisms have different time scales, with coping related to the fast and short-lasting catecholamine response and defence related more to the pituitary-adrenal response. Since cortisol is regarded as suppressive, dampening the acute stress response, defence mechanisms have most impact on the psychological phenomena occurring when coping is ineffective. The immediate defence mechanisms seem also to affect catecholamines. Interaction between these behavioral strategies, therefore, may have biological consequences not only for the immediate endocrine responses, but also for the interaction between the endocrine response patterns. The consequences of this complex psychobiological feedback circuit are unknown, but open new perspectives and explanations for genetic and environmental influences on personality development.

One common view of psychological defences is that they serve to keep 'threat' unconscious, and therefore are assumed to be related to low activation or stress responses. However, studies of EEG correlates of perceptual defence (Olff and Sterman, 1991) suggest that there is more activation, rather than less, in subjects with high perceptual defence scores. The same conclusion can be drawn from studies on defence and cardiovascular parameters (Olff, 1991). The large noradrenaline response to a laboratory stressor for high defence subjects also indicates a high activation level for subjects high on perceptual defence, at least for the sympathetic response. Usually, high sympathetic activation is accompanied by activation of the pituitary adrenal axis. However, the lower values of cortisol and ACTH for these subjects may be dependent upon a purely psychological mechanism: that of defence. Consequently, the activation may last longer because it is not dampened by glucocorticoids. Therefore, defence may be related to increased, rather than decreased, activation. This is schematically shown in Figure 1.2.

1.13 SIMILARITIES AND DIFFERENCES BETWEEN THE STRESS RESPONSE IN MAN AND ANIMALS

To judge the state of a subject involves using questionnaires in human studies, or measuring physiological changes in both humans and other animals. Apart from this obvious response difference, the similarities in the responses of different species are more apparent than any differences. We seem to be dealing with essential brain mechanisms, which is to be expected if we accept that this is the alarm system of a self-regulating organism.

The filtering mechanisms may be regarded as logical functions when defined in terms of stimulus and response expectancies, and there is really no reason why there

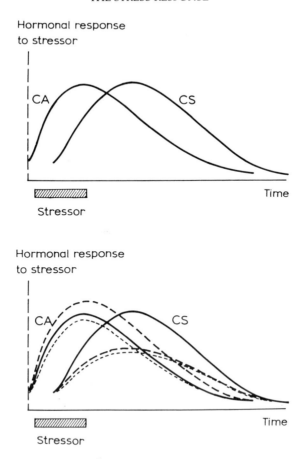

Figure 1.2 Hypothetical curves for catecholamine (CA) and cortisol (CS) responses to a stressor. Dotted lines show subjects who are coping instrumentally, and broken lines show those using defence mechanisms. The continuous line represents an 'average' response. Instrumental copers are shown with the same initial catecholamine reaction as the average responders, but a smaller cortisol response because of their positive response-outcome expectancy. Defenders have a higher than average response for the initial activation, but they also have a low cortisol response which is a result of the defence mechanism itself. Their catecholamine response, and other aspects of the activation that cortisol was supposed to turn off, may continue longer than 'normal'. CA: catecholamines; CS: cortisol.

should be phylogenetic leaps in the logical functions of the brain. Stimulus and response expectancies exist, of course, in animals as well as humans, as is evident from the conditioning literature. Classical conditioning corresponds to stimulus expectancy, instrumental conditioning to response expectancy (Bolles, 1972). But the cognitive defence mechanisms that we have dealt with have only been identified and operationalized in humans and rely heavily on verbal reports. We do not know how to verify or exclude the existence of similar mechanisms in animals.

There is also no apparent *a priori* reason why the response itself should be different in man and animals. The brainstem activation system appears rather early in phylogeny. 'Cortical' electroencephalographic activation patterns appear before the neocortex is developed, and has been described in bone fish (Enger, 1957); bone fish have also developed a reticular formation (see Pickering, 1981). In agreement with the positions outlined in this chapter, fish show a stress response which has the basic behavioural, electroencephalographic, vegetative and endocrine components described for mammals (Pickering, 1981). In mammals, the electroencephalographic, vegetative, endocrine and immunological consequences of the alarm response seem basically similar across species. The levels reached, and molecular details, may differ, and there are significant sex differences in many species, including humans. The behavioural characteristics of the orienting response, for instance, seem to differ mainly with regard to which sensory system is the most important for the species: cats look around, rats sniff. There are also obvious species differences in response repertoire and in species specific 'action patterns'. Nevertheless, basic logical principles remain constant across species.

The situational characteristics which elicit stress responses in animals and humans obviously differ. However, the consensus is that the animal research by, for instance, Bohus et al. (1987), Bohus and Croiset (1990), Koolhaas and Bohus (1989), Dantzer (1989), Murison et al. (1989) and Glavin et al. (1991), is revealing basic psychosomatic patterns which also exist in humans. When the emphasis is on the logical aspects of the situations producing psychosomatic consequences, we conclude that general principles are revealed which are common to man and other animals. The relations between load, control, feedback, and psychosomatic consequences in man and animals are summarized in Figure 1.3.

One of the pyramids summarizes extensive work in human behavioural epidemiology (Karasek and Theorell, 1990). The other pyramid is based on work by Weiss (1972), and is consistent with a large body of data from many researchers. The basic assumption in this chapter is that a similar pyramid may be drawn whether the vertical axis indicates sustained tonic activation or sustained stress responses. Figure 1.3 suggests that the relationship between load, control and health problems is a general biological phenomenon, expressing basic neurobiological principles. These relationships are mediated *via* the activation (or stress response) mechanisms, and express fundamental principles for the logical operations of central nervous systems.

1.14 CONCLUDING REMARKS

The purpose of the stress response is to drive the organism to eliminate the sources of the response by invoking appropriate behaviour. The peak of the response is a mobilization of resources described by many early physiologists concerned with homeostasis and appreciated as an important, albeit limited, element in improving psychological performance. It is only when these limits are surpassed, and only in very complex situations, that stress may be an inadequate way of responding to a crisis.

The later stage of the acute stress response is characterized mainly by the more slowly reacting hormones, such as cortisol. This part of the total stress response has suppression of the body's response to stress as its main function. The 'tail' of the acute response, therefore, may be regarded as a homeostatic device, itself

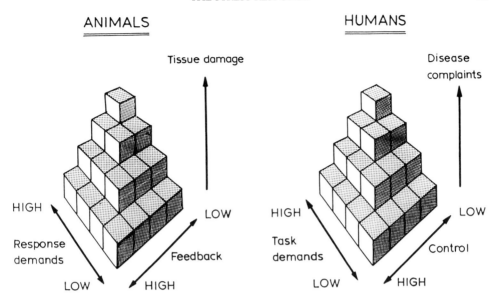

Figure 1.3 The relationship between stress defined as load (stressor, task or performance demands), control possibilities or feedback from the responses, and the development of psychosomatic disease in experimental animals (left) and health complaints (right). The left pyramid summarizes extensive work in several species, and is based on an initial figure published by Weiss (1972). The right pyramid summarizes extensive work in human behavioural epidemiology (Karasek and Theorell, 1990). The basic assumption in this paper is that a similar pyramid may be drawn, with the vertical axis indicating sustained tonic activation, or sustained stress responses and that these relationships express basic and fundamental biological principles common across species. (Modified from Ursin and Murison, 1983.)

reestablishing the physiological balance. The physiological response seems to turn itself off. This makes it difficult to monitor the sustained activation in anything but pathophysiological consequences.

The behaviour elicited by the stress also aims to turn itself off. It is a homeostatic mechanism, with its main effect being to eliminate the sources of stress. The response is general, it is uncomfortable and it serves as a general alarm system. It elicits somatic and behavioural responses which tend to eliminate this imbalance, just like any other homeostatic mechanism.

Sustained activation may occur when the organism is without control and information, or when there is reason to expect negative events. It occurs whenever the organism has an expectancy of negative events, or negative response-outcome expectancies, or no response-outcome expectancies (hopelessness and helplessness). The stress state which results from a lack of coping is not dangerous in itself, therefore. The alarm system is there for a purpose and the alarm *per se* does not produce pathological changes. It is only when prolonged and sustained that the homeostatic elements in the response may be surpassed. If this takes place, disease might occur in the somatic locus with least resistance.

Our discussion of the stages and levels of activation in relation to defence or coping is tentative. Further research requires repeated measurement of several indicators of activation. Until the resources are available to fund this kind of design, the exact psychological and physiological mechanisms during the stressful encounter will remain obscure.

ACKNOWLEDGEMENTS

The research reported here was supported in part through grants by the Norwegian Research Council for Science and the Humanities, the Royal Norwegian Research Council for Science and Industrial Research, and the Legacy of Grete Harbitz.

1.15 REFERENCES

Anisman, H. (1978) Neurochemical changes elicited by stress Behavioral correlates, in H. Anisman and G. Bignami (eds.), *Psychopharmacology of Aversively Motivated Behavior*, Plenum, New York

Berger, H. (1930) Uber das Elektroenkephalogramm des Menschen. II. *J. Physiol. Neurol.* **40:** 60–179

Bohus, B., Benus, R.F., Fokkema, D.S. *et al.* (1987) Neuroendocrine states and behavioral and physiological stress responses. *Prog. Brain Res.* **72:** 57–70

Bohus, B. and Croiset, G. (1990) Neuropeptides, behavior, and the autonomic nervous system, in D. de Wied (ed.), *Neuropeptides: Basics and Perspectives*, Elsevier, Amsterdam, pp 255–76

Bolles, R.C. (1972) Reinforcement, expectancy and learning. *Psychological Rev.* **79:** 394–409

Charvat, J., Dell, P. and Folkow, B. (1964) Mental factors and cardiovascular disorders. *Cardiologia*, 44: 124–41

Coover, G.D., Ursin, H. and Levine, S. (1973) Plasma corticosterone levels during active-avoidance learning in rats. *J. Comp. Physiol. Psychol.* **82:** 170–4

Coover, G.D., Ursin, H. and Murison, R. (1983) Sustained activation and psychiatric illness, in H. Ursin, and R. Murison, (eds.), *Biological and Psychological Basis of Psychosomatic Disease*, Pergamon Press, Oxford.

Coover, G.D., Murison, R., Sundberg, H. *et al.* (1984) Plasma corticosterone and meal expectancy in rats: Effects of low probability cues. *Physiol. Behav.* **33:** 179–84

Dantzer, R. (1989) Neuroendocrine correlates of control and coping, in A. Steptoe and A. Appels (eds.), *Stress, Personal Control, and Health*, Wiley, Chichester, pp 277–94

Davidson, J., Smith, E.R. and Levine, S. (1978) Testosterone, in H. Ursin, E. Baade, and S. Levine, (eds.), *Psychobiology of Stress. A Study of Coping Men*, Academic Press, New York, pp 57–62

Ellertsen, B., Johnsen, T.B. and Ursin, H. (1978) Relationship between the hormonal responses to activation and coping, in H. Ursin, E. Baade, and S. Levine, (eds.), *Psychobiology of Stress: A Study of Coping Men*, Academic Press, New York, pp 105–22

Endresen, I.M. and Ursin, H. (1991) The relationship between psychological defence, cortisol, immunoglobulins, and complements, in M. Olff, G. Godaert, and H. Ursin, (eds.), *Quantification of Human Defence Mechanisms*, Springer-Verlag, Berlin, pp 262–72

Enger, P.S. (1957) The electroencephalogram of the codfish (*Gadus callarias*). *Acta Physiol. Scand.* **39:** 55–72

Frankenhaeuser, M. (1980) Psychobiological aspects of life stress, in S. Levine, and H. Ursin, (eds.), *Coping and Health*, Plenum Press, New York

Glavin, G.B., Murison, R., Overmier, J.B. *et al.* (1991) The neurobiology of stress ulcers. *Brain Res. Rev.* **16:** 301–43

Glass, D.C. (1977) Stress, behavior patterns, and coronary disease. *American Scientist*, **65:** 177–87

Gleser, G.C. and Ihilivich, D. (1969) An objective instrument for measuring defense mechanisms. *J. Consult. Clin. Psychol.* **33:** 51–60

Goldman, L., Coover, G.D. and Levine, S. (1973) Bidirectional effects of reinforcement shifts on pituitary-adrenal activity. *Physiol. Behav.* **10:** 209–14

Haan, N. (1978) *Coping and Defending,* Academic Press, New York

Hamilton, V., Bower, G.H. and Fryda, N.H. (eds.) (1987) *Cognition, Motivation, and Affect: A Cognitive Science View,* Nijhoff, Dordrecht

Hansen, J.R., Støa, K.F., Blix, A.S. and Ursin, H. (1978) Urinary levels of epinephrine and norepinephrine in parachutist trainees, in H. Ursin, E. Baade, and S. Levine, (eds.), *Psychobiology of Stress. A Study of Coping Men,* Academic Press, New York, pp 63–74

Henry, J.P. and Meehan, J.P. (1981) Psychosocial stimuli, physiological specificity, and cardiovascular disease, in H. Weiner, M.A. Hofer and A.J. Stunkard (eds.), *Brain, Behaviour and Bodily Disease,* Raven Press, New York

Hockey, G.R.J. (1986) A state control theory of adaptation to stress and individual differences in stress management, in G.R.J. Hockey, A.W.K. Gaillard, and M.G.H. Coles, (eds.), *Energetics and Human Information Processing,* Nijhoff, Dordrecht, pp 285–98

Jenkins, C.D. (1982) Psychosocial risk factors for coronary heart disease, *Acta Medica Scand.* **660 (suppl.),** 123–36

Jones, B.E. (1981) Understanding the physiological correlates of a behavioral state as a constellation of events, *Behav. Brain Sci.* **4:** 482–3

Karasek, R.A. and Theorell, T. (1990) *Healthy Work. Stress, Productivity, and the Reconstruction of Working Life,* Basic Books, New York

Kavelaars, A., Ballieux, R.E. and Heijnen, C.J. (1990) Differential effects of b-Endorphin on CAMP levels in human peripheral blood mononuclear cells. *Brain, Behav. Immunity,* **4,** 171–9

Koolhaas, J. and Bohus, B. (1989) Social control in relation to neuroendocrine and immunological parameters, in A. Steptoe and A. Appels (eds.), *Stress, Personal Control, and Health,* Wiley, Chichester, pp 295–305

Kragh, U. (1960) The Defense Mechanism Test: A new method for diagnosis and personnel selection. *J. Appl. Psychol.* **44:** 303–9

Lazarus, R.S. and Folkman, S. (1984) *Stress, Appraisal, and Coping,* Springer-Verlag, New York

Lazarus, R.S. and Launier (1978) Stress-related transactions between persons and environment, in: L.A. Pervin and I.M. Lewis (eds.), *Perspectives in Interactional Psychology,* Plenum, New York

Levenson H. (1974) Activism and powerful others: distinction within the concept of internal–external control. *J. Personal. Assess.* **38,** 377–83

Levine, S. (1969) An endocrine theory of infantile stimulation, in A. Ambrose, (ed.), *Stimulation in Early Infancy,* Academic Press, London, pp 45–63

Levine, S. (1983) A psychological approach to the ontogeny of coping, in: N. Garmezy, and M. Rutter, (eds.), *Stress, Coping and Development in Children,* McGraw Hill, New York

Levine, S., Madden, J., Conner, R.L., Moskal, J.R. and Anderson, D.C. (1973) Physiological and behavioral effects of prior aversive stimulation (preshock) in the rat. *Physiol. Behav.* **10:** 467–71

Levine, S. and Coover, G. D. (1976) Environmental control of suppression of the pituitary-adrenal system. *Physiology and Behavior,* **17:** 35–7

Levine, S. and Ursin, H. (1991) What is stress? in M.R. Brown, C. Rivier and G. Koob (eds.), *Neurobiology and Neuroendocrinology of Stress,* Marcel Decker, New York

Lindsley, D.B. (1951) Emotion, in S.Stevens (ed.), *Handbook of Experimental Psychology,* Wiley, New York

Modigh, K. (1974) Effects of social stress on the turnover of brain catecholamines and 5-hydroxytryptamine in mice. *Acta Pharmacol. Toxicol.,* **34,** 97–105

Moruzzi, G. and Magoun, H.W. (1949) Brain stem reticular formation and activation of the EEG. *Electroenceph. Clin. Neurophysiol.* **1:** 455–73

Munck, A., Guyre, P.M. and Holbrook, N.J. (1984) Physiological functions of glucocorticoids in stress and their relation to pharmacological actions. *Endocrine Rev.* **5:** 25–44

Murison, R., Overmier, J.B., Hellhammer, D.H. and Carmona, M. (1989) Hypothalamo-pituitary-adrenal manipulations and stress ulcerations in rats. *Psychoneuroendocrinology,* **14:** 331–8

Myhre, G., Ursin, H. and Hansen, I. (1981) Corticosterone and body temperature during acquisition of social hierarchy in the captive willow ptarmigan (*Lagopus l. lagopus*). *Zeitschrift für Tierpsychologie*, **57**: 123–30

Olff, M. (1991) *Defense and coping: Self-reported health and psychobiological correlates*. PhD thesis, University of Utrecht

Olff, M., Brosschot, J.F., Godaert, G.L.R. *et al.* (1991) Defence and coping: Effects on hormones and immunology in a laboratory stress situation, in M. Olff (ed.), *Defense and coping: Self-reported health and psychobiological correlates*. PhD thesis, University of Utrecht

Olff, M., Ursin, H. and Sterman, M.B. (1992). Perceptual defense and EEG correlates. *Biofeedback and Self-regulation*, **17**: 233

Overmier, J.B. and Murison, R. (1991) Juvenile and adult footshock stress modulate later adult gastric pathophysiological reactions to restraint stresses in rats. *Behav. Neurosci.* **105**: 246–52

Pickering, A.D. (ed.) (1981) *Stress and Fish*, Academic Press, London

Plutchik, R., Kellerman, H. and Conte, H. R. (1979) A structural theory of ego defenses and emotions, in C.E. Izard (ed.), *Emotions and Psychopathology*, Plenum, New York

Pribram, K. and McGuiness, D. (1975) Arousal, activation and effort in the control of attention. *Psychol. Rev.* **82**: 116–49

Rose, R.M., Poe, R.O. and Mason, J.W. (1967) Observations on the relationship between psychological state, 17-OHCS excretion, and epinephrine, norepinephrine, insulin, BEI, estrogen and androgen levels during basic training. *Psychosom. Med.* **29**: 544

Schreurs P. J. G., Tellegen, B., Van De Willige, G. and Brosschot J. F. (1988) *De Utrechtse Coping Lijst: Handleiding*, Swets en Zeitlinger, Lisse

Selye, H. (1936) A syndrome produced by diverse nocuous agents. *Nature*, **138**: 32

Selye, H. (1974) *Stress Without Distress*, J.B. Lippincott, Philadelphia

Stanton, M.E., Patterson, J.M. and Levine, S. (1985) Social influences on conditioned cortisol secretion in the squirell monkey. *Psychoneuroendocrinology*, **10**: 125–34

Strømme, S.B., Wikeby, P.C., Blix, A.S. and Ursin, H. (1978) Additional heart rate, in: H. Ursin, E. Baade and S. Levine (eds.), *Psychobiology of Stress. A Study of Coping Men*, Academic Press, New York, pp 83–9

Ursin, H. (1980) Personality, activation, and somatic health. A new psychosomatic theory, in: S. Levine, and H. Ursin, (eds.), *Coping and Health*, Plenum Press, New York, pp 259–79

Ursin, H. (1988) Expectancy and activation: An attempt to systematize stress theory, in: D. Hellhammer, I. Florin and H. Weiner (eds.), *Neurobiological Approaches to Human Disease*, Hans Huber, Toronto

Ursin, H., Baade, E. and Levine, S. (eds.) (1978) *Psychobiology of Stress. A Study of Coping Men*, Academic Press, New York

Ursin, H. and Murison, R. (eds.) (1983) *Biological and Psychological Basis of Psychosomatic Disease*, Pergamon, Oxford

Ursin, H., Mykletun, R., Tønder, O. *et al.* (1984) Psychological stress-factors and concentrations of immunoglobulins and complement components in humans, *Scand. J. Psychol.* **25**, 340–7

Vanderwolf, C.H. and Robinson, T.E. (1981) Reticulo-cortical activity and behavior. A critique of the arousal theory and a new synthesis. *The Brain and Behav. Sci.* **4**: 459–514

Værnes, R.J. (1982) The Defense Mechanism Test predicts inadequate performances under stress. *Scand. J. Psychol.* **23**, 37–43

Værnes, R.J., Ursin, H., Darragh, A. and Lambe, R. (1982) Endocrine response patterns and psychological correlates. *J. Psychosom. Res.* **26**, 123–31

Weiss, J.M. (1972) Influence of psychological variables on stress-induced pathology, in: *Ciba Foundation Symposium: Physiology, Emotion and Psychosomatic Illness*, Elsevier, Amsterdam

2

The role of life events in the aetiology of depressive and anxiety disorders

DEPARTMENT OF SOCIAL POLICY AND SOCIAL SCIENCE
ROYAL HOLLOWAY AND BEDFORD NEW COLLEGE
UNIVERSITY OF LONDON
EGHAM HILL, UK

ISBN 0–12–663370–3

2.1 INTRODUCTION

2.1.1 BACKGROUND TO LIFE-EVENT RESEARCH

The study of life events has developed because of the part they appear to play in bringing about disease in human beings. In psychiatry, interest can be traced back to figures like Adolf Meyer at the turn of the century, although systematic enquiry began some 50 years later (e.g. Hinkle and Wolf, 1957; Holmes and Rahe, 1967; Brown and Birley, 1968; Paykel *et al.*, 1969). For obvious reasons, research has been almost entirely non-experimental. Moreover, it has not, on the whole, been theoretically driven. Where it has, ideas have not always been securely based. An example is the influential assumption by Holmes and Rahe (1967), derived from Selye's work, that change in habit, rather than the threat or meaning of life events, was involved in the genesis of disease. We will see that recent research suggests just the reverse: that it is unpleasantness or threat, rather than life change, that is usually involved. Furthermore, current work underlines the importance of going beyond general concepts, such as change and threat, to consider the specific meaning of events in terms of categories such as loss, danger and challenge (Lazarus, 1966; Brown and Harris, 1989).

Despite the scope of such an agenda, life-event research may convey to the newcomer a degree of parochialism, with fierce controversies over methodological issues. But the tolerant reader may accept that this combativeness reflects the importance of the theoretical and practical implications of the effects that are being studied. As well-based empirical generalizations have emerged, the relative isolation of life events research has declined. An example is the recent work on life events in the relapse of schizophrenic patients (Leff, 1991). Such replicable findings call for greater integration into the mainstream of clinical work. Life-events research has implications for research originating from quite different intellectual roots. For instance, in *Human Nature and Suffering*, Gilbert (1989) reviews work in sociobiology, ethology, biology, psychology and clinical sciences, and focuses on the importance of threats to basic biosocial goals, particularly those concerning care giving, care taking, competition and cooperation. Such threats, he argues, disturb central nervous system function with profound implications for psychopathology and illness in general. He makes the point that pursuit of such goals will be coordinated within general arousal systems which facilitate goal-directed behaviour and the converse, disengagement. Gilbert, in fact, discusses several such negative systems, each proposed to have a different phylogenetic root. One stems from carryovers of the reptilian brain and relates to the 'yielding' subroutine that follows defeat; it can involve changes that last a considerable time, and from which an animal may never fully recover. It has been seen as the basis of 'learned helplessness' in animals (Seligman, 1975; Hellhammer, 1983) which is commonly, albeit speculatively, related to profound depression in humans. Such 'defeat states' are distinguished from submissive acts of a subordinate to a dominant animal which are related to the later limbic brain area and more linked to social anxieties.

So far, progress in establishing definite links between psychosocial and biological processes has been modest. The bulk of biological research has sought to elucidate *correlates* of disorders. However, this does not necessarily take us far in aetiological terms. Perhaps most useful are biological markers that reflect vulnerability to onset or relapse of a particular disorder. While there are promising developments, nothing has yet been firmly established in the area of depression (Kupfer, 1991). The area of genetics is the most troublesome at present. As Bronson (1987) notes, it is all too easy to assume that a biological system is entirely inherited. But it seems more likely that genetic influences for most depressive and anxiety disorders will work indirectly through their influence on vulnerability to relevant stressors. That is, the environment will play a major role in the expression of any genetic influence, including individuals' capacity to create their own environment. Individual differences in propensity to derailment may reflect genetically-based vulnerability and protection, developmentally-based risk and protection, the influence of current circumstances, or all three. Certainly a disorder can follow a stressor so quickly that it appears that the event is the sole cause. However, this kind of simple link may be rare. Current research underlines the critical mediating role of other background psychosocial factors and I will argue that life-event research provides the opportunity to examine their role.

2.1.2 VULNERABILITY AND PROTECTIVE FACTORS

By the mid-1970s the need to distinguish life events from background vulnerability factors that modify a person's response to them had been recognized (Brown *et al.*, 1975). Vulnerability was defined in terms of an increased risk once a life event had occurred. For example, low self-esteem would contribute to depressive onset only in the presence of a major loss (e.g. bereavement). This formulation implies the presence of an interaction between the event and the vulnerability factor, and there has been considerable controversy about the statistical procedures necessary to test for this (Tennant and Bebbington, 1978; Brown and Harris, 1978b). While the issues are not entirely settled (e.g. Rutter and Pickles, 1992), there is widespread acceptance of the likely importance of such interactive effects. There is also little dispute about the intellectual gains that have stemmed from efforts to distinguish provoking agents (e.g. life events) from ongoing states of vulnerability (e.g. poor support). However, the distinction between events and background factors might be counterproductive if rigidly applied when dealing with processes occurring over a substantial period of time. The loss of a husband (i.e. event), even when not producing a clinically relevant disorder, can lead to social isolation and lowered self-esteem (i.e. vulnerability) and, in turn, to a further crisis (i.e. event) of a seriously threatening nature. This might involve, for example, a series of 'risky' sexual relationships which are entered upon in an effort to cope with the distress associated with the vulnerability.

Protective factors have typically been seen as the mirror-image of vulnerability: support *versus* lack of support, high *versus* low self-esteem, and so on. However, it is now clear that protective and vulnerability factors often need to be considered separately. A woman can lack support from a husband (vulnerability), but receive it from a close friend (protection) or she can have both low self-esteem (defined by negative comments about self) and high self-esteem (defined by positive comments). Later, research examples will be given that suggest that, in neither instance, would it be useful to see one as the mirror image of the other where clinical depression is concerned.

2.2 MEASUREMENT ISSUES

2.2.1 WHAT SHOULD BE MEASURED?

2.2.1.1 *Subjective reports*

Some of the muddle that has arisen over use of the term 'stress' can be avoided by recognizing that both external stimuli and internal states are involved, and that there is a need to deal with how each translates into the other. However, it does not follow that the best way forward, in either a methodological or theoretical sense, will be to accept what a person says about this. If we simply accept a person's account of his or her response to an event (even assuming that this is accurate) it will be difficult to study ways in which the experience of stress is affected by other factors. Suppose, for example, in a large cross-sectional survey, we wish to see how far social support reduces distress created by loss of a job. The task would be compromised because some instances of job loss will have been excluded as support had eliminated their reported stressfulness. Therefore, to focus on *experience* would risk defeating a major aim of stress research. Furthermore, if we rely on what the respondent tells us about his or her reaction to an event it will be difficult to rule out contamination from the very illness we wish to explain. This stems particularly from 'effort after meaning', whereby a person's report that a stressful event has occurred merely reflects the need to find some explanation for the occurrence of the disorder (Bartlett, 1932; Brown, 1974).

2.2.1.2 *Checklist methods*

Unfortunately, equally serious problems arise from an approach which bypasses such issues by assuming that all events of a given class, such as a job loss, are equally stressful. This may be appropriate for some particularly traumatic events, but it is now clear that most categories of event will contain experiences that vary a great deal in their meaning; for example, was the respondent dismissed, has he another job or any savings, does he have support? Nonetheless, the almost universally-used 'checklist' approach to recording events has invariably employed the same weight or score to denote the stressfulness of 'events' such as job loss. The shortcomings of such instruments do not end here, however. The fact that the respondent is merely asked to check whether a class of event has occurred means that it is impossible to date events with any accuracy. The approach also has low reliability in terms of both (1) agreement when subjects are re-interviewed about the same events, and (2) agreement with ratings of panels of judges about the stress expected to be associated with particular classes of event. In general, empirical findings have been inconsistent and ambiguous and there is growing acceptance that the approach has failed in terms of both accuracy and sensitivity (Dohrenwend *et al.*, 1987; Brown, 1989a; Raphael *et al.*, 1991).

Despite such shortcomings, checklist instruments are still widely used. A technical reason for the continued use of such instruments is that most do not deal with the impact of particular events, but with an overall score based on summing events occurring in a given period of time. In this sense most research has been one step removed from the problem of event variability. Another consequence of the assumption of additivity has been the way it has diverted concern from the issue of meaning. Notwithstanding, there is, as yet, no empirical support for the assumption that it is the total impact of a number of events that is critical – at least in the case of psychiatric disorders. Research with depression, in fact, suggests quite the reverse: that it is the impact of one severe event, or one series of highly related events, that is usually critical (Brown, 1989b). An alternative approach is based on the use of flexible, semi-structured interviews with rating scales that deal with the complexity of verbal and vocal aspects of what is being said. The two approaches are radically different and, although the more structured interview has its uses, this chapter will underline its grave shortcomings, at least for stress research.

2.2.2 CONTEXTUAL RATINGS

What appears to be needed is a way of characterizing events in terms of *potential* stressfulness. A distinctive approach to describing events, which takes into account the context in which they occur, has attempted to achieve this by encouraging respondents to talk at length about circumstances surrounding particular events. In dealing with this material the approach has moved from reliance on what the respondent said he or she felt to characteristics of the event itself. The investigator rates the *likely* meaning of an event in the light of the person's wider social context. For example, in the case of a birth, relevant factors would include housing (will there be overcrowding?), financial insecurity (will she have to stop work?), career plans (will it mean giving up further education?), and routine matters (is there a grandmother willing to care for the child during the day?). In this way, the event's meaning is gauged in terms of its probable significance in the light of the plans and identities of the person involved.

The Life Events and Difficulties Schedule (LEDS) uses this approach and has three main stages. The first involves the investigator using a series of standard questions (with unlimited probes) to establish whether or not particular events have occurred. The LEDS includes only events about which there has been prior agreement that they are likely to produce fairly marked positive or negative emotion. Although there is bound to be an arbitrary element to such criteria, the important point is that the examples are sufficiently extensive to minimize the need for *ad hoc* decisions by the investigator. Moreover, standards will be the same for target and comparison groups.

The second stage involves the interviewer in rating likely threat by collecting contextual material. This is done both from the perspective of the day it occurred (short-term threat) and some 10 to 14 days later (long-term threat). The third stage involves the rating process itself. The approach to rating attempts to systematize the notion of *verstehen* (or 'understanding') stemming from German social science (Jaspers, 1962; Schutz, 1971; Weber, 1964). Because there is a risk of bias at this point, another check is used. At a fourth stage, certain key scales, including threat, are rated again at consensus meetings. Here the interviewer does not disclose information about how the respondent said she had reacted, or whether she developed a disorder. It is this consensus rating that is used in all analyses.

As already noted, life events occurring at a particular time do not constitute the only relevant adversity. A woman may live in unpleasant and overcrowded housing, but not experience an event in the year covered by an enquiry (Pearlin *et al.*, 1981). The LEDS records material about difficulties provided that they have lasted at least four weeks. The procedure is similar to that for events and there is again an extensive manual dealing with definitions and examples.

The inclusion rules and the contextual approach to rating threat meet, to some degree, all the problems so far discussed. Since, for example, in the rating of contextual threat, there is no attempt to assess what was felt, the possibility of missing a *potential* stressor because it was not experienced as threatening should be avoided. Because of this requirement, the approach deliberately rules out taking account of factors such as current social support and coping style when considering context, even though these might well influence the degree of threat caused by an event. This is done because of the need to know whether these factors *separately* influence a person's response to events. Contextual ratings ignore these potential influences and restrict attention to recent behaviour relevant to prior plans and purposes. There is, then, in a sense, a sixth stage where wider factors such as degree of support are taken into account to try to explain why some people do, while others do not, succumb to disorder following a particular life event. Contextual ratings are therefore an overture to more broadly based considerations of possible aetiological processes.

2.3 DIAGNOSTIC ISSUES

The issue of establishing whether or not a psychiatric disorder is present has faced similar problems with data collection. For some of the same reasons, the research to be described has utilized one of the semi-structured psychiatric interviews developed over the last 20 years. The key starting point of such instruments has been the systematic definition of individual symptoms and the development of diagnostic rules which take account of clusters of symptoms. Their construction has been guided a good deal by psychiatric hospital experience. The use of such instruments to determine thresholds for clinically significant conditions has contributed crucially to the plausibility of epidemiological research. The threshold used for the 'Bedford College' surveys reviewed in this chapter identifies 'cases' having syndromes comparable to those of women seen in psychiatric out-patient clinics. It also distinguishes 'borderline cases' having symptoms that are not sufficiently typical, frequent or intense to be rated as cases but which are more than isolated symptoms. In the general population, diagnoses of 'case' depression and 'case' anxiety are the most common.

The surveys to be described have largely concerned women and were based on the 9th edition of the Present State Examination (PSE). The criteria chosen to delineate 'caseness' of depression were that women had to have 'depressed mood' and at least four of 10 other core symptoms such as weight loss, poor concentration, self-deprecation and anergia.

Depressive episodes at a caseness level tend to be self-limiting even without treatment: in a recent survey of mothers living in Islington, an inner-city area of north London, about 10% of episodes arising in the previous year resolved within 13 weeks and about half within six months. Nonetheless, as many as a quarter lasted for at least one year. This leaves out those already depressed so that, at any one time, half the women with a depressive condition will have had it for at least one year. Therefore, despite the episodic nature of much depression, 'chronicity' is an important public health issue.

Many studies (Nolen-Hoeksema, 1987) have reported less depression in men than in women, but the size of the difference is not clear. For example, men are often more difficult to interview and their clinical picture is often complicated by alcoholism or drug addiction. Given the difficulty and cost of collecting detailed material, many community surveys concerned with aetiological issues have concentrated on women. Nevertheless, research suggests that much the same broad categories of psychosocial factors are at work for men as for women (e.g. Bebbington *et al.*, 1981a,b; Bolton and Oatley, 1987; Eales, 1988).

The definition of types of depression presents a major problem. Depressive disorders, particularly those seen by psychiatrists, can differ markedly in clinical terms. While there is general agreement that the rare bipolar conditions need to be treated separately, there remains much uncertainty about the bulk of depressed patients. There is a tradition of distinguishing at least two further categories: 'neurotic' *versus* 'psychotic', 'reactive' *versus* 'endogenous', 'non-melancholic' *versus* 'melancholic', and so on. The basic idea has been that an endogenous category can be distinguished by the clustering of certain symptoms such as 'retardation', 'anergia', 'anhedonia' and 'weight loss'; but lurking at the heart of the distinction has been the idea that some depression (neurotic, non-endogenous, reactive, non-melancholic) is more often provoked by environmental stress and is more reactive to the environment once the depression is underway. There can be no doubt about the reality of clinical differences and the importance of taking them into account. Beyond this, however, there are a host of unsettled issues from a life-event perspective, especially about the 'reactive'/'endogenous' distinction – a matter to which I will return (Monroe, 1990; Young *et al.*, 1987).

Like nearly all work within psychiatry, the approach to diagnosis throughout this review will be firmly categorical. This contrasts with a good deal of other work, especially in psychology, that treats clinical syndromes in dimensional terms. Goldberg and Huxley (1992) are the latest of a long line of commentators who have strongly argued for dealing with anxiety and depression in dimensional terms. Fortunately, it is highly unlikely that a categorical approach is misleading. It may be helpful to recognize that the dispute is not about whether or not depression is a dimensional phenomenon; most of it almost certainly is. The issue is whether, by dealing with it in one or the other way, important empirical generalizations are masked. It is doubtful that a number of key findings to be discussed would have emerged if analyses had been carried out in dimensional terms. It is also the case that a dimensional approach can lead to some highly misleading errors in interpretation (Brown *et al.*, 1991).

2.4 DEPRESSION AND LIFE EVENTS

2.4.1 ONSET OF DEPRESSION

As already noted, contextual ratings of threat have been at the heart of aetiological enquiries using the LEDS. Threatening events of one kind or another have been shown to be important in the development and course of a range of physical and psychiatric disorders (Harris, 1989a) such as schizophrenia (Brown and Birley, 1968; Day, 1989), self-poisoning (Farmer and Creed, 1989), gastrointestinal disorder (both organic and functional) (Creed, 1989; Craig, 1989), multiple sclerosis (Grant *et al.*, 1989) and menstrual disorders (Harris, 1989b).

Only a small minority of life events recorded by the LEDS appear to be capable of provoking a depressive disorder. In technical terms they are those rated as 'moderate' or 'marked' on a 4-point scale of long-term (referring to the situation some 10–14 days after the occurrence of the event) threat or unpleasantness and which are focused on the subject. In our terminology, they are labelled as 'severe' (Brown and Harris, 1978a, 1989). To a lesser extent, major difficulties (i.e. marked difficulties that have lasted two years or more, excluding those which are threatening purely in terms of physical health) have also been shown to relate to onset. Table 2.1 summarizes results from twelve studies using the LEDS. The first, carried out in Camberwell in south London, is typical (see first row). The interview covered an average period of 38 weeks before onset (or before interview for those not depressed) and the figures suggest a considerable association of onset with severe events and, to a lesser extent, with major difficulties.

Some have sought to play down the significance of such results because only about one in five women who experience a severe event go on to develop depression. Andrews and Tennant (1978), in response to the Camberwell results, even suggested that life events are unlikely to have clinical or preventive importance, since they explain only a small proportion of the variance in occurrence of the disorder. However, they failed to point out that this situation holds for almost all aetiological agents in medicine; for instance, heavy smoking explains less than 1% of the variance of lung cancer, even though most of the cancers occur among heavy smokers (Cooke, 1987). A more informative index is population attributable risk which gives the proportion of instances of the disorder associated with the putative causal agent, allowing for the fact that this will at times be due to chance (Markush, 1977). When this index is used there are substantial associations between severe life events and depression. The result of the Outer Hebrides study shown in Table 2.1 is typical: 84% of the cases of depression had a provoking agent (i.e. a severe event or major difficulty) with a population attributable risk (PAR) of 73%, although with only 12% of variance explained (see Brown and Harris, 1986; Cooke, 1987 for further details). Such estimates are, if anything, likely to be conservative. Some events not rated as severely threatening, such as the loss of a pet dog, are at times closely associated with onset, suggesting they *were* experienced as severely threatening.

While findings concerning severe events have been broadly consistent, with estimates of PAR ranging between 50% and 80%, there has been more variability in estimates of the role of major difficulties. Consequently, their causal status is less securely established. A fully satisfactory aetiological case will require better understanding of the causal processes involved. One possibility is that certain quite minor events can bring home to a person what she is missing because of the major difficulty that she is experiencing and, in this way, precipitate onset: for instance, a sister's happy engagement may, for a woman with a major marital problem, serve to underline her own unhappy situation (Brown and Harris, 1978a).

There has been a general agreement about the critical importance of 'loss' for depression (e.g. Bibring, 1953; Beck, 1967; Paykel, 1974; Finlay-Jones and Brown,

Table 2.1 Summary of population studies using LEDs of women in the 18–65 age range, giving relationship of severe events and major difficulties to onset of caseness of depression

Studies (random sample unless stated)	Duration of period studied	Onset cases			Noncases
		Severe event[a]	Major difficulty[a]	Severe event or major difficulty[a]	Severe event or major difficulty[a]
Brown and Harris (1978a): Camberwell	38/52 (weeks)	25/37 (68)	18/37 (49)	33/37 (89)	115/382 (30)
Brown and Prudo (1981): Lewis, Outer Hebrides, Scotland	1 year	11/16 (69)	6/16 (38)	13/16 (81)	42/171 (25)
Costello (1982): Calgary, Alberta, Canada	1 year	18/38 (47)	20/38 (53)	—	—
Campbell, Cope, and Teasdale (1983): Oxford (workingclass with at least one child)	1 year[b]	6/11 (55) 5/12 (42)	6/11 (55) 5/12 (42)	10/11 (91) 9/12 (75)	21/60 (35) 17/52 (33)
Cooper and Sylph (1973): London (general practice)	3/12 (months)	16/34 (47)	—	—	—
Finlay-Jones and Brown (1981): London (general practice)	1 year	27/32 (84)	6/32 (19)	27/32 (84)	32/119 (27)
Martin et al. (1989): Manchester (pregnant women)	1 year	13/14 (93)	4/14 (29)	13/14 (93)	25/64 (39)
Brown, et al. (1986a): Islington (workingclass with at least one child)	6/12[b] (months)	29/32 (91) 25/33 (76)	15/32 (47) 14/33 (42)	30/32 (94) 28/33 (85)	92/271 (34) 107/323 (33)
Parry and Shapiro (1986): Sheffield (workingclass with at least one child)	1 year	12/20 (60)	3/20 (15)	14/20 (70)	62/172 (36)
Bebbington, Hurry, Tennant, and Sturt (1984): Camberwell	10/12 (months)	—	—	13/21 (62)	45/131 (34)
Totals		212/312 (68)	102/279 (38)	218/261 (84)	558/1745 (32)

Note. All study locations were in England unless otherwise specified. Chronic cases of depression were excluded. Average population attributable risk = 73.1%
[a] Percentages of those with at least one event/difficulty are in parentheses
[b] Sample seen on two occasions 12 months apart; first line refers to follow-up period, and second, to anterior period. The proportion with a severe event among non-cases not shown in table was 19%

1981; Miller and Ingham, 1983; Dohrenwend *et al.*, 1986). This was confirmed in a study of severe events before an onset of depression among women living in an inner-city area of London, Islington. The study was longitudinal and dealt with working-class women with children, a fifth of whom were single mothers. Women were contacted one year after they had first been seen, to explore the reason for any onset of depression during the intervening period. The follow-up interview gathered information about events occurring in the year, together with an account of the support received (Brown *et al.*, 1985; Brown *et al.*, 1986a). Twenty-nine of the 32 women developing depression had important loss experiences. The impact of severe events was so substantial that major difficulties did not contribute once they were taken into account; thus in what follows only events will be discussed.

For 12 of the 29 depressed women, the severe event presented a threat to their identity as a wife or mother and about which they could do very little. For most of them, it was part of a long history of failure and disappointment in at least one of these roles. One single mother lived with three children in a comfortable flat. Her only son had left home after a quarrel and she had had no contact with him over the previous year; this had, in fact, led to an episode of depression from which she had largely recovered. A second episode occurred when she learned that he faced a criminal charge. She conveyed a sense of loss and disappointment "That he should turn out like that. I think I made a mistake. It is the result of my actions . . . There is no way I can help him now." She dwelt on the effects of her divorce ten years before, and her difficulty in obtaining custody of her children. She also conveyed a second failure in her attempts to obtain further qualifications for her job by studying full-time.

A second set of eight women had a more diverse set of experiences, but all appeared to feel imprisoned in a non-rewarding and deprived setting. The event itself underlined how little they could do about extracting themselves. Five of the eight had events concerning poor housing or debt, or both. However, there were usually wider ramifications: one woman (a single mother) lived with an extremely hyperactive child; another (again a single mother) was pregnant by a man who had let her down when she had been left homeless after her flat had been set on fire (the severe event). Of the remaining three events in this second group, two concerned severe physical handicap, one involving a husband and the other the woman herself. The eighth woman was unusual in the sense that background circumstances leading to the event were not blatantly depressing; when seen first she reported feeling trapped in a dull and unrewarding marriage. The subsequent event was a love affair going badly wrong. These women were distinguished by the fact that they appeared to be trapped; the event itself seemed to bring this home to them. However, the two groups certainly overlap and the dominant theme for both was one of loss, failure and disappointment.

The final nine women all had grounds to feel rejected, at least for a time. Six were involved in the same themes of failure and disappointment. One, for instance, had originally conveyed how she disliked her unmarried status, how she found it too much responsibility, how lonely she was and how she wished to marry. During the following year she became close to a man: the 'event' was to discover that he was going out with another woman. Another had a similar experience; a further two had trouble with sisters with whom they had been very close; another's husband left home and another suffered the death of a child in circumstances that might have indicated some failure on her part. The remaining three women experienced a death (mother, husband and friend, respectively), but there was no obvious reason for them to feel in any way responsible or rejected. For one of these last three there appeared to be a puzzling element of 'over-reaction': she had an excellent marriage and a wide circle of friends. There does not seem, however, to be much doubt about the conclusion to be drawn: the events for the most part presented an integral threat to core aspects of the sense of identity and self-worth of the women. Almost all could be seen to reflect being stuck in a deprived or punishing situation that could not be changed; a result consistent with the argument of the original Camberwell enquiry: that a general feeling of hopelessness was often critical in mediating between event and depression (Brown and Harris, 1978a: see also Abramson *et al.*, 1989).

The idea of loss alone is not enough, therefore. Even if broadened beyond its everyday usage, it does not convey the sense of personal failure and defeat, or the idea that any way forward is blocked. Such ideas regarding depression are hardly new. Many have discussed the related ideas of helplessness and hopelessness (e.g. Bibring, 1953; Beck, 1967; Lazarus, 1966; Melges and Bowlby, 1969) and the well known learned helplessness theory developed on the basis of animal research is highly relevant (Seligman, 1975, Chapter 6). Unger (1984) discusses psychiatric disorder in terms of fundamental human needs to experience relationships which confirm a distinctive self-identity, and also to enter into some kind of acceptance of one's own character. In light of these, he discusses the paradigmatic experiences of 'blockage' and 'loss'. Loss is linked to disbelief in one's ability to reaffirm and reconstruct one's identity in the absence of the relations that have been destroyed. Here I have suggested that loss of valued ideas about self or another is enough. Blockage refers to a person's doubts about her or his power to escape or change a setting which has become a prison. For the Islington women, the life event before depression often appeared critical in the sense of both blockage and loss.

Since related deprivations and difficulties had often gone on for years, the role of life events also underlines the remarkable capacity of humans to *avoid* depression. It is the sudden 'bringing home' to a person of the implications of a situation that often appears to be critical.

2.4.2 GOING BEYOND CONTEXTUAL RATINGS

The Islington survey was designed to go beyond confirmation of the aetiological role of severe events and, in particular, to explore the reason why only one in five women experiencing such an event developed depression. The next section discusses this in terms of vulnerability and protective factors, but I deal first with the possibility that break-down follows a severe event that is particularly threatening. This uses detailed information collected about each woman at the time of first contact and characterizes each severe event in additional ways. The Self Evaluation and Social Support (SESS) interview-based instrument used to get the necessary background data is described elsewhere (Brown *et al.*, 1990a,b).

As already explained, contextual ratings take into account relevant plans, purposes and commitments insofar as these can be *indirectly* assessed by biographical material. However, the detailed information about each woman collected by the SESS at the time of the first Islington interview allowed a *direct* assessment of 'commitment', and this in turn could be used in making a second-stage assessment of the threat. Six life domains were discussed: marriage, motherhood, housework, employment, extra-household activities and close ties outside the home. A four-point rating of a woman's commitment was made for each. Four-fifths of the 303 women at risk of developing depression in the follow-up year had at least one marked commitment, with an average of 1.5 when all six domains are considered. It was relatively straightforward to 'match' each subsequent severe event with domains of marked commitment, e.g. a husband's infidelity matching a woman's marked commitment to marriage (Brown *et al.*, 1987). If only Islington women with a severe event are considered, matching trebled the risk of depression. The same procedure was followed, with similar results, for marked difficulties that had lasted at least six months. A third exercise, involving role conflict produced relatively few new onsets over and above those for matching D-events and in the final analysis the two types are combined. Results are summarized in Table 2.2, which shows an overall fourfold increase in risk for any woman with a matching, compared with a non-matching, severe event.

Table 2.2 Onset among 130 Islington women with a severe event in Islington by whether it 'matched' certain first interview measures

A.	Severe event matching prior difficulty of 6/12 or more (D-event)	% onset
	YES	46 (16/35)
	NO	14 (13/95)
	p <0.001	22 (29/130)

B.	Severe event matching prior commitment (C-event)	% onset
	YES	40 (16/40)
	NO	14 (13/90)
	p <0.01	22 (29/130)

C.	Either severe event matching prior difficulty, role conflict or prior commitment (i.e. D/R-event or C-event).	% onset
	YES	36 (25/70)
	NO	7 (4/60)
	p <0.001	22 (29/130)

The much greater risk of depression in women with matching commitment confirms our assumptions about the significance of plans and purposes that provide the underlying rationale for contextual threat ratings. The idea of 'matching' has wide ramifications. It has, for example, recently been used to help explain why mothers in full-time work are at greater risk of depression than those not employed, and why those in part-time employment are at the lowest risk (Brown and Bifulco, 1990). It seems likely that those in full-time work are at increased risk because of the way severe events can match critical aspects of employment status. Some women appear to be at risk because their marked commitment leaves them vulnerable to events (such as discovery of glue sniffing by a child) that underline the role conflict they experience about their employment. A second group of women have, at best, modest commitment, seeing full-time work as necessary to keep their household afloat or to improve its living standards. They appear to be particularly vulnerable when a severe event brings home to them that their sacrifices have been undervalued or in vain: for example, a woman who had been working excessive hours to save for a deposit for a house finding out her husband had been feckless and irresponsible with these hard won savings. While this kind of interpretation may only just be acceptable in scientific terms, the underlying associations are reasonably soundly based.

2.4.3 REMAINING CONTROVERSIES

It would be inappropriate, however, to give an impression that life-event research is without controversy. There is still a good deal of conflict about the issue, with opposed 'hard' and 'soft' positions being strongly argued. A 'hard' position is taken by an influential group working in New York who wish to keep quite distinct the factors that influence how threatening an event is: for example, career plans influencing the threat of a pregnancy. It is argued that these should be considered in auxiliary analyses (Dohrenwend et al., 1987). This position underestimates the associations between life events and disorder. Indeed, ratings made with the instrument developed by the New York group appear to show hardly any correlation with those based on a LEDS rating of long-term threat (Brown, 1989a) and have so far produced only modest associations with onset of depression. The best result has been obtained by isolating 'incidents verified to be fateful and disruptive', for which 27% of onset cases and 10% of the non-depressed subjects were positive (Shrout et al., 1989).

By contrast to the New York group, a 'soft' position is taken by Lazarus (1991). In his argument for a subjective approach, the intermediate position taken by the contextual ratings of the LEDS is not considered. He sees emotion (and stress) in terms of a transaction between person and environment. He argues that this has a bearing on personal goals which are brought to the encounter and for which the environmental conditions are relevant. Emotions (and stress) are engendered around the business of realising these goals as well as managing the demands, constraints and opportunities presented by the environment. Such a view, with which I am entirely sympathetic, leads him, however, to an unwarranted stand regarding the need for subjective measurement of people's experience. He ignores the grave shortcomings that it has shown in aetiological research as discussed above (and as exemplified in his work on minor life events, or 'hassles'; Brown, 1989c). He argues that a measure must aim to be maximally accurate, but he fails to recognize that there may be certain advantages to be derived from some forms of 'inaccuracy', as argued for contextual ratings. It follows that there can be no general solution. The scientific question concerns the *degree* to which subjective elements are allowed into the research process.

2.5. VULNERABILITY TO DEPRESSION

It will be recalled that vulnerability factors are defined by their ability to increase risk only in the presence of a provoking agent. The search for such factors will be greatly hampered by an ineffective life-event instrument. This is because, to the degree that events are 'missed', a vulnerability factor will show up as acting independently as a risk factor. Where the LEDS is concerned, it has been useful to go on to ask why most people experiencing a severe event do not develop depression; in fact, as many as two-thirds do not do so even in response to a matching severe event (Table 2.2).

2.5.1 PARTICULAR VULNERABILITIES

The Camberwell enquiry concluded that low self-esteem might explain why the background vulnerability factors that were isolated (such as little intimacy with husband and having three or more children under 15 living at home) raised risk in the presence of a provoking agent (Brown and Harris, 1978a). In Islington, low self-esteem, when judged from negative comments that a woman made about herself, did appear to play such a role. Among those with a severe event, risk of onset was increased almost threefold when low self-esteem had been present at the time of the first interview (Table 2.3; Brown et al., 1986a; see also Ingham et al., 1986; Miller et al., 1987). However, the exact role of low self-esteem cannot be settled without taking into account other relevant factors and, in particular, the quality of social support. The need for such a wider view is shown by two puzzling findings of the Islington survey. First, whereas a measure of self-esteem based on negative comments was related to onset, another based on positive comments was quite unrelated. Secondly, and more surprisingly, not one of a series of positive measures of marriage was related to the risk of onset despite the fact that almost all negative measures dealing with marriage were highly related.

Table 2.3 Low self-esteem at first interview and onset of depression in the follow-up year among 303 Islington women

Low self-esteem	Severe event (% onset)	
	Yes	No
Yes	34 (17/50)	6 (2/31)
No	15 (12/80)	1 (1/142)
	p <0.02	ns

The reason for these curious results turned out to relate to the quality of support occurring in the follow-up year. For married women, support at the time of the first interview did not guarantee help in a subsequent crisis; such discontinuity was particularly likely to relate to depression. It is as though there were two relevant events: the crisis itself and the fact that the anticipated support was not forthcoming. Positive measures of the marriage failed to predict depression because women who were 'let down' were more likely to be positive in this way at the earlier first interview. Because being 'let down' was highly related to onset, the predictive power of any such positive measure was suppressed in a statistical sense. It was only when those 'let down' were excluded that the expected link between positive measures and low risk emerged (Brown et al., 1986a; Brown, 1992). Table 2.4 illustrates this. For the sake of simplicity it deals with married women and only considers support received from a husband.

Table 2.4 Confiding in husband at first interview, crisis support during the follow-up year and onset of depression among those with a provoking agent (97 married Islington women)

Confiding in husband at first interview	Crisis support from husband during follow-up (% with onset)		
	Yes	No	Total
Yes	'a' 4 (1/28)	'b' 37 (7/19) 'let down'	17 (8/47)
No	'c' 25 (2/8)	'd' 24 (10/42)	24 (12/50)
	(Crisis support not known for 4)		

The large difference in terms of onset between cells 'b' ('let down') and 'a' (not 'let down') demonstrates statistical suppression, i.e. when cells 'a' and 'b' are combined, and the undifferentiated measures of support at first interview considered, first interview support is not linked to depression (see the right-hand column).

The data also illustrate the need for protective and vulnerability factors to be considered separately. Women receiving inadequate support at times 1 and 2 were, like those 'let down', at high risk (compare cells 'd' and 'b'). However, a further result (not shown in the table) dramatically changes the implications of this. Among those without support from a husband on either occasion (i.e. cell 'd'), those receiving support from another core tie (usually a woman friend or close relative) had practically no chance of developing depression. Both protection and vulnerability therefore appear to be at work. Sadly, such protection did not hold for those 'let down', i.e. cell 'b'. This is, of course, one more example of how life events can be used to build complex causal propositions.

It is also of interest that single mothers were hardly ever 'let down'. Support from close ties predicted extremely well those who would get help in a crisis in the following year. The problem for single mothers was that many of them (about half) lacked effective support on both occasions.

2.5.2 AN OVERALL CAUSAL MODEL OF DEPRESSION

By taking account of factors such as matching events and the quality of support in the follow-up year, a reasonably convincing causal model with a good deal of explanatory power has been created. Since most of the component factors were established retrospectively at the follow-up interview, some might question its validity (Brown *et al.*, 1990c). However, an equally impressive result is obtained by taking account only of factors measured at first interview: in particular, negative psychological factors (negative evaluation of self or low grade symptoms of anxiety or depression) and negative environmental factors (negative interaction in the home and, for single mothers, lack of a core close tie). While only a fifth of the 303 women at risk of developing depression had *both* risk factors at first interview, three-quarters of the subsequent onsets occurred among them. The fact that ratings made before any onset related as well as those taken after onset suggests that measurement artefacts have not been seriously involved. The first interview risk factors therefore need to be seen in terms of event-production *and* vulnerability (see Figure 2.1).

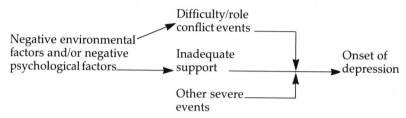

Figure 2.1 Schematic causal diagram linking first interview measures with those collected at follow-up

It is also of note that severe events alone rarely relate to onset. Few women developed depression without at least one of the first interview risk factors. In other words, multifactorial models are essential. Finally, although the associations are impressive, half of those in the highest risk group (with both background risk factors) did *not* develop depression following a severe event. There is therefore plenty of room for other factors, and there is no reason why risk defined in biological terms should not form a part of any final model.

2.5.3 MODEL BUILDING

The analyses so far will have conveyed the fluid quality of the factors involved. If environmental circumstances change, risk may change. Rutter (1990), to underline this point, uses the terms 'process' or 'mechanism' rather than 'variable' or 'factor':

"this is because any one variable may act as a risk factor in one situation but a vulnerability factor in another. There has been much unhelpful dispute in the literature on the supposed buffering effect of social support, because most investigators have assumed that the vulnerability (or protection) lies more in the variable than the process."

I have employed instead the notion of 'model' to convey the importance of exploring such processes. I prefer to retain the term 'factor' to describe the elements of the model as it is doubtful whether we can hope to do much more at present than plot the occurrence of such factors at distinct times. I have used 'process' for attempts to interpret the links between the various factors. But the message is the same: 'factors' can change and, at any one time, a single 'factor' may serve several functions.

The Islington results also point to the need to cover a range of factors in any model, and the study of these is likely to demand familiarity with a number of disciplines. Since such collaboration is likely to mean that more material will have to be collected, it is fortunate that such model building does not necessarily require large numbers of subjects. On the assumption that important causal effects have been uncovered, the greater sensitivity and accuracy of interview-based ratings make it possible to use a relatively small number of subjects. But the development of an effective model will almost certainly require a series of studies, each gradually pushing back into the complex of causal factors. The research outlined here started with the isolation of proximal stressors and, in the course of a series of studies, sought to isolate relevant vulnerability and protective factors. It should be clear that the success of this, and particularly the success of any test of vulnerability, has crucially depended on establishing time order. Major studies involving life events which ignore this vital requirement can still be found (e.g. Phelan *et al.*, 1991).

A common problem is that chronic conditions are muddled with those of more recent onset. Provoking events for chronic disorders will usually not be documented because they are outside the study period. Therefore, a given variable may be ruled out as a vulnerability factor because it occurs often without a severe life event (Brown, 1981). It is therefore necessary to ask quite distinct questions about onset and course, not only to deal with this kind of possibility, but to account for more complex situations. These include the possibility that 'stressor' and 'vulnerability' factors not only independently *perpetuate* depression, but synergistically *provoke* it.

Finally, as already implied, it will be necessary to take into account relevant biological processes. One possible explanation for the role of ongoing difficulties in the onset of depression is that they lead to more or less irreversible changes in certain biological substrates. This, for example, could involve sensitisation and kindling phenomena seen in animals and which might be relevant to human depression. In essence, with repeated depressive episodes, the biochemical and physiological processes involved in a depressive illness could be more easily triggered by the external environment (Post *et al.*, 1986). This could come about in a number of ways. Ongoing support may excite a system (say 5-hydroxytryptamine function) and thereby increase resilience (Deakin, 1990; Deakin *et al.*, 1990; Goldberg and Huxley, 1992), while ongoing difficulties and deprivations may attentuate it. Both processes would be compatible with a vulnerability model. In some circumstances, the adverse biological changes, although increasing risk, may still require a provoking severe event.

2.6 PSYCHIATRIC DEPRESSED PATIENTS AND LIFE EVENTS

So far I have concentrated on depressive episodes occurring in the general population, albeit defined in terms of the level of severity typically seen in out-patient practice. The associations between events and depression among patients have proved to be substantial (see Table 2.5), but somewhat lower than in the general population. This (and the somewhat greater variability of the findings) is hardly surprising. In depressive conditions treated on an in-patient basis, clinical experience suggests that some truly 'endogenous' disorders are almost certainly present. However, systematic enquiry has consistently failed to demonstrate major differences in the experience of life events between patients regarded as having 'endogenous' or 'non-endogenous' ('neurotic') depression. Some studies have shown modest differences and there is an overall tendency for endogenous groups (defined in clinical terms) to contain a somewhat lower proportion with an event, but the basic negative finding predominates (see Katschnig *et al.*, 1986, for a recent review). This still leaves the issue of why *clinically* defined subgroupings of depression are at best only modestly differentiated in terms of the presence of provoking life events.

Table 2.5 Summary of recent studies of life events and onset of depression among psychiatric patients

	Number of subjects		Percentage with at least one event			
Source	Patients	Comparison group	Patients (%)	Comparison group (%)	Period covered	Type of event
Paykel *et al.* (1969): New Haven, CT, USA	185	185	44	17	6 months	'Undesirable'
Brown and Harris, (1978a): London	114	382	61	19	8 months	'Severe threat'
Barrett (1979): Boston	130	—	58	—	6 months	'Undesirable'
Glassner, Haldipur and Dessauersmith (1979): New York	25[b]	25	56	16	12 months	'Major role loss'
Fava, Munari, Pavan, and Kellner (1981): Padua, Italy	40	40	73	30	6 months	'Uncontrollable'
Benjaminsen (1981): Odense, Denmark	89	—	63	—	6 months	'Severe loss'
Vadher and Ndetei (1981): Nairobi, Kenya	30	40	67	8	12 months	'Severe threat'
Murphy (1982): London	100[c]	168	48	23	12 months	'Severe threat'
Katschnig *et al.* (1981): Vienna	176[d]	—	42[d]	—	12 months	'Severe threat'
Bebbington, Tennant, moderate and Hurry (1981b): London	45[e]	257	18	10	3 months	'Marked an independent threat'
Perris (1984a): Sweden	38	—	62	—	12 months	'Exit event'
Calloway and Dolen (1989): North London	72	—	39	—	6 months	'Severe threat'
Brown, Harris and Hepworth (unpublished): North London	128	—	64	—	6 months	'Severe threat'[a]
Lora and Fava (1992): Milan	46	74	87	19	12 months	'Severe threat'[a]

[a] Either independent or possibly independent events
[b] Manic-depressive, bipolar sample
[c] Elderly sample aged 65–87 years
[d] Special analysis for this chapter
[e] Includes some anxiety states

A possible explanation arises from a finding that has been even more variable and uncertain. A number of studies of depressed patients have reported that the presence of a stressor is less common with successive episodes of depression (Angst,

1966; Dolan *et al.*, 1985; Ezquia *et al.*, 1987; Ghaziuddin *et al.*, 1990). However, Perris (1984a,b) found only small differences, and my own research group has failed to establish such an association. Nonetheless, the idea of some kind of increased sensitivity after a first attack is an interesting one and a recent analysis has taken up this idea in the context of the reactive-endogenous issue. In a study of depressed patients in north London, the presence of a severe event in the six months before onset proved to be high among a non-melancholic/non-psychotic depressive group whether or not the episode was the first one or not (Brown *et al.*, unpublished). When turning to the group that could be expected to be 'endogenous', the proportion was equally high among those with a first onset melancholic/psychotic condition, but much lower among those who had had a prior episode (see row 1, Table 2.6). Something appears to be occurring that would go some way towards explaining the puzzling inconsistencies in the research literature about endogenous conditions. It suggests that whether or not a particular melancholic/psychotic group differs from the neurotic group will depend on the relative proportion of first onsets. Such patients might, for instance, be much less common in a tertiary treatment centre than in an ordinary psychiatric service. Of course, if correct, the result raises a set of further questions. Does the first onset among the 'endogenous' group result in some kind of 'scarring' that leaves a person open to 'spontaneous' episodes? What would be the biological basis for this? It will now be clear that one of the roles of life-event research is to raise just such questions.

Table 2.6 Ordinal number of adult episodes of depression by 'melancholic/psychotic' division and presence of a severe event in 6/12 of onset

	Melancholic/psychotic score (% severe event)			
	Least	Most	Odds ratio	P
1st onset	72 (26/36)	59 (10/17)	1.82	ns
Non-1st onset	74 (42/57)	22 (4/18)	9.80	$p \leqslant 0.01$

2.7 ANXIETY AND LIFE EVENTS

Much less attention has been devoted to enquiries about another common disorder, anxiety, despite general agreement that, like depression, it is especially common in women. Anxiety needs some consideration in the present review if only because of its frequent overlap with depression (Wittchen, 1988; Wittchen *et al.*, 1991).

The DSM-III-R manual (American Psychiatric Association, 1987) distinguishes panic disorder, agoraphobia, generalized anxiety disorder, social phobia and simple phobia, and allows for the presence of more than one of these. This helps highlight the conceptual and practical problems raised by the overlap of depressive and anxiety conditions, i.e. their comorbidity. In the past, it has been common to give priority to one of the two conditions (usually depression). However, there has been a recent move away from such arbitrary rules. This breaking from hierarchial diagnostic rules has allowed the issue of comorbidity in depression and anxiety to be systematically tackled.

A useful start is provided by the accumulating evidence that severely threatening life events play as important a role in onset of anxiety disorder as in depression. Finlay-Jones, studying attenders at a general practice surgery in the Regent's Park area of London, found that just over two-thirds of women with a recent onset had had an accompanying severe event (Finlay-Jones and Brown, 1981; Finlay-Jones, 1989). Of particular interest was his introduction of the *second-order* level of meaning. This has already been touched upon with regard to 'loss'; 'loss' was distinguished from 'danger', the latter defined in terms of the degree of unpleasantness of a specific future crisis consequent upon an event. As predicted, severe events involving loss occurred more commonly before depression whereas those involving danger more often before anxiety. Consistent with this, mixed onsets of anxiety and depression showed an increase in both loss and danger, either in terms of distinct events or by one event containing significant aspects of loss and danger (e.g. death of a husband causing loss and leaving his widow severely in debt thereby also causing a danger of future credit crisis). The Islington survey, where anxiety conditions were classified in terms of DSM III-R criteria, has confirmed this result (Brown, 1993).

I will return to the issue of comorbidity when discussing recovery (see section 2.8) and the impact of childhood experience (section 2.9). At this point I have used the issue of comorbidity to illustrate the important role of quality of meaning for understanding different, albeit overlapping, psychiatric conditions and, with this, the need to move beyond general threat ratings.

2.8 RECOVERY, EVENTS AND DIFFICULTIES

Depression can take a chronic course (one quarter of all new onsets persist for at least 12 months), and anxiety disorders are even more likely to do so. It is therefore necessary to consider the effect of psychosocial factors on the course of a disorder, as well as on its onset. The original Camberwell survey underlined the role of ongoing difficulties in chronicity. Surtees and Ingham (1980) have used the LEDS to show that the occurrence of threatening events was related to the failure to recover in treated patients. Several other studies have also suggested a significant role for psychosocial factors in chronicity (e.g. Huxley *et al.*, 1979; Mann *et al.*, 1981; Tennant *et al.*, 1981). More recently a case has been made that, in Islington, psychosocial factors not only play a role in improvement and recovery in depressive conditions, but that they appear to do so in terms of processes that are the mirror-image of those producing onset. Two LEDS contextual measures have proved important: fresh-start events and difficulty-reduction events (the latter subsequently subsumed under the notion of relief events). Fresh starts are directly focused on the subject and involve hope of something better, either by resolving a LEDS difficulty or the amelioration of a state of deprivation. A single mother, for example, dissatisfied at her limited range of social contacts (that would not as such be rated as a LEDS difficulty), may acquire a boyfriend. By contrast, difficulty-reduction (or relief events) refers to a reduction in a LEDS difficulty without any implications of a 'fresh-start': for example, a noisy, intrusive neighbour moving away.

In the Islington study two-thirds of women improving or recovering from an episode of depression lasting at least three months had either a fresh-start event or a difficulty-reduction event in the prior three months. This proportion is much higher than would be expected in any one three-month period and even higher than their rate (36%) during the total time of the episode throughout the study period (three years average; Brown et al., 1988). The effect did not hold for episodes of less than three months.

A more recent analysis of the same data suggests that much the same effects are present for anxiety conditions defined in DSM-III-R terms (Brown et al., 1992b). However, an association only emerges when 'positive' events are considered that could anchor the person in a new role, or renew an old one. In both instances, there are implications for increased security: for example, owning a house for the first time which involves moving from a run-down inner-city area to a new estate in the suburbs, a second pregnancy, a new job which, although not doing much to reduce a financial difficulty, is required if the woman is not to live on social security. While fresh-start and anchoring are often aspects of the same event, events relating to recovery (and onset) tend to have different meanings according to whether depression or anxiety is involved: those involving depression more often relate to hope (or lack of it) and those involving anxiety to security (or lack of it). The fact that an event will often contain implications for both dimensions rules out the emergence of a clear-cut result; nonetheless, the patterning of results across a number of studies makes a reasonable *prima facie* case for specific processes relating particular meanings to particular diagnostic categories.

2.9 LIFE EVENTS OVER THE LIFE SPAN

Many experiences of childhood and adolescence can be viewed as life events. The aetiological model of the original Camberwell enquiry, for example, included loss of mother before the age of 11 due to either death or a separation of at least one year. This was associated with a raised risk of depression in adulthood once a provoking agent had occurred. Along with this went a greater overall prevalence of chronic depressive conditions (Brown et al., 1977). While there has been a good deal of controversy surrounding these findings (Crooke and Eliot, 1980; Tennant et al., 1980; Harris et al., 1986, 1987), two further population studies have produced equally clear results. An enquiry in Walthamstow first screened women in the general population as a means of providing a large number of instances of such a loss, and these women were again found to have a much higher rate of current depression in adult life. This was also found for losses between the ages of 11 and 17 (Brown et al., 1986b; Harris et al., 1986, 1987). A high rate of adult depression also occurred among Islington women losing a mother between age 11 and 17, again suggesting that the original cut-off at age 11 needed to be amended (Bifulco et al., 1987).

The aetiological processes involved are undoubtedly complex, with recent research emphasizing the role of untoward experiences *after* the loss, rather than the loss itself, as being critical (Harris et al., 1986, 1987; Bifulco et al., 1987). Future research will need to trace a woman's history in detail, from the early loss itself to later depression. Already there is suggestive evidence that, in terms of the life span as a whole, intervening experiences can reduce, as well as increase, risk among those with adverse childhoods. Initial attempts to chart the relevant chain of circumstances suggest that certain early life experiences are particularly likely to be associated with a raised risk of provoking agents and vulnerability factors in adult life (Harris *et al.*, 1986, 1987). Especially important in setting a woman on an adverse life-course trajectory was the quality of replacement parental care after the loss. If this was judged inadequate (in terms of an index of parental indifference and lax control, termed *lack of care*), the risk of current depression was doubled; this held true irrespective of whether there had been a loss of mother (Bifulco *et al.*, 1987).

Another factor which plays a critical mediating role is the experience of premarital pregnancy. This often traps women in relationships which they might not otherwise have chosen and, like lack of care, may in time become a source of provoking agents: for example, housing and financial problems consequent upon a couple starting a family too young, or marital difficulties with undependable partners. In Walthamstow, such women also emerged as less upwardly mobile in terms of social class. In interpreting such experiences, a conveyor belt of adversities was outlined along which some women moved from one crisis to another, starting with lack of care in childhood and passing *via* premarital pregnancy to current working-class status, lack of social support, and high rates of provoking agents. It would be short-sighted, however, to attribute such a chain of circumstances solely to environmental factors. Although it was often hard to see how they could have left this conveyor belt once their childhood had located them on it, a more personal element, as well as bad luck, undoubtedly could play an important role.

More recently, this early index has been expanded to include sexual and physical abuse which, while correlated with lack of care, independently increase risk (Bifulco *et al.*, 1992, submitted). The resulting index of *childhood adversity* has been found in the Islington survey material to relate also to a much increased risk of DSM-III-R anxiety conditions other than simple phobia and mild agoraphobia (Brown and Harris, 1992; Brown *et al.*, 1993a and b). There is a much increased rate among those with marked agoraphobia, generalized anxiety disorder and social phobia. Other research reports have indicated a raised risk of panic disorder among women with significant adversity in childhood (e.g. Raskin *et al.*, 1982; Faravelli *et al.*, 1985; Tweed *et al.*, 1989).

The Islington analysis also throws further light on the puzzling question of comorbidity of anxiety and depression. A substantial amount of this was explained by the presence of childhood adversity because this led to an increased prevalence of both disorders. If replicated, this result is another example of the way life events can illuminate complex clinical issues.

2.10 CONCLUDING REMARKS

Ideally, this review would have covered research in a range of disciplines that have shown some interest in life events, although it is frustrating that almost all of it has been based on questionnaire-type measures. But there are signs of a change. In cognitive psychology, for example, there is some recognition that cognitions are typically responses to external events and that psychopathology is not necessarily just a consequence of an individual 'overreacting' because of 'faulty' cognitions.

Once a role for the environment is seriously considered, many new avenues for research open up. What about seeing treatment as a potentially fresh-start event and this as part of the explanation for the powerful placebo effects that can occur? How far do psychologically based therapies (or drug related ones for that matter) work through their ability to facilitate 'positive' life events? How far is the notorious length of full psychoanalysis related to the fact that 'support' within the therapeutic setting tends to discourage crucial 'delogjamming' and 'fresh-start' decisions in everyday life?

Perhaps most notable of all developments is the growing realisation that events (and particularly interpersonal difficulties) in childhood and adolescence play a significant role in adult psychopathology. Part of the importance here of life-event studies resides in its emphasis on establishing time order. Results of the kind reported have been convincing to the degree that they have managed to juxtapose life events and onset (as well as relapse and recovery). This is something made possible by intensive interviews and narrative accounts. It now looks as though it will be necessary to do this for the whole life span. This need can be seen in the many genetically inspired family studies, which have concentrated on the extent to which there is a raised rate of psychiatric disorder in first degree relatives of probands and the extent to which disorders 'breed' true. While the main thrust is genetic in spirit, there is a recognition that, so far, findings could equally be explained by psychosocial factors. One fundamental problem has been that psychosocial adversity experienced by offspring of a psychiatrically ill parent is so common that it is extremely difficult to sort out the contributions of environment and heredity. One way forward, and perhaps, at the moment, the only feasible way, is to plot the temporal relation between both psychopathology and relevant life experiences. Did a child's experience of persistent sexual abuse predate a mother's depression? If it did not, how far did the experience of abuse relate to change in behaviour on the child's part and to what degree on the mother's? For example, did her care become more ineffective after the onset of her depression? How far were both the mother's depression and the child's experiences part of a wider context of adversity: the violent behaviour of the father, for instance?

Despite immense effort in biological research in recent years we are still, at best, on the edge of discovering critical risk factors in depression or anxiety. Relating life events to biological factors has hardly begun: one study in particular points to the way in which biological research can be illuminated by the study of life events and difficulties. This examined the role of LEDS-type events on the function of the hypothalamic-pituitary-adrenal axis and the hypothalamic-pituitary-thyroid axis in depressed and non-depressed men and women (Calloway and Dolan, 1989), and showed the degree to which severe events can influence both types of endocrine function. However, enough has been achieved to encourage collaborative research. Unfortunately, since more data are needed for each subject, there is bound to be increased pressure to use simple and cheap measures. Research reports are therefore likely to continue to be largely characterized by a startling gap between the sophistication of the statistical manipulation of data and mediocrity of measurement. While correction of this imbalance will require some rethinking on the part of research worker and funding body, it is difficult not to be optimistic about new developments in life-event research over the next decade.

ACKNOWLEDGEMENTS

Much of the research described in this chapter has been supported by the Medical Research Council. All of it has been carried out in close collaboration with colleagues who have shared closely in the development of both the research and ideas. I am particularly indebted to Tirril Harris who made valuable comments on an early draft of this chapter.

2.11 REFERENCES

Abramson, L.Y., Metalsky, G.I. and Alloy, A.B. (1989) Hoplessness depression: A theory-based subtype of depression. *Psychological Rev.*, **96**: 358–72

American Psychiatric Association (1987) *Diagnostic and Statistical Manual of Mental Disorders (Third edition – revised)*, American Psychiatric Association, Washington

Andrews, G. and Tennant, C. (1978) Editorial: Life event stress and psychiatric illness. *Psychol. Med.*, **8**: 545–9

Angst, J. (1966) *Atiologie und Nosologie Endogener Depressiver Psychosen*, Springer-Verlag, Berlin

Barrett, J.E. (1979) The relationship of life events to the onset of neurotic disorders, in J.E. Barrett (ed.), *Stress and Mental Disorders*, Raven Press, New York, pp 87–109

Bartlett, F. (1932) *Remembering: A Study of Experimental and Social Psychology*, Cambridge University Press, UK

Bebbington, P.E., Hurry, J., Tennant, C. *et al.* (1981a) Epidemiology of mental disorders in Camberwell. *Psychol. Med.*, **11**: 561–79

Bebbington, P.E., Tennant, C. and Hurry, J. (1981b) Adversity and the nature of psychiatric disorder in the community. *J. Affect. Disord.*, **3**: 345–66

Bebbington, P.E., Hurry, J., Tennant, C. and Sturt, E. (1984) Misfortune and resilience: a community study of women. *Psychol. Med.*, **14**: 347–63

Beck, A.T. (1967) *Depression: Clinical, Experimental and Theoretical Aspects*, Staples Press, London

Benjaminsen, S. (1981) Stressful life events preceding the onset of neurotic depression. *Psychol. Med.*, **11**: 369–78

Bibring, E. (1953) Mechanisms of depression, in P. Greenacre (ed.), *Affective Disorders: Psychoanalytic Contributions to their Study*, International Universities Press, New York

Bifulco, A., Brown, G.W. and Harris, T.O. (1987) Childhood loss of parent and adult psychiatric disorder: the Islington Study. *J. Affect. Disord.*, **12**: 115–28

Bifulco, A., Harris, T. and Brown, G.W. (1992) Mourning or early inadequate care – reexamining the relationship of maternal loss in childhood with adult depression and anxiety. *Develop. and Psychopath.*, **4**: 433–49

Brown, G.W. (1993a) Life events and affective disorders: replications and limitations *Psychom. Med.*, **55**; 248–259.

Brown, G.W. and Harris, T.O. (1989) *Life Events and Illness*, Unwin and Hyman, London

Brown, G.W. and Harris, T.O. (1992) Aetiology of anxiety and depressive disorders in an inner-city population: 1: early adversity. *Psychol. Med.* (in press)

Brown. G.W. and Prudo, R. (1981) Psychiatric disorder in a rural and an urban population. 1. Aetiology of depression. *Psychol. Med.*, **11**: 581–99

Brown, G.W., Bhrolchain, M.N. and Harris, T. (1975) Social class and psychiatric disturbance among women in an urban population. *Sociology*, **9**: 225–54

Brown, G.W., Harris, T.O. and Copeland, J.R. (1977) Depression and loss. *Br. J. Psychiatr.*, **130**: 1–18

Brown, G.W., Craig, T.K.J. and Harris, T.O. (1985) Depression: disease or distress? Some epidemiological considerations. *Br. J. Psychiatr.*, **147**: 612–22

Brown, G.W., Bifulco, A., Harris, T. and Bridge, L. (1986a) Life stress, chronic psychiatric symptoms and vulnerability to clinical depression. *J. Affect. Disorders*, **11**: 1–19

Brown, G.W. Harris, T.O. and Bifulco, A. (1986b) The long term effects of early loss of parent, in M. Rutter, C.E. Izard and P.B. Read (eds.), *Depression in Young People*, Guilford Press, New York, pp 251–96

Brown, G.W., Bifulco, A. and Harris, T. (1987) Life events, vulnerability and onset of depression: Some refinements. *Br. J. Psychiatr.*, **150**: 30–42

Brown, G.W., Adler, Z. and Bifulco, A. (1988) Life events, difficulties and recovery from chronic depression. *Br. J. Psychiatr.*, **152**: 487–98

Brown, G.W., Andrews, B., Bifulco, A. and Veiel, H. (1990a) Self-esteem and depression: 1. Measurement issues and prediction of onset. *Social Psychiatr. and Psychiatric Epidemiol.*, **25**: 200–9

Brown, G.W., Bifulco, A. and Andrews, B. (1990b) Self-esteem and depression: 3. Aetiological issues. *Social Psychiatr. and Psychiatric Epidemiol.*, **25**: 235–43

Brown, G.W., Bifulco, A. and Andrews, B. (1990c) Self-esteem and depression: 4. Effect on course and recovery. *Social Psychiatr. and Psychiatric Epidemiol.*, **25**: 244–9

Brown, G.W., Harris, T.O. and Lemyre, L. (1991) Now you see it, now you don't – some considerations on multiple regression, in D. Magnusson, L.R. Bergman, G. Rudinger and B. Torestad (eds.), *Problems and Methods in Longitudinal Research: Stability and Change*, Cambridge University Press, Cambridge, pp 67–94

Brown, G.W., Harris, T.O. and Eales, M.J. (1993b) Aetiology of anxiety and depressive disorders in an inner-city population: 2: comorbidity. *Psychol. Med.* **23**: 155–65

Brown, G.W., Lemyre, L. and Bifulco, A. (1992b) Social factors and recovery from anxiety and depressive disorders: a test of the specificity. *Br. J. Psychiatr.* **161**: 44–54

Calloway, P. and Dolan, R. (1989) Endocrine changes and clinical profiles in depression, in G.W. Brown and T.O. Harris, (eds.), *Life Events and Illness*, Guilford Press, New York

Campbell, E., Cope, S. and Teasdale, J. (1983) Social factors and affective disorder: an investigation of Brown and Harris's model. *Br. J. Psychiatr.*, **143**: 548–53

Cooke, D.J. (1987) The significance of life events as a cause of psychological and physical disorder, in B. Cooper (ed.), *Psychiatric Epidemiology: Progress and Prospects*, Croom Helm, London, pp 67–70

Cooper, B. and Sylph, J. (1973) Life events and the onset of neurotic illness: an investigation in general practice. *Psychol. Med.* **3**: 421–35

Costello, C.G. (1982) Social factors associated with depression: a retrospective community study. *Psychol. Med.*, **3**: 421–35

Craig, T.K.J. (1989) Abdominal pain, in G.W. Brown and T.O. Harris (eds.), *Life Events and Illness*, Guilford Press, New York, pp 213–31

Creed, F. (1989) Appendectomy, in, G.W. Brown and T.O. Harris (eds), *Life Events and Illness*, Guilford Press, New York

Crooke, T. and Eliot, J. (1980) Parental death during childhood and adult depression: a critical review of the literature. *Psychol. Med.* **87**: 252–9

Day, R. (1989) Schizophrenia, in G.W. Brown and T.O. Harris (eds), *Life Events and Illness*, Guilford Press, New York, pp 113–37

Deakin, J.W. (1990) Serotonin subtypes and affective disorder, in C. Idzidowski and P. Cowen (eds.), *Serotonin – Sleep and Mental Disorder*, Blackwell Scientific, Oxford

Deakin, J.W., Pennell, I., Upadhaya, A. and Lofthouse, R. (1990) A neuroendocrine study of 5-HT function in depression: evidence for biological mechanisms of endogenous and psychosocial causation. *Psychopharmacol.*, **101**: 85–92

Dohrenwend, B.P., Shrout, P.E., Link, B.G. *et al.* (1986) Overview and initial results from a risk-factor study of depression and schizophrenia, in J.E. Barrett and R.M. Rose (eds.), *Mental Disorders in the Community*, Guilford Press, New York, pp 184–215

Dohrenwend, B.P., Link, B.G., Kern, R. *et al.* (1987) Measuring life events: the problem of variability within event categories, in B. Cooper (ed.), *Psychiatric Epidemiology: Progress and Prospects*, Croom Helm. London, pp 103–19

Dolan, R.J., Calloway, S.P., De Souza, F.V.A. and Wakeling, A. (1985) Life events depression and hypothalamic-pituitary-adrenal axis function. *Br. J. Psychiatr.*, **147**: 429–33

Eales, M.J. (1988) Affective disorders in umemployed men. *Psychol. Med.*, **18**: 935–46

Ezquiaga, E., Gutierrez, J.L.A. and Lopez, A.G. (1987) Psychosocial factors and episode number in depression. *J. Affective Disord.*, **12**: 135–8

Faravelli, C., Webb, T., Ambonetti, A. *et al.* (1985) Prevalance of traumatic early life events in 31 agoraphobic patients with panic attacks. *Am. J. Psychiatr.*, **142**: 1493–4

Farmer, R. and Creed, F. (1989) Life events and hostility in self poisoning. *Br. J. Psychiatr.*, **154**: 390–5

Fava, G.A., Munari, F., Pavan, L. and Kellner, R. (1981) Life events and depression. *J. Affect. Disord.*, **3**: 159–65

Finlay-Jones, R. (1989) Anxiety, in G.W. Brown and T.O. Harris (eds.), *Life Events and Illness*, Guilford Press, New York, pp 95–112

Finlay-Jones, R. and Brown, G.W. (1981) Types of stressful life events and the onset of anxiety and depressive disorders. *Psychol. Med.*, **11**: 803–15

Ghaziuddin. M., Ghaziuddin, N. and Stein, G. (1990) The role of life events in the onset of depression: First versus recurrent episodes. *Canad. J. Psychiatr.*, **35**: 239–42

Gilbert, P. (1989) *Human Nature and Suffering*, Lawrence Erlbaum, London and New York

Glassner, B., Haldipur, C.V. and Dessauersmith, J. (1979) Role loss and working-class manic depression. *J. Nerv. and Mental. Dis.*, **167**: 530–41

Goldberg, D. and Huxley, P. (1992) *Common Mental Disorders: A Bio-Social Model*, Routledge, London and New York

Grant, I., McDonald, W.I., Patterson, T. and Trimble, M.R. (1989) Multiple sclerosis, in G.W. Brown and T.O. Harris, (eds.), *Life Events and Illness*, Guilford Press, New York, pp 295–311

Harris, T.O. (1989a) Physical illness: an introduction, in G.W. Brown and T.O. Harris (eds.), *Life Events and Illness*, Guilford Press, New York, pp 119–212

Harris, T.O. (1989b) Disorders of menstruation, in G.W. Brown and T.O. Harris (eds.), *Life Events and Illness*, Guilford Press, New York, pp 261–94

Harris, T.O., Brown, G.W. and Bifulco, A. (1986) Loss of parent in childhood and adult psychiatric disorder. The Walthamstow study 1. The role of lack of adequate parental care. *Psychol. Med.*, **16**: 641–59

Harris, T.O., Brown, G.W. and Bifulco, A. (1987) Loss of parent in childhood and adult psychiatric disorder. The Walthamstow study 2. The role of inadequate substitute care. *Psychol. Med.*, **17**: 163–83

Hellhammer, D. (1983) Learned helplessness – an animal model revisited, in *The origins of Depression: Current Concepts and Approaches*, J. Angst (ed.), Springer-Verlag, Berlin

Hinkle, L.E. and Wolff, H.G. (1957) The nature of man's adaptation to his total environment and the relation of this to illness. *Arch. Int. Med.*, **99**: 442–60

Holmes, T.H. and Rahe, R.H. (1967) The social readjustment rating scale. *J. Psychosomat. Res.*, **11**: 213–18

Huxley, P.G., Goldberg, D.P., Macguire, G.P. and Kincey, V.A. (1979) The prediction of the course of minor psychiatric disorders. *Br. J. Psychiatr.*, **135**: 535–43

Ingham, J.G., Kreitman, N.B., Miller, P.McC. Sashidharan, S.P. and Surtees, P.G. (1986) Self-esteem, vulnerability and psychiatric disorder in the community. *Br. J. Psychiatr.*, **148**: 375–85

Jaspers, K. (1962) *General Psychopathology* (J. Hoenig and M.W. Hamilton, transl.), Manchester University Press, Manchester

Katschnig, H., Brandl-Nebehay, A., Fuchs-Robetin, G., Seelig, P., Eichberger, G., Strobl, R. and Sint, P.P. (1981) in *Lebensverandernde Ereignisse, Psychosoziale Dispositionen und Depressive Verstimmungszustande*, Psychiatrische Universitatsklinik Wein, Abteilung fur Sozialpsychiatrie und Dokumentation, Vienna

Katschnig, H., Pakesch, G. and Egger-Zeidener, E. (1986) Life stress and depressive sub-types: a review of present diagnostic criteria and recent research results, in H. Katschnig, (ed.), *Life Events and Psychiatric Disorders: Controversial Issues*, Cambridge University Press, Cambridge, pp 201–45

Kupfer, D.J. (1991) Biological markers of depression, in J.P. Feighner and W.F. Boyer (eds.), *Perspectives in Psychiatry: Volume 2: Diagnosis of Depression*, Wiley, Chichester, UK, pp 79–98

Lazarus, R.S. (1966) *Psychological Stress and the Coping Processes*, McGraw Hill, New York

Lazarus, R.S. (1991) Cognition and motivation in emotion. *Am. Psychologist*, **46**: 352–67

Leff, J. (1991) Schizophrenia: social influences on onset and relapse, in D.H. Bennet and H.L. Freeman (eds.), *Community Psychiatry*, Churchill Livingston, UK, pp 198–214

Lora, A. and Fava, E. (1992) Provoking agents, vulnerability factors and depresssion in an Italian setting: a replication of Brown and Harris' model. *J. Affective Disord.* **24:** 227–36

Mann, A.H., Jenkins, R. and Belsey, E. (1981) The twelve-month outcome of patients with neurotic illness in general practice. *Psychol. Med.,* **11:** 535–50

Markush, R.E. (1977) Levin's attributable risk statistic for analytic studies and vital statistics. *Am. J. Epidemiol.,* **105:** 401–7

Martin. C.J., Brown, G.W, Goldberg, D.P. and Brockington, I.F. (1989). Psychosocial stress and puerperal depression. *J. Affect. Disord.,* **16:** 283–93

Melges, F.T. and Bowlby, J (1969) Types of hopeless in psychopathological processes. *Arch. Gen. Psychiatr.,* **20:** 690–9

Miller, P.M. and Ingham, J.G. (1983) Dimensions of experience. *Psychol. Med.,* **13:** 417–29

Miller, P.McC., Ingham, J.G., Kreitman, N.B. *et al.* (1987) Life events and other factors implicated in onset and in remission of psychiatric illness in women. *J. Affect. Disord.,* **12:** 73–88

Monroe, S.M. (1990) Psychosocial factors in anxiety and depression, in J.D. Maser and C.R. Cloninger, (eds.), *Comorbidity of Mood and Anxiety Disorders,* American Psychiatric Press Inc., Washington, pp 463–97

Murphy, E. (1982) Social origins of depression in old age. *Br. J. Psychiatr.,* **141:** 135–42

Nolen-Hoeksema, S. (1987) Sex differences in unipolar depression: evidence and theory. *Psychol. Bull.,* **101:** 259–82

Parry, G. and Shapiro, D.A. (1986) Social support and life events in working-class women. *Arch. Gen. Psychiatr.,* **43:** 315–23

Paykel, E.S. (1974) Recent life events and clinical depression, in I.K.E. Gunderson and R.D. Rahe (eds.), *Life Stress and Illness,* Charles, C. Thomas, Illinois, pp 134–63

Paykel, E.S., Myers, J.K., Dienelt, M.N. *et al.* (1969) Life events and depression: a controlled study. *Arch. Gen. Psychiatr.,* **21:** 753–60

Pearlin, L.I., Lieberman, M.A., Menaghan, E.G. and Mullan, J.T. (1981) The stress process. *J. Health and Soc. Behav.,* **22:** 337–56

Perris, H. (1984a) Life events and depression: Part 2. Results in diagnostic subgroups, and in relation to the recurrence of depression. *J. Affect. Disord.,* **7:** 25–36

Perris, H. (1984b) Life events and depression: Part 3. Relation to severity of the depressive syndrome. *J. Affect. Disord.,* **7:** 37–44

Phelan, J., Schwartz, J.E., Bromet, E.J. *et al.* (1991) Work stress, family stress and depression in professional and managerial employees. *Psychol. Med.,* **21:** 999–1012

Post, R.M., Rubinow, D.R. and Ballenger, J.C. (1986) Conditioning and sensitisation in the longitudinal course of affective illness. *Br. J. Psychiatr.,* **149:** 191–210

Raphael, K.G., Cloitre, M. and Dohrenwend, B.P. (1991) Problems of recall and misclassification with checklist methods of measuring stressful life events. *Health Psychol.,* **10:** 62–74

Raskin, M., Harmon, V.S., Peeke, H.V. *et al.* (1982) Panic and generalized anxiety disorders. *Arch. Gen. Psychiatr.,* **39:** 687–9

Rutter, M. (1990) Psychosocial resilience and protective mechanisms, in J. Rolf, A.S. Masten, D. Cicchetti, K.H. Nuechterlein and S. Weintraub (eds.), *Risk and Protective Factors in the Development of Psychopathology,* Cambridge University Press, Cambridge pp 181–214

Rutter, M. and Pickels, A. (1991) Person-environment interactions: concepts, mechanisms and implications for data analysis, in T.D. Wachs and R. Plomin (eds.) *Conceptualisation and Measurement of Organism-Environment Interaction.* pp. 105–41, American Psychological Association, Washington DC

Schutz, A. (1971) Concept and theory formulation in the social sciences, in *A. Schutz, Collected Papers, Volume 1,* Nijhoff, The Hague, pp 48–98

Seligman, M.E.P. (1975) *Helplessness on Depression, Development and Health,* W.H. Freeman, San Francisco

Shrout, P.E., Link, B.G., Dohrenwend, B.P. *et al.* (1989) Characterizing life events as risk factors for depression: the role of fateful loss events. *J. Abnormal Psychol.,* **98:** 460–7

Surtees, P.G. and Ingham, J.G. (1980) Life stress and depressive outcome: application of a dissipation model to life events. *Social Psychiatr.,* **15:** 21–31

Tennant, C. and Bebbington, P. (1978) The social causation of depression: a critique of the work of Brown and his colleagues. *Psychol. Med.*, **8:** 565–75

Tennant, C., Bebbington, P. and Hurry, J. (1980) Parental death in childhood and risk of adult depressive disorders: a review. *Psychol. Med.*, **10:** 289–99

Tennant, C., Bebbington, P. and Hurry, J. (1981) The short-term outcome of neurotic disorders in the community: the relation of remission to clinical factors and to 'neutralising' life events. *Br. J. Psychiatr.*, **139:** 213–20

Tweed, L.J., Schoenback, V.J., George, L.K. and Blazer, D.G. (1989). The effects of childhood parental death and divorce on six-month history of anxiety disorders. *Br. J. Psychiatr.*, **154:** 823–8

Unger, R.M. (1984) *Passion: An Essay on Personality*, The Free Press, New York

Vadher, A. and Ndetel, D.M. (1981) Life events and depression in a Kenyan setting. *Br. J. Psychiatr.*, **139:** 134–7

Weber, M. (1964) *The Theory of Social and Economic Organisation* (T. Parson, transl.), Collier-MacMillan, London

Wittchen, H.V. (1988) Natural course and spontaneous remissions of untreated anxiety disorders: results of the Munich follow-up study (MFS) In I. Hand and H.V. Wittchen (eds.), *Panic and Phobia 2. Treatments and Variables Affecting Course and Outcome*, Springer-Verlag, Heidelberg, pp 3–17

Wittchen, H.V., Essav, C.A. and Krieg, J.L. (1991) Anxiety disorders: similarities and differences of comorbidity in treated and untreated groups. *Br. J. Psychiatr.*, **159** (Suppl. 12): 22–33

Young, M.A., Keller, M.B., Lavori, P.W. *et al.* (1987) Lack of stability of the RDC endogenous subtype in consecutive episodes of major depression. *J. Affect. Disord.*, **12:** 139–44

Part Two
Clinical effects of stress

3
Stress and psychiatric disorder: Reconciling social and biological approaches

PAUL GLUE[1], DAVID NUTT AND NICK COUPLAND

PSYCHOPHARMACOLOGY UNIT
SCHOOL OF MEDICAL SCIENCES
UNIVERSITY OF BRISTOL
BRISTOL, BS8 1TD, UK

3.1 INTRODUCTION

The term 'stress' may be used in two ways in psychiatry: it may be used to identify events or circumstances that are perceived adversely ('stressors'), or to describe the state induced by such events or circumstances (the 'stress reaction'). Studies into the social antecedents or consequences of stress have concentrated on the former meaning, whereas biological studies into the physiology and neurochemistry of stress are intimately involved with the latter interpretation of the term. In this chapter we will try to reconcile both approaches in an overview of the role of stress in psychiatric disorder, by reviewing clinical findings and integrating these with proposed

---

[1] Present address: Schering-Plough Research Institute, Kenilworth, New Jersey, USA

ISBN 0–12–663370–3

biological mechanisms. This approach is not without its difficulties, as clinical studies in this area are almost exclusively social in orientation, whereas work on biological mechanisms of stress is predominantly preclinical.

3.2 STRESS, SYMPTOMS AND PSYCHIATRIC DIAGNOSES

It is important to note that there are no symptoms that are unique or specific to stress-related or reactive-type disorders; virtually any psychiatric symptom can occur (e.g. depressed or anxious mood, delusions or hallucinations, changes in behaviour, etc.). However, it is possible to generalize about responses to stress, in terms of type of symptoms and their duration. Transient emotional or cognitive symptoms may be observed in normal people in response to stressful circumstances, such as a major relationship problem or an accident. These occur shortly after the stressful event, and would appear to an observer to be in proportion to the severity of the event. Symptoms reported most commonly include changes in mood (e.g. anxiety, depression or irritability), poor concentration, physical symptoms (e.g. muscle tension, abdominal pain), or alterations in behaviour (e.g. social withdrawal, insomnia). The occurrence of depressive or anxiety symptoms in response to stress is not synonymous with the development of depressive or anxiety disorders. The former occur extremely frequently, and might be regarded as part of human experience; the latter occur relatively infrequently despite the association with stress. Factors that may contribute to the development of stress-induced psychiatric disorders are discussed in section 3.4.

The importance of stress in the genesis of some disorders has been emphasized in two ways in the US psychiatric classification system, the revised third edition of the Diagnostic and Statistical Manual (DSM-III-R) (American Psychiatric Association, 1987). This manual is based on an atheoretical and descriptive approach to psychiatric disorders, but the exceptions are several categories which incorporate 'stress' or 'reactive' aspects into the diagnostic label. These include: Brief Reactive Psychosis, Posttraumatic Stress Disorder, Adjustment Disorders, and childhood disorders such as Separation Anxiety Disorder and Reactive Attachment Disorder. Such diagnoses are made on the basis of symptoms arising shortly after the occurrence of stressful circumstances. Diagnoses are made on the basis of not only symptoms, but also of their duration. If stress-related symptoms do not last longer than a few days, they would be classified as acute reactions to stress. Symptoms that persist for up to six months would be termed an adjustment disorder. A particularly intense response to exceptionally stressful circumstances (e.g. torture, a major natural disaster), is called posttraumatic stress disorder (PTSD) (see Chapter 15).

The second way in which the role of stress can be incorporated into a diagnosis is through the use of additional codes. The DSM-III-R is a multiaxial classification system which permits supplementary information to be recorded, as well as formal diagnostic labels. Stress-related information can be detailed on Axis IV, a Severity of Psychosocial Stressors Scale. As well as outlining a number of examples of stressful events or circumstances, this scale also distinguishes between acute and enduring stresses, and allows them to be rated. A broadly similar approach has been adopted by the International Classification of Diseases (ICD-10) diagnostic system (Mezzich 1988).

3.3 CLINICAL ASPECTS OF STRESS

To examine this topic, studies in this area have been divided into those studying the stressful effects of environmental factors, and those looking at aspects of individuals' responses to the environment or to a particular event. As will be discussed below, it should be noted that any identified factor may be simultaneously stress-promoting and stress-relieving. For instance, the financial and social benefits of employment may be negated by having to carry out boring or distasteful duties; similarly, marriage may be a support or a stress depending on the quality of the relationship. This makes interpretation of research findings difficult, as frequently only the presence or absence of factors are described, rather than what the presence or absence of a factor may represent to a particular subject.

3.3.1 EXTERNAL DEMANDS AND SUPPORTS

The association between adverse life experiences and psychiatric disorder has been recognized for centuries. For instance, the French psychiatrist Pinel would routinely ask new patients: "Have you suffered vexation, grief, or reversal of fortune?" (Rutter, 1985). Investigating this link has been a major focus of social psychiatry, and a range of evaluation techniques have been developed, including the study of life events. Although this topic is reviewed in Chapter 2, it will be mentioned here briefly for the sake of completeness.

Early subjective epidemiological studies into the role of life events noted that episodes of illness clustered around periods of life change (for a review see Wolff, 1962). Holmes and Rahe (1967) attempted to standardize life change events by giving them a numerical weighting which reflected their apparent severity. These researchers found increased rates of all types of illness in subjects with high life event scores, whether these arose from unpleasant events (e.g. bereavement, unemployment) or pleasant ones (job promotion, birth of a child) (Rahe *et al.*, 1970). Investigations into life events have been modified in several ways. For instance, the distinction has been made between life events that are clearly independent of an illness (e.g. losing a job because a factory is closed down) and those that might be related (e.g. losing a job because one's work performance was impaired by a developing depressive illness). Illnesses are dated as accurately as possible, and a time limit is set on the duration over which events are recalled, to reduce memory distortion. The type of event is classified not only in terms of severity, but also in terms of nature (e.g. loss or threat events). To reduce observer bias, information gathering is performed in a consistent way, using structured interviewing techniques.

There is a general consensus that an excess number of life events occur in the months before the onset of a wide range of illnesses. While the greatest body of work has demonstrated an association with depressive illness (Paykel *et al.*, 1969; Brown and Harris, 1978; Shrout *et al.*, 1989), there is an association with onset and relapse of schizophrenia (Brown and Birley, 1968; Jacobs et al., 1974; Jacobs and Myers, 1976; Leff and Vaughan, 1981; Ventura et al., 1989), panic disorder (Roy-Byrne et al., 1986), minor anxiety and depressive symptoms (Cooper and Sylph, 1973) (for a review see Brown and Harris, 1989). There is also an association with physical disorders including multiple sclerosis, abdominal pain and menorrhagia (Brown and Harris, 1989). There has also been considerable interest in type of life event and subsequent illness. For example, 'loss' events (e.g. bereavement) have been reported to occur more commonly before depressive or anxiety depressive disorders, while anxiety disorders are more commonly preceded by 'danger' events (e.g. assault) (Finlay-Jones and Brown, 1981).

Another way of studying the role of stress on psychiatric disorder has been to look at the influence of patterns of relationships on illness relapse. The majority of this work has examined family communication patterns in schizophrenia, using a standardized assessment of family interactions (Vaughan and Leff, 1976). Several groups have reported that relapse rates for schizophrenia are highest in patients whose family members or friends display high 'expressed emotion', this is, they make hostile or critical comments to the patient, are emotionally overinvolved, and spend more than 35 h per week in face-to-face contact (Brown et al., 1962; Leff and Vaughan, 1981; for a review see, Vaughan, 1989). Subsequently, several groups have reduced relapse rates by altering family interactions (Falloon et al., 1982, 1985; Leff et al., 1982, 1989; Tarrier et al., 1989). This has been achieved using a number of different therapeutic approaches, including teaching problem-solving techniques, emphasizing non-critical ways of communication, reducing time spent together, and education. Expressed emotion has also been shown to influence the course of illness in other conditions such as bipolar illness (Priebe et al., 1989), anorexia nervosa (Szmukler et al., 1987), dementia (Orford et al., 1987) and mental handicap (Greedharry, 1987).

The influence of living circumstances has also been closely studied, especially in depression. In a classic study, working-class women who were looking after young children, did not work outside the home, had no-one close to confide in, and had lost their mothers by death or separation before the age of 11, were found to have higher rates of depression (Brown and Harris, 1978). In the majority of the studies in this area, the main variable studied in determining susceptibility to stress is the presence or absence of social support.

3.3.2 INDIVIDUAL NEEDS AND RESOURCES

As well as stressful or adverse circumstances inducing symptoms or a disorder in an individual, another group of factors to consider are those influencing individuals' susceptibility to stress, or their responses to it.

Rates of stress-related psychiatric illness may be influenced by factors which increase the overall likelihood of being exposed to stressors. In a recent epidemiological survey on rates of traumatic life events, those subjects who were more likely to be exposed to such events were male, of low education, with high self-ratings of extraversion, high rates of childhood behaviour problems, and high rates of family psychiatric or drug and alcohol problems (Breslau et al., 1991). A similar link has been has been proposed for individuals who experience early parental loss and subsequent depression: that a mood disorder *per se* is not transmitted, but rather characteristics or behaviour that predispose to depression are promoted (McGuffin *et al.*, 1988). For instance, an association has been made between lack of care during childhood due to maternal death and increased risk of early premarital pregnancy, which also increases the risk of marriage to an unsuitable partner and the subsequent relationship difficulties this entails (Brown *et al.*, 1986; Harris *et al.*, 1987). It has been reported that such individuals have higher rates of immature, impulsive, hostile, dependent or manipulative features and also have a greater risk of using alcohol and other drugs in adult life (Akiskal, 1989). Such personality features might produce relationship problems or contribute to life events that might then lead to a depressive or other illness. This is confirmed by other studies showing that depressed or anxious patients with personality disorders tend to experience more life stress and have poorer responses to treatment than those without such features (Sargant and Dally, 1962; Pfohl *et al.*, 1984; Black *et al.*, 1988; Reich, 1988).

The association of personality factors and physical illness has also been closely studied. Patients with Type A behaviour patterns (aggressiveness, hostility, time urgency and competitive striving for achievement) have increased rates of cardiovascular disease compared with those without such features (Rosenman *et al.*, 1975; Haynes *et al.*, 1980). Type A personality features have also been associated with high ratings of neuroticism and psychiatric problems in general (Irvine *et al.*, 1982; Bass, 1984).

Low self-esteem has been closely associated with both anxiety and depression (Brown and Harris, 1978; Ingham *et al.*, 1986). The direction of this association is unclear, with some authors suggesting that it is a consequence of depression or anxiety, while others see it as one component of a depression-prone personality (Bagley and Evan-Wong, 1975; Altman and Wittenborn, 1980; Lewinsohn *et al.*, 1981). Despite the intuitive attractiveness of the concept of self-esteem, it is still poorly defined and cannot be reliably measured (for a review see Robson, 1988).

The literature on genetic aspects of stress is limited and almost entirely derived from animal studies. This area is of particular interest as it may shed light on the reasons for individual differences in response to stress. Different strains of mice show different behavioural and brain biochemical responses to footshock stress (Zacharko *et al.*, 1987; Shanks and Anisman, 1988). However more detailed analysis is complicated by the fact that behavioural responses within mice strains vary depending on the type of stress paradigm (for reviews see Anisman and Zacharko, 1990, 1992). Studies into familial aspects of stress have shown that first degree relatives of depressed patients experience more life events than those of control subjects, even when events relating to the depressed patient are excluded (McGuffin *et al.*, 1988). This raises the interesting possibility that there are familial influences on the tendency to experience problems.

Specific enquiries have been made into the ways in which people cope with stress, most frequently using checklists of possible coping thoughts or behaviours, to assess subjects' responses to a recent stressful incident (Lazarus and Folkman, 1984; see Coyne and Downey, 1991). Generally, the use of coping strategies is found to be positively associated with measures of distress or depression, with virtually no strategies correlating with lower scores on these measures (Coyne *et al.*, 1981; Parker and Brown, 1982). While these findings might offer the pessimistic prospect that coping is of little benefit in reducing stress or preventing its consequences, some caution is needed in interpreting this research. For instance, it may be that coping strategies are applied less effectively in depressed compared with non-depressed subjects. It has been suggested that coping checklists may not be sensitive enough to assess competence in coping, or to take into account environmental factors that may influence coping skills (for further discussion, see Coyne and Downey, 1991).

Finally, an alternative approach to this topic has been to examine factors that promote resistance to stress, or 'invulnerability' factors. This topic has been comprehensively reviewed recently (Rutter, 1985) and readers are referred to this article. Some summary points can be drawn from this review. An individual's resistance to stress is relative rather than absolute, and the degree to which an individual can deal with stress is variable over time rather than a fixed quality. As has been discussed above in connection with vulnerability factors, the concept of stress resistance is also based around individual and environmental factors.

3.4 BIOLOGICAL MEDIATORS OF STRESS

The mechanisms by which stressful events or circumstances produce psychological or physical responses are complicated. In this section we identify several neurochemical systems which are involved in the perception of stress, and in subsequent responses to it. Since these systems are reviewed in greater detail in later chapters (11, 12 and 13), our main purpose is to develop an integrated hypothesis to account for the biological responses to stress and the way in which these are manifest clinically.

The essence of our conceptualization is given in Figure 3.1. We suggest that there are three levels at which neurotransmitters are involved in the stress response; perception, immediate response and prolonged manifestations. These ideas are based on well-established principles that neurotransmitters can be distinguished on the basis of speed of response, duration of action and range of activity. Inhibitory amino acids (e.g. γ-aminobutyric acid; GABA), and excitatory amino acids (e.g. glutamate) have specific, localized short-term effects. Monoamine transmitters (noradrenaline, dopamine and 5-hydroxytryptamine), acetylcholine and histamine produce slower but more durable responses that appear to modulate the actions of the primary amino acid transmitters. Neuropeptides (e.g. corticotropin-releasing factor (CRF) and endorphins) produce even slower responses that tend to be more widespread and long-lasting (Cooper et al., 1991). We hypothesize that different aspects of stress vulnerability and response can be conceptualized by examining responses and interactions of these three groups of neurotransmitters.

3.4.1 STRESS PERCEPTION

Stress perception is probably modulated through fast-acting amino acid mechanisms. The primary sensory transmitter is thought to be the excitatory amino acid (EAA) glutamate. This acts at two sorts of receptors; those defined by a preferential response to α-amino-3-hydroxy-5-methylisoxazole 4-proprionic acid

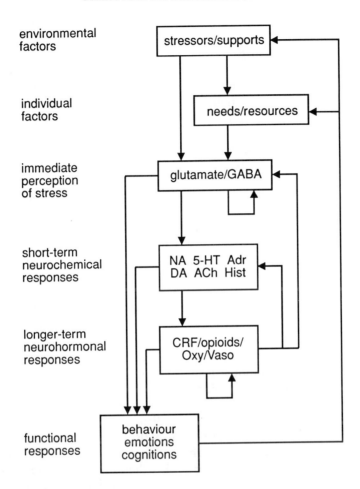

Figure 3.1 Three levels of biological response to stress. ACh: acetylcholine; Adr: adrenaline; DA: dopamine; 5-HT: 5-hydroxytryptamine; Hist: histamine; CRF: corticotropin-releasing factor; Oxy: oxytocin; Vaso: vasopressin

(AMPA) and those selective for N-methyl-D-aspartate (NMDA) (Watkins *et al.*, 1990). The former mediate the immediate excitatory synaptic throughput, whereas the latter modulate this in a sophisticated way, in that co-activation of both receptors can lead to processes such as long-term potentiation, a physiological process that is associated with learning (Morris *et al.*, 1986). Thus glutamate transmission is necessarily involved in the registration of threat and the behavioural response to it. The finding that interference with either sort of glutamate receptor can produce amnesia in animals suggests that the learned component of stress (e.g. memory of catastrophes, flashbacks) may be mediated by glutaminergic processes. In addition, we would suggest that the dissociative states seen following acute severe stress might be caused by a disruption of EAA transmission either in the first relay stations (e.g. the thalamus and lateral geniculate) or in the cortex. Similar states can be produced by drugs such as phencyclidine which block the NMDA-gated ion channel (Javitt and Zukin, 1991; Domino, 1992). Stress-induced dissociation might be construed as an adaptive response that attempts to minimize the most disturbing aspects of stress by preventing sensory information from reaching consciousness.

There is growing evidence that drugs, which have been used for years to reduce stress, act in part through blocking EAA transmission. Thus both barbiturates and ethanol reduce EAA-mediated ion fluxes (Hoffman et al., 1989; Lovinger et al., 1989; see Nutt, 1991; Weight et al., 1992). Moreover, at least in the case of ethanol, there is data showing that, on chronic administration, the NMDA receptor is upregulated (Grant et al., 1990). Thus the stress of alcohol withdrawal is probably mediated in part through increased NMDA-receptor activation which may also lead to a kindling of these responses (Glue and Nutt, 1990; see below).

GABA is the main inhibitory neurotransmitter in the central nervous system (CNS). GABAergic interneurones are widely distributed and modulate pre- and postsynaptic activity of most grey matter neurones (for more details see Chapter 12). In particular, GABA inhibits a number of neurotransmitter systems which amplify stress responses, especially the glutamate system. Based on clinical and research observations, it might be hypothesized that abnormalities of GABA inhibition would lead to inappropriately heightened awareness of, or response to, stress. There is some preliminary evidence to support this, in that the overarousal, vigilance and fearful anticipation of anxiety is associated with reduced sensitivity of GABA receptors (see below). This section will concentrate mainly on anxiety disorders, as there have been very few studies of GABA sensitivity in depression. Although GABA receptor sensitivity cannot be tested directly in man, there are several studies examining responses to benzodiazepines and barbiturates (see Glue et al., 1992b), as these drugs bind to sites on the GABAa receptor molecule (Nutt, 1990). Patients with the severe anxiety state called 'panic disorder' have been shown to have reduced sensitivity to benzodiazepine agonists (Roy-Byrne et al., 1990), and there appear to be similar abnormalities in normal subjects with high levels of trait anxiety (Glue et al., 1992a). This confirms earlier studies demonstrating reduced sensitivity to barbiturates in anxious patients (Shagass, 1954; Shagass and Naiman, 1956).

We have suggested that many of the phenomena seen in those patient groups that are particularly sensitive to stress could be due to altered benzodiazepine receptor function. This alteration is manifest as a shift in receptor function towards 'inverse agonism': such a receptor shift means that full agonists at the benzodiazepine receptor would have only partial agonist activity, antagonists would become partial inverse agonists (i.e. have the opposite pharmacological effects from the agonists) and the activity of inverse agonists would be enhanced (see Nutt, 1990; Nutt et al., 1990b). There is now strong clinical evidence to support this theory, as demonstrated by the reduced sensitivity to benzodiazepine agonists (see above) and the markedly anxiogenic effects of the benzodiazepine antagonist, flumazenil, in patients with panic disorder (Nutt et al., 1990b).

It is tempting to speculate that the exaggerated sensitivity of anxious patients to reading lists of threatening (as opposed to non-threatening) words is due to heightened perception of stress-inducing stimuli (Mathews and MacLeod, 1985; see Eysenck, 1991). Our present conceptualization of the role that benzodiazepine receptor dysfunction may play in anxiety disorders is presented diagrammatically in Figure 3.2. While this might explain why anxious subjects are more sensitive to stressful stimuli, other mechanisms are probably involved in their exaggerated responses to stress. One of the best studied neurotransmitter systems in the 'fight or flight' response to stress is the noradrenergic system (see below). Although there are probably abnormalities in the inhibitory controls for this system, such as the inhibitory α2-adrenoceptor, the paroxysmal overactivity of noradrenergic neurones thought to occur during panic attacks may be due in part to reduced GABA receptor function at the level of the cell body or terminal regions (Nutt and Lawson, 1992).

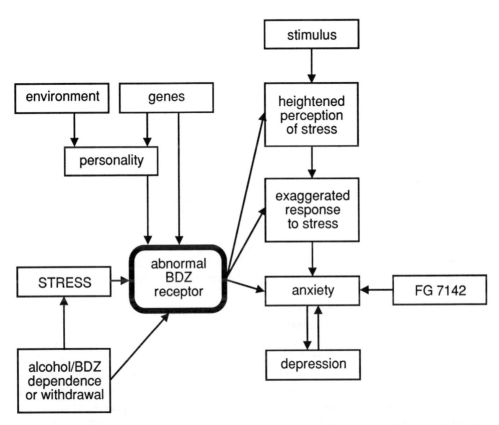

Figure 3.2 The role of the benzodiazepine receptor in anxiety: relative contributions of genetic, environmental and pharmacological factors. BDZ: benzodiazepine; FG 7142: (N-methyl-β)-carboline-3-carboxamide

Drugs that enhance GABA function are of immediate benefit in reducing anxiety. Barbiturates, ethanol and benzodiazepines all enhance GABA function, the first two by direct effects on the receptor-regulated chloride channel conductance, the latter by facilitating GABA activity. There is limited evidence that patients are relatively resistant to some of the adverse actions of these drugs (e.g. the amnestic, sedating or performance-impairing properties; see Glue *et al.*, 1992b) which further supports the idea that a receptor dysfunction may underlie these conditions. Tolerance to many of the actions of GABA-acting tranquillizers often occurs after chronic administration, probably due to reduced GABA function. For instance, this can be demonstrated by reduced effects of benzodiazepine agonists (Roy-Byrne *et al.*, 1991; see Glue, 1991). The arousal and anxiety which occurs during withdrawal from these drugs reflects the loss of inhibitory control. We have suggested that a change in benzodiazepine receptor setpoint may underlie aspects of tolerance (Nutt, 1990; Nutt *et al.*, 1990b; Glue *et al.*, 1992b). This change is the same as that thought to be present in panic disorder, which could explain the high frequency of panic attacks in benzodiazepine withdrawal.

3.4.2 SHORT-TERM MODULATION OF STRESS RESPONSES

The best understood of the short-term neurochemical responses to stress are those involving catecholamines. A complicated pattern of peripheral and central autonomic changes and adrenal medullary secretion has been shown to underlie the behavioural and cardiovascular changes associated with defence reactions or alerting responses, the 'fight or flight' syndrome. Release of noradrenaline and adrenaline is dependent upon the circumstances and behavioural response. For instance, noradrenaline release is most commonly associated with anger and motor activity ('physical stress'), while adrenaline release is associated with anticipation, fear or unpredictability ('psychological stress') (Frankenhaeuser, 1975; Trap-Jensen *et al.*, 1982; but see Chapter 11). Elevated plasma, urinary and CSF levels of catecholamines, or their metabolites, have been reported in a range of conditions associated with anxiety such as panic disorder (Ko *et al.*, 1983; Villacres *et al.*, 1987), alcohol withdrawal (Nutt *et al.*, 1988) and in normal subjects undergoing unpleasant emotional arousal (e.g. public speaking; Dimsdale and Moss, 1980).

In addition to peripheral measures, substantial changes in central adrenergic activity have been demonstrated, including stress-induced increases in noradrenergic neuronal firing rates (Abercrombie and Jacobs, 1987), noradrenaline turnover (Iimori *et al.*, 1982) and changes in adrenoceptor density in several brain regions (for reviews see Krystal, 1990, and Stanford, 1990). Studies in the latter area have confirmed that cortical, α1-, α2- and β-adrenoceptor numbers are reduced, as are second messenger (adenosine 3′, 5′-cyclic monophosphate; cyclic AMP) reponses to agonist stimulation (Krystal, 1990). α2-Adrenoceptors appear to be most sensitive in response to stress, as downregulation may occur after mild stress, in the absence of β-adrenoceptor changes (Stanford and Salmon, 1989). The net result of this would be to increase noradrenergic reactivity because of the loss of autoreceptor inhibition of neuronal activity. The reduction in β-adrenoceptor density might then be regarded as a response to the elevated noradrenaline levels due to the heightened noradrenergic neuronal activity (see Figure 3.3).

If noradrenergic systems are so sensitive to stress what role(s) does their activation play? One theory is that the coactivation of noradrenergic inputs into the forebrain with primary sensory inputs is necessary for appropriate registration and response to environmental change (Aston-Jones *et al.*, 1984). It is now thought that primary sensory afferents may send collaterals into the locus coeruleus that result in parallel activation of the two ascending systems. Thus the degree of noradrenergic coactivation

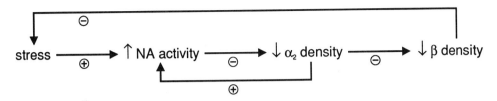

Figure 3.3 Stress and adrenoceptor changes

could contribute to the emotional colouring or tone of the sensory input, either directly or through a change in arousal (see Figure 3.4). Insufficient noradrenergic activation leads to inattention, as may occur in attention deficit disorder and various dementias. On the other hand, excessive input may contribute to the exaggerated attentional bias seen in anxiety disorders (see above), as well as the pronounced encoding of traumatic stimuli, particularly images, in posttraumatic stress disorder (PTSD).

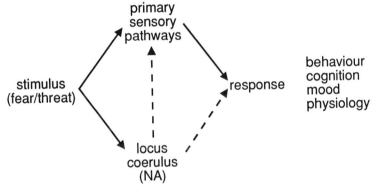

Figure 3.4 Parallel activation of sensory (———) and noradrenergic (– – –) pathways in response to threat. NA: noradrenaline

These concepts can be used to explain phenomena in a variety of psychiatric disorders ranging from schizophrenia through to dementia. The condition that has been investigated most deeply is panic disorder. A fair body of evidence suggests that paroxysms of noradrenergic overactivity produce many of the symptoms of panic attacks (Nutt, 1990; Nutt et al., 1990a). The basis of the dysregulation has been thought to be at the level of the inhibitory α2-adrenoceptor, although more recent data suggests that this may not be the complete story (Nutt and Glue, 1991). A dysregulation of noradrenergic responses to sensory stimuli may be one way of explaining why trivial stressors (that may be at or below the very limit of perception) provoke panic in susceptible individuals (Figure 3.4) (for a review see Nutt and Lawson, 1992).

If noradrenergic activation is necessary for the registration of, and response to stressors, what is the anatomical location of the comparator, the structure that determines whether or not sensory inputs represent stress or threat? One possibility is the locus coeruleus, either alone or in conjunction with the nuclei paragigantocellularis and prepositus hypoglossi (Hsiao and Potter, 1990). Activation by collaterals of ascending sensory inputs could explain stress responses in these brain regions to excessive stimulation, such as loud noise or sharp pain (see above). This cannot explain responses to stressors which depend on more sophisticated associative or psychological appraisal, such as conditioned fear, nonreward (frustration), loss, etc. The best conceptualization of these issues is that of Gray (1987), who postulates that the septohippocampal system is critical in controlling the anxiogenic response to such stimuli. His theory also postulates a permissive role for noradrenaline in these processes. Perhaps the best explanation of stress responses is that comparison and appraisal of stressors occurs as a result of cortical integration of sensory, noradrenergic and limbic inputs. The septohippocampal system in this model co-ordinates retrieval of memories which are then compared in the cortex, the emotional tone of which is determined by the level of concurrent noradrenergic activity.

The second role for noradrenaline is in co-ordinating stress responses by activation of corticotropin-releasing factor (CRF) neurones. CRF neurones are innervated by noradrenergic neurones (Kitazawa et al., 1987), and noradrenaline stimulates CRF release (Alonso et al., 1986). There are also CRF afferents to the locus coeruleus (Al-Damluji et al., 1987), and CRF injections into this area produce excessive and irregular firing patterns and stimulate sympathetic nervous system activity (Brown et al., 1982; Valentino et al., 1983). The role of CRF in stress responses is discussed briefly below, and in greater detail, in Chapter 13.

The third role of noradrenaline is in adaptation to stress. Attention has been drawn to the similarity between the actions of stress and antidepressants in downregulating β-adrenoceptors (Stone, 1979). Stone has suggested that behavioural adaptation to stress involves downregulation of β-adrenoceptors, and that the stress-protective effects of antidepressants are therefore through a similar mechanism (Stone, 1983).

Noradrenergic processes are influenced by many other transmitters. Of particular relevance are those already mentioned. Stimulation of the locus coeruleus is in part under EAA control whereas GABA-ergic and opioid inputs provide inhibition (Bird and Kuhar, 1977; Cedarbaum and Aghajanian, 1976; Cherubini et al., 1988; see Aston-Jones et al., 1990). These inputs provide an explanation for the interactive process outlined in Figure 3.1. Moreover, it is possible that the antistress effects of benzodiazepines are mediated in part through their ability to attenuate stress-induced activation of noradrenergic systems. A number of studies have suggested that benzodiazepines have a preferential ability to prevent the increase in noradrenaline turnover and release caused by stress (Taylor and Laverty, 1969; Ida et al., 1985; Rossetti et al., 1990). There are also clinical data supporting this concept; for instance, the increased turnover produced by surgical procedures is attenuated by benzodiazepines (Liu et al., 1977; Dionne et al., 1984), although the ability of benzodiazepines to lower basal indices of noradrenaline release is controversial (Nutt, 1986; see Glue et al., 1992b).

3.4.3 LONG-TERM MODULATION OF STRESS RESPONSES

Several neuropeptides have central roles in the facilitation or inhibition of stress responses. The best known are corticotropin-releasing factor (CRF), the opioid peptides (β-endorphin, the enkephalins and dynorphins), and the posterior pituitary hormones, oxytocin and vasopressin. The role of other peptides is less well researched, but neurotensin, cholecystokinin, neuropeptide Y, substance P, thyrotropin-releasing hormone, somatostatin and others have all been implicated in stress responses.

The role of CRF as the principal mediator and coordinator of endocrine, behavioural and autonomic responses to stress is described in detail in Chapter 13, and will not be discussed at length here. The purpose of this chapter is to highlight the integration of central neurotransmitter systems in response to stress, and thus the interactions between the central role of CRF and the modulatory roles of other neurotransmitters will be emphasized. Substances that inhibit and enhance secretion of CRF, adrenocorticotropic hormone (ACTH) and cortisol are summarized in Figure 3.5.

Neurotransmitters implicated in perception of stress (e.g. GABA) and in the modulation of short-term responses to it (e.g. noradrenaline, 5-hydroxytryptamine) affect the release of CRF (see Delbende *et al.*, 1992). The release of CRF sets in motion

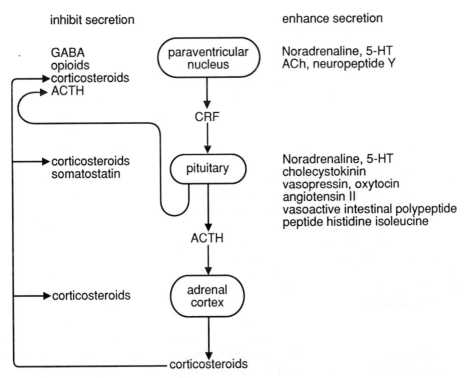

Figure 3.5 Modulation of corticosteroid responses to stress: substances that inhibit or enhance secretion of CRF, ACTH and cortisol

several positive and negative feedback loops, or may be associated with the release of other substances which amplify its biological effects. One well-described positive feedback interaction is that between CRF and noradrenergic pathways. CRF stimulates noradrenergic neurones in the locus coeruleus (Al-Damluji et al., 1987) which in turn may activate CRF neurones. In addition to these, there are also response-amplification mechanisms. For instance, vasopressin is released at the same time as CRF and, in combination, these produce a much greater release of ACTH than does either substance alone (Rivier et al., 1984; see Delbende et al., 1992). Several negative feedback mechanisms have been identified, such as the inhibition of CRF release by ACTH and cortisol, or the inhibitory effects of opioids on CRF release (see Chapter 13).

The fact that there are multiple interactions between CRF and other neurotransmitter pathways, some of which are facilitatory and some inhibitory, may provide some insight into the diversity of responses to stressors. Under normal conditions, release of CRF will be modulated by other pathways or 'counterregulated' (for a review see McEwen et al., 1992a). In conditions where there are defects in the modulatory pathways, effective control does not occur, leading to excessive release of CRF. This may have three effects: stress-induced changes may be exaggerated via a series of positive feedback loops in the modulatory pathways; kindling-like changes may be promoted (Ehlers et al., 1983); or it may led to neuronal death in certain brain regions (Uno et al., 1989; McEwen et al., 1992a, 1992b). Such long-term changes may underlie the clinical observation that the development of certain clinical disorders follows a kindling-like pattern, with episodes increasing in severity and frequency over time (see Klein, 1981, and Nutt, 1990).

Although there are many influences on CRF release, three examples will be used to show how abnormalities in modulatory pathways may contribute to abnormal CRF responses, pathological sensitivity or responses to stress and excessive vulnerability to stress. Presynaptic α2-adrenoceptors provide inhibitory feedback control for noradrenergic neurones (Starke, 1981). As described above, stress has been shown to reduce the density of these receptors. Clinically, there is reduced functional activity of this receptor in depression and panic anxiety (for a review see Glue and Nutt, 1988). Thus it is possible that stress-induced changes in α2-adrenoceptors would lead to a reduction in the inhibitory control over noradrenergic neuronal firing, so that activity in the positive feedback pathway between CRF and noradrenaline would be increased, leading ultimately to excessive CRF responses and clinical symptoms. Even after remission of these illnesses, underactivity of α2-adrenoceptors appears to persist (Mitchell et al., 1988; Middleton, 1992), which would leave patients vulnerable to relapse.

Abnormalities in the activity of GABA receptors have been reported in anxiety and alcoholism (see section 3.4.1). While the literature for depression is less clear, there is some evidence to implicate underactivity of GABA pathways, and preclinical studies suggest that GABA agonists may have an antidepressant effect (Lloyd and Morselli, 1987). The pharmacological effect of benzodiazepine receptor subsensitivity would be to reduce inhibitory control of CRF secretion, or of other factors that might promote it (e.g. noradrenergic pathways). The functional consequences of this might be to provoke exaggerated, inappropriate or prolonged responses to stressful stimuli. Subsequent treatment with benzodiazepines or self-medication with alcohol could tend to aggravate this process by causing a further subsensitivity of benzodiazepine receptors.

Benzodiazepine receptors are affected in turn by alterations in corticosteroid levels. Adrenalectomy leads to an increase in brain benzodiazepine receptors, and this effect is reversed by corticosteroid administration (de Souza et al., 1986; Acuna et al., 1990). Stress decreases benzodiazepine receptor numbers in various brain regions, and this effect is blocked by adrenalectomy (Weizman et al., 1989, 1990). Thus stress-induced hypercortisolism reduces benzodiazepine receptor function through a mechanism that is dependent on functioning adrenals. This might explain the association noted in earlier clinical studies between plasma cortisol concentrations and anxiety levels (Sachar, 1970; Berger, 1980), or the observation that anxiety can be produced by exogenously administered hydrocortisone (Levitt et al., 1963).

A third example of abnormalities in modulatory pathways contributing to CRF dysregulation are those involving opioid receptors. There is considerable colocalization of CRF and opioid peptides in a number of subcortical areas, especially those involved with autonomic function (de Souza and Appel, 1991; Lind and Swanson, 1984). They appear to have complementary effects, in that CRF stimulates the release of opioids, while opioids inhibit the release of CRF (Nikolarakis et al., 1987, 1989). Studies on opioid/CRF interactions have concentrated on anorexia nervosa (this is discussed in detail in Chapter 13) and depression (Berger and Nemeroff, 1987; Schmauss and Emrich, 1988). Although there is no strong evidence to suggest that endogenous opioid levels are altered in depression, full and partial opioid agonists have shown marked, if inconsistent, antidepressant effects (Schmauss and Emrich, 1988). More convincing evidence of opioid receptor abnormalities in depression come from agonist challenge studies. In normal subjects morphine and fentanyl reduce plasma cortisol levels, but this effect is reversed or absent in severe depression (Zis et al., 1985; Matussek and Hoehe, 1989). The latter authors have also presented preliminary data showing that cortisol responses to morphine normalize after effective antidepressant treatment (Zis et al., 1985). These studies suggest that, in depression, there is a reduction of the normal inhibitory effect of opioid receptors on the hypothalamic-pituitary-adrenal axis, and this reduction appears to normalize after successful antidepressant treatment. Whether the loss of inhibitory opioid control is a cause or effect of the hypercortisolism is unclear, however the functional consequences are the loss of another inhibitory control over CRF release.

3.5 CONCLUDING REMARKS

This chapter has explored the association of stress and psychiatric disorder from several perspectives, attempting to evaluate social and psychological findings in the light of recent psychopharmacological discoveries. Biological responses to stress may be usefully examined at three levels, from perception of stress to short- and longer-term adaptive responses. There are different but interactive neurotransmitter systems mediating each of these three levels of response, and recent advances in our understanding of these systems provides some insight into how these may be involved in psychiatric illness arising from stress. Several mechanisms are also described whereby genetic or environmental stressors may produce changes in neurotransmitter activity. One of the central mediators of normal stress responses is CRF, which interacts dynamically with a number of these pathways. Excessive or prolonged responses to stress may be caused by abnormal function of these pathways, leading to CRF overactivity and further dysequilibrium. The pattern of abnormal activity in the modulatory pathways may determine the nature of the illness arising in response to stress.

3.6 REFERENCES

Abercrombie, E.D. and Jacobs, B.L. (1987) Single unit response of noradrenergic neurons in the locus coeruleus of freely moving cats. II. Adaptation to chronically presented stressful stimuli. *J. Neurosci.*, **7**: 2837–43

Acuna, D., Fernandez, B., Gomar, M.D., del Aguila, C.M. and Castillo, J.L. (1990) Influence of the pituitary-adrenal axis on benzodiazepine receptor binding to rat cerebral cortex. *Neuroendocrinology*, **51**: 97–103

Akiskal, H.S. (1989) New insights into the nature and heterogeneity of mood disorders. *J. Clin. Psychol.*, **50**: 6–12

Al-Damluji, S., Perry L., Tomlin, S. *et al.* (1987) Alpha-adrenergic stimulation of corticotropin secretion by a specific central mechanism in man. *Neuroendocrinology*, **45**: 68–76

Alonso, G., Szafarczyk, A., Balmefrezol, M. and Assenmacher, I. (1986) Immuno-cytochemical evidence for stimulatory control by the ventral noradrenergic bundle of parvo-cellular neurons of the paraventricular nucleus secreting corticotropin releasing hormone and vasopression in rats. *Brain Res.*, **397**: 297–307

Altman, J.H. and Wittenborn, J.R. (1980) Depression-prone personality in women. *J. Abnormal Psychol.*, **89**: 303–8

American Psychiatric Association (1987) *Diagnostic and Statistical Manual of Mental Disorders (3rd Edn).* American Psychiatric Association Press, Washington, DC

Anisman, H. and Zacharko, R.M. (1990) Multiple neurochemical and behavioral consequences of stressors: implications for depression. *Pharmacol. Ther.*, **46**: 119–36

Anisman, H. and Zacharko, R.M. (1992) Resistance to stress. Multiple neurochemical, behavioral and genetic factors. *J. Psychopharmacol.*, **6**: 8–10

Aston-Jones, G., Foote, S.L. and Bloom, F.E. (1984) Anatomy and physiology of locus coeruleus neurones: functional implications, in M.G. Ziegler and C.R. Lake (eds.), *Norepinephrine: Clinical Aspects*, Williams and Wilkins, Baltimore, pp 92–116

Aston-Jones, G., Shipley, M.T., Ennis, M., Williams, J.T. and Pieribone, V.A. (1990) Restricted afferent control of locus coeruleus neurones revealed by anatomical, physiological and pharmacological studies, in D.J. Heal and C.A. Marsden (eds.), *The Pharmacology of Noradrenaline in the Central Nervous System*, Oxford University Press, Oxford, pp 187–247

Bagley, C. and Evan-Wong, L. (1975) Neuroticism and extraversion in responses to Coopersmith Self Esteem Inventory. *Psychol. Reports*, **36**: 253–4

Bass, C. (1984) Type A behaviour in patients with chest pain: test-retest reliability and psychometric correlates of Bortner scale. *J. Psychosomatic. Res.*, **28**: 289–300

Berger, F.M. (1980) Effect of antianxiety drugs on fear and stress. *Behav. Sci.*, **25**: 315–25

Berger, P.A. and Nemeroff, C.B. (1987) Opioid peptides in affective disorders, in H.Y. Meltzer (ed.), *Psychopharmacology: The Third Generation of Progress*, Raven Press, New York, pp 637–46

Bird, S.J. and Kuhar, M.J. (1977) Iontophoretic application of opiates to the locus coeruleus. *Brain Res.*, **122**: 523–33

Black, D.W., Bell, S., Hulbert, J. and Nasrallah, A. (1988) The importance of Axis II disorders in patients with major depression: a controlled study. *J. Affect. Dis.*, **14**: 115–22

Breslau, N., Davis, G.C., Andreski, P. and Peterson, E. (1991) Traumatic events and posttraumatic stress disorder in an urban population of young adults. *Arch. Gen. Psychiat.* **48:** 216–22

Brown, G.W. and Birley, J.L.T. (1968) Crises and life changes and the onset of schizophrenia. *J. Health. Soc. Behav.,* **9:** 203–24

Brown, G.W. and Harris, T.D. (1978) *Social Origins of Depression: A Study of Psychiatric Disorder in Women.* Free Press, New York

Brown, G.W. and Harris, T. (1989) *Life Events and Illness.* Unwin Hyman, London

Brown, G.W., Monck, E.M., Carstairs, G.M. and Wing, J.K. (1962) Influence of family life on the course of schizophrenic illness. *Br. J. Prev. Soc. Med.,* **16:** 55–68

Brown, M.R., Fisher, L.A., Rivier, J. *et al.* (1982) Corticotropin-releasing factor: effects on the sympathetic nervous system and on oxygen transport. *Life Sci.,* **30:** 207–10

Brown, G.W., Bifulco, A., Harris, T. and Bridge, L. (1986) Life stress, chronic sub-clinical symptoms and vulnerability to clinical depression. *J. Affect. Dis.,* **11:** 1–9

Cedarbaum, J.M. and Aghajanian, G.K. (1976) Noradrenergic neurons of the locus coeruleus: inhibition of epinephrine and activation by the alpha-antagonist piperoxane. *Brain Res.,* **112:** 413–19

Cherubini, E., North, R.A. and Williams, J.T. (1988) Synaptic potentials in rat locus coeruleus neurones. *J.Physiol.,* **406:** 431–42

Cooper, B. and Sylph, J. (1973) Life events and the onset of neurotic illness: an investigation in general practice. *Psychol. Med.,* **3:** 421–35

Cooper, J.R., Bloom, F.E. and Roth, R.H. (1991) *The Biochemical Basis of Neuropharmacology.* Oxford University Press, Oxford

Coyne, J.C., Aldwin, C. and Lazarus, R.S. (1981) Depression and coping in stressful episodes. *J. Abnormal. Psychol.,* **90:** 439–47

Coyne, J.C. and Downey, G. (1991) Social factors and psychopathology: stress, social support and coping processes. *Ann. Rev. Psychol.,* **42:** 401–25

Delbende, C., Delarue, C., Lefebvre, H. *et al.* (1992) Glucocorticoids, transmitters and stress. *Br. J. Psychiat.,* **160** (suppl 15): 24–34

de Souza, E.B. and Appel, N.M. (1991) Distribution of brain and pituitary receptors involved in mediating stress responses, in M.R. Brown, G.F. Koob and C. Rivier (eds.), *Stress: Neurobiology and Neuroendocrinology,* Marcel Dekker, New York, pp 91–117

de Souza, E.B., Goeders, N.E. and Kuhar, M.J. (1986) Benzodiazepine receptors in rat brain are altered by adrenalectomy. *Brain Res.,* **381:** 176–81

Dimsdale, J.E. and Moss, J. (1980) Plasma catecholamines in stress and exercise. *J. Am. Med. Assn.,* **243:** 340–2

Dionne, R.A., Goldstein, D.S. and Wirdzek, P.R. (1984) Effects of diazepam premedication and epinephrine-containing local anaesthetic on cardiovascular and catecholamine responses to oral surgery. *Anesth. Analges.,* **63:** 640–6

Domino, E. (1992) Chemical dissociation of human awareness forms on non-competitive NMDA receptor antagonism. *J. Psychopharmacol.,* **6:** 418–24

Ehlers, C., Henricksen, S., Wang, M. *et al.* (1983) Corticotropin releasing factor produces increases in brain excitability and convulsive seizures in rats. *Brain Res.,* **278:** 332–6

Eysenck, M. (1991) Cognitive factors in clinical anxiety: potential relevance to therapy, in M. Briley and S.E. File (eds.), *New Concepts in Anxiety,* Macmillan, London, pp 418–33

Falloon, I.R.H., Boyd, J.L., McGill, C.W. *et al.* (1982) Family management in the prevention of relapse in schizophrenia. *New Engl. J. Med.,* **306:** 1437–40

Falloon, I.R.H., Boyd, J.L., Williamson, M. *et al.* (1985) Family management in the prevention of morbidity of schizophrenia. *Arch. Gen. Psychiat.,* **42:** 887–96

Finlay-Jones, R.A. and Brown, G.W. (1981) Types of stressful life event and the onset of anxiety and depressive disorders. *Psychol. Med.,* **11:** 803–15

Frankenhaeuser, M. (1975) Experimental approach to the study of catecholamines and emotion, in L. Levi (ed.), *Emotions: Their Parameters and Measurement,* Raven Press, New York, pp 209–34

Glue, P. (1991) The pharmacology of saccadic eye movements. *J. Psychopharmacol.,* **5:** 377–87

Glue, P. and Nutt, D.J. (1988) Clonidine challenge testing of alpha-2-adrenoceptor function in man: the effects of mental illness and psychotropic medication. *J. Psychopharmacol.*, **2:** 119–37

Glue, P. and Nutt, D.J. (1990) Overexcitement and disinhibition: dynamic neurotransmitter interactions in alcohol withdrawal. *Br. J. Psychiat.*, **157:** 491–9

Glue, P., Wilson, S.J., Ball, D. and Nutt, D.J. (1992a) Benzodiazepine sensitivity in subjects with high and low trait neuroticism. *Biol. Psychiat.*, **31:** 149A

Glue, P., Wilson, S.J. and Nutt, D.J. (1992b) in C. Hallstrom (ed.), *Benzodiazepine Dependence,* Oxford University Press, Oxford (in press)

Grant, K.A., Valverius, P., Hudspith, M. and Tabakoff, B. (1990) Ethanol withdrawal seizures and the NMDA receptor complex. *Eur. J. Pharmacol.*, **176:** 289–96

Gray, J.A. (1987) The neuropsychology of emotion and personality, in S.M. Stahl, S.D. Iverson and E.C. Goodman (eds.), *Cognitive Neurochemistry,* Oxford University Press, Oxford

Greedharry, D. (1987) Expressed emotion in the families of the mentally handicapped: a controlled study. *Br. J. Psychiat.*, **150:** 400–2

Harris, T., Brown, G.W. and Bifulco, A. (1987) Loss of parent in childhood and adult psychiatric disorder: the role of social class and premarital pregnancy. *Psychol. Med.*, **17:** 163–83

Haynes, S.G., Feinleib, M. and Kannel, W.B. (1980) The relationship of psychological factors to coronary heart disease in the Framingham study. III. Eight year incidence of coronary heart disease. *Am. J. Epidemiol.*, **111:** 37–58

Holmes, T. and Rahe, R.H. (1967) The Social Readjustment Rating Scale. *J. Psychosomatic Res.*, **11:** 213–18

Hoffman, P.L., Rabe, C.S., Moses, F. and Tabakoff, B. (1989) N-methyl-D-aspartate receptors and ethanol. *J. Neurochem.*, **52:** 1937–40

Hsiao, J.K. and Potter, W.Z. (1990) The mechanism of action of antipanic drugs, in J.C. Ballenger (ed.), *Clinical Aspects of Panic Disorder,* Wiley Liss, New York, pp 297–317

Ida, Y., Tanaka, M., Tsuda, A., Tsujimaru, S. and Nagasaki, N. (1985) Attenuating effects of diazepam on stress-induced increase in noradrenaline turnover in specific brain regions of rats. *Life Sci.*, **37:** 2491–7

Iimori, K., Tanaka, K., Kohno, Y. *et al.* (1982) Psychological stress enhances noradrenaline turnover in specific brain regions in rats. *Pharmacol. Biochem. Behav.*, **16:** 637–40

Ingham, J.G., Kreitman, N.B. and Miller, P.McC. (1986) Self esteem, vulnerability and psychiatric disorder in the community. *Br. J. Psychiat.*, **148:** 375–85

Irvine, J., Lyle, R.C. and Allon, R. (1982) Type A personality as psychopathology: personality correlates and an abbreviated scoring system. *J. Psychosomatic. Res.*, **26:** 183–9

Jacobs, S. and Myers, J. (1976) Recent life events and acute schizophrenic psychosis: a controlled study. *J. Nerv. Mental. Dis.*, **162:** 75–87

Jacobs, S., Prusoff, B.A. and Paykel, E.S. (1974) Recent life events in schizophrenia and depression. *Psychol. Med.*, **4:** 444–52

Javitt, D.C. and Zukin, S.R. (1991) Recent advances in the phencyclidine model of schizophrenia. *Am. J. Psychiat.*, **148:** 1301–8

Kitazawa, S., Shioda, S. and Nakai, Y. (1987) Catecholaminergic innervation of neurons containing corticotropin-releasing factor in the paraventricular nucleus of the rat hypothalamus. *Acta Anatomica*, **129:** 337–43

Klein, D.F. (1981) Anxiety reconceptualized, in D.F. Klein and J.G. Rabkin (eds.), *Anxiety: New Research and Changing Concepts,* Raven Press, New York, pp 235–63

Ko, G.N., Elsworth, J.D., Roth, R.H. *et al.* (1983) Panic-induced elevation of plasma MHPG levels in phobic-anxious patients: effects of clonidine and imipramine. *Arch. Gen. Psychiat.*, **40:** 425–30

Krystal, J.H. (1990) Animal models for posttraumatic stress disorder, in E.L. Giller (ed.) *The Biological Assessment and Treatment of PTSD,* American Psychiatric Association Press, Washington, DC, pp 3–26

Lazarus, R.S. and Folkman, S. (1984) *Stress, Appraisal and Coping.* Springer-Verlag, New York

Leff, J. and Vaughan, C. (1981) The role of maintenance therapy and relatives' expressed emotion in relapse of schizophrenia: a two year follow up. *Br. J. Psychiat.*, **139:** 102–4

Leff, J.P., Kuipers, L., Berkowitz, R et al. (1982) A controlled trial of social intervention in the families of schizophrenic patients. *Br. J. Psychiat.*, **141**: 121–34

Leff, J.P., Berkowitz. R., Shavit, N. et al. (1989) A trial of family therapy versus a relatives' group for schizophrenia. *Br. J. Psychiat.*, **154**: 58–66

Levitt, E.E., Persky, H., Brady, J.P. and Fitzgerald, J.A. (1963) The effect of hydrocortisone infusion on hypnotically induced anxiety. *Psychosomatic Med.*, **25**: 158–61

Lewinsohn, P.M., Steinmetz, J.L., Larson, D.W. and Franklin, J. (1981) Depression-related cognitions: antecedent or consequence. *J. Abnormal. Psychol.*, **90**: 213–19

Lind, R.W. and Swanson, L.W. (1984) Evidence for corticotropin releasing factor and Leu-enkephalin in the neural projection from the lateral parabrachial nucleus to the median preoptic nucleus: a retrograde transport, immunohistochemical double labelling study in the rat. *Brain Res.*, **321**: 217–24

Liu, W.S., Bidwai, A.V., Lunn, J.K. and Stanley, T.H.E. (1977) Urine catecholamine excretion after large doses of fentanyl, fentanyl and diazepam, diazepam and pancuronium. *Can. Anaesth. Soc. J.*, **24**: 371–9

Lloyd, K.G. and Morselli, P.L. (1987) Psychopharmacology of GABAergic drugs, in H.Y. Meltzer (ed.), *Psychopharmacology: The Third Generation of Progress*, Raven Press, New York, pp 183–95

Lovinger, D.M., White, G. and Weight, F.F. (1989) Ethanol inhibits NMDA-activated ion current in hippocampal neurons. *Science*, **243**: 1721–4

Mathews, A. and MacLeod, C. (1985) Selective processing of threat cues in anxiety states. *Behav. Res. Ther.*, **23**: 563–9

Matussek, N. and Hoehe, M. (1989) Investigations with the specific mu-opiate receptor agonist fentanyl in depressive patients: growth hormone, prolactin, cortisol, noradrenaline and euphoric responses. *Neuropsychobiology*, **21**: 1–8

McEwen, B.S., Angul, J., Cameron, H. et al. (1992a) Paradoxical effects of adrenal steroids on the brain: protection versus degeneration. *Biol. Psychiat.*, **31**: 177–99

McEwen, B.S., Gould, E.A. and Sakai, R.R. (1992b) The vulnerability of the hippocampus to protective and destructive effects of glucocorticoids in relation to stress. *Br. J. Psychiat.*, **160** (suppl 15): 18–23

McGuffin, P., Katz, R. and Bebbington, P. (1988) The Camberwell Collaborative Depression Study. III. Depression and adversity in the relatives of depressed probands. *Br. J. Psychiat.*, **152**: 775–82

Mezzich, J.E. (1988) On developing a psychiatric multiaxial schema for ICD-10. *Br. J. Psychiat.*, **152** (suppl 1): 38–43

Middleton, H.C. (1990) An enhanced hypotensive response to clonidine can still be found in panic patients despite psychological treatment. *J. Affect. Dis.* **4**: 213–19

Mitchell, P.B., Bearn, J.A., Corn, T.H.E. and Checkley, S.A. (1988) Growth hormone response to clonidine after recovery in patients with endogenous depression. *Br. J. Psychiat.*, **152**: 34–38

Morris, R.G.M., Anderson, E., Lynch, G.S. and Baudry, M. (1986) Selective impairment of learning and blockade of long-term potentiation by an N-methyl-D-aspartate receptor antagonist, AP5. *Nature*, **319**: 774–6

Nikolarakis, K., Pfeiffer, A., Stalla, G.K. and Herz, A. (1987) The role of CRF in the release of ACTH by opiate agonists and antagonists in the rat. *Brain Res.*, **421**: 373–6

Nikolarakis, K., Pfeiffer, A., Stalla, G.K. and Herz, A. (1989) Facilitation of ACTH secretion by morphine is mediated by activation of CRF containing neurons and sympathetic neuronal pathways. *Brain Res.*, **498**: 385–8

Nutt, D.J. Interactions between diazepam and noradrenergic function in man. *Human Psychopharmacol.*, **1**: 35–40

Nutt, D.J. (1990) The pharmacology of human anxiety. *Pharmacol. Ther.*, **47**: 233–66

Nutt, D.J. (1991) The three S's: subtyping subunits in Sardinia. GABAergic synaptic transmission: molecular, pharmacological and clinical aspects. *J. Psychopharmacol.*, **5**: 426–36

Nutt, D.J. and Glue, P. (1991) Imipramine in panic disorder. 2. Effects on α2-adrenoceptor function. *J. Psychopharmacol.*, **5**: 135–41

Nutt, D.J. and Lawson, C.W. (1992) Panic attacks. A neurochemical overview of models and mechanisms. *Br. J. Psychiat.*, **160**: 165–78

Nutt, D.J., Glue, P., Molyneux, S. and Clarke, E. (1988) Alpha-2-adrenoceptor activity in alcohol withdrawal: a pilot study of the effects of i.v. clonidine in alcoholics and normals. *Alcohol: Clin. Exp. Res.*, **12**: 14–18

Nutt, D.J., Glue, P. and Lawson, C.W. (1990a) The neurochemistry of anxiety. *Prog. Neuropsychopharmacol. Biol. Psychiat.*, **14**: 737–52

Nutt, D.J., Glue, P., Lawson, C. and Wilson, S. (1990b) Evidence for altered benzodiazepine receptor sensitivity in panic disorder: effects of the benzodiazepine antagonist flumazenil. *Arch. Gen. Psychiat.*, **47**: 917–25

Orford, J., O'Reilly, P. and Goonatilleke, A. (1987) Expressed emotion and perceived family interaction in the key relatives of elderly patients with dementia. *Psychol. Med.*, **17**: 963–70

Parker, G. and Brown, L. (1982) Coping behaviours that mediate between life events and depression. *Arch. Gen. Psychiat.*, **39**: 1386–91

Paykel, E.S., Myers, J.K., Dienelt, M.N. *et al.* (1969) Life events and depression: a controlled study. *Arch. Gen. Psychiat.*, **21**: 753–7

Pfohl, B., Stangl, D. and Zimmerman, M. (1984) The implications of DSM-III-R personality disorders for patients with major depression. *J. Affect. Dis.*, **7**: 309–18

Priebe, S., Wildgrube, C. and Muller-Oerlinghausen, B. (1989) Lithium prophylaxis and expressed emotion. *Br. J. Psychiat.*, **154**: 396–9

Rahe, R., Gunderson, E.K.E. and Arthur, R.J. (1970) Demographic and psychosocial factors in acute illness reporting. *J. Chronic. Dis.*, **23**: 245–55

Reich, J.H. (1988) DSM-III personality disorders and the outcome of treated panic disorder. *Am. J. Psychiat.*, **145**: 1149–52

Rivier, C., Bruhn, T. and Vale, W. (1984) Effect of ethanol on the hypothalamic-pituitary-adrenal axis in the rat: role of corticotropin-releasing factor (CRF). *J. Pharmacol. Exp. Ther.*, **229**: 127–31

Robson, P.J. (1988) Self esteem – a psychiatric view. *Br. J. Psychiat.*, **153**: 6–15

Rosenman, R.H., Brand, R.J., Jenkins, C.D. *et al.* (1975) Coronary heart disease in the Western Collaborative Group study. Final follow up experience of 8½ years. *J. Am. Med. Assn.*, **233**: 872–7

Rossetti, Z.L., Postas, C., Pani L. *et al.* (1990) Stress increases noradrenaline release in the rat frontal cortex: prevention by diazepam. *Eur. J. Pharmacol.*, **176**: 229–31

Roy-Byrne, P.P., Geraci, M. and Uhde, T.W. (1986) Life events and the onset of panic disorder. *Am. J. Psychiat.*, **143**: 1424–7

Roy-Byrne, P.P., Cowley, D.S., Greenblatt, D.J. *et al.* (1990) Reduced benzodiazepine sensitivity in panic disorder. *Arch. Gen. Psychiat.*, **47**: 534–8

Roy-Byrne, P.P., Cowley, D.S., Ritchie, J. *et al.* (1991) Benzodiazepine sensitivity in panic disorder: effects of alprazolam treatment. *Biol. Psychiat.*, **29**: 55A

Rutter, M. (1985) Resilience in the face of adversity. Protective factors and resistance to psychiatric disorder. *Br. J. Psychiat.*, **147**: 598–611

Sachar, E.J. (1970) Psychological factors relating to activation and inhibition of the adrenocortical stress response in man: a review. *Prog. Brain Res.*, **32**: 316–24

Sargant, W. and Dally, P. (1962) Treatment of anxiety states by antidepressant drugs. *Br. Med. J.*, **i**: 6–9

Schmauss, C. and Emrich, H.M. (1988) Narcotic antagonist and opioid treatment in psychiatry, in R.J. Rodgers and S.J. Cooper (eds.), *Endorphins, Opiates and Behavioural Processes*, Wiley, Chichester, pp 327–51

Shagass, C. (1954) Sedation threshold. Method for estimating tension in psychiatric patients. *Electroenceph. Clin. Neurophysiol.*, **6**: 221–33

Shagass, C. and Naiman, J. (1956) The sedation threshold as an objective index of manifest anxiety in psychoneurosis. *J. Psychosom. Res.*, **1**: 49–57

Shanks, N. and Anisman, H. (1988) Stressor-provoked behavioral changes in six strains of mice. *Behav. Neurosci.*, **102**: 894–905

Shrout, P.E., Link, B.G, Dohrenwend, B.P. *et al.* (1989) Characterising life events as risk factors for depression; the role of fateful loss events. *J. Abnorm. Psychol.*, **98**: 460–7

Stanford, S.C. (1990) Central adrenoceptors in response and adaptation to stress, in D.J. Heal and C.A. Marsden (eds.), *The Pharmacology of Noradrenaline in the Central Nervous System*, Oxford University Press, Oxford, pp 379–422

Stanford, S.C. and Salmon, P. (1989) Neurochemical correlates of behavioural responses to frustrative nonreward in the rat: implications for the role of central noradrenergic neurones in behavioural adaptation to stress. *Exp. Brain Res.*, 75: 133–8

Starke, K. (1981) Presynaptic receptors. *Ann. Rev. Pharmacol. Toxicol.*, 21: 7–30

Stone, E.A. (1979) Subsensitivity to norepinephrine as a link between adaptation to stress and antidepressant therapy: an hypothesis. *Res. Commun. Psychol. Psychiat. Behav.*, 4: 241–55

Stone, E.A. (1983) Problems with current catecholamine hypotheses of antidepressant agents: speculations leading to a new hypothesis. *Behav. Brain Sci.*, 6: 535–77

Szmukler, G.I., Berkowitz, R., Eiser, I. *et al.* (1987) Expressed emotion in individual and family settings: a comparative study. *Br. J. Psychiat.*, 151: 174–8

Tarrier, N., Barrowclough, C., Vaughan, C.E. *et al.* (1989) Community management of schizophrenia. A two year follow up of a behavioural intervention with families. *Br. J. Psychiat.*, 154: 625–8

Taylor, K.M. and Laverty, R. (1969) The effect of chlordiazepoxide, diazepam and nitrazepam on catecholamine metabolism in regions of the rat brain. *Eur. J. Pharmacol.*, 8: 296–301

Trap-Jensen, J., Carlsen, J.E., Hartling, O.J. *et al.* (1982) Beta-adrenoceptor blockade and psychic stress in man. A comparison of the acute effects of labetolol, metoprolol, pindolol and propranolol on plasma levels of adrenaline and noradrenaline. *Br. J. Clin. Pharmacol.*, 13: 391 S–5 S

Uno, H., Ross, T., Else, J. *et al.* (1989) Hippocampal damage associated with prolonged and fatal stress in primates. *J. Neurosci.*, 9: 1705–11

Valentino, R.J., Foot, S.L. and Aston-Jones, G. (1983) Corticotropin-releasing factor activates noradrenergic neurones of the locus coeruleus. *Brain Res.*, 270: 363–7

Vaughan, C.E. (1989) Annotation: expressed emotion in family relationships. *J. Child Psychol. Psychiat.*, 30: 13–22

Vaughan, C.E. and Leff, J., (1976) The measurement of expressed emotion in the families of psychiatric patients. *Br. J. Social Clin. Psychol.*, 15: 157–65

Ventura, J., Nuechterlein, K.H., Lukoff, D. and Hardesty, J.P. (1989) A prospective study of stressful life events and schizophrenic relapse. *J. Abnorm. Psychol.*, 98: 407–11

Villacres, E.C., Hollifield, M., Katon, W.J. *et al.* (1987) Sympathetic nervous system activity in panic disorder. *Psychiat. Res.*, 21: 313–21

Watkins, J.C., Krogsgaard-Larsen, P. and Honore, T. (1990) Structure activity relationships in the development of excitatory amino acid receptor agonists and competitive antagonists. *Trends Pharmacol. Sci.*, 11: 25–33

Weight, F.F., Aguayo, L.G., White, G. *et al.* (1992) Ethanol and excitatory amino acid receptors. *Adv. Biochem. Psychopharmacol.* (in press)

Weizman, R., Weizman, A., Kook, K.A. *et al.* (1989) Repeated swim stress alters brain benzodiazepine receptors measured *in vivo*. *J. Pharmacol. Exp. Ther.*, 249: 701–7

Weizman, A., Weizman, R., Kook, K.A. *et al.* (1990) Adrenalectomy prevents the stress-induced decrease in *in vivo* [3H] Ro 15–1788 binding to GABA-A receptors in the mouse. *Brain Res.*, 519: 347–50

Wolff, H.G. (1962) A concept of disease in man. *Psychosomatic. Med.*, 24: 25–30

Zacharko, R.M., Lalonde, G., Kasian, M. and Anisman, H. (1987) Strain-specific effects of inescapable shock on intracranial self-stimulation from the nucleus accumbens. *Brain Res.*, 426: 164–8

Zis, A.P., Haskett, R.F., Albala, A.A. *et al.* (1985) Opioid regulation of hypothalamic-pituitary-adrenal axis in depression. *Arch. Gen. Psychiat.*, 42: 383–6

4

Stress and the cardiovascular system: Central and peripheral physiological mechanisms

Béla Bohus and Jaap M. Koolhaas

Department of Animal Physiology
University of Groningen
Haren
The Netherlands

75

ISBN 0–12–663370–3

4.1 INTRODUCTION

The cardiovascular system signals stressful influences of the psychosocial and/or physical environment on humans and other animals. The prominent British physiologist Sherrington (1952) maintained that natural emotional states also profoundly change cardiovascular function:

"yet heightened beating of the heart, blanching or flushing of the blood vessels, . . . all these are prominent characters in the pantomime of natural emotion".

The stressful event may have an immediate influence on the function of the heart and blood vessels, therefore, or it may affect heart rate and blood pressure *via* cognitive (learning and memory) processes over a long period of time.

Walter Bradford Cannon was the first to study physiological responses to directly threatening environmental influences (Cannon, 1929a). Cannon's description of the 'fight or flight' response as the natural reaction to threat in animals, including humans, has remained a keystone of our understanding of the physiology and pathology of stress. This reaction pattern is characterized by activation of the sympatho-adrenal system, *via* a mass-discharge, and inhibition of the parasympathetic nervous system (Cannon, 1929b). The sympathetic mass-discharge has long been considered to be responsible for the 'classical' cardiovascular signs of stress such as the pressor response (increase in blood pressure) and tachycardia (increase in heart rate). It also inhibits the enteric limb of the parasympathetic nervous system, thereby reducing gastrointestinal activity during stress. The idea of sympathetic mass-discharge has only recently been challenged as a result of extensive physiological and neurobiological research on nervous regulation of cardiovascular function (Jänig and McMachlan, 1992).

Although Cannon can be considered the father of the physiology of stress, the concept itself was introduced a few years later by Hans Selye (1935). The concept of non-specificity of bodily responses to diverse noxious stimuli, primarily *via* the activation of the adrenal cortex, led endocrinologists to study mechanisms of adaptation. More than a quarter of a century later, the non-specific nature of stress was challenged by Mason (1968a,b): his and others' research drew attention to the necessity of experiencing novelty or distress on an individual basis to elicit adrenal cortical or sympatho-adrenal activation.

That stress cannot be viewed as a unidimensional reaction has also been advocated by Henry (Henry and Stephens, 1977). Henry's studies of social stress in mice showed diverse stress responses ranging from sympathetic mass-discharge to adreno-cortical and parasympathetic hyperactivation. The pattern of the stress response in any case depends on an individual's appreciation of the environment which is, in turn, determined by the social status of that individual. In our own view, the non-specificity *versus* specificity of stress response patterns depends upon the controllability and/or predictability of the environment, coping strategies (which are either genetically determined or acquired during the life-span), the duration and nature of the stressors, and finally, the relevant systems (e.g. neuroendocrine, cardiovascular, gastrointestinal) and their state (resting or activated; physiological or pathological; Bohus *et al.*, 1987a). This interactional view, that follows recent and influential psychological theories of stress (e.g. Levine and Ursin, 1991), led to the recognition of two, equally successful, coping patterns in rodents: active and passive behavioural coping. These have distinct physiological characteristics which determine successful adaptation to certain environments but, at the same time, leave the latent possibility of failure in others (Koolhaas and Bohus, 1989, 1991).

Surprisingly, the concept of stress has remained practically 'undiscovered' by cardiovascular physiologists. The absence of proper instruments for cardiovascular studies in conscious animals or humans is only a partial excuse. Bard's observation, as early as in 1928, that integration of the flight or fight response requires an intact hypothalamus (Bard, 1928), was largely ignored until fairly recently. Hilton and colleagues (see Hilton, 1982) constructed a detailed map of areas in the forebrain, hypothalamus, mid-brain and the medulla oblongata involved in behavioural and cardiovascular regulation in defence. The results showed a remarkable similarity between cardiovascular reaction patterns elicited by hypothalamic (or forebrain) stimulation in anaesthetized cats and the natural reactions occurring during defence in this species (Zanchetti *et al.*, 1972). These studies somewhat shifted the interest to investigations of brain circuits involved in cardiovascular responses to diverse stressors. It was then recognized that more than one autonomic neuronal network exists which, together with (neuro)endocrine systems, control cardiac and vascular reaction to various stressors.

This chapter is devoted primarily to presenting a view on the physiological mechanisms of control of cardiovascular responses to natural and experimental stress in animals, particularly in rats. Although cardiovascular studies are not always easy in this species, the vast amount of knowledge on the behavioural, neurobiological and neuroendocrine characteristics of the stress response justifies the use of rats for cardiovascular stress research as well. However, when extrapolating this knowledge to humans, differences in the cardiovascular function between rodents and humans should be borne in mind (Herd, 1991). Our aim here is to provide an integrated view of different levels of the control of the cardiovascular stress response, both between and within organ systems. Accordingly, attention will be given to central and peripheral nervous and endocrine systems. As far as recent knowledge permits, the mechanisms are discussed at system, cellular and molecular levels. Finally, recent knowledge concerning probable mechanisms of cardiovascular stress pathology is summarized, and the influence of pathological states, such as hypertension, on stress responsiveness is discussed. To focus on an integrative view means that this chapter cannot provide a complete picture. Fortunately, several recent and excellent reviews may help the reader to obtain more details.

4.2 PHYSIOLOGY AND PATHOLOGY OF CARDIOVASCULAR RESPONSES TO STRESS

Despite an obvious cardiovascular response to stress, the number of studies providing a complete picture of cardiac and vascular function is rather limited. In many studies changes in heart rate were used to indicate a cardiac response to stress. Technically, this measure, with the aid of electrocardiographic (ECG) recordings either through wire leads or by radiotelemetry, is relatively easy even in small rodents such as rats. Measurement of blood pressure *via* chronically implanted (aortic) catheters in the rat has been mastered only by a few laboratories and mostly for restricted periods. The majority of studies apply a non-invasive technique (tail-cuff) to follow changes in resting blood pressure. The use of carnivores is rather problematic due to problems with standardization of stress procedures. Primates would be the ideal subjects for such studies but their use is understandably restricted. Measurement of other important cardiovascular indices, such as organ blood flow or cardiac output seems technically feasible, but it has not aroused much interest among stress researchers.

4.2.1 CARDIOVASCULAR RESPONSES TO SINGLE OR REPEATED STRESS

The cardiovascular reaction pattern during defence induced naturally or by electrical stimulation of the brain was the first stress response to be studied in detail in experimental animals. The distinctive feature of this pattern is that it appears, in part, to be dissimilar to other sympathetically induced cardiovascular changes (e.g. sympathetic mass-discharge). Sympathetic mechanisms seem to be responsible for vasoconstriction in the splanchnic, renal and cutaneous vascular beds. In contrast, vasodilatation occurs in skeletal muscles (Hilton, 1982). Interestingly, in the conscious baboon, bilateral lesions in a site in the hypothalamus that elicits the defence pattern abolishes muscle vasodilatation leaving vasoconstriction intact (Smith *et al.*, 1980). This vasodilatation is effected by sympathetic cholinergic nerves and is differentially regulated in various species due to differences in the distribution of these nerves (Bolme *et al.*, 1970). Blockade of vasoconstrictor nerves does not affect vasodilatation in carnivores, ruminants and primates, whereas the response is inhibited by the cholinergic blocker, atropine, in rodents. A dissimilar pattern of cardiovascular activation was described by Smith *et al.* (1979) while studying cardiovascular responses in monkeys in relation to a conditioned emotional response (CER). An increase in blood pressure, heart rate and terminal aortic flow and a decrease in renal flow occur during a CER. Our own studies in the rat, using an electrical stimulation model of stress and arousal, indicate that activation of the ascending reticular activating system in the mesencephalon induces abrupt arrest of ongoing behavioural activity in awake animals, or wakening of a sleeping animal, followed by attentional/orientational movements. This type of behaviour is accompanied by an increase in systolic and diastolic blood pressure and a *decrease* in heart rate, i.e. a pressor response with bradycardia. The cardiovascular response also develops in rats under urethane anaesthesia suggesting that the behavioural and cardiovascular responses are not dependent on each other (Bohus *et al.*, 1989). Collectively, these studies suggest the existence of more than one cardiovascular stress pattern. Neurobiological mechanisms underlying these responses support this hypothesis as discussed later in this chapter.

4.2.1.1 Stress and hypertension

It is impossible to review all the studies in rats, dogs and monkeys suggesting elevation of blood pressure during stress. Briefly, the majority show a different degree of elevation of blood pressure and tachycardia in response to routine laboratory stressors, such as painful electric footshock, immobilization, noise or more 'delicate' stress procedures such as avoidance conditioning (Henry and Stephens, 1977; Kelleher et al., 1981; McCarty, 1983; Smith and DeVito, 1984). The conditioning procedures allow a precise analysis of behaviour which parallels the cardiovascular response. The behavioural measure might be related to the cognitive component (i.e. what is learned) or to the motor component (i.e. the expression of what is learned). In addition, the nature of the stressful stimulus and the adopted coping strategy are important determinants of the hypertensive response. Fokkema and Koolhaas (1985) showed that social defeat in the male rat induces a more pronounced increase in mean blood pressure than does social victory. This difference exists both during the confrontation and in conditioned situations in which body contact with the dominant animal is prevented. Korte et al. (1992a) demonstrated that the increase in mean blood pressure in a CER paradigm requiring passive behavioural coping is much more pronounced (20 mm Hg) than that in rats trained in the active coping paradigm of defensive burying (average 10 mm Hg). This finding suggests that greater physical effort does not necessarily mean a larger hypertensive response to the stressor. Rather, 'control' may be the variable that determines the magnitude of the response.

Other research into the effect of stress upon blood pressure has focused on whether chronic stress can lead to hypertension. Many 'classical' stress procedures such as immobilization, escapable or inescapable electric footshock, noise or flashing light given continuously or discontinuously over several months induce a sustained elevation of blood pressure in rats and monkeys (Rosencrans et al., 1966; Lamprecht et al., 1973; Herd et al., 1969; Gamallo et al., 1988). In general, these stressors cause a moderate (borderline) increase in blood pressure. However, male offspring of female spontaneously hypertensive rats (SHR) and male Wistar–Kyoto (WKy) normotensive rats develop spontaneous borderline hypertension. These hybrid rats show a chronic and persisting hypertension (185 mm Hg or higher) in a stressful behavioural conditioning paradigm (Lawler et al., 1980a, b; 1981; 1984a, b). Cardiac output, vascular resistance and increased sympathetic drive of the hybrids, in comparison to normotensive WKys, may contribute to the hypertension (Hubbard et al., 1986). An interesting model for stress-induced hypertension in Wistar rats was developed by genetic selection (Markel, 1985 cited by Naumenko et al., 1989). The selection criterion is an exaggerated increase in blood pressure during immobilization. Whereas the 15th generation has a mean blood pressure of 148 mm Hg, the 19th generation displays a value of 178 mm Hg (Naumenko et al., 1989).

Chronic social stressors (i.e. species specific natural stimuli) also induce hypertension in rodents and primates. The sustained elevation of blood pressure may reach 175 mm Hg; the effects depend on social status. For example, Henry et al. (1975) found that mice living for at least nine months in a complex population cage develop irreversible hypertension. According to Ely and Henry (1971), the order of dominance within the group determines the magnitude of the blood pressure elevation: the dominant animal is more hypertensive than the subordinates. Dominant macaques have also higher blood pressure than subordinates (Shively and Kaplan, 1984). Social isolation may also act as a stressor to induce hypertension. Gardiner and Bennett (1977) reported reversible blood pressure elevation in rats throughout a three-week isolation period. Carlier et al. (1988) showed that isolation with brief daily grouping induces hypertension after six weeks. In contrast with these studies, isolated Wistar rats show an average elevation of 33 mm Hg of blood pressure after only seven days (Naranjo et al., 1986). That isolation-induced hypertension depends on other factors as well is shown by the study of Gardiner and Bennett (1977): all rats housed in glass metabolic cages developed hypertension, whereas about 50% of conventionally housed, but isolated, rats remain normotensive.

The importance of the complexity of the social and non-social environment in stress-induced hypertension is suggested by several studies. Ely (1981) demonstrated that disturbances in an established social order within a group of mice living in population cages increase the incidence of cardiovascular pathology. The experiments of Henry et al. (1975) have been replicated with rats in complex population cages (Harrap et al., 1984). This particular social stress fails to induce hypertension, but the appearance of gastric ulceration in one study indicates that individuals experienced appreciable stress (Harrap et al., 1984). In other studies, normotensive Wistars were used as intruders into an established colony. After six weeks, elevated blood pressure was observed in these rats. A recent paper reports elevated blood pressure in group-housed normotensive male Wistar rats, up to a systolic level of 135 mm Hg, following switching of some rats to another conventional cage (Szilagyi, 1991); this procedure is also effective in SHRs (Ely and Weigand, 1983). However, in our hands, prevention of the establishment of a social hierarchy in group-housed rats, by daily changes in the composition of the group, fails to induce substantial changes in blood pressure (Korte and Bohus, unpublished).

That individual differences which predetermine the social position of an animal are important for the development of mild sustained elevation in blood pressure is shown by studies in this laboratory (Fokkema, 1986). Rats that are more offensive and defensive in social interaction occupy the higher ranks in a rat colony that is established in a large open space (4 m²) with small 'nest' boxes. A three-month colony aggregation fails to influence the blood pressure of the single dominant. Subdominants showing many attempts to challenge the dominant's position are mildly hypertensive and show increased blood pressure reactivity to a single social challenge following the three-month colony experience. Subordinate rats remain normotensive and show low reactivity of blood pressure to social challenge. Observations of Fokkema and Koolhaas (1985) and Fokkema et al. (1988) in dynamic social interactions show coherent relations between behaviour, blood pressure response and endocrine changes. The more competitive a rat is during social interaction (victory and defeat), the higher is the increase in both blood pressure and the ratio of plasma noradrenaline/adrenaline. In a recent study in baboons, individuals with higher behavioural and endocrine indices of stress in a conflict situation also showed a greater systolic blood pressure elevation (Turkkan, 1991).

Together, these data suggest that development of hypertension during chronic stress requires a complex set of social and environmental stimuli. Comparison of the data by Henry *et al.* (1975) and Fokkema and Koolhaas (1985) suggests that the effort to maintain a social position may be of primary aetiological importance in development of sustained hypertension. Obviously, social disorder increases the probability of developing hypertension, but this factor cannot be considered as the only determinant of high blood pressure.

4.2.1.2 Heart rate response to stress

The cardiovascular system is generally viewed as a coordinated physiological entity. However, experiments with diverse stressors demonstrate a preferentially high reactivity of the heart rate to changes in the environment. This is a short latency response effected *via* the autonomic nervous system. The reactivity of blood pressure is less and of somewhat longer latency. Since handling, a procedure that is often unavoidable when starting to record heart rate, immediately increases heart rate in all species, it is difficult to obtain basal heart rates in freely moving animals subjected to an experimental environment. Heart rate is also increased by restriction of movement resulting from wire contact with the recording equipment. Nevertheless, the ratio between motionless and self-grooming states remain the same (Bohus, 1974). These results obtained in the rat can be replicated in baboons (Adams *et al.*, 1988).

The stress-induced increase in heart rate and reduction of interbeat interval is usually measured as the shortening of the 'R-R interval' (i.e. the time elapsed between two consecutive R waves of the ECG). This change occurs with very short latency. The magnitude of the increase is profoundly influenced by the motor activity accompanying the stress. The issue of cardiac-somatic coupling (Obrist, 1981) is of primary importance in evaluating stressor effects on heart rate. It should be noted, however, that the duration of the stress-induced tachycardia nearly always exceeds that of the stress (Gärtner *et al.*, 1980; Gomez *et al.*, 1989; Roozendaal *et al.*, 1992a). The severity of stress determines the magnitude of the tachycardia in sheep and it correlates with other indices of stress such as elevation of plasma cortisol (Harlow *et al.*, 1987; Baldock and Sibly, 1990). Finally, tachycardia as a conditioned stress response occurs across a range of species (Liang *et al.*, 1979; Gomez *et al.*, 1989).

Hofer (1970) was the first to show that a decrease in heart rate may also occur in response to stress. This bradycardia, induced by sudden noise or threatening visual stimuli, is usually accompanied by behavioural immobility. Sudden silence superimposed on background noise also leads to immediate behavioural arrest and bradycardia in the rat (Hagan and Bohus, 1984; Steenbergen *et al.*, 1989; Buwalda *et al.*, 1991; 1992a). The bradycardiac stress response is greater in rats on a reversed feeding schedule, i.e. if a normal food intake rhythm in the dark is shifted to the light period (Steenbergen *et al.*, 1989). In addition, the magnitude of the bradycardiac response to such a mild stress is diminished with aging (Buwalda *et al.*, 1991).

A marked decrease in heart rate is displayed in the rat in conditioned emotional stress situations in which there is a conflict between two stressors (e.g. natural avoidance of a brightly lit area *versus* footshock punishment in the preferred dark area; Bohus, 1974). A similar bradycardiac stress response is observed, particularly in the early phase of exposure, when male rats are forced into areas in which a brief inescapable footshock was experienced at least a day earlier (Bohus, 1985; Nyakas *et al.*, 1990; Korte *et al.*, 1992a). These bradycardiac responses are imposed on a tachycardiac pre-shock baseline of the freely moving animal. This response is typical for young and young adult male rats and for females in the dioestrus phase of the oestrus cycle. Females in oestrus or old adult, aged and senescent male rats fail to show this response (De Loos *et al.*, 1979; Bohus, 1985; Nyakas *et al.*, 1990). Although forced exposure to the former shock compartment is accompanied by immobility, a cardiac-somatic coupling is, by and large, absent in the conflict situation. For example, approach of the former punishment area (i.e. a somatomotor, activity results in a marked phasic decrease in heart rate instead of an expected tachycardia (Bohus, 1977). This suggests a large impact of the cognitive component of stress on heart rate.

Although the unconditioned stress response to inescapable footshock is tachycardia, a conditioned cardiac response in CER paradigms is often bradycardia both in the rat and rabbit (e.g. Bohus, 1973). Finally, decreased heart rate occurs in male tree shrews defeated by a male conspecific and taking a subordinate position. This response is opposite to the marked tachycardiac reactions of the victorious (sub)dominant male (von Holst *et al.*, 1983).

Few studies have been devoted to determining heart rate changes as a consequence of long-term stress. Socially isolated rats show a temporary increase in resting heart rate. A remarkable dissociation occurs between heart rate and blood pressure. In contrast to heart rate the hypertensive response fails to habituate to isolation (Gardiner and Bennett, 1977). Dissociation of the blood pressure and heart rate response of rats to the chronic psychosocial stress of being transferred between cages occurs as well: the hypertensive response is not accompanied by tachycardia or bradycardia (Szilagyi, 1991). The tachycardiac response to handling and exposure to a novel environment fails to habituate provided that the rats are connected to a wire lead to record the ECG. This finding can be interpreted as persistence of (mild) stressor effects due to relative restraint of the animals. Rats bearing a miniature radiotransmitter to record the ECG show more rapid heart rate habituation both between and within repeated exposures (Bohus, 1974). Reports from Adams *et al.* (1988) support these findings. In contrast, Konarska *et al.* (1989) failed to observe differences in basal heart rate of control and stressed rats exposed daily to footshock, restraint or repeated cold swim for 26 days. Heart rate reactivity to a conditioned emotional stressor in male rats is an increasing tendency for tachycardia following a three-month period of social stress by colony aggregation (Bohus, Schoemaker and Koolhaas, unpublished results). Finally, a recent report by Eisermann (1992) shows that wild European rabbits living in a semi-natural environment for more than 1500 days display chronic alterations in heart rate which are dependent on social rank. Subordinate individuals show a chronically elevated heart rate compared to the dominant ones. This relationship is opposite to the one which would be expected on the basis of invested effort (aggressive encounters, motor activity, etc.) to attain a dominant position. However, the long term effort to maintain dominance is minimal since the position is accepted by subordinates. Therefore, in the rabbit, unlike in tree shrews (von Holst, 1986) or monkeys (Cherkovich and Tatoyan, 1973), the heart rate adapts to social position. If the social position of the dominants is constantly challenged such adaptation may not occur (Ely, 1981; Manuck et al., 1983).

4.2.2 STRESS AND CARDIAC ARRHYTHMIAS

The strong impact of emotional (psychic) stressors on the autonomic nervous system ensures that psychogenic factors are important precursors of somatic events leading to cardiac dysrhythmias and coronary or non-coronary types of sudden death syndrome. This statement appears to be valid for both humans (Kamarck and Jennings, 1991) and other animals (Hughes and Lynch, 1978). Richter (1957) was the first to demonstrate sudden death immediately after immersion in water of wild rats with their vibrissae removed. The rats display marked bradycardia and their psychic state can be described as hopeless. Lown et al. (1973) presented evidence that psychic stress in the dog increases the electric instability of the heart and consequent vulnerability for fatal ventricular fibrillation. Dogs anticipating electric shocks show tachycardia and a high degree of ventricular excitability (Liang et al., 1979). Corley et al. (1973) reported arrhythmias and abnormal ECG recordings from squirrel monkeys exhibiting tachycardia in response to restraint and avoidable electric shock.

The first observations by Lown et al. (1973) have been replicated and extended with moderate consistency. In these studies ventricular vulnerability was characterized by the intensity of the electric current required to induce ventricular fibrillation or repetitive extrasystoles when applied to the heart (Matta et al., 1976a). It was shown that restraint and avoidance conditioning reduce the repetitive extrasystole threshold in dogs. The same stress induces spontaneous ventricular fibrillation following experimental coronary occlusion (Corbolan et al., 1974; Verrier and Lown, 1977). Interestingly, adaptation to physical restraint before occlusion reduces the arrhythmogenic consequences of acute stress in pigs (Skinner et al., 1975). According to Kamarck and Jennings (1991) these data suggest that acute rather than chronic stress is associated with changes in arrhythmic potentials.

Although attention is paid mainly to ventricular tachycardia and fibrillation as triggering events for sudden cardiac death, both in animals and humans (Kamarck and Jennings, 1991) bradyarrhythmia (slow, abnormal rhythm) may also be the basis of fatal disturbance in rhythmogenesis. Bradyarrhythmia is observed in rats exposed to conflict stress or to a conditioned emotional stress of fear of inescapable shock (Bohus, 1985; Nyakas et al., 1990). The latter study also shows that the incidence of extrasystoles in stressed rats increases with age. The studies with young and young adult rats suggest that, depending on the stressful situation, parasympathetic activation may be equally important in predisposition to sudden cardiac death. Aging is an extra factor increasing the probability of fatal arrhythmia due to decreased parasympathetic activity in stress (Nyakas et al., 1990; Buwalda et al., 1992a, b).

4.2.3 PATHOLOGICAL LESIONS OF THE CARDIOVASCULAR SYSTEM AND CHRONIC STRESS

Selye (1976) proposed that, in the case of failure to adapt physiologically to stressful stimuli, non-reversible organ pathologies may develop. This, often designated as the exhaustion phase of stress, is probably the consequence of a sustained increased secretion of stress hormones, especially catecholamines and corticosteroids. One of the prominent vascular pathologies is arteriosclerosis. It is maintained that sustained peripheral vasoconstriction leads to thickening of the periarterial smooth muscle and loss of elasticity (Schneiderman, 1983). Chronic social disorder in a colony which leads to irreversible high blood pressure is followed by pathological alterations in the coronary arteries and the aorta (Henry et al., 1971; Henry and Stephens, 1977). The alterations manifest themselves as mucopolysaccharide and muscle protein deposition that leads to thickening of those vessels. In addition, replacement of elastin and collagen in the intima of the vessels by fibrous tissue results in a loss of elasticity.

Stress contributes to the development of atherosclerosis in cynomolgus monkeys provided with high or low cholesterol diet (Manuck et al., 1983, 1989; Kaplan et al., 1983). The stress was caused by frequent changes in the composition of the social group for a period of 21 months, and led to a high incidence of aggressive behaviour and extreme submissivity. Housing previously group-aggregated pigs in solitary confinement also leads to arteriosclerosis in the coronary arteries and the aorta (Ratcliffe et al., 1969). Housing male chickens in heterogenous sex groups leads to coronary pathologies, whereas arteriosclerosis is absent in all-male groups. In females kept in mixed-sex groups coronary atherosclerosis also develops, but it takes twice as long to do so (32 versus 16 meals; Ratcliffe and Snyder, 1967). Von Eiff and Piekarski (1977) proposed that oestrogens have a protective effect against cardiovascular pathology in humans. The reader is reminded that oestrogen also suppresses the bradycardiac response to stress in female rats (De Loos et al., 1979; Bohus, 1985).

In addition to atherosclerosis, focal myocardial necrosis and fibrosis were described in several species of mammals following chronic social stress (Henry and Stephens, 1977), crowding (Weber and van der Walt, 1973, 1975), surgery (Weber et al., 1973), electric footshock (Hall et al., 1960; Corley et al., 1973; Miller et al., 1978), restraint and immobilization (Johansson and Jönsson, 1977; Tapp et al., 1989). Myocardial necrosis due to coronary arteriosclerosis occurs in both male and female chickens kept in heterosexual groups (Ratcliffe and Snyder, 1967). This pathology is absent in all-male groups or individually housed birds. Chronic signalled, but irregular, escapable footshock in the rat induces venous congestion and disturbances in the myocardium together with lipid deposition in the intima layer of coronary vessels in the rat (Bassett and Cairncross, 1975, 1977). Finally, renal pathology, independent of blood pressure changes, also occurs in socially stressed mice (Henry and Stephens-Larson, 1985), wild rats (Barnett et al., 1975) and in tree shrews (von Holst et al., 1983).

4.3 THE NEUROBIOLOGY OF STRESS AND CARDIOVASCULAR REGULATION

Relatively few, and mostly recent, studies have focused directly on the central and peripheral nervous and neuroendocrine mechanisms which are involved in the organization of cardiovascular stress responses in physiological and pathological conditions. The variety of response patterns and their pathological alterations suggest the existence of alternative nervous and endocrine networks in the neurobiology of cardiovascular stress. This section presents a personal view on the few direct and numerous indirect pieces of evidence for these alternative networks in the brain and periphery.

4.3.1 CENTRAL NERVOUS REGULATION OF CARDIOVASCULAR STRESS RESPONSES

4.3.1.1 Central autonomic networks, blood pressure and heart rate

During the last decade the classical idea about higher nervous, particularly brainstem, clusters controlling cardiovascular function has undergone substantial change. First of all, the renaissance of neuroanatomy through development of sensitive tracing techniques has allowed us to map networks of functional significance. These studies are supported by electrophysiological observations using delicate techniques of single cell recording with electrical or chemical stimulation of areas of interest. Secondly, the development of immunohistochemical techniques has allowed chemical characterization of the neural networks and morphological identification and quantification of receptors for neurotransmitters and neuromodulators. Thirdly, the rapidly growing fields of cellular and molecular biology are helping to identify cellular mechanisms underlying the physiology and pathology of the cardiovascular system. Finally, observations on unanaesthetized organisms, particularly in 'real-life' stress situations, are invaluable in the integration of knowledge about regulatory mechanisms and their disturbances.

Most authors suggest that two levels of organization exist. These effect reflex adjustment of the end organs on the one hand, and integrative control on the other (e.g. Loewy, 1990). The reflex circuit is simple, with few neurones. The nucleus tractus solitarius (NTS) is a major recipient of baroreceptor information which is then conveyed *via* interneurones to brainstem nerve cells which subsequently project either directly or indirectly to parasympathetic (vagal) and sympathetic outflow neurones. This reflex organization is, by and large, governed by a central network of forebrain and hypothalamic neurones that are concerned with specific patterns of cardiovascular function. This network receives external sensory and emotional information *via* cortical and limbic connections. In addition the network receives information about the acute state of the viscera, including the cardiovascular system, *via* the NTS. Through ascending neurones, this may consequently affect adaptive integrated behaviour. It may also convey the visceral message in chronic pathological states like hypertension. Besides an organ-specific input and internal interconnections (e.g. Jordan and Spyer, 1986; Loewy, 1990) the NTS is a major source of fibres ascending to hypothalamic and limbic-forebrain structures. In turn, the NTS receives major reciprocal monosynaptic afferents from these sites (Veening *et al.*, 1984; Van Giersbergen *et al.*, 1992). Part of the ascending visceral information is bisynaptic with a synaptic connection in the lateral parabrachial nucleus. This nucleus also receives bodily information (e.g. plasma electrolytes, hormones) *via* the area postrema and projects monosynaptically to the hypothalamus, limbic forebrain and cortical areas (Saper and Loewy, 1980; Fulwiler and Saper, 1984; Loewy, 1990). The direct monosynaptic projections originating from the medial parabrachial nucleus reach the frontal, infralimbic and insular regions of the cerebral cortex. Not yet clear is the difference in the visceral/hormonal information carried by the ascending neurones of the NTS and the parabrachialis. An important aspect of the parabrachial afferentation is to connect the brainstem with cerebrocortical areas *via* diverse thalamic nuclei. The cortical projections include the parietal, cingulate and anterior limbic areas. Accordingly, the cortical connections provide an extra dimension to the visceral/hormonal (stress) information reaching higher brain areas. The parabrachial nuclei, as a target of higher brain neurones is clearly important in the descending organization of the cardiovascular stress response.

The locus coeruleus in the brainstem is frequently mentioned in relation to both stress and cardiovascular regulation. This nucleus, which contains noradrenergic cells of the ascending noradrenergic bundle system, receives ascending projections from lower brainstem areas such as the rostral ventrolateral medulla, a major cardiovascular control centre (see Aston-Jones *et al.*, 1986). It receives multisynaptic inputs from splanchnic, pelvic and vagal nerves (Svensson and Thoren, 1979; Elam *et al.*, 1984). A role for the locus coeruleus in cardiovascular function is suggested by, among others, the fact that chemostimulation of the nucleus leads to an increase in blood pressure and heart rate. This effect is eliminated by local injection of 6-hydroxydopamine, a selective neurotoxin that destroys catecholaminergic cells and abolishes noradrenergic neurotransmission (Sved and Felsten, 1987). In addition, peptidergic stimulation of the locus coeruleus (e.g. by vasopressin) also increases blood pressure (Berecek *et al.*, 1984). The locus coeruleus does not project directly to the spinal autonomic preganglionic nuclei of the intermediolateral cell column, the vagal motor nuclei or the NTS (Jones and Yang, 1985; Thor and Helke, 1988). It is therefore assumed that the cardiovascular response is produced *via* ascending neurones to the lateral hypothalamus (LH) and to the bed nucleus stria terminalis (BNST) (Jones and Yang, 1985). In view of the extensive noradrenergic innervation of the limbic forebrain (Aston-Jones *et al.*, 1984), and the involvement of the LH and BNST in the central autonomic circuit, it is likely that activation of the locus coeruleus induces cardiovascular changes *via* these central sites.

The validity of such a view of the organization of the cardiovascular stress response is supported by our own findings. Electrical stimulation of the cuneiform nucleus of the mesencephalon induces diffuse activation of the cerebral cortex, synchronization of the electrical activity of the hippocampus (Moruzzi and Magoun, 1949), increase in systolic and diastolic blood pressure and decrease in heart rate (Versteeg *et al.*, 1982) and behavioural arousal of sleeping rats or immediate arrest of ongoing behavioural activity with orientational head movements (Korte and Bohus, unpublished). This nucleus shows extensive ascending projections, of which those to the LH, BNST, the central nucleus of the amygdala (CEA) and the paraventricular nucleus of the hypothalamus (PVN) are of importance here. Descending projections reach, among other regions, the locus coeruleus, the NTS, the parabrachial complex and the dorsal motor nucleus of the vagus nerve (DMVN) (Korte *et al.*, 1992b). Transections destroying the ascending projections diminish the pressor response by about half (Bohus *et al.*, 1983). This indicates that at least a part of the response is caused *via* the hypothalamic and limbic sites. Transection of the vagal afferent nerve blocks the bradycardiac response suggesting a direct vagal activation by the nucleus. The descending sympathetic projections run, by and large, through the parabrachial-Kolliker-Fuse nuclei complex and are mainly responsible for the pressor response (Korte *et al.*, 1992b).

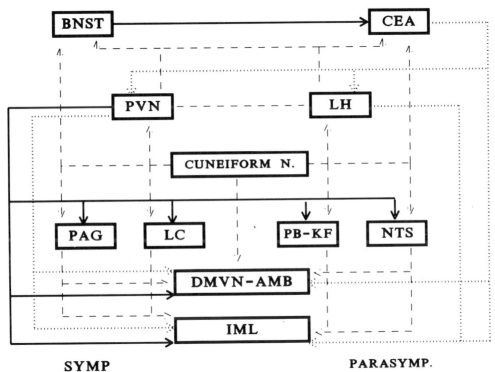

Figure 4.1 Schematic representation of the autonomic network in the brain which is involved in a sympathetically or a parasympathetically governed cardiovascular response to various stressors: the role of vasopressin and corticotropin releasing factor (CRF) containing neurones. BNST: bed nucleus of the stria terminalis; CEA: central amygdaloid nucleus; PVN: paraventricular nucleus of the hypothalamus; LH: lateral hypothalamic nucleus; PAG: periaqueductal gray; LC: locus coeruleus; PB-KF: parabrachial and Kolliker-Fuse nuclei; NTS: nucleus tractus solitarius; DMVN-AMB: dorsal motor nucleus of the vagus nerve and the ambiguous nucleus; IML: intermedio-lateral column of the spinal cord. Solid lines: vasopressinergic neurones; dashed lines: corticotropin releasing factor containing neurones; broken lines: other ascending and descending neurones

The forebrain nuclei such as the CEA, the BNST, the LH and the PVN have been suggested to be part of a central autonomic circuit (Loewy, 1990, 1991). Besides the reciprocal connections to the NTS, parabrachial nucleus and the locus coeruleus, these nuclei project monosynaptically to the DMNV and nucleus ambiguous which represents the major relay station of the parasympathetic outflow. The projections of the PVN also include the sympathetic preganglionic neurones of the intermediolateral column of the spinal cord. In addition, the forebrain areas are interconnected by extensive projection systems (e.g. Swanson and Sawchenko, 1983; Veening et al., 1984; Holstege et al., 1985; Luiten et al., 1987). Together, the forebrain and the brainstem form an autonomic circuit by which the neural network conveys reciprocal information via monosynaptic and multisynaptic pathways.

There is ample direct and indirect evidence to suggest the involvement of the forebrain autonomic circuit in stress-related cardiovascular regulation. It is, however, as yet unclear which effects are regulated by the monosynaptic and which by polysynaptic pathways. Hilton and his associates were the first to demonstrate the involvement of the amygdala in the cardiovascular defence response (see Hilton, 1982). Electrical stimulation of the amygdala produces defence reactions that are similar to those induced from the hypothalamus (Stock et al., 1979). Chemostimulation of the CEA and adjacent areas, with unilateral injection of the cholinomimetic, carbachol, or bilateral administration of noradrenaline, induces an increase in blood pressure and bradycardia in the rat. The behavioural reaction to carbachol is immobility and/or freezing (Ohta et al., 1991). The excitatory amino acid glutamate also induces a pressor response (Iwamoto et al., 1987). Electrolytic lesions of the CEA inhibit the bradycardiac response to conditioned emotional stressors of fear of shock in the rabbit (Kapp et al., 1979) and of the conditioned social stress of defeat or conditioned fear in the rat (Roozendaal et al., 1990). It inhibits the blood pressure increase caused by fear of stress (LeDoux et al., 1988) and the acute tachycardiac response to inescapable footshock in the rat (Roozendaal et al., 1991). In addition, it attenuates the exaggerated pressor and vasoconstrictor response to noise stress in spontaneously hypertensive rats (Galeno et al., 1984). The role of BNST in stress-induced cardiovascular changes is unclear. The highly comparable projection pattern of the CEA and BNST (e.g. Weller and Smith, 1982; Holstege et al., 1985) implies the involvement of both nuclei in autonomic regulation.

The involvement of various hypothalamic regions in the regulation of cardiovascular function has been frequently suggested (Smith et al., 1980). However, the role of the PVN and LH in stress-related autonomic function is practically unknown. LH lesion prevents the hypertensive effects of fear conditioning (LeDoux et al., 1988). Adrenergic chemostimulation or inhibition of these nuclei may differentially affect circulating levels of the catecholamines, adrenaline and noradrenaline in physical exercise (Scheurink and Steffens, 1990). This suggests that the various parts of the sympatho-adrenal system are activated differentially rather than en masse by forebrain areas. In a recent report Nakamura et al. (1992) showed that the blood pressure increase during discriminative learning involving aversive stimuli is diminished by microinjection of an α_1-adrenergic blocker into the PVN of the rat. Glutaminergic chemostimulation of the PVN increases adrenal nerve activity, but decreases blood pressure and renal nerve activity (Katafuchi et al., 1988).

Accordingly, the chemical diversity of these nuclei may contribute to a dynamic control of the various stress functions, including cardiovascular alterations. A remarkable chemical diversity characterizes all these forebrain nuclei and also brainstem nuclei such as the NTS and the parabrachialis nucleus: besides classical neurotransmitters, a large number of peptidergic transmitters in each nucleus effect the reciprocal connections within the autonomic circuit (Loewy, 1990; Gray, 1991; van Giersbergen *et al.*, 1992). The interactive role of transmitters, neuropeptides and peripheral hormones at the level of the PVN is already documented for control of macronutrient intake and metabolism (Leibowitz, 1992). Such complex interactions in cardiovascular regulation remain to be shown.

The periaqueductal gray area (PAG) of the midbrain periventricular gray also deserves attention. Since Hilton's original observation (see Abrahams *et al.*, 1960), many studies have reinforced the view that this region is involved in the integration of behavioural and cardiovascular reactions to threatening stressful stimuli. Electrical stimulation of the PAG elicits behaviours and cardiovascular changes that resemble natural defence responses (Bandler, 1988). Extensive studies suggest that such stimulation affects an intrinsic brainstem circuit and not descending fibres from the forebrain autonomic area (Bandler *et al.*, 1991). The PAG shows viscerotopic and somatotopic organization of the modulation of specific cardiovascular parameters in specific visceral and somatic regions (Bandler *et al.*, 1991). The modulation may be both excitatory or inhibitory in character (Bandler *et al.*, 1991; Van der Plas and Maes, unpublished). As far as the innervation of the PAG is concerned, it receives vast projections from the forebrain autonomic nuclei such as the CEA, BNST, and the hypothalamic areas (Holstege *et al.*, 1985; Holstege, 1987; Luiten *et al.*, 1987). These projections reach those PAG areas that have extensive topographically organized outputs to the lower brainstem. It is therefore possible that part of the higher nervous autonomic (and behavioural) information is further topographically organized within the PAG. However, not all integrated information that evokes specific response patterns necessarily reaches the PAG. Cardiovascular changes evoked by the stimulation of the cuneiform nucleus are unaffected by the transection of the afferents (Korte *et al.*, 1992b).

Taken together, neuroanatomical and physiological studies show that certain key areas of the limbic forebrain and hypothalamus form a network that regulates autonomic nervous activity *via* brainstem and spinal outputs. Although a limbic autonomic network seems common to different responses, different output systems mediate different responses: fight/flight is sympathetically dominated while conservation/withdrawal is parasympathetically dominated leading to two types of cardiovascular stress response (Bohus *et al.*, 1990b; Bohus, 1993).

4.3.1.2 *Central nervous influences on autonomically mediated cardiac arrhythmias and coronary circulation*

Evidence suggests a central nervous mechanism in the development and expression of cardiac arrhythmias. An association of arrhythmias and myocardial lesions with severe brain damage was reported by Natelson (1985). Electrical stimulation of the LH produces arrhythmia both in intact animals (Satinsky *et al.*, 1971) and in dogs with experimentally induced coronary occlusion (Garvey and Melville, 1969). Arrhythmias also appear after termination of the electrical stimulation of the anterior-medial hypothalamus. That vagotomy or atropine-induced blockade of vagal influences eliminates the arrhythmia in the cat suggests the involvement of the parasympathetic nervous system (Tashiro *et al.*, 1988). In contrast, the effect of posterior hypothalamic stimulation on ventricular fibrillation in the dog is abolished by blockade of adrenergic transmission with β-adrenergic receptor antagonists (Verrier *et al.*, 1975). This latter finding suggests a sympathetic involvement. Finally, cardiac ventricular fibrillation and myocardial ischaemia in the pig following stress can be prevented by cryogenic blockade of frontocortical-brainstem efferents projecting to sympathetic outflow stations (Skinner and Reed, 1981) or intracerebral administration of the β-adrenergic antagonist, propranolol (Skinner *et al.*, 1975; Skinner and Reed, 1981). Taken together, the evidence supports a role for central nervous mechanisms in the development and expression of cardiac arrhythmias and myocardial lesions. One cannot yet construct a picture of the network involved in this action. However, the involvement of the LH suggests that the forebrain autonomic circuit may be involved.

4.3.2 AUTONOMIC NERVOUS OUTPUT AND CARDIOVASCULAR RESPONSES TO STRESS

The integrated neuronal message from the autonomic network of the brain activates both the sympathetic and parasympathetic divisions of the autonomic nervous system. Whereas the sympathetic outflow may be regulated at various levels of the spinal cord, the parasympathetic output is controlled at the level of the preganglionic vagal output from the DMVN and/or the nucleus ambiguous at the lower medullary level. The sympathetic outflow reaches the preganglionic cells in the intermediolateral column of the spinal cord either *via* the lower medullary vasomotor control regions like the rostral ventrolateral nuclei, the A5 (noradrenergic) cell group or the raphe (5-hydroxytryptaminergic) nuclei, or through direct projections from the hypothalamic and parabrachial nuclei or the NTS (Strack *et al.*, 1989). The preganglionic neurones then leave the central nervous system and reach the target areas *via* postganglionic fibres. The preganglionic cells integrate the signals descending from the lower medulla and the higher autonomic centres together with those arising from segmental primary afferents (Jänig and McLachlan, 1992).

4.3.2.1 *Sympatho-adrenal output, blood pressure and heart rate during stress*

Whereas it has long been maintained that the parasympathetic system affects the function of various target organs in a rather discrete way, the emphasis on mass-discharge has been the leading idea about stress-induced sympathetic function. The discrete regulation of parasympathetic function *via* the vagus nerve is related to the anatomical organization of the input to, and the output from, the preganglionic cells of the DMVN and the nucleus ambiguous (e.g. Loewy and Spyer, 1990). In contrast, since Cannon (1928a, b), a neuroendocrine concept has ruled the sympathetic view of the autonomic nervous system. Recent knowledge of the discrete anatomical organization of the input to preganglionic sympathetic cells, however, suggests a precise and differentiated regulation of the sympathetic limb. In addition, integrative properties of preganglionic neurones and ganglia and the modulation of discharge of postganglionic neurones contribute to the more discrete regulation of their output (Jänig and McLachlan, 1992). Besides the anatomical topography, the colocalization of neuropeptides with the classical transmitters, noradrenaline or acetylcholine, is important for coding the specific output (e.g. Furness et al., 1992). Recent evidence suggests that release of the various sympathetic messengers from the nerve endings may be a differentially regulated and discrete process. This view challenges earlier concepts suggesting that target cells of the autonomic effectors are simply 'bathed' in neurotransmitter (Hirst et al., 1992; Jänig and MacLachlan, 1992).

The release of adrenaline from the adrenal medulla and of noradrenaline from the adrenal medulla and/or sympathetic nerve terminals represents the classical sympathetic stress response. A third catecholamine, dopamine, found in the adrenals and autonomic ganglia is less affected by stress (Popper et al., 1977). The physiological effects of catecholamines en masse can explain practically all changes occurring during the flight/fight stress response: cardiac output is augmented by means of increased heart rate and cardiac contractility; venous return is increased due to peripheral vasoconstriction and redistribution of blood from sites such as the gastrointestinal organs and spleen.

A large body of evidence suggests that various forms of stress of both acute and chronic nature increase circulating catecholamines in different species (McCarty et al., 1981; McCarty, 1983; see also Chapter 11). Besides the intensity of stressors (Konarska et al., 1989; Natelson et al., 1981), factors such as acute or chronic exposure, habituation and conditioning, controllability and predictability of the stressor, individual variation, sex and age all affect the levels of circulating catecholamines and the ratio of adrenaline and noradrenaline release (e.g. De Turck and Vogel, 1980; Swenson and Vogel, 1983; Livezey et al., 1985; De Boer et al., 1989; Sachser and Lick, 1989). All these data support the view that stress does not cause a simple mass-discharge of catecholamines. Catecholamine release alone may explain a number of features of the cardiovascular response to stress. However, we have to bear in mind recent findings, discussed above, concerning the transmission of signals to the targets (see also Hirst et al., 1992; Jänig and McLachlan, 1992). Therefore, one should be cautious in interpreting changes in circulating catecholamines as indicators of cardiovascular pathology.

It is generally maintained that acute changes in blood pressure caused by stress are the result of a balance between constrictor and vasodilatatory action of catecholamines via the sympathetic nerve endings or through the general circulation. The constriction is due to a stimulation of α_1 adrenoceptors in the arterial vasculature, whereas the vasodilatation is due to β_2-adrenoceptors with the greatest density in the skeletal muscle vasculature (Vanhoutte, 1981). Cardiovascular adjustment during stress is the result of noradrenaline and adrenaline release and their subsequent actions at these two receptor sites (Smith et al., 1979; Carobrez et al., 1983; Sakaguchi et al., 1983). Studies by Kirby and Johnson (1990) suggest that the reduction in blood pressure mediated by β_2-adrenoceptor-induced vasodilation is greater in rats with increased sympatho-adrenal activity, such as in spontaneous hypertensive rats, by comparison with normotensive controls. This blunting of the stressor response produced by circulating adrenaline may have two consequences. First of all, there is a shift of flow to the muscle vascular bed (fight/flight reaction). Secondly, the full magnitude of the stress response is blunted. This may reflect adaptation to the metabolic demands of the given stressful situation.

The final issue to be discussed in relation to the involvement of the sympathetic nervous system in cardiovascular stress responses is the role of peptides colocalized with the classical transmitter noradrenaline. Most attention has focused on neuropeptide Y (NPY). NPY is colocalized with noradrenaline in postganglionic neurones innervating the cardiovascular system (Ekblad et al., 1984). NPY is considered to potentiate vasoconstrictor effects of noradrenaline (Zukovska-Grojec et al., 1985), but circulating noradrenaline also potentiates the actions of NPY (Wahlestedt et al., 1990). The cardiovascular effects of NPY differ from those of noradrenaline. It causes vasoconstriction in coronary and cerebral vessels, vascular regions where noradrenaline is weakly active, and induces long-term cardiodepression in the rat (Edvinsson et al., 1983; Rudehill et al., 1986; Zukowska-Grojec et al., 1987b). NPY is released during intense stress and contributes to haemodynamic adjustments (Zukowska-Grojec et al., 1987a; Zukowska-Grojec and Vaz, 1988). Stress of handling or cold water exposure induces greater pressor and tachycardiac responses in male than in the female rats which may be ascribed to the high release of NPY which is seen in males, but not in females (Zukowska-Grojec et al., 1991).

All these findings show that autonomic outputs can be regulated at levels ranging from the spinal cord to the nerve endings and their associated synaptic receptors. Functional differentiation is provided by selective activation of autonomic output at the sympathetic and the vagal parasympathetic systems. In addition, the message conveyed by central drive mechanisms may reach different targets through the same pathways by using different, or different combinations of, messengers.

4.3.2.2 The sympathetic nervous system, cardiac dysrhythmias and stress

A number of studies present evidence that stress-induced changes in rhythmogenesis are associated with alterations in sympathetic function. A reduction in the threshold of repetitive extrasystoles is correlated with an increase in levels of circulating adrenaline and noradrenaline in the dog (Liang et al., 1979). β-Adrenoceptor blockade reduces the incidence of arrhythmias due to stress (Matta et al., 1976a; Verrier and Lown, 1977; Rosenfeld et al., 1978). Surgical removal of the sympathetic stellate ganglion also partially protects against the occurrence of stress-induced arrhythmias (Verrier and Lown, 1977). Stellectomy reduces spontaneous arrhythmias and mortality after myocardial ischaemia in the dog (Schwartz and Stone, 1980), whereas sympathetic stimulation increases the incidence of ventricular fibrillation in dogs with coronary occlusion (Euler et al., 1985).

A remarkable lateral asymmetry can be discerned in sympathetic-cardiac interaction. Stimulation of the left-side of the stellate ganglion is more effective than stimulation of the right-side in increasing ventricular vulnerability (Brooks et al., 1977). An increased vulnerability to fibrillation occurs after cold blockade of the right, but not the left, sympathetic nerve in dogs with coronary occlusions (Schwartz et al., 1976). In contrast, removal of the left stellate ganglion increases fibrillation threshold and reduces spontaneous arrhythmias following myocardial ischaemia in dogs with coronary occlusions (Kliks et al., 1975; Schwartz and Stone, 1980). Lane and Schwartz (1987) proposed that the activation of left and right cerebral hemispheres may exert differential control of the heart, particularly in strong emotional states. This hypothesis attempts to build a bridge between emotion, cerebral lateralization, arrhythmia and sudden cardiac death. Although direct evidence supporting this attractive hypothesis is not available, some indirect evidence shows that it cannot be discarded: lateralized control of immune function is well documented, for example (Neveau, 1992).

Whereas the above evidence demonstrates direct neural influences on the heart, other data suggest that environmental stress may influence coronary vascular tone, thereby effecting oxygen delivery to the heart. Psychological stressors, causing myocardial ischaemia, may serve as priming processes to induce fatal dysrhythmias. Loud noise, cold water and conditioned fear of a painful shock all induce changes in coronary tone (Bergamaschi et al., 1973a, b; Bergamaschi and Longoni, 1973; Billman and Randall, 1980a, b). Sympathetic influences on the vascular tone may be both vasoconstrictor (α_1-adrenergic) or vasodilator (β_1-adrenergic), whereas the parasympathetic effects are vasodilatory (Shepherd and Vanhoutte, 1985; Young and Vatner, 1986). Interestingly, stress-related coronary vasomotor changes in dogs are related to the ischaemic state of the heart muscles. Whereas stress provoking anger increases coronary blood flow, heart rate and blood pressure before coronary occlusion, this procedure reduces coronary flow, increases arterial resistance and causes ischaemic changes in the ECG; hypertension and tachycardia remain unaffected. Accordingly, a coronary vasomotor change is influenced by the pre-existing condition of the heart and may eventually lead to a shift in adrenergic receptor mechanisms regulating vasomotor tone.

An additional mechanism by which coronary blood supply may be reduced by psychological stress is platelet aggregation. Environmental stress such as cold water or electric shock increases platelet aggregation in the coronary arteries of the cat (Haft and Fani, 1973a, b). The autonomic mediation may be through α-adrenoceptor mechanisms. Platelet aggregation is potentiated by catecholamines (Ardlie et al., 1985; Larsson et al., 1989), probably via α-adrenergic binding sites on the platelets.

4.4 CHEMICAL NEUROBIOLOGY OF THE CARDIOVASCULAR STRESS RESPONSE

In the previous section the neuroanatomical basis of the regulation of cardiovascular stress responses was discussed. Some examples of the importance of messengers, particularly within the peripheral autonomic nervous system, were presented. It was further emphasized that reciprocal, often monosynaptic, connections between the forebrain and brainstem autonomic areas form a network. This may be of special importance in the regulation of the various cardiovascular and neuroendocrine responses to stress. In this respect, neuropeptides may play a decisive role (Gray, 1991; Sawchenko, 1991; van Giersbergen et al., 1992). Although the major changes have been reported in the case of corticotropin releasing factor (CRF) and thyrotropin releasing hormone (TRH), acute and chronic stress also alter secretion of neurotensin, cholecystokinin, substance P, vasopressin and somatostatin (see Bissett, 1991). CRF has been regarded as the main stress messenger in the brain inducing behavioural, autonomic and neuroendocrine alterations resembling the stress response (Dunn and Berridge, 1990; Owens and Nemeroff, 1991; Chapter 12). Although it may appear rather reductionist to assign a central role to a single peptide in stress, it is likely that, together with vasopressin, CRF plays a vital role in the regulation of various forms of cardiovascular stress response (Bohus *et al.*, 1990a,b; Bohus, 1993).

4.4.1 CORTICOTROPIN RELEASING FACTOR

CRF is classically considered as the hypothalamic messenger activating the pituitary-adrenal response to stress. CRF-containing neurones in the medial parvocellular part of the hypothalamic PVN projecting to the median eminence represent the neuroanatomical locus of this neuroendocrine response (Sawchenko, 1991). In addition, CRF neurones from the PVN innervate the parasympathetic DMVN and the preganglionic cells of the spinal sympathetic limb. An extensive network of CRF axon terminals and synaptic contacts with CRF cell bodies (Silverman *et al.*, 1989) may ensure an integrated neuroendocrine response to stress. The LH, CEA and BNST, structures of the forebrain autonomic network, also contain CRF cell bodies, and project to the parabrachial nuclei and the DMVN (Moga and Gray, 1985; Gray and Magnuson, 1987). This extends the idea that this peptidergic system serves multiple autonomic mechanisms. In addition, CRF immunoreactive cell bodies and terminals can be localized in lower brainstem and spinal autonomic areas like the NTS, DMVN and the intermedio-lateral column of the spinal cord (IML; see Fisher, 1991). The brain regions with CRF cell bodies and axon terminals also contain specific receptors for this peptide (De Souza, 1987). Finally, changes in the hypothalamic and extrahypothalamic concentrations of CRF occur following acute and chronic stress (e.g. Chappell *et al.*, 1986).

The action of intracerebroventricularly administered CRF on the cardiovascular system is to increase mean blood pressure, heart rate, and stroke volume and to decrease total peripheral resistance (see Brown, 1991; Fisher, 1991). The increase in blood pressure is due to the increased cardiac output (Brown, 1991), and is blocked by the sympathetic ganglionic blocker, chlorisondamine (Fisher *et al.*, 1983). This indicates a sympathetic nervous involvement. Intracerebroventricular CRF was reported to increase plasma catecholamine concentration as well. Since administration of the cholinergic blocker, methylatropine, fails to affect CRF-induced tachycardia, the higher heart rate may be ascribed to withdrawal of cardiac vagal tone by the peptide (Fisher, 1988).

Local CRF injection in the CEA (Brown and Gray, 1988) or LH (Diamant, 1991) is effective in increasing blood pressure and heart rate in the rat. CRF may act through multiple sites within the autonomic network activating the cardiovascular system (Fisher, 1991). Plasma concentrations of adrenaline and noradrenaline are not uniformly increased following local brain injections of CRF. Whereas no consistent changes in adrenaline concentration occur, noradrenaline levels increase, particularly following injection of the peptide into the dorsal hypothalamus, zona incerta, ventromedial hypothalamus and LH (Brown, 1986). It is also maintained that CRF does not cause an *en masse* release of noradrenaline from all noradrenergic nerve endings. Noradrenaline turnover, taken as the indicator of neuronal activation, is increased in the kidney, but decreased in the pancreas and brown fat tissue (Brown and Fisher, 1985). Others report an increase in the sympathetic activity of brown adipose tissue following intracerebroventricular injection of CRF (Arase et al., 1988).

These and other data (see Fisher, 1991 and later in this section) strongly suggest that CRF-induced changes in cardiovascular function are the result of sympatho-medullary activation and actions at target organs. These result in a differentiated response mimicking the cardiovascular components of a fight/flight like stress reaction. There are, however, some observations that suggest a more complex picture. A small amount of CRF, which induces a substantial increase in blood pressure and heart rate in the freely moving rat, increases plasma noradrenaline concentration substantially whereas changes in adrenaline levels are minimal (Korte et al., 1993). In addition, the increase in heart rate is highly correlated with rats' motor activity. This finding suggests that heart rate changes are the consequence of metabolic demand during motor activity. However, it is suggested that CRF-induced changes in cardiovascular function are independent of locomotor activity (Overton and Fisher, 1989). Recent findings also show that different fragments of CRF are responsible for its behavioural (motor activity) and cardiovascular effects (Diamant, 1991). That an increase in the release of catecholamines cannot explain all the cardiovascular alterations observed following intracerebroventricular injection of CRF is supported by the finding that the peptide-induced pressor response is absent in anaesthetized rats whereas the tachycardiac response is unaffected (Kurosawa et al., 1986). Acute adrenalectomy, which removes the source of adrenal medullary catecholamines, leaves the pressor and tachycardiac effects of CRF unaltered (Fisher et al., 1983). This suggests regulation via the postganglionic sympathetic nerves.

Collectively, these findings assume a parallel, but possibly independent action of CRF on behavioural and physiological processes. Multiple sites of action in the brain would not necessarily mean that separate loci are involved in the diverse actions. Instead, the available findings suggest that neurochemically diverse, parallel mechanisms from the same sites mediate the various reactions. Intracerebroventricular CRF increases plasma renin activity, but the consequent increase in circulating levels of angiotensin II is not responsible for the pressor reactions: catopril, which blocks the enzyme which converts angiotensin I to its active form, angiotensin II, does not modify the cardiovascular effects of CRF (Fisher et al., 1983). Like various stressors, centrally administered CRF inhibits cardiovascular baroreceptor function in the rat (Fisher, 1988). This may be due to the suppression of incoming baroreceptor information or the inhibition of vagal parasympathetic output. CRF does not exclusively affect baroreceptor input (Overton et al., 1990), but it suppresses vagal output, probably via peptidergic neurones that terminate on the vagal motor nuclei. CRF neurones from the CEA innervate the DMVN (Gray, 1991). Modification of CEA output by local neuropeptide administration affects heart rate via CRF-mediated actions (Wiersma et al., unpublished).

CRF may be physiologically involved in the central regulation of cardiovascular responses to stress. Central administration of the CRF receptor antagonist α-helical CRF blocks the action of stressors like ether exposure, insulin-induced hypoglycaemia or haemorrhage on circulating catecholamine levels (Brown *et al.*, 1985, 1986). Treadmill exercise induced increases in blood pressure, heart rate, iliac blood flow and mesenteric vascular resistance are also attenuated by the α-helical CRF antagonist peptide (Kregel *et al.*, 1990). Unfortunately, to our knowledge, no data are available on the effects of the antagonist on cardiovascular responses to psychosocial stressors.

There are some indirect suggestions that CRF may operate in chronic alterations of cardiovascular function. CRF levels in the brain are changed in the spontaneously hypertensive rats (SHR; Hashimoto *et al.*, 1985). In addition, electrical lesions of the hypothalamic PVN (Ciriello *et al.*, 1984) or the CEA (Galeno *et al.*, 1984) appear to retard the development of hypertension of SHR rats. Both nuclei are the loci of CRF cell bodies in hypertension of SHR rats. Both nuclei are the loci of CRF cell bodies in the brain (Fisher, 1991). It is of interest that intracerebral CRF induces a fall in blood pressure in the spontaneously hypertensive rats, in contrast to the pressor response in normotensive controls. This decreased blood pressure is in contrast to the exaggerated adrenaline response to CRF in these rats (Brown *et al.*, 1988). Remarkably, various stressors also cause a hypotensive response in rats with genetic (spontaneous) or experimental hypertension (see later in this chapter). According to Fisher (1991) the increased adrenaline release of SHRs, interacting with vasodilatory β_2-adrenoceptors, might be the cause of this paradoxical response.

Taken together, the data suggest that CRF may be an important messenger of cardiovascular responses in situations in which the primary response is fight/flight behaviour with sympathetic activation. CRF conveys the cardiovascular fight/flight message by a selective activation of the sympathetic outflow and the inhibition of parasympathetic (vagal) output. It is remarkable that there is a close parallelism between the CRF network and the autonomic neuronal network in the brain.

4.4.2 VASOPRESSIN

Vasopressor (increase in blood pressure) and antidiuretic (reduced renal outflow) actions have been long regarded as the main physiological functions of vasopressin: its corticotropin (ACTH) releasing activity, recognized in the 1950s, indicated a putative role for this peptide in stress. Although the discovery of CRF (Vale *et al.*, 1983) implied that vasopressin is not the major releasing factor in the HPA, its physiological significance in ACTH release is still recognized (Gillies *et al.*, 1982). Vasopressin is released from the posterior lobe of the pituitary gland by several stimuli that are regarded as stressors, but its role as a peripheral stress hormone is still elusive. Interestingly, the release of vasopressin into the cerebral ventricular system has been reported by several groups (Rodriguez, 1976; Laczi *et al.*, 1984). Oxytocin, a structurally related nonapeptide with milk-ejecting and uterus-contracting activity, is often regarded as a stress-related peptide (e.g. Lang *et al.*, 1983).

Vasopressin and oxytocin were the first peptides found to be synthesized in the CNS. The magnocellular cells of the hypothalamic supraoptic nucleus and PVN were the first recognized sites of synthesis. The neurohormones, together with their carrier peptides, are transported to the posterior pituitary. In addition, fibres from the PVN also run to the median eminence where their nerve terminals end on the cranial branches of the hypophyseal portal vessels (Swanson and Sawchenko, 1983). These axons form the neuroendocrine branch of the peptide system involved in the endocrine stress response. These pathways appear to represent only a small part of the neuronal network using vasopressin as a chemical messenger. Vasopressinergic cells of the parvocellular PVN project to the lower brainstem and spinal autonomic nuclei such as the parabrachial nuclei, the NTS, the DMVN and the sympathetic cells of the IML. Oxytocinergic innervation of the autonomic brainstem and forebrain nuclei from the PVN is also known. Vasopressinergic cell bodies are also located in the forebrain areas such as the BNST which projects, among other sites, to the CEA. In addition, scattered cell bodies can be recognized in the locus coeruleus (see Buijs, 1983; Sawchenko, 1991). The actions of argine-8-vasopressin, the nonapeptide found in the brains of most rodents and higher species including humans, are mediated by cell membrane receptors (Jard, 1985). The 1A subtype of peripheral vasopressin receptors is responsible for a series of neural actions of the peptide. In addition, vasopressin may act through oxytocin receptors in the brain: vasopressin acts as an agonist at these receptors, while oxytocin has the opposite effects (De Wied, 1991).

The evidence that vasopressin may act as a messenger/modulator of various cardiovascular stress responses has been reviewed recently (Bohus et al., 1990a,b; Versteeg et al., 1993). Briefly, intracerebroventricular vasopressin attenuates the pressor response induced by electrical stimulation of the cuneiform area of the mesencephalic reticular formation. This action is shared by oxytocin and can be evoked from brain sites such as the dorsal hippocampus or the PAG of the brainstem (Bohus et al., 1983). Unfortunately, no data are yet available about the action of peptides on blood pressure to psychosocial stress. Neither is their exact mode of action known. In non-stressed rats, vasopressin administration intracerebroventricularly or to local areas such as the locus coeruleus, has been reported to increase mean blood pressure (Berecek and Swords, 1990), but there are also negative reports (Versteeg et al., 1993). Microinjection of vasopressin into the NTS changes baroreceptor gain with an inverted U-shaped dose-response relation.

More knowledge is accumulating on the effect of vasopressin on heart rate in various stress situations (Bohus et al., 1990a,b). A bradycardiac response to the emotional stress of conditioned fear of inescapable footshock is intensified by vasopressin administered either peripherally or intracerebroventricularly. Vasopressin also facilitates bradycardiac responses to mild stress of a non-aversive nature such as sudden reduction of background noise or novelty (Hagan and Bohus, 1984; Buwalda et al., 1992a, b). The intensified and/or prolonged cardiac stress response is accompanied by a pronounced immobility reaction (Buwalda et al., 1992a, b). This putative parasympathetic effect of vasopressin is probably mediated by a direct activation of cardiac vagal output, rather than a modification of baroreceptor function. However, inhibition of sympathetic outflow may also contribute. Vasopressin administration appears to block novelty-induced elevation of plasma noradrenaline levels (Buwalda, Nyakas, Koolhaas, Bohus, unpublished). The CEA may be only one site where vasopressin (and oxytocin) affects the central drive of vagal output. Local microinjection of vasopressin in the CEA induces bradycardia in non-stressed rats provided that low doses of the peptide are given. This action is probably mediated by the central vasopressin receptor. High doses of vasopressin and various doses of oxytocin cause tachycardia, behavioural activation and a rise in plasma corticosterone. The tachycardiac action of the peptides can be blocked by an oxytocin antagonist, indicating the involvement of oxytocin receptors (Roozendaal et al., 1992a). A remarkable action of the peptides can be observed in emotionally stressed rats of a genetically inbred line. Rats selected for low active avoidance behaviour (Roman Low Avoidance rats: 'RLA') display immobility and marked bradycardia in reaction to inescapable stress. Low doses of vasopressin microinjected into the CEA intensify this stress reaction: high doses of vasopressin and oxytocin inhibit both reactions. Both peptides are ineffective in rats selected for high active avoidance behaviour (Roman High Avoidance rats: 'RHA'). The RHA rats show only tachycardia and low immobility in the stress situation (Roozendaal et al., 1992b).

Taken together, the data suggest that vasopressin is an important messenger in the cardiovascular stress response in situations where orientation and/or immobility and vagal cardiac activation is the predominant stress response. The mechanism of action is most probably an enhancement of the central drive of vagal output to the heart. Remarkably, other vagal functions such as the meal-induced release of insulin in the rat are not affected by vasopressin (Buwalda et al., 1991). In addition, inhibition of sympathetic outflow may also play a role. The remarkable coincidence of the vasopressinergic network with the autonomic neuronal network in the brain suggests that vasopressin may be the chemical messenger of a conservation/ withdrawal type stress reaction. Central vasopressin may also activate sympathetic outflow and could be an important factor in the pathogenesis of hypertension (Berecek and Swords, 1990). However, this is seemingly difficult to reconcile with the data presented here. It is possible that under certain circumstances there is a shift in receptor sensitivity that results in sympathetic activation: the existence of such a mechanism is shown by Roozendaal et al. (1992a). It is worth mentioning that vasopressin content in a number of brain areas of stroke-prone spontaneously hypertensive rats (SP-SHRs; Lang et al., 1981) and in the hypothalamus of SHRs is lower than in controls (Negro-Vilar and Saavedra, 1980). Intracerebral vasopressin decreases blood pressure and heart rate in SP-SHRs (Imai et al., 1987). These data suggest a protective effect of vasopressin against hypertension.

4.4.3 NEUROPEPTIDES AND CARDIAC ARRHYTHMIAS

The tachycardiac action of CRF, and the bradycardiac effects of vasopressin may be important in the development of disorders of heart rhythm. Unfortunately, data are not yet available about this important aspect of peptide action. Most interest has been focused on the involvement of endogenous opioids in arrhythmogenesis (Verrier and Carr, 1991). The chemically and neuroanatomically rather diverse opioid systems play an important role in the stress response (Amir et al., 1980) and in cardiovascular regulation (Holaday, 1983; van Giersbergen et al., 1992). Opioid involvement in the development of hypertension is also frequently suggested (see Szilagyi, 1988; Versteeg et al., 1993). In addition, interaction between centrally-acting antihypertensive drugs and endogenous opioids, particularly in the NTS, is of importance to therapy (see van Giersbergen et al., 1992). The effects of opioids on circulation and rhythmogenesis manifest themselves in interactions with catecholamines both centrally and peripherally (Holaday, 1983; Randich and Maixner, 1984; Feuerstein, 1985). The three classes of endogenous opioids (the endorphins, enkephalins and dynorphins) are localized, among other sites, in the limbic and brainstem nuclei involved in the central autonomic circuit (e.g. Khachaturian et al., 1985; van Giersbergen et al., 1992). In the peripheral nervous system, endogenous opioid cells are localized in sympathetic ganglia (e.g. Hervonen et al., 1981), the catecholamine containing cells of the carotid body and the adrenal medulla (Varndell et al., 1982). The complexity of the system arising from differential processing of the precursor molecules and differential localization is increased by the fact that more than one receptor type mediates the actions of the various opioids (Höllt, 1986).

From the early 1950s it was known that opioids such as morphine increase parasympathetic and decrease sympathetic tone. Increased vulnerability to repetitive extrasystoles in dogs exposed to an aversive environment is prevented by pretreatment with morphine (DeSilva et al., 1978). Although the major component is a vagally mediated action, withdrawal of sympathetic tone or direct myocardial effects may also play a protective role. A different type of opioid action is demonstrated in models for arrhythmogenesis using digitalis intoxication. Behavioural stress enhances digitalis-induced ventricular arrhythmia (Natelson and Cagin, 1981). A low dose of the opiate antagonist, naloxone, (4 mg/kg) abolishes such deleterious effects of stress (Natelson et al., 1982). Since the parasympathetic blocker, atropine, also blocks the effects of stress, endogenous opioids acting *via* the vagal nerve might be responsible for the increased arrhythmogenesis (Cagin and Natelson, 1981). In contrast, it was reported that much lower doses of naloxone (0.01–0.1 mg/kg) increase digitalis-induced arrhythmia (Rabkin and Roob, 1986). Morphine administration suppresses digitalis-induced arrhythmia in hypokalaemic, but not in normokalaemic guinea pigs. Intracerebroventricular administration of a mixed μ–δ opioid receptor antagonist enhances digitalis arrhythmia. These results were interpreted as suggesting a μ-receptor mediated protection, and a δ-receptor mediated provocation of arrhythmia (Rabkin et al., 1987). In the rat, the opioid antagonist, naloxone, diminishes ventricular arrhythmias induced by coronary ligation (Fagbemi et al., 1982). At the same time, the partial opioid agonist, meotazinol, reduces the incidence and severity of arrhythmias in rats following coronary artery occlusion (Fagbemi et al., 1983). In contrast, naloxone fails to show an antiarrhythmogenic action in pigs (Bergey and Beil, 1983) and dogs (Pinto et al., 1989).

Taken together, opioids may play a role in the mechanism of stress-induced changes in arrhythmogenesis. However, it is impossible at present to draw any general conclusions. First of all, due to the complexity of the opioidergic systems and their receptors, the use of more specific agonists and antagonists seems to be obligatory. In addition, one has to consider the differences between the aetiology of arrhythmias induced by different drugs or mechanical interventions and their interaction with stressors. For example, the influence of opioid agonists on rhythmogenesis in relation to haemorrhage involves primarily the baroreceptor reflex arc, whereas enhanced vagal influences are of importance in the case of acute myocardial ischaemia (Saini et al., 1988,1989).

4.5 PATHOLOGICAL STATES AND THE CARDIOVASCULAR STRESS RESPONSE

It is generally accepted that pathological conditions, such as hypertension or obesity, and behavioural factors, such as emotionality or personality, represent high risk for coronary disease and sudden cardiac death. It is also suggested that stress and coronary disease are practically inseparable events (Kamarck and Jennings, 1991). Although interest is focused on the relation between stress, behaviour and hypertension in animal models, particularly in relation to spontaneous (genetic) hypertension in selected rat strains (e.g. McCarty, 1983), less attention is paid to stress, rhythmogenesis and pathological conditions (Kamarck and Jennings, 1991). This section deals first with some recent and surprising findings. These concern stress and blood pressure responses in rats with chronically high blood pressure. Secondly, the relation between stress, arrhythmia, and pathophysiological and behavioural states is considered and attention addressed to the possible mechanisms involved.

4.5.1 EXPERIMENTAL AND GENETIC HYPERTENSION AND HYPOTENSIVE RESPONSES TO PSYCHOSOCIAL STRESS

As already mentioned, rats with spontaneous (genetic) hypertension display exaggerated sympathetic responses to stressful stimuli as determined from sympathetic nerve activity or levels of adrenaline and noradrenaline in the plasma (Lundin et al., 1984; Kirby et al., 1989). In contrast to what might be expected, acute footshock fails to produce a greater pressor response in the hypertensive SHR as compared to their WKy controls (Kirby et al., 1987, 1989). Activation of β_2-adrenoreceptors may be responsible for blunting the pressor response to stress (Kirby and Johnson, 1990). This suggests a complex regulation of blood pressure in stress, particularly in pathological conditions. Although this interpretation suggests that a peripheral mechanism is involved, one cannot exclude the role of central neuronal and behavioural factors. For instance, whereas footshock delivered to a single rat increases postshock blood pressure, this parameter decreases when two rats receive the same shock together. Also, paired rats show species-specific threat and attack postures, whereas single rats display escape attempts (Williams and Eichelman, 1971). The activity of the sympathetic neurones is important: rat strains with low sympathetic activity fail to show the hypotensive response (Lamprecht et al., 1974). Selective destruction of sympathetic nerves changes the hypotensive response into a hypertensive one in paired rats, whereas the hypertensive response in single animals remains unaffected. Conversely, adrenalectomy, which also removes the source of medullary catecholamines, reverses the hypertensive response leaving the hypotensive one unaltered (Williams et al., 1979).

Subsequent studies showed that social stress of defeat by a dominant rat increases postfight mean blood pressure, but baseline values one day after defeat are significantly lower than before the stress (Fokkema and Koolhaas, 1985). Postfight blood pressure in hypertensive Dahl salt-sensitive rats also show a decreased response as compared to normotensive salt-insensitive animals (Adams and Blizard,

1987). Since the strain of rats used in the experiments of Fokkema and Koolhaas (1985) appears to be hypertension-prone in chronic social stress experiments (Fokkema, 1986), the next studies were designed to investigate the influence of social and non-social stressors on baseline blood pressure of rats suffering from hypertension of various origins (Bohus *et al.*, 1988). Repeated social stress of defeat by a dominant rat results in lower morning baseline blood pressure in SHRs. This decrease develops gradually and lasts for several days following the termination of stress. The normotensive WKY controls show a mild hypertension instead. Rats of Wistar strain with hypertension caused by subcutaneous deoxycorticosterone implants and 1% salt in drinking water (DOC/salt hypertension) also display a lasting fall of blood pressure after social stress, whereas this does not change in normotensive controls. Stress of exposure to swimming at room temperature in groups of three rats (social swimming) results in a similar fall of baseline blood pressure in hypertensive SHRs which also lasts for several days after termination of stress. This effect of stress affects hypertensive animals regardless of the cause of this disorder: social swimming resulted in decreased blood pressure in SP-SHRs and in rats with DOC/salt or renal hypertension. Furthermore, various forms of non-social stress cause the same decrease in blood pressure as does social stress: solitary swimming, single inescapable, escapable and avoidable footshock in rats with DOC/salt hypertension all cause a similar alteration in blood pressure. Remarkably, exercise of swimming at 32°C causes a 'stress-like' increase in circulating corticosterone, adrenaline and noradrenaline, together with a short hypertensive and tachycardiac response (Scheurink and Steffens, 1990), yet this fails to alter baseline blood pressure. This indicates that psychosocial stressors induce a paradoxical reaction in blood pressure regulation in the pathological state of hypertension. None of the factors concerned with coping with stress, such as controllability and predictability of the stressor, seem to play a role in this phenomenon.

The mechanism(s) by which the various stressors cause this phenomenon is/are not yet known. In certain cases, such as rats with genetic hypertension or rats prone to mild social hypertension, an increase in sympathetic nerve activity may be an essential prerequisite for the fall in blood pressure. However, this does not hold true for DOC/salt and renal hypertensive rats. It is not unreasonable to assume a role for a baroreceptor-mediated mechanism. Remarkably, electrolytic lesion of the CEA in the autonomic network of the forebrain prevents development of the decrease in blood pressure (Bohus *et al.*, 1988). This lesion does not alter baroreceptor function (Roozendaal *et al.*, 1990), but it does block tachycardiac and neuroendocrine consequences of stressors, such as inescapable footshock (Roozendaal *et al.*, 1991). It is therefore likely that the occurrence of a complex stress response, or one of the components of the response, is essential for the long lasting phenomenon. It is also found that endogenous opioids may play a preventive role: administration of the opiate antagonist, naloxone, is followed by a more marked fall in blood pressure (Bohus *et al.*, 1988). Finally, the significance of this phenomenon also remains to be determined. Preliminary observations suggest that the bradycardiac response to superimposed stress of fear of inescapable footshock is prevented by the fall in blood pressure (Nyakas and Bohus, unpublished).

Taken together, these data suggest that pathological states of the cardiovascular system, such as hypertension, dramatically change the blood pressure response to single or repeated stressful stimuli. The prognostic value of these reactions remains to be determined.

4.5.2 BEHAVIOURAL AND PATHOPHYSIOLOGICAL STATES: EFFECTS ON CARDIAC STRESS REACTIONS AND RHYTHMOGENESIS

As mentioned before, the cardiac response to certain forms of stress, such as fear of an inescapable footshock or sudden changes in the rats' environment, results in bradycardia. This bradycardiac reaction is often accompanied by bradyarrhythmic changes in the heart (Bohus, 1985). In young adult, male Wistar rats a significant negative correlation exists between exploratory behaviour in a complex novel environment and the cardiac response during the CER. Rats with greater exploratory activity show less marked bradycardiac stress response. Rats that explore less display more marked bradycardiac response with occasional bradyarrhythmia. These responses can be ascribed to marked parasympathetic (vagal) activation. The negative correlation is preserved in old and senescent rats (24–25 and 33 months of age, respectively), but the sympathetic-parasympathetic balance is shifted towards the sympathetic direction, probably due to a decreased vagal drive (Nyakas et al., 1990). A negative correlation between territorial offensive behaviour and the cardiac response to emotional stress exists in male rats of the pigmented TMD-S3 strain which are territorially aggressive: the more aggressive the rats in their own territory, the less the bradycardiac stress reaction. This suggests that the sympathetic-parasympathetic balance is shifted in the sympathetic direction in aggressive rats (Bohus et al., unpublished data).

Genetic selection for behavioural characteristics in active avoidance behaviour also affects cardiac stress responsiveness. Roman High Avoidance (RHA) rats learn an active avoidance response rapidly, whereas the Roman Low Avoidance (RLA) show freezing instead of avoidance or escape (Driscoll and Bättig, 1982). The RLA rats show a marked bradycardia with bradyarrhythmia and immobility behaviour in response to conditioned fear. The RHA rats fail to show bradycardia and immobility (Roozendaal et al., 1992b). It seems, therefore, that the central drive of the vagus nerve is very powerful in RLA rats, whereas the parasympathetic influence on the heart of the RHA rats is practically absent in emotional stress. This suggestion is supported by results of studies of the cardiac and behavioural reaction of the Roman Avoidance strains to a sudden change in environment. As mentioned already, the mild stress of sudden silence after low intensity background noise induces a marked bradycardia, behavioural arrest, and immobility in Wistar rats. RLA rats show a more marked and prolonged bradycardia with occasional bradyarrhythmia and prolonged immobility. Bradycardia and immobility are absent in RHA rats. Instead, a mild increase in sympathetic effects is visible (Bohus and Balkan, unpublished).

Although the importance of hypertension and obesity as risk factors in cardiac dysrhythmias and sudden cardiac death is generally acknowledged, we are not yet aware of experimental evidence concerning stress and these pathological conditions. Our recent observations in male rats reinforce the notion suggested by human epidemiological studies (Bohus and Nyakas, in preparation). Young adult (approx. four months old) hypertensive SP-SHR rats of the German–Heidelberg line (mean systolic blood pressure 190 mm Hg) fail to display significant changes in cardiac rate in response to conditioned fear. Surprisingly, their normotensive WKy controls (mean systolic blood pressure 132 mm Hg) show a slight tachycardiac response. Three month old male SHR rats of the Dutch–Zeist line with mild hypertension (mean systolic blood pressure 155 mm Hg) display a bradycardia and occasional arrhythmia in response to the strong emotional stressor. In the normotensive WKy controls the bradycardia in response to emotional stress is absent. Of the Wistar strains, DOC/salt hypertensive rats (mean systolic blood pressure 173 mm Hg *versus* 128 mm Hg in controls) and renal hypertensive rats with a mean systolic blood pressure of 191 mm Hg (controls 131 mm Hg) fail to show a bradycardiac response to emotional stress. Borderline renal hypertensive rats (mean blood pressure 157 mm Hg) show a bradycardiac stress response like their normotensive controls. These findings suggest that the vagal cardiac stress mechanism is impaired in hypertensive rats independently of the cause of their high blood pressure. Analysis of data for individual rats shows that those with a systolic blood pressure higher than 180 mm Hg fail to show bradycardia: these rats show occasional tachyarrhythmia and extrasystoles. Accordingly the degree, rather than the origin of high blood pressure, is of importance for vagal regulation. The relative shift in autonomic balance towards a sympathetic hyperactivity may then be responsible for the pathology of rhythmogenesis. Interestingly, the normotensive WKy controls show a vagal response deficit with a shift to sympathetic activation. Since WKy-lines derive from inbreeding, one should always take care to use proper control animals in stress studies.

The role of obesity as a risk factor has been investigated recently in rats with genetically determined obesity (Zucker ob/ob males) in comparison to their lean monozygous (ob/-) controls. The lean stressed rats do exhibit the bradycardiac response to emotional stress, whereas the obese animals do not (Nyakas *et al.*, in press). Accordingly, in the obese rats the autonomic balance involved in the cardiac stress response is shifted to the sympathetic direction probably due to a disturbance of vagal input to the heart.

Taken together, the risk factors of behaviour, hypertension and obesity contribute to the development of changes in cardiac rhythmogenesis. The observations favour the view that, in the rat, disturbance of (central) vagal drive is the primary cause of rhythmogenesis. Although sudden cardiac death does not occur in the current experimental model, it does seem to be suitable for investigating the pathomechanisms of fatal dysrhythmia.

4.6 CONCLUDING REMARKS

That stress of various types profoundly affects the function of the cardiovascular system is, in physiological terms, self-explanatory. It is more remarkable that there are different patterns of responses that are not physiologically directly self-explanatory. Influential psychophysiological theories on stress and cardiovascular responses suggested the metabolic relevance of sympathetically controlled cardiac and circulatory function. In contrast, parasympathetic (vagal) responses were considered to be metabolically irrelevant and non-adaptive (Obrist, 1981). The data presented here advocate the view that both sympathetically and parasympathetically dominated responses represent relevant patterns of physiological importance, and reflect different parts of the complex behavioural, neuroendocrine and physiological stress response. Physiologically speaking, one pattern reflects the need to expend energy (fight/flight), whereas the other requires energy conservation (conservation/withdrawal).

Increased attention is being focused on individual differences in response to environmental challenge, and the correlations between various measures of behavioural, physiological, and endocrine responses in animals and humans (e.g. Koolhaas et al., 1983, 1986; Henry and Stephens-Larson, 1985; Dantzer et al., 1988; Cools et al., 1990; Turkkan, 1991). Our own observations suggest that two general strategies can be recognized in rodents. These can be designated as active and passive coping and serve to adapt to environmental challenges (Koolhaas et al., 1986; Bohus et al., 1987a, b; Koolhaas and Bohus, 1991). The characteristics of the two strategies are summarized in Table 4.1. Briefly, an active strategy means a high demand for environmental control (fight/flight), but less flexibility in behavioural change. A passive strategy is reflected by a lower demand for active control (conservation/withdrawal). This strategy is characterized by high behavioural plasticity according to environmental demand. These behavioural strategies have physiological and neuroendocrine characteristics. An imbalance of autonomic reactivity, reflected by disturbed cardiovascular functioning is one result of the two strategies. An active behavioural strategy is accompanied by more sympathetic reactivity, whereas a passive strategy is associated with more parasympathetic reactivity. In our view, both strategies represent successful coping with

Table 4.1 Behavioural, neuroendocrine and physiological responses characteristic of active or passive strategies to cope with environmental challenge in male rats and mice. ↑ High response level; ↓ low response level

Behaviour	Active coping	Passive coping
Exploration	↑	↓
Avoidance	↑	↓
Escape	↑	↓
Immobility	↓	↑
Offence	↑	↓
Defence	↑	↓
Attention	↓	↑
Spatial orientation	↓	↑
Stereotypy	↑	↓
Temporal orientation	↓	↑
Neuroendocrinology/physiology		
Adrenaline	↑	↓
Noradrenaline	↑	↓
Corticosterone	↓	↑
Testosterone	↑	↓
Prolactin	↑	↓
Blood pressure	↑	↑
Heart rate	↑	↓

environmental demands. Active and passive coping should be viewed as a continuum with preferential use of one or the other strategy depending on individuals and environmental demands. Genetic selection results in extremes in the use of the two strategies. Remarkably, selection for one characteristic (e.g. behaviour or blood pressure) results in coupled changes in both characteristics in the following generations (Bohus *et al.*, 1987b). Finally, individual differences in pathology during active or passive strategies can result in disturbances or failure to cope.

The data reviewed suggest that different networks in the brain co-ordinate cardiovascular responses to stressors demanding active or passive strategies. Within individuals, one or other network may function preferentially. As shown in Figure 4.1, the circuit serving the passive cardiovascular coping response, together with the PAG of the midbrain, is neuroanatomically identical to the circuit of the active or defence network. The functional difference is in the preferential use of CRF or vasopressin as a peptidergic messenger within the circuit: CRF conveys active, and vasopressin, passive messages. Since CRF and vasopressin represent only part of the complex neuropeptide system within the circuit, it is an inviting prospect to determine whether galanin, neurotensin, substance P, calcitonin, somatostatin and, most importantly, oxytocin are also involved.

The data reviewed here show clearly that stress is an additional risk factor for various cardiovascular pathologies in animals. It is also clear that individual differences in the preferential use of active or passive coping strategies may predispose to particular types of pathology in the heart and/or the circulatory system. In addition, the pathological state also influences animals' stress responsiveness. It must be emphasized that stress alone merely increases animals' vulnerability to certain pathologies. This may in part be due to innate coping abilities of the experimental animals.

It is also clear that stress is a comparable risk factor in animals and humans. Accordingly, animal models can serve to test hypotheses that cannot be investigated easily in humans. This is especially relevant to stress management, in particular with the aid of drugs. Increased interest is being focused on anti-stress properties of drugs acting primarily in the CNS. Further research is essential to determine which level of organization (central autonomic circuit, brainstem regulatory centres, sympathetic or vagal outflow to the target organ) is the best target for effective therapy of stress-related diseases of the cardiovascular system.

ACKNOWLEDGEMENTS

The studies reported here were supported in part by several grants of the Saal van Zwanenberg Foundation, Dutch Heart Foundation, the Foundation for Medical Sciences of the Dutch Research Organization NWO, and by a special grant from the School of Sciences, University of Groningen. The editorial assistance of Mrs. J. Poelstra is gratefully acknowledged.

4.7 REFERENCES

Abrahams, V.C., Hilton, S.M. and Zbrozyna, A.W. (1960) Active muscle vasodilatation produced by stimulation of the brain stem: its significance in the defence reaction. *J. Physiol.*, **154**: 491–513

Adams, N. and Blizard, D.A. (1987) Defeat and cardiovascular response. *Psychol. Rec.*, **37**: 349–68

Adams, M.R., Kaplan, J.R., Manuck, S.B. *et al.* (1988) Persistent sympathetic nervous system arousal associated with tethering in cynomolgus macaques. *Lab. Anim. Sci.*, **38**: 279–81

Amir, S., Brown, Z.W. and Amit, Z. (1980) The role of endorphins in stress: evidence and speculations. *Neurosci. Biobehav. Rev.*, **4**: 77–86

Arase, K., York, D.A., Shimizu, H. *et al.* (1988) Effects of corticotropin-releasing factor on VMH-lesioned obese rats. *Am. J. Physiol.*, **25**: R751–6

Ardlie, N.G., McGuiness, J.A. and Garrett, J.J. (1985) Effect on human platelets of catecholamines at levels achieved in the circulation. *Atherosclerosis*, **58**: 251–9

Aston-Jones, G., Foote, S.L. and Bloom, F.E. (1984) Anatomy and physiology of locus coeruleus neurons. Functional implications, in M. Ziegler and C. Lake (eds.), *Frontiers in Clinical Neuroscience*, Williams and Wilkins, Baltimore, pp 92–116

Aston-Jones, G., Ennis, M., Pieribone, V.A. *et al.* (1986) The brain nucleus locus coeruleus: Restricted afferent control of a broad efferent network. *Science*, **234**: 734–7

Baldock, N.M. and Sibly, R.M. (1990) Effects of handling and transportation on the heart rate and behaviour of sheep. *Appl. Anim. Behav. Sci.*, **28**: 15–39

Bandler, R. (1988) Brain mechanisms of aggression as revealed by electrical and chemical stimulation: Suggestion of a central role for the midbrain periaqueductal grey region, in A. Epstein and A. Morrison (eds.), *Progress in Psychobiological Physiology Psychology vol. 13*, Academic Press, New York, pp 67–154

Bandler, R., Carrive, P. and Zhang, S.P. (1991) Integration of somatic and autonomic reactions within the midbrain periaqueductal grey: Viscerotopic, somatotopic and functional organization. *Progr. Brain Res.*, **87**: 269–305

Bard, P. (1928) A diencephalic mechanism for the expression of rage with special reference to the sympathetic nervous system. *Am. J. Physiol.*, **84**: 490–515

Barnett, S.A., Hocking, W.E. *et al.* (1975) Socially-induced renal pathology of captive wild rats (*Rattus villosissimus*). *Agress. Behav.*, **1**: 123–33

Bassett, J.R. and Cairncross, K.D. (1975) Morphological changes induced in rats following prolonged exposure to stress. *Pharmacol. Biochem. Behav.*, **3**: 411–20

Bassett, J.R. and Cairncross, K.D. (1977) Changes in the coronary vascular system following prolonged exposure to stress. *Pharmacol. Biochem. Behav.*, **6**: 311–18

Berecek, K.H. and Swords, B.H. (1990) Central role of vasopressin in cardiovascular regulation and the pathogenesis of hypertension. *Hypertension*, **16**: 213–24

Berecek, K.H., Olpe, H.R. *et al.* (1984) Microinjections of vasopressin into the locus coeruleus of conscious rat. *Am. J. Physiol.*, **247**: H675–81

Bergamaschi, M. and Longoni, A.M. (1973) Cardiovascular events in anxiety: Experimental studies in the conscious dog. *Am. Heart J.*, **86**: 385–94

Bergamaschi, M., Caravaggi, A.M., Mandelli, V. and Shanks, R.G. (1973a) The role of beta-adrenoceptors in the coronary and systemic hemodynamic responses to emotional stress in conscious dogs. *Am. Heart J.*, **86**: 216–26

Bergey, J.L. and Beil, M.E. (1983) Antiarrhythmic evaluation of naloxone against acute coronary occlusion-induced arrhythmias in pigs. *Eur. J. Pharmacol.*, **90**: 427–31

Billman, G.E. and Randall, D.C. (1980a) Classic aversive conditioning of coronary blood flow in mongrel dogs. *Pavlov. J.*, **15**: 93–101

Billman, G.E. and Randall, D.C. (1980b) Mechanisms mediating the coronary vascular response to behavioral stress in the dog. *Circ. Res.*, **48**: 214–23

Bissett, G. (1991) Neuropeptides involved in stress and their distribution in the mammalian central nervous system, in J.A. McCubbin, P.G. Kaufmann and C.B. Nemeroff (eds.), *Stress, Neuropeptides, and Systemic Disease*, Academic Press Inc., New York, pp 55–71

Bohus, B. (1973) Pituitary-adrenal influences on avoidance and approach behavior of the rat. *Progr. Brain Res.*, **39:** 407–20

Bohus, B. (1974) Telemetered heart rate responses of the rat during free and learned behaviour. *Biotelemetry*, **1:** 193–201

Bohus, B. (1977) Pituitary neuropeptides, emotional behaviour and cardiac responses. *Progr. Brain Res.*, **47:** 277–88

Bohus, B. (1985) Acute cardiac responses to emotional stressors in the rat; the involvement of neuroendocrine mechanisms, in J.F. Orlebeke, G. Mulder and L.J.P. van Doornen (eds.), *Psychophysiology of Cardiovascular Control. Models, Methods, and Data*, Plenum Press, New York, London, pp 131–50

Bohus, B. (1993) Physiological functions of vasopressin in behavioural and autonomic responses to stress, in P. Burbach and D. de Wied (eds.), *Brain Functions of Neuropeptides*, Parthenon Publishing, Carnforth, pp 15–40

Bohus, B., Versteeg, C.A.M., De Jong, W. *et al.* (1983) Neurohypophyseal hormones and central cardiovascular control. *Progr. Brain Res.*, **60:** 445–57

Bohus, B., Koolhaas, J.M., Nyakas, C. *et al.* (1987a) Physiology of stress: a behavioral view, in P.R. Wiepkema and P.W.M. van Adrichem (eds.), *Biology of Stress in Farm Animals: An Integrative Approach*, pp 57–70

Bohus, B., Benus, R.F., Fokkema, D.S. *et al.* (1987b) Neuroendocrine states and behavioral and physiological stress responses. *Progr. Brain Res.*, **72:** 57–70

Bohus, B., Koolhaas, J.M., Nyakas, C. and Hindriks, F. (1988) Stress-induced fall in blood pressure in hypertensive rats: studies on environmental, neural and neuroendocrine factors. *Soc. Neurosci. Abs.*, **19:** 422

Bohus, B., Koolhaas, J.M., Luiten, P.G.M. *et al.* (1989) Vasopressin and related peptides: involvement in central cardiovascular regulation, in F.P. Nijkamp and D. De Wied (eds.), *Neuropeptides, Brain and Hypertension*, Elsevier, Amsterdam, pp 99–110

Bohus, B., Koolhaas, J.M., Korte, S.M. *et al.* (1990a) Behavioural physiology of serotonergic and steroid-like anxiolytics as antistress drugs. *Neurosci. Biobeh. Rev.*, **14:** 529–34

Bohus, B., Koolhaas, J.M., Nyakas, C. *et al.* (1990b) Neuropeptides and behavioural and physiological stress response: the role of vasopressin and related peptides, in S. Puglisi-Allegra and A. Oliverio (eds.), *Psychobiology of Stress*, Kluwer, Dordrecht, pp 103–23

Bolme, P., Novotny, J., Uvnas, B. and Wright, P.G. (1970) Species distribution of sympathetic cholinergic vasodilator nerves in skeletal muscle. *Acta Physiol. Scand.*, **78:** 60–4

Brown, M. (1986) Corticotropin releasing factor: central nervous system sites of action. *Brain Res.*, **199:** 10–14

Brown, M.R. (1991) Neuropeptide-mediated regulation of the neuroendocrine and autonomic responses to stress, in J.A. McCubbin, P.G. Kaufmann and C.B. Nemeroff (eds.), *Stress, Neuropeptides, and Systemic Disease*, Academic Press, New York, pp 73–93

Brown, M.R. and Fisher, L.A. (1985) Corticotropin releasing factor: Effects on the autonomic nervous system and visceral system. *Fed. Proc.*, **44:** 243–8

Brown, M.R. and Gray, T.S. (1988) Peptide injections into the amygdala of conscious rats: Effects on blood pressure, heart rate and plasma catecholamines. *Regul. Pept.*, **21:** 93–106

Brown, M.R., Fisher, L.A., Webb, V. *et al.* (1985) Corticotropin-releasing factor: A physiologic regulator of adrenal epinephrine secretion. *Brain Res.*, **328:** 355–7

Brown, M.R., Gray, T.S. and Fisher, L.A. (1986) Corticotropin-releasing factor receptor antagonist: Effects on the autonomic nervous system and cardiovascular function. *Regul. Pept.*, **16:** 231–329

Brown, M.R., Hauger, R. and Fisher, L.A. (1988) Autonomic and cardiovascular effects of corticotropin-releasing factor in the spontaneously hypertensive rat. *Brain Res.*, **441:** 33–40

Buijs, R.M. (1983) Vasopressin and oxytocin – their role in neurotransmission. *Pharmacol. Ther.*, **22:** 127–41

Buwalda, B., Koolhaas, J.M. and Bohus, B. (1992a) Behavioral and cardiac responses to mild stress in young and aged rats: Effects of amphetamine and vasopressin. *Physiol. Behav.*, **51:** 211–16

Buwalda, B., Nyakas, C., Koolhaas, J.M. *et al.* (1992b) Vasopressin prolongs behavioral and cardiac responses to mild stress in young but not in aged rats. *Physiol. Behav.*, **52:** 1127–31

Buwalda, B., Strubbe, J.H., Hoes, M.W.M. and Bohus, B. (1991) Reduced preabsorptive insulin response in aged rats: differential effects of amphetamine and arginine-vasopressin. *J. Auton. Nerv. Syst.*, **36:** 123–8

Cagin, N.A. and Natelson, B.H. (1981) Cholinergic activation produces psychosomatic digitalis toxicity. *J. Pharmacol. Exp. Ther.*, **218:** 709–11

Cannon, W.B. (1929a) *Bodily Changes in Pain, Hunger, Fear and Rage*, M.A. Branford, Boston

Cannon, W.B. (1929b) Organization for physiological homeostasis. *Physiol. Rev.*, **9:** 399–431

Carlier, P.G., Crine, A.F., Yema, N.M. and Rorive, G.L. (1988) Cardiovascular structural changes induced by isolation-stress hypertension in the rat. *J. Hypertens. Suppl.*, **6:** S112–15

Carobrez, A.P., Schenberg, L.C. and Graeff, F.G. (1983) Neuroeffector mechanisms of the defense reaction in the rat. *Physiol. Behav.*, **31:** 439–44

Chappell, P.B., Smith, M.A., Kilts, C.D. *et al.* (1986) Alterations in corticotropin-releasing factor-like immunoreactivity in discrete rat brain regions after acute and chronic stress. *J. Neurosci.*, **6:** 2908–14

Cherkovich, G.M. and Tatoyan, S.K. (1973) Heart rate (radiotelemetrical registration) in macaques and baboons according to dominant-submissive rank in a group. *Folia Primatol.*, **20:** 265–73

Ciriello, J., Kline, R.L., Zhang, T.X. and Caverson, M.M. (1984) Lesions of the paraventricular nucleus alter development of spontaneous hypertension in the rat. *Brain Res.*, **310:** 355–9

Corbolan, R., Verrier, R. and Lown, B. (1974) Psychological stress and ventricular arrhythmias during myocardial infarction in the conscious dog. *Am. J. Cardiol.*, **34:** 692–6

Cools, A.R., Brachten, R., Heeren, D. *et al.* (1990) Search after neurobiological profile of individual-specific features of Wistar rats. *Brain Res. Bull.*, **24:** 49–69

Corley, K.C., Shiel, F.O'M, Mauck, H.J.P. and Greenhoot, J. (1973) Electrocardiographic and cardiac morphological changes associated with environmental stress in squirrel monkeys, *Psychosom. Med.*, **35:** 361–4

Dantzer, R., Terlouw, C., Tazi, A. *et al.* (1988) The propensity for schedule-induced polydipsia is related to differences in conditioned avoidance behaviour and in defense reactions in a defeat test. *Physiol. Behav.*, **45:** 269–73

De Boer, S.F., Van der Gugten, J. and Slangen, J.L. (1989) Plasma catecholamine and corticosterone responses to predictable and unpredictable noise stress in rats. *Physiol. Behav.*, **45:** 789–95

De Loos, W.S., De Jong, W., Bohus, B. and De Wied, D. (1979) Reduction of heart rate reaction to emotional stress by ovarian hormones. *J. Endocrinol.*, **81:** 138P–9P

DeSilva, R.A., Verrier, R.L. and Lown, B. (1978) The effects of psychological stress and vagal stimulation with morphine on vulnerability to ventricular fibrillation (VF) in the conscious dog. *Am. Heart J.*, **95:** 197–203

De Souza, E.B. (1987) Corticotropin-releasing factor receptors in the rat central nervous system: characterization and regional distribution. *J. Neurosci.*, **7:** 88–100

De Turck, K.H. and Vogel, W.H. (1980) Factors influencing plasma catecholamine levels in rats during immobilization. *Pharmacol. Biochem. Behav.*, **13:** 129–31

De Wied, D. (1991) Effects of neurohypophyseal hormones and related peptides on learning and memory processes, in R.C.S. Frederickson, J.L. McGaugh and D.L. Felten (eds.), *Peripheral Signaling of the Brain*, Hogrefe and Huber, Toronto, pp 335–50

Diamant, M. (1991) *Neuropeptides, autonomic stress responses and behavior*. PhD Thesis, University of Utrecht

Driscoll, P. and Bättig, K. (1982) Behavioral, emotional and neurochemical profiles of rats selected for extreme differences in active, two-way avoidance performance, in I. Lieblich (ed.), *Genetics of the Brain*, Elsevier, Amsterdam, pp 95–123

Dunn, A.J. and Berridge, C.W. (1990) Physiological and behavioral responses to corticotropin-releasing factor administration: is CRF a mediator of anxiety or stress responses? *Brain Res. Rev.*, **15:** 71–100

Edvinsson, L., Emson, P., McCulloch, J. *et al.* (1983) Neuropeptide Y: cerebrovascular innervation and vasomotor effects in the cat. *Neurosci. Lett.*, **43:** 79–84

Eiserman, K. (1992) Long-term heart rate responses to social stress in wild european rabbits: predominant effect of rank position. *Physiol. Behav.*, **52:** 33–6

Ekblad, E., Edvinsson, L., Wahlestedt, C. *et al.* (1984) Neuropeptide Y coexists and cooperates with noradrenaline in perivascular nerve fibers. *Regul. Pept.*, **8:** 225–35

Elam, M., Yao, T., Svensson, T.H. and Thoren, P. (1984) Regulation of locus coeruleus neurons and splanchnic sympathetic nerves by cardiovascular afferents. *Brain Res.*, **390:** 281–7

Ely, D.L. (1981) Hypertension, social rank, and aortic arteriosclerosis in CBA/J mice. *Physiol. Behav.*, **26:** 655–61

Ely, D.L. and Henry, J.P. (1971) Effects of social role upon the blood pressure of individual male mice. *Fed. Proc.*, **30:** 265

Ely, D.L. and Weigand, J. (1983) Stress and high sodium effects on blood pressure and brain catecholamines in spontaneously hypertensive rats. *Clin. Exp. Hypertens.*, **A5:** 1559–87

Euler, D.E., Nattel, S., Spear, J.F. *et al.* (1985) Effect of sympathetic tone on ventricular arrhythmias during circumflex coronary occlusion. *Am. J. Physiol.*, **249:** H1045–50

Fagbemi, O., Lepran, I., Parratt, J.R. and Szekeres, L. (1982) Naloxone inhibits early arrhythmias resulting from acute coronary ligation. *Br. J. Pharmacol.*, **76:** 504–6

Fagbemi, O., Kane, K.A., Lepran, I. *et al.* (1983) Antiarrhythmic actions of meptazinol, a partial agonist at opiate receptors in acute myocardial ischaemia. *Br. J. Pharmacol.*, **78:** 455–60

Feuerstein, C. (1985) The opioid system and central cardiovascular control: Analysis of controversies. *Peptides (N.Y.)*, **6:** 51–8

Fisher, L.A. (1988) Corticotropin-releasing factor: central nervous system effects on baroreflex control of heart rate. *Life Sci.*, **42:** 2645–9

Fisher, L.A. (1991) Corticotropin-releasing factor and autonomic-cardiovascular responses to stress, in J.A. McCubbin, P.G. Kaufmann and C.B. Nemeroff (eds.), *Stress, Neuropeptides, and Systems Disease*, Academic Press, New York, pp 95–118

Fisher, L.A., Jessen, G. and Brown, M.R. (1983) Corticotropin-releasing factor (CRF): Mechanism to elevate mean arterial pressure and heart rate. *Regul. Pept.*, **5:** 153–61

Fokkema, D.S. (1986) *Social behavior and blood pressure. A study of rats.* PhD Thesis, University of Groningen

Fokkema, D.S. and Koolhaas, J.M. (1985) Acute and conditioned blood pressure changes in relation to social and psychosocial stimuli in rats. *Physiol. Behav.*, **34:** 33–8

Fokkema, D.S., Smit, K., Van der Gugten, J. and Koolhaas, J.M. (1988) A coherent pattern among social behavior, blood pressure, corticosterone and catecholamine measures in individual male rats. *Physiol. Behav.*, **42:** 485–98

Fulwiler, C.E. and Saper, C.B. (1984) Subnuclear organization of the efferent connections of the parabrachial nucleus in the rat. *Brain Res. Rev.*, **7:** 229–59

Furness, J.B., Bornstein, J.C., Murphy, R. and Pompolo, S. (1992) Role of peptides in transmission in the enteric nervous system. *Trends in Neurosci.*, **15:** 66–71

Galeno, T.M., Van Hoesen, G.W. and Brody, M.J. (1984) Central amygdaloid nucleus lesion attenuates exaggerated hemodynamic responses to noise stress in the spontaneously hypertensive rat. *Brain Res.*, **291:** 249–59

Gamallo, A., Alario, P., Villanua, M.A. and Nava, M.B. (1988) Effects of chronic stress in the blood pressure in the rat: ACTH administration. *Horm. Metab. Res.*, **20:** 336–8

Gardiner, S.M. and Bennett, T. (1977) The effects of short-term isolation on systolic blood pressure and heart rate in rats. *Med. Biol.*, **55:** 325–9

Gärtner, K., Büttner, D., Döhler, K. *et al.* (1980) Stress response of rats to handling and experimental procedures. *Lab. Anim.*, **14:** 267–74

Garvey, H.L. and Melville, K.I. (1969) Cardiovascular effects of lateral hypothalamic stimulation in normal and coronary-ligated dogs. *J. Cardiovasc. Surg.*, **10:** 377–85

Gillies, G., Linton, E. and Lowry, P. (1982) Corticotropin releasing activity of the new CRF is potentiated several times by vasopressin. *Nature*, **299:** 355–7

Gomez, R.E., Büttner, D. and Cannata, M.A. (1989) Open field behaviour and cardiovascular responses to stress in normal rats. *Physiol. Behav.*, **45:** 767–9

Gray, T.S. (1991) Amygdala: Role in autonomic and neuroendocrine responses to stress, in J.A. McCubbin, P.G. Kaufmann and C.B. Nemeroff (eds.), *Stress, Neuropeptides, and Systemic Disease*, Academic Press, New York, pp 37–53

Gray, T.S. and Magnuson, D.J. (1987) Neuropeptide neuronal efferents from the bed nucleus of the stria terminalis and central amygdaloid nucleus to the dorsal vagal complex in the rat. *J. Comp. Neurol.*, **262**: 365–74

Haft, J.I. and Fani, K. (1973a) Intravascular platelet aggregation in the heart induced by stress. *Circulation*, **47**: 353–8

Haft, J.L. and Fani, K. (1973b) Stress and the induction of intravascular platelet aggregation in the heart. *Circulation*, **48**: 164–9

Hagan, J.J. and Bohus, B. (1984) Vasopressin prolongs bradycardiac response during orientation. *Behav. Neural. Biol.*, **41**: 77–83

Hall, C.E., Cross, E. and Hall, O. (1960) Amyloidosis and other pathologic changes in mice exposed to chronic stress. *Tex. Rep. Biol. Med.*, **18**: 205–13

Harlow, H.J., Thorne, E.T., Willams, E.S. *et al.* (1987) Adrenal responsiveness in domestic sheep (*Ovis aries*) to acute and chronic stressors as predicted by remote monitoring of cardiac frequency. *Can. J. Zool.*, **65**: 2021–7

Harrap, S.B., Louis, W.J. and Doyle, A.E. (1984) Failure of psychosocial stress to induce chronic hypertension in the rat. *J. Hypertens.*, **20**: R82–90

Hashimoto, K., Hattori, T., Murakami, K. *et al.* (1985) Reduction in brain immunoreactive corticotropin-releasing factor (CRF) in spontaneously hypertensive rats. *Life Sci.*, **36**: 643–7

Henry, J.P. and Stephens, P.M. (1977) *Stress, Health and the Social Environment. A Sociobiological Approach to Medicine*, Springer-Verlag, Berlin

Henry, J.P. and Stephens-Larson, P. (1985) Specific aspects of stress on disease processes, in G.P. Moberg (ed.), *Animal Stress*, Amer. Physiol. Ass., Bethesda, Maryland, pp 161–75

Henry, J.P., Ely, D.L., Stephens, P.M. *et al.* (1971) The role of psychosocial factors in the development of arteriosclerosis in CBA mice. *Atherosclerosis*, **14**: 203–18

Henry, J.P., Stephens, P.M. and Santisteban, G.A. (1975) A model of psychosocial hypertension showing reversibility and progression of cardiovascular complications. *Circ. Res.*, **36**: 156–64

Herd, J.A. (1991) Cardiovascular response to stress. *Physiol. Rev.*, **71**: 305–30

Herd, J.A., Morse, W.H., Kelleher, R.T. and Jones, L.G. (1969) Arterial hypertension in the squirrel monkey during behavioural experiments. *Am. J. Physiol.*, **217**: 24–9

Hervonen, A., Linnoila, I., Pickel, V.M. *et al.* (1981) Localization of met- and leu-enkephalin immunoreactivity in nerve terminals in human paravertebral sympathetic ganglia. *Neuroscience*, **6**: 323–30

Hilton, S.M. (1982) The defence-arousal system and its relevance for circulatory and respiratory control. *J. Exp. Biol.*, **190**: 159–74

Hirst, G.D.S., Bramich, N.J., Edwards, F.R. and Klemm, M. (1992) Transmission at autonomic neuroeffector junctions. *Trends in Neurosci.*, **15**: 40–6

Hofer, M.A. (1970) Cardiac and respiratory function during sudden prolonged immobility in wild rodents. *Psychosom. Med.*, **32**: 533–647

Holaday, J.W. (1983) Cardiovascular effects of endogenous opiate systems. *Ann. Rev. Pharmacol. Toxicol.*, **23**: 541–94

Höllt, V. (1986) Opioid peptide processing and receptor selectivity. *Ann. Rev. Pharmacol. Toxicol.*, **26**: 59–77

Holstege, G. (1987) Some anatomical observations on the projections from the hypothalamus to brainstem and spinal cord: an HRP and autoradiographic tracing study in the cat. *J. Comp. Neurol.*, **260**: 98–126

Holstege, G., Meiners, L. and Tan, K. (1985) Projections of the bed nucleus of the stria terminalis to the mesencephalon, pons, and medulla oblongata in the cat. *Exp. Brain Res.*, **58**: 370–91

Hubbard, J.W., Cox, R.H., Sanders, B.J. and Lawler, J.E. (1986) Changes in cardiac output and vascular resistance during behavioral stress in the rat. *Am. J. Physiol.*, **251**: R82–90

Hughes, C.W. and Lynch, J.J. (1978) A reconsideration of psychological precursors of sudden death in infrahuman animals. *Am. Psychol.*, **33**: 419–29

Imai, Y., Abe, K., Sasaki, S. *et al.* (1987) Hypotensive and bradycardiac effects of centrally administered vasopressin in stroke-prone spontaneously hypertensive rats. *Hypertension*, **10**: 346–9

Iwamoto, J., Chida, K., LeDoux, J.E. (1987) Cardiovascular responses elicited by stimulation of neurons in the central amygdaloid nucleus in awake but not anesthetized rats resemble conditioned emotional responses. *Brain Res.*, **418:** 183–8

Jänig, W. and McLachlan, E.M. (1992) Characteristics of function-specific pathways in the sympathetic nervous system. *Trends in Neurosci.*, **15:** 475–81

Jard, S. (1985) Vasopressin receptors. *Front. Horm. Res.*, **13:** 89–104

Johansson, G. and Jönsson, G. (1977) Myocardial cell damage in the porcine stress syndrome. *J. Comp. Pathol.*, **87:** 67–74

Jones, B.E. and Yang, T.Z. (1985) The efferent projections from the reticular formation and the locus coeruleus studied by anterograde and retrograde axonal transport in the rat. *J. Comp. Neurol.*, **242:** 56–92

Jordan, D. and Spyer, K.M. (1986) Brainstem integration of cardiovascular and pulmonary afferent activity. *Progr. Brain Res.*, **67:** 295–314

Kamarck, T. and Jennings, J.R. (1991) Biobehavioral factors in sudden cardiac death. *Psychol. Bull.*, **109:** 42–75

Kaplan, J.R., Manuck, S.B., Clarkson, T.B. *et al.* (1983) Social stress and atherosclerosis in normocholesterolemic monkeys. *Science*, **220:** 733–5

Kapp, B.S., Frysinger, R.C., Gallagher, M. and Haselton, J.R. (1979) Amygdala central nucleus lesions: Effect on heart rate conditioning in the rabbit. *Physiol. Behav.*, **23:** 1109–17

Katafuchi, T., Oomura, Y. and Kurosawa, M. (1988) Effects of chemical stimulation of paraventricular nucleus on adrenal and renal nerve activity in rats. *Neurosci. Lett.*, **86:** 195–200

Kelleher, R.T., Morse, W.H. and Herd, J.A. (1981) Pharmacological studies of behavioral influences on cardiovascular function. *Neurosci. Biobehav. Rev.*, **5:** 325–34

Khachaturian, J., Lewis, M.E., Schäfer, M.K.-H. and Watson, S.J. (1985) Anatomy of CNS opioid systems. *Trends in Neurosci.*, **8:** 111–19

Kirby, R.F. and Johnson, A.K. (1990) Role of β_2-adrenoceptors in cardiovascular response of rats to acute stressors. *Am. J. Physiol.*, **258:** H683–8

Kirby, R.F., Callahan, M.V. and Johnson, A.K. (1987) Regional vascular responses to an acute stressor in spontaneously hypertensive and Wistar-Kyoto rats. *J. Autonom. Nerv. Syst.*, **20:** 185–8

Kirby, R.F., Callahan, M.V., McCarty, R. and Johnson, A.K. (1989) Cardiovascular and sympathetic nervous system responses to acute stress in borderline hypertensive rats. *Physiol. Behav.*, **46:** 309–13

Kliks, B.R., Berguss, M.J. and Abildskov, J.A. (1975) Influence of sympathetic tone on ventricular fibrillation threshold during experimental coronary occlusion. *Am. J. Cardiol.*, **36:** 46–9

Konarska, M., Stewart, R.E. and McCarty, R. (1989) Sensitization of sympathetic-adrenal medullary responses to a novel stressor in chronically stressed laboratory rats. *Physiol. Behav.*, **46:** 129–35

Koolhaas, J.M. and Bohus, B. (1989) Social control in relation to neuroendocrine and immunological responses, in A. Steptoe and A. Appels (eds.), *Stress, Personal Control and Health*, John Wiley, Brussels, pp 295–305

Koolhaas, J.M. and Bohus, B. (1991) Coping strategies and cardiovascular risk: a study of rats and mice, in A. Appels, J. Groen and J. Koolhaas (eds.), *Behavioral Observations in Cardiovascular Research*, Swets and Zeitlinger, Amsterdam, pp 45–58

Koolhaas, J.M., Schuurman, T. and Fokkema, D.S. (1983) Social behavior of rats as a model for the psychophysiology of hypertension, in T.M. Dembroski and G. Blümchen (eds.), *Biobehavioral Bases of Coronary Heart Disease*, Karger, Basel, pp 391–400

Koolhaas, J.M., Fokkema, D.S., Bohus, B. and Van Oortmerssen, G.A. (1986) Individual differences in blood pressure reactivity and behavior of male rats, in T.H. Schmidt, T.M. Dembroski and G. Blümchen (eds.), *Biological and Psychological Factors in Cardiovascular Disease*, Springer-Verlag, Berlin, pp 517–26

Korte, S.M., Buwalda, B., Bouws, G.A.H. *et al.* (1992a) Conditioned neuroendocrine and cardiovascular stress responsiveness accompanying behavioral passivity and activity in aged and in young rats. *Physiol. Behav.*, **51:** 815–22

Korte, S.M., Jaarsma, D., Luiten, P.G.M. and Bohus, B. (1992b) Mesencephalic cuneiform nucleus and its ascending and descending projections serve stress-related cardiovascular responses in the rat. *J. Autonom. Nerv. Syst.*, **41**: 157–76

Korte, S.M., Bouws, G.A.H. and Bohus, B. (1993) Central actions of corticotropin-releasing hormone (CRH) on behavioral, neuroendocrine and cardiovascular regulation: brain corticoid receptor involvement. *Horm. Behav.* (in press)

Kregel, K.C., Overton, J.M., Seals, D.R. *et al.* (1990) Cardiovascular responses to exercise in the rat: role of corticotropin-releasing factor. *J. Appl. Physiol.*, **68**: 361–7

Kurosawa, M., Sato, A., Swenson, R.S. and Takahaski, Y. (1986) Sympatho-adrenal medullary functions in response to intracerebroventricularly injected corticotropin-releasing factor in anesthetized rats. *Brain Res.*, **367**: 250–7

Laczi, F., Gaffori, O., Fekete, M. *et al.* (1984). Levels of arginine-vasopressin in cerebrospinal fluid during passive avoidance behavior in rats. *Life Sci.*, **34**: 2385–91

Lamprecht, F., Williams, R.B. and Kopin, J.J. (1973) Serum dopamine-beta-hydroxylase during development of immobilisation-induced hypertension. *Endocrinology*, **92**: 953–6

Lamprecht, F., Eichelman, B.S., Williams, R.B. *et al.* (1974) Serum dopamine-beta-hydroxylase activity and blood pressure response of rat strains to shock-induced fighting. *Psychosom. Med.*, **36**, 298–303

Lane R.D. and Schwartz, G.E. (1987) Induction of lateralized sympathetic input to the heart by the CNS during emotional arousal: a possible neurophysiologic trigger of sudden cardiac death. *Psychosom. Med.*, **49**: 274–84.

Lang, R.E., Heil, J.W.E., Ganten, D. *et al.* (1983) Oxytocin unlike vasopressin is a stress hormone in the rat. *Neuroendocrinology*, **37**: 314–16

Lang, R.E., Rascher, W., Unger, Th. and Ganten, D. (1981) Reduced content of vasopressin in the brain of spontaneously hypertensive as compared to normotensive rats. *Neurosci. Lett.*, **23**: 199–202

Larsson, P.T., Hjemdahl, P., Olsson, G., *et al.* (1989) Altered platelet function during mental stress and adrenaline infusion in humans: Evidence for an increased aggregability in vivo as measured by filtragometry. *Clin. Sci.*, **76**: 369–76

Lawler, J.E., Barker, G.J., Hubbard, J.W. and Allen, M.T. (1980a) The effects of conflict on tonic levels of blood pressure in the genetically borderline hypertensive rat. *Psychophysiol.*, **17**: 363–70

Lawler, J.E., Barker, G.F., Hubbard, J.W. and Schaub, P.G. (1980b) Pathophysiological changes associated with stress-induced hypertension in the borderline hypertensive rat. *Clin. Sci.*, **50**: 307s–10s

Lawler, J.E., Barker, G.F., Hubbard, J.W. and Schaub, R.G. (1981) Effects of stress on blood pressure and cardiac pathology in rats with borderline hypertension. *Hypertension*, **3**: 496–505

Lawler, J.E., Barker, G.F., Hubbard, J.W. *et al.* (1984a) Blood pressure and plasma renin activity responses to chronic stress in the borderline hypertensive rat. *Physiol. Behav.*, **32**: 101–5

Lawler, J.E., Cox, R.H., Barker, G.F. *et al.* (1984b) Cardiovascular, plasma catecholamine and corticosterone responses to alterations in the predictability of electric shock in the rat. *Physiol. Psychol.*, **12**: 227–32

LeDoux, J.E., Iwata, J., Cichetti, P. and Reis, D.J. (1988) Different projections of the central amygdaloid nucleus mediate autonomic and behavioral correlates of conditioned fear. *J. Neurosci.*, **8**: 2517–29

Leibowitz, S.F. (1992) Neurochemical-neuroendocrine systems in the brain controlling macronutrient intake and metabolism. *Trends in Neurosci.*, **15**: 491–7

Levine, S. and Ursin, H. (1991) What is stress, in M.R. Brown, G.F. Koob and C. Rivier (eds.), *Stress. Neurobiology and Neuroendocrinology*, Marcel Dekker, New York, pp 3–21

Liang, B., Verrier, R.L., Melman, J. and Lown, B. (1979) Correlation between circulating catecholamine levels and ventricular vulnerability during psychological stress in conscious dogs. *Proc. Soc. Exp. Biol. Med.*, **161**: 266–9

Livezey, G.T., Miller, J.M. and Vogel, W.H. (1985) Plasma norepinephrine, epinephrine and corticosterone stress responses to restraint in individual male and female rats, and their correlations. *Neurosci. Lett.*, **62**: 51–6

Loewy, A.D. (1990) Central autonomic pathways, in A.D. Loewy and K.M. Spyer (eds.), *Central Regulation of Autonomic Functions*, Oxford University Press, New York, Oxford, pp 88–103

Loewy, A.D. (1991) Forebrain nuclei involved in autonomic control. *Progr. Brain Res.*, **87:** 253–68

Loewy, A.D. and Spyer, K.M. (1990) Vagal preganglionic neurons, in A.D. Loewy and K.M. Spyer (eds.), *Central Regulation of Autonomic Functions*, Oxford University Press, New York, pp 68–87

Lown, B., Verrier, R. and Corbalan, R. (1973) Psychological stress and threshold for repetitive ventricular response. *Science*, **182:** 834–6

Luiten, P.G.M., ter Horst, G.J. and Steffens, A.B. (1987) The hypothalamus, intrinsic connections and outflow pathways to the endocrine system in relation to the control of feeding and metabolism. *Progr. Neurobiol.*, **28:** 1–54

Lundin, S., Ricksten, S.E. and Thoren, P. (1984) Interaction between 'mental stress' and baroreceptor reflexes concerning effects on heart rate, mean arterial pressure and renal sympathetic activity in conscious spontaneously hypertensive rats. *Acta Physiol. Scand.*, **120:** 273–81

Manuck, S.B., Kaplan, J.R. and Clarkson, T.B. (1983) Behaviorally induced heart rat reactivity and atherosclerosis in cynomoyglus monkeys. *Psychosom. Med.*, **45:** 95–108

Manuck, S.B., Kaplan, J.R., Adams, M.R. and Clarkson, T.B. (1989). Behaviourally-elicited heart rate reactivity and atherosclerosis in female cynomolgus monkeys (*Macaca fascicularis*) *Psychosom. Med.*, **51:** 306–18

Mason, J.W. (1968a) A review of psychoendocrine research on the pituitary-adrenal cortical system. *Psychosom. Med.*, **30:** 567–607

Mason, J.W. (1968b) A review of psychoendocrine research on the sympathetic-adrenal medullary system. *Psychosom. Med.*, **30:** 631–53

Matta, R.J., Lawler, J.E. and Lown, B. (1976a) Ventricular electrical instability in the conscious dog: Effect of psychological stress and beta adrenergic blockade. *Am. J. Cardiol.*, **38:** 594–8

Matta, R.J., Verrier, R.L. and Lown, B. (1976b) Repetitive extra systole as an index of vulnerability to ventricular fibrillation. *Am. J. Physiol.*, **230:** 1469–73

McCarty, R. (1983) Stress, behaviour and experimental hypertension, *Neurosci. Biobehav. Rev.*, **7:** 493–502

McCarty, R., Kvetnansky, R. and Kopin, J.J. (1981) Plasma catecholamines in rats: daily variations in basal levels and increments in response to stress. *Physiol. Behav.*, **26:** 27–31

Miller, D.G., Gilmour, R.F., Grossman, Z.D. *et al.* (1978) Myocardial uptake of Tc-99m skeletal agents in the rat after experimental induction of microscopic foci of injury. *J. Nucl. Med.*, **18:** 1005–9

Moga, M.M. and Gray, T.S. (1985) Evidence for corticotropin-releasing factor, neurotensin and somatostatin in the neural pathway from the central nucleus of the amygdala to the parabrachial nucleus. *J. Comp. Neurol.*, **241:** 275–84

Moruzzi, G. and Magoun, H.W. (1949) Brain stem reticular formation and activation of the EEG. *Electroenceph. Clin. Neurophysiol.*, **1:** 455–73

Nakamura, K., Ono, T., Fukuda, M. and Uwano, T. (1992) Paraventricular neuron chemosensitivity and activity related to blood pressure control in emotional behavior. *J. Neurophysiol.*, **67:** 255–64

Naranjo, J.R., Urdin, M.C., Borrell, J. and Fuentes, J.A. (1986) Evidence for a central but not adrenal, opioid mediation in hypertension induced by brief isolation in the rat. *Life Sci.*, **38:** 1923–30

Natelson, B.H. (1985) Neurocardiology: An interdisciplinary area for the 80s. *Arch. Neurol.*, **42:** 178–84

Natelson, B.H. and Cagin, N.A. (1981) The role of shock predictability during aversive conditioning in producing psychosomatic digitalis toxicity. *Psychosom. Med.*, **43:** 191–7

Natelson, B.H., Tapp, W.N., Adamus, J.E. *et al.* (1981) Humoral indices of stress in rats. *Physiol. Behav.*, **26:** 1049–54

Natelson, B.H., Cagin, N.A., Turfts, M. *et al.* (1982) Bidirectional effect of naloxone on emotionally conditioned digitalis toxicity. *Psychosom. Med.*, **44:** 397–400

Naumenko, E.V., Markel, A.L., Amstislavsky, S.F. and Dygalo, N.N. (1989) Brain adrenergic mechanisms of adrenocortical regulation in rats with inherited stress-induced arterial hypertension, in G.R. van Loon, R. Kvetnansky, R. McCarty and J. Axelrod (eds.), *Stress, Neurochemical and Humoral Mechanisms*, Gordon and Breach, New York, pp 453–60

Negro-Vilar, A. and Saavedra, J.M. (1980) Changes in brain somatostatin and vasopressin levels after stress in spontaneously hypertensive and Wistar-Kyoto rats. *Brain Res. Bull.*, **5**: 353–8

Neveau, P.J. (1992) Asymmetrical brain modulation of the immune response. *Brain Res. Rev.*, **17**: 101–7

Nyakas, C., Prins, A.J.A. and Bohus, B. (1990) Age-related alterations in cardiac response to emotional stress: relations to behavioural reactivity in the rat. *Physiol. Behav.*, **47**: 273–80

Nyakas, C., Balkan, B., Steffens, A.B. and Bohus, B. (1993) Cardiac and behavioral response of obese and lean Zucker rats to emotional stress. *Physiol. Behav.* (in press)

Obrist, P.A. (1981) *Cardiovascular Psychophysiology: A Perspective*, Plenum, New York

Ohta, H., Watanabe, S. and Ueki, S. (1991) Cardiovascular changes induced by chemical stimulation of the amygdala in rats. *Brain Res. Bull.*, **26**: 575–81

Overton, J.M. and Fisher, L.A. (1989) Central nervous system actions of corticotropin-releasing factor on cardiovascular function in the absence of locomotor activity. *Regul. Pept.*, **25**: 314–34

Overton, J.M., Davis-Gorman, G. and Fisher, L.A. (1990) Central nervous system cardiovascular actions of CRF in sinoaortic denervated rats. *Am. J. Physiol.*, **258**: R596–601

Owens, M.J. and Nemeroff, C.B. (1991) Physiology and pharmacology of corticotropin-releasing factor. *Pharmacol. Rev.*, **43**: 425–73

Pinto, J.M.B., Kirby, D.A. and Verrier, R.L. (1989) Abolition of clonidine's effects on ventricular refractoriness by naloxone in the conscious dog. *Life Sci.*, **45**: 413–20

Popper, C.W., Chiuch, C.C. and Kopin, J.J. (1977) Plasma catecholamine concentrations in unanaesthetized rats during sleep, wakefulness, immobilization and after decapitation. *J. Pharm. Exp. Ther.*, **202**: 144–8

Rabkin, S.W., Einzig, S. and Benditt, D.G. (1987) Morphine suppression of digitalis-induced arrhythmias. *Arch. J. Phar.*, **289**: 267–77

Rabkin, S.W. and Roob, O. (1986) Effect of the opiate antagonist naloxone on digitalis-induced cardiac arrhythmias. *Eur. J. Pharmacol.*, **130**: 47–55

Randich, A. and Maixner, W. (1984) Interactions between cardiovascular and pain regulatory systems. *Neurosci. Biobehav. Rev.*, **8**: 343–67

Ratcliffe, H.L., Luginbuhl, H., Schnarr, W.R. and Chacko, K. (1969) Coronary arteriosclerosis in swine: evidence of a relation to behaviour. *J. Comp. Physiol. Psychol.*, **68**: 385–92

Ratcliffe, H.L. and Snyder, R.L. (1967) Arteriosclerotic stenosis of the intramural arteries of chickens: further evidence of a relation to social factors. *Br. J. Exp. Path.*, **48**: 357–65

Richter, C.P. (1957) On the phenomenon of sudden death in animals and man. *Psychosom. Med.*, **19**: 191–8

Rodriguez, E.M. (1976) The cerebrospinal fluid as a pathway in neuroendocrine integration. *J. Endocrinol.*, **1**: 407–37

Roozendaal, B., Koolhaas, J.M. and Bohus, B. (1990) Differential effect of lesioning of the central amygdala on the bradycardiac and behavioral response of the rat in relation to conditioned social and solitary stress. *Behav. Brain Res.*, **41**: 39–48

Roozendaal, B., Koolhaas, J.M. and Bohus, B. (1991) Attenuated cardiovascular, neuroendocrine, and behavioral responses after a single footshock in central amygdaloid lesioned male rats. *Physiol. Behav.*, **50**: 771–5

Roozendaal, B., Schoorlemmer, G.H.M., Wiersma, A. *et al.* (1992a) Opposite effects of central amygdaloid vasopressin and oxytocin on the regulation of conditioned stress responses in male rats. *Ann. N.Y. Acad. Sci.*, **652**: 460–1

Roozendaal, B., Wiersma, A., Driscoll, P. *et al.* (1992b) Vasopressinergic modulation of stress responses in the central amygdala of the Roman high-avoidance and low-avoidance rat. *Brain Res.*, **596**: 35–40

Rosencrans, J.A., Watzman, N. and Buckley, J.P. (1966) The production of hypertension in male albino rats subjected to experimental stress. *Biochem. Pharmacol.*, **15**: 1707–18

Rosenfeld, J., Rosen, M.R. and Hoffman, B.F. (1978) Pharmacological and behavioral effects on arrhythmias that immediately follow abrupt coronary occlusion: A canine model of sudden coronary death. *Am. J. Cardiol.*, **41**: 1075–82

Rudehill, A., Sollevi, A., Franco-Cereceda, A. and Lundberg, J.M. (1986) Neuropeptide Y (NPY) and the pig heart: Release and coronary vasoconstrictor effects. *Peptides 7*: 821–6

Sachser, N. and Lick, C. (1989) Social stress in guinea-pigs. *Physiol. Behav.*, **46**: 137–44

Saini, V., Carr, D.A., Hagestad, E.L. *et al.* (1988) Antifibrillatory mechanism of the narcotic agonist fentanyl. *Am. Heart J.*, **115**: 598–605

Saini, V., Carr, D.B. and Verrier, R.L. (1989) Comparative effects of the opioids fentanyl and byprenorphine on ventricular vulnerability during acute coronary artery occlusion. *Cardiovasc. Res.*, **23**: 1001–6

Sakaguchi, A., LeDoux, J.E. and Reis, D.J. (1983) Sympathetic nerves and adrenal medulla: contributions to cardiovascular-conditioned emotional responses in spontaneously hypertensive rats. *Hypertension*, **5**: 728–38

Saper, C.B. and Loewy, A.D. (1980) Efferent connections of the parabrachial nucleus in the rat. *Brain Res.*, **197**: 291–317

Satinsky, J., Kosowsky, B., Lown, B. and Kerzner, J. (1971) Ventricular fibrillation induced by hypothalamic stimulation during coronary occlusion. *Circulation*, **44** (Suppl. 2): II–60

Sawchenko, P.E. (1991) A tale of three peptides: corticotropin-releasing factor-, oxytocin-, and vasopressin-containing pathways mediating integrated hypothalamic responses to stress, in J.A. McCubbin, P.G. Kaufmann and C.B. Nemeroff (eds.), *Stress, Neuropeptides, and Systemic Disease*, Academic Press, New York, pp 3–17

Scheurink, A.J.W. and Steffens, A.B. (1990) Central and peripheral control of sympathoadrenal activity and energy metabolism in rats. *Physiol. Behav.*, **48**: 909–20

Schneidermann, N. (1983) Behaviour, autonomic function and animal models of cardiovascular pathology, in T.M. Dembrowski, T.H. Schmidt and G. Blümchen (eds.), *Biobehavioural Bases of Coronary Heart Diseases*, Karger, Basel, pp 304–64

Schwartz, P.J. and Stone, H.L. (1980) Left stellectomy in the prevention of ventricular fibrillation caused by acute myocardial ischemia in conscious dogs with anterior myocardial infarction. *Circulation*, **62**: 1256–65

Selye, H. (1935) A syndrome produced by diverse noxious agents. *Nature Lond.*, **138**: 32–3

Selye, H. (1976) *The Stress of Life*, McGraw-Hill, New York

Shepherd, J.T. and Vanhoutte, P.M. (1985) Spasm of the coronary arteries: causes and consequences (the scientist's viewpoint). *Mayo Clin. Proc.*, **60**: 33–46

Sherrington, C.S. (1952) Reflexes as adapted reactions: bodily resonance of emotions, in *The Integrative Action of the Nervous System. Lecture VII*, Cambridge University Press, London, p 257

Shively, C. and Kaplan, J. (1984) Effects of social factors on adrenal weight and related physiology of *Macaca fascicularis*. *Physiol. Behav.*, **33**: 777–82

Silverman, A.-J., Hou-Yu, A. and Chen, W.-P. (1989) Corticotropin-releasing factor synapses within the paraventricular nucleus of the hypothalamus. *Neuroendocrinology*, **49**: 291–9

Skinner, J.E. and Reed, J.C. (1981) Blockade of frontocortical-brainstem pathway prevents ventricular fibrillation of ischemic heart. *Am. J. Physiol.*, **240**: H156–63

Skinner, J.E., Lie, J.T. and Entman, M.J. (1975) Modification of ventricular fibrillation latency following coronary artery occlusion in the conscious pig. The effects of psychological stress and beta-adrenergic blockade. *Circulation*, **51**: 656–67

Smith, O.A., Astley, C.A., DeVito, J.L. *et al.* (1980) Functional analysis of hypothalamic control of the cardiovascular responses accompanying emotional behavior. *Fed. Proc.*, **39**: 2487–94

Smith, O.A. and DeVito, J.L. (1984) Central neural integration for the control of autonomic responses associated with emotion. *Ann. Rev. Neurosci.*, **7**: 43–65

Smith, O.A., Hohimer, A.R., Astley, C.A. and Taylor, D.J. (1979) Renal and hindlimb vascular control during acute emotion in the baboon. *Am. J. Physiol.*, **236**: R198–205

Steenbergen, J.M., Koolhaas, J.M., Strubbe, J.H. and Bohus, B. (1989) Behavioural and cardiac responses to a sudden change in environmental stimuli: effect of forced shift in food intake. *Physiol. Behav.*, **45**: 729–33

Stock, G., Stumpf, H., Schlor, K.H. (1979) Absence of sympathetic cholinergic vasodilation in cats during early stages of affective behavior elicited by stimulation of central amygdala, postero-lateral hypothalamus and locus coeruleus. *Clin. Sci.*, **57**: 205s–8s

Strack, A.M., Sawyer, W.B., Hughes, J.H., Platt, K.B. and Loewy, A.D. (1989) CNS cell groups regulating the sympathetic nervous outflow to adrenal gland as revealed by transneuronal cell body labeling with pseudorabies virus. *Brain Res.*, **491**: 276–96

Sved, A. and Felsten, G. (1987) Stimulation of the locus coeruleus decreases arterial pressure. *Brain Res.*, **414**: 199–232

Svensson, T.H. and Thoren, P. (1979) Brain noradrenergic neurons in the locus coeruleus: Inhibition by blood volume load through vagal afferents. *Brain Res.*, **172**: 174–8

Swanson, L.W. and Sawchenko, P.E. (1983) Hypothalamic integration: organization of the paraventricular and supraoptic nuclei. *Ann. Rev. Neurosci.*, **6**: 269–324

Swenson, R.M. and Vogel, W.H. (1983) Plasma catecholamine and corticosterone as well as brain catecholamine changes during coping in rats exposed to stressful footshock. *Pharmacol. Biochem. Behav.*, **16**: 689–93

Szilagyi, J.E. (1988) Endogenous opiate modulation of baroreflexes in normotensive and hypertensive rats. *Am. J. Physiol.*, **1255**: H987–91

Szilagyi, J.E. (1991) Psychosocial stress elevates blood pressure via an opioid dependent mechanism in normotensive rats. *Clin. Exper. Hyper. – Theory and Practice*, **A13**: 1383–94

Tapp, W.N., Natelson, B.H. Creighton, D. *et al.* (1989) Alprazolam reduces stress-induced mortality in cardiomyopathic hamsters. *Pharmacol. Biochem. Behav.*, **32**: 331–6

Tashiro, N., Hirata, K., Maki, A. and Nakao, H. (1988) Cardiac arrhythmias induced in cats by stimulation of the anteriomedial hypothalamus. *Int. J. Psychophysiol.*, **6**: 231–40

Taylor, J., Weyers, P., Harris, N. and Vogel, W.H. (1989) The plasma catecholamine stress response is characteristic for a given animal over a one-year period. *Physiol. Behav.*, **46**: 853–6

Thor, K.B. and Helke, C.J. (1988) Catecholamine-synthesizing neuronal projections to the nucleus tractus solitarii of the rat. *J. Comp. Neurol.*, **264**: 265–80

Turkkan, J.S. (1991) Individual behavioral and neuroendocrine changes are correlated with blood-pressure elevations during conflict alone and combined with high-salt diet in baboons. *Psychobiol.*, **19**: 161–7

Vale, W., Rivier, C., Brown, M.R. *et al.* (1983) Chemical and biological characterization of corticotropin releasing factor. *Recent Prog. Horm. Res.*, **39**: 245–70

Vanhoutte, P.M. (1981) Introductory remarks: alpha- and beta-adrenergic receptors and the cardiovascular system. *J. Cardiovasc. Pharmacol.*, **3** (suppl. 1): S1–S13

Van Giersbergen, P.L.M., Palkovits, M. and de Jong, W. (1992) Involvement of neurotransmitters in the nucleus tractus solitarii in cardiovascular regulation. *Physiol. Rev.*, **72**: 789–824

Varndell, I.M., Tapia, F.J., De Mey, J. *et al.* (1982) Electron immunocytochemical localization of enkephalin-like material in catecholamine-containing cells of the carotid body, the adrenal medulla, and in pheochromocytomas of man and other mammals. *J. Histochem. Cytochem.*, **30**: 682–90

Veening, J.G., Swanson, L.W. and Sawchenko, P.E. (1984) The organization of projection from the central nucleus of the amygdala in brainstem sites involved in central autonomic regulation: a combined retrograde transport-immunohistochemical study. *Brain Res.*, **303**: 337–57

Verrier, R.L. and Carr, D.B. (1991) Stress, opioid peptides, and cardiac arrhythmias, in J.A. McCubbin, P.G. Kaufmann and C.B. Nemeroff (eds.), *Stress, Neuropeptides, and Systemic Disease*, Academic Press, New York, pp 409–27

Verrier, R.L. and Lown, B. (1977) Effect of left stellectomy on enhanced cardiac vulnerability induced by psychological stress. *Circulation*, **56** (Suppl. 3): III–80

Verrier, R.L., Calvert, A. and Lown, B. (1975) Effect of posterior hypothalamic stimulation on ventricular fibrillation threshold. *Am. J. Physiol.*, **228**: 923–7

Versteeg, D.H.G., Petty, M.A., Bohus, B. and De Jong, W. (1993) The central nervous system and hypertension: the role of catecholamines and neuropeptides, in D Ganten (ed.), *Experimental and Genetic Models of Hypertension*, Elsevier, Amsterdam, in press

Versteeg, C.A.M., Bohus, B. and De Jong, W. (1982) Attenuation by arginine- and desglycinamide-lysine-vasopressin of a centrally evoked pressor response. *J. Auton. Nerv. Syst.*, **5**: 253–62

Von Eiff, A.W. and Piekarski, C. (1977) Stress reactions of normotensives and hypertensives and the influence of female sex hormones on blood pressure regulation. *Progr. Brain Res.*, **47**: 289–99

Von Holst, D. (1986) Vegetative and somatic compounds of tree shrews' behavior. *J. Auton. Nerv. Syst. Suppl.*, 657–60

Von Holst, D., Fuchs, E. and Stohr, W. (1983) Physiological changes in male *Tupaia belangeri* under different types of social stress, in T.H. Dembroski, T.H. Schmidt and G. Blümchen (eds.), *Biobehavioural Bases of Coronary Heart Diseases*, Karger, Basel, pp 382–90

Wahlestedt, C., Hakanson, R., Vaz, C.A. and Zukowska-Grojec, Z. (1990) Norepinephrine and neuropeptide Y: vasoconstrictor cooperation in vivo and in vitro. *Am. J. Physiol.*, **258**: R736–42

Weber, H.W. and Van der Walt, J.J. (1973) Cardiomyopathy in crowded rabbits: a preliminary report. *S. Afr. Med. J.*, **47**: 1591–5

Weber, H.W. and Van der Walt, J.J. (1975) Cardiomyopathy in crowded rabbits. *Recent. Adv. Stud. Card. Struct. Metab.*, **6**: 441–7

Weber, H.W., Van der Walt, J.J. and Greef, M.J. (1973) Spontaneous cardiomyopathies in chacma baboons. *Recent Adv. Stud. Card. Struct. Metab.*, **2**: 361–75

Weller, K.L. and Smith, D.A. (1982) Afferent connections to the bed nucleus of the stria terminalis. *Brain Res.*, **232**: 255–70

Williams, R.B. and Eichelman, B.S. (1971) Social setting: influence on the physiological response to electric shock in the rat. *Science* **174**: 613–14

Williams, R.B., Eichelman, B.S. and Ng, L.K.Y. (1979) The effects of peripheral chemosympathectomy and adrenalectomy upon blood pressure responses of the rat to footshock under varying conditions: evidence for behavioral effects on patterning of sympathetic nervous system responses. *Psychophysiol.*, **16**: 89–93

Young, M.A. and Vatner, S.F. (1986) Regulation of large coronary arteries. *Circ. Res.*, **59**: 579–96

Zanchetti, A., Baccelli, E., Mancia, G. and Ellison, G.D. (1972) Emotion and the cardiovascular system in the cat, in *Physiology. Emotion and Psychosomatic Illness: CIBA Found. Symp. Vol. 8*, Ass. Sci. Publ., Amsterdam, pp 201–19

Zukowska-Grojec, Z. and Vaz, C.A. (1988) Role of neuropeptide Y (NPY) in cardiovascular responses to stress. *Synapse*, **2**: 293–8

Zukowska-Grojec, Z., Bayorh, M.A. and Haass, M. (1985) Neuropeptide Y and peptide YY mediate noradrenergic vasoconstriction and modulate sympathetic responses. *Regul. Pept.*, **15**: 99–106

Zukowska-Grojec, Z., Konarska, M. and McCarty, R. (1987a) Differential plasma catecholamine and neuropeptide Y responses to acute stress in rats. *Life Sci.*, **42**: 1615–24

Zukowska-Grojec, Z., Marks, B.R. and Haass, M. (1987b) Neuropeptide Y is a potent vasoconstrictor and a cardiodepressant in the rat. *Am. J. Physiol.*, **253**: H1234–9

Zukowska-Grojec, Z., Shen, G.H., Capraro, P.A. and Vaz, C.A. (1991) Cardiovascular, neuropeptide Y, and adrenergic responses in stress are sexually differentiated. *Physiol. Behav.*, **49**: 771–7

5

Stress and the cardiovascular system: A psychosocial perspective

ANDREW STEPTOE

DEPARTMENT OF PSYCHOLOGY
ST. GEORGE'S HOSPITAL MEDICAL SCHOOL
UNIVERSITY OF LONDON
LONDON, UK

ISBN 0–12–663370–3

5.1 INTRODUCTION

Cardiovascular responses are central to the pattern of physiological adjustment during stressful transactions. Increases in heart rate and stroke volume and modifications in the regional distribution of blood flow lead to raised blood pressure under many conditions of behavioural challenge. Other cardiovascular responses include increased platelet adhesion and aggregation, the mobilization of lipids, and modifications in the electrophysiological stability of the myocardium, leading in vulnerable individuals to cardiac arrhythmia. Cardiovascular stress responses are relevant to a number of anxiety disorders such as panic and phobia (Steptoe and Johnston, 1991). The emphasis in this chapter will be on the relevance of cardiovascular stress responses to the major clinical disorders of the cardiovascular system, in particular coronary (ischaemic) heart disease, and hypertension (high blood pressure).

Coronary heart disease is among the leading causes of premature death in the developed world. The underlying pathology is atherosclerosis of the coronary arteries, leading to restriction of blood flow (ischaemia) to the heart muscle. Coronary atherosclerosis has a prolonged aetiology commencing in childhood for many people, but typically comes to attention only at advanced stages with manifestations such as angina pectoris (ischaemic chest pain), myocardial infarction and sudden cardiac death. The major risk factors for coronary heart disease include age, high blood pressure, elevated serum cholesterol concentration and cigarette smoking. Hypertension itself is common in the adult population. Prevalence levels depend on the criterion for the definition of hypertension, but in the British Regional Heart Study it was found that 26% of middle-aged men had diastolic pressures above 90 mmHg (Shaper et al., 1988). Raised blood pressure is also an important risk factor for cerebral stroke and can lead to other disorders such as heart failure.

The role of psychosocial factors in the development and maintenance of cardiovascular disorders has been the subject of considerable research, and several volumes have reviewed the field (e.g. Steptoe, 1981; Schmidt et al., 1986). Books have also been published on specific aspects such as the role of the work environment (Karasek and Theorell, 1990), Type A behaviour (Dembroski et al., 1978; Houston and Snyder, 1988), anxiety (Byrne and Rosenman, 1990), anger and hostility (Chesney and Rosenman, 1985); the methodology of stress investigations (Schneiderman et al., 1989); and behavioural interventions (Blanchard et al., 1988; Ornish, 1990). To this must be added numerous chapter, journal issues and individual reviews. It would be impossible to compress this literature into a short space. Instead, this chapter will focus on a series of major themes that have emerged in recent work, and will trace these themes through the range of research strategies that are relevant to cardiovascular disorders. The emphasis will be on human research and on psychosocial parameters, since animal research and physiological processes are discussed elsewhere in this volume.

Before addressing these themes, it is necessary to consider two preliminary issues. The first concerns the pathways that may be responsible for mediating the association between psychosocial factors and cardiovascular disorders, and the second concerns the range of research strategies with which the association can be studied.

5.2 MEDIATING PATHWAYS

There are three broad processes through which psychosocial factors may be associated with cardiovascular responses and disorders.

5.2.1 PSYCHOPHYSIOLOGICAL PROCESSES

Psychophysiological processes involve stimulation from the brain leading to modifications in autonomic and neuroendocrine function. Some influences may be chronic, resulting from intermittent or sustained stimulation over a long period of time; much research on hypertension and coronary heart disease points to this possibility. Acute but intense stimulation may also be relevant, particularly in triggering sudden cardiac events. Classic physiological stress responses may be responsible for mediating emotional and behavioural influences on health, and are the prime focus for much research on stress and the cardiovascular system. However, before their involvement in clinical settings can be confirmed, it is necessary to rule out the alternatives described below.

5.2.2 BEHAVIOURAL PROCESSES

The second set of mediating processes involves behaviours such as cigarette smoking, vigorous exercise, dietary habits and alcohol consumption, that are known to influence coronary heart disease and hypertension. These behaviours have complex determinants, but there is evidence that they can be influenced by psychosocial stress and may form part of the psychological coping process (e.g. DeFrank *et al.*, 1987; McCann *et al.*, 1990). Moreover, they can interact with autonomic and neuroendocrine processes, producing synergistic responses. It is therefore possible that some associations between psychosocial factors and cardiovascular disorders can be accounted for by behaviour changes, without the need to invoke psychophysiological stress responses at all.

5.2.3 COGNITIVE PROCESSES RELATED TO SYMPTOM APPRAISAL AND REPORTING

People differ in their perception of symptoms and in their decisions about when to consult medical services. Negative affectivity, a non-specific state of dysphoria and dissatisfaction, is associated with symptoms and increased physician consultation (Watson and Pennebaker, 1989). In response to adverse life experience, people may also complain of more illness and seek medical attention. The importance of these processes for psychosocial studies of cardiovascular function is that many disorders (notably hypertension and angina) are frequently first recognized in response to complaints on the part of patients. Many other people with equally severe cardiovascular problems remain undiagnosed in the population, since they are not predisposed to somatic complaints. This means that clinical investigations of psychosocial factors in cardiovascular disease can be biased towards finding associations with psychological disturbance, because samples may contain an excess of patients with high negative affect.

This argument can be illustrated by research on personality and coronary heart disease. Booth-Kewley and Friedman (1987) have proposed, on the basis of meta-analysis, that coronary heart disease is associated with a variety of negative psychological traits including depression, anxiety and anger/hostility. This conclusion has been discounted by many investigators, notably Stone and Costa (1990), who suggest that these characteristics reflect the more general trait of negative affectivity which is in turn associated with subjective diagnoses such as angina. Several prospective studies have now shown that neuroticism (a marker of negative affectivity) is related prospectively with angina but not with 'hard' cardiac end points such as myocardial infarction and cardiac death (e.g. Almada *et al.*, 1991; Shekelle *et al.*, 1991). Consequently, many associations between personality and coronary heart disease may be spurious.

Similar arguments can be put forward in relation to high blood pressure. Hypertension is frequently diagnosed when blood pressure is recorded as part of the physician's response to non-specific health complaints (Barlow *et al.*, 1977). Telling a person that they have hypertension can also have adverse psychological consequences; psychological studies of diagnosed hypertensives may therefore be compromised in that they will be biased towards finding higher levels of psychological disturbance (and its antecedents) in patients compared with controls. This process is also relevant to psychophysiological studies of cardiovascular response, as is evident from the work of Rostrup and co-workers (1992; Rostrup *et al.*, 1990). These investigators identified young men with high blood pressure during medical tests for military service, and randomized them to 'informed' and 'uninformed' conditions. One year later, the informed or aware group were invited for a second examination, having been told that their blood pressure was elevated at screening, while the uninformed group was simply asked for a second test. At this second session, resting blood pressure was significantly higher in the informed than uninformed groups, and their plasma adrenaline and heart rate responses to the cold pressor test were larger. The two groups also differed on self-report measures, with the aware subjects being less assertive.

It is important to bear these alternative mediating pathways in mind when evaluating the strength of evidence linking psychosocial factors with cardiovascular responses.

5.3 LEVELS OF STUDY

The data relating psychosocial factors and stress responses with cardiovascular disorders comes from a range of sources. Each level of study has advantages and limitations. It is important to recognize that the convergence of different strands of evidence is necessary for evaluating arguments relating stress with cardiovascular responses.

5.3.1 STUDIES OF AUTONOMIC, NEUROENDOCRINE AND NEUROCHEMICAL PROCESSES

These studies are crucial for identifying mechanisms that may be relevant in human disease. Without defined biological mechanisms, the force of evidence from other levels of investigation is inevitably weakened.

5.3.2 ANIMAL STUDIES

Experimental studies in animals can document the ways in which cardiovascular pathology may arise through exposure to different forms of behavioural environment, and have been used to demonstrate the effects of stress on atherosclerosis, ventricular fibrillation, hypertension and other processes. In interpreting animal studies, it is important to resist the assumption that, just because behavioural factors can lead to cardiovascular pathology under controlled experimental conditions, the association is necessarily important in the clinical context.

5.3.3 STUDIES OF PATHOPHYSIOLOGY

Pathophysiological studies of cardiovascular disease help to identify the metabolic and regulatory processes that may be disturbed by central nervous system stimulation. Biochemical studies of hypertensives, for example, have shown that heightened noradrenaline spillover from the kidneys and heart is characteristic of the early stages of the disorder (Esler *et al.*, 1991). Enhanced sympathetic activity has also been documented in borderline hypertensives using microneurographic recordings from sympathetic fibres (Anderson *et al.*, 1989). The relevance of such findings is that one of the most plausible explanations for disturbed sympathetic activity is enhanced activation of central stress pathways.

5.3.4 MENTAL STRESS TESTING

Acute reactions to stress in humans are studied extensively in the cardiovascular context. The pattern of response has been described in detail by Hjemdahl (1990) and Herd (1991). Under conditions which engage subjects in active efforts to cope (as when presented with difficult problem-solving tasks), there is typically sympathetically-induced tachycardia accompanied by increased stroke volume, vasoconstriction in the renal and splanchnic tissues and vasodilatation in skeletal muscle and adipose tissue. In most people, the result is a decrease in systemic vascular resistance, but this is more than offset by the increase in cardiac output, so arterial blood pressure rises. Mental stress testing is used not only to investigate haemodynamic changes, but also ischaemic responses and changes in platelet function (Rozanski *et al.*, 1988; Larsson *et al.*, 1989).

   Mental stress testing in the laboratory has been a subject of extensive debate, and has been vigorously criticized and defended (Parati *et al.*, 1991; Steptoe and Vögele,

1991). The advantage of mental stress testing is that cardiovascular responses can be monitored under carefully controlled conditions, so that hypotheses may be tested using genuine experimental designs. The disadvantages are that responses are short-lived and that they may not always be representative of function outside the laboratory.

5.3.5 FIELD STUDIES

A relatively new domain of cardiovascular stress research has arisen through the development of ambulatory or portable measurement equipment. This has enabled heart rate, blood pressure and even parameters of left ventricular function and catecholamine levels to be monitored while people go about their everyday lives. When measures of the psychosocial environment are also taken, these techniques provide valuable additional information about natural covariations between demands, coping and cardiovascular activity (e.g. Johnston et al., 1990; Theorell et al., 1991a). Other types of field study used less elaborate measures, but have assessed the chronic effects of exposure to environmental stressors such as crowding (e.g. Ostfeld et al., 1987).

5.3.6 CLINICAL AND EPIDEMIOLOGICAL STUDIES

The final level of study involves investigations of psychosocial factors related to the actual occurrence of cardiovascular pathology. This approach itself includes a wide variety of research strategies, from studies of personality, life events and other personal experiences to population-based surveys related to social status, employment and social support. It can be argued that without evidence at this level, no amount of experimental or pathophysiological data can prove the role of psychosocial factors in cardiovascular disorders. On the other hand, such studies are rarely conclusive because of the impossibility of random assignment of people to different experiences, and the difficulty of measuring and controlling for every potential confounding factor.

It is, of course, necessary to take account of evidence from all levels of study. Moreover, the relevance of stress can seldom be examined in isolation. More frequently, stress processes interact with other factors and predispositions in rendering the person more or less vulnerable to cardiovascular disorder.

5.4 SOCIAL ISOLATION AND SOCIAL INTEGRATION

The influence of social isolation and integration on cardiovascular responses provides the first important research theme for this chapter. Over the past 20 years, a number of prospective epidemiological studies have been reported, showing a negative association between social integration and mortality (see House et al., 1988). For example, several thousand adults aged 35–69 at initial examination were followed up for 10–12 years in Tecumseh, Michigan (House et al., 1982). 13% of men and 6.1% of women died over this period, coronary heart disease figuring strongly as a cause. Composite measures of social relationships and social activities were inversely related to mortality after controlling for age, smoking and various biomedical risk factors. The role of social isolation was also clear in a three-year follow-up of patients with acute myocardial infarction (Ruberman *et al.*, 1984). After controlling for other prognostic factors, men who were socially isolated (seeing friends or relations infrequently and not belonging to any social organizations), and who also had experienced relatively high levels of recent life stress (family separations, financial difficulties, etc.), had more than four times the risk of death compared with those who were low on these factors.

Social isolation has an influence on cardiovascular parameters such as blood pressure. A study of elderly Swedish men found that the extent to which subjects were socially anchored in formal and informal social groups was negatively correlated with blood pressure level after controlling for social class, marital status, alcohol, smoking, body weight and physical activity (Hanson *et al.*, 1988). A follow-up over 20 years of a secluded order of Italian nuns showed that blood pressure rose with age to a lesser extent in this stable, socially integrated religious group compared with population controls (Timio *et al.*, 1988). Several investigators have also examined blood pressure changes in migrants moving from isolated stable cultures to developed countries in which social ties and structures are less defined (e.g. Salmond *et al.*, 1989). Although increases in body weight and salt intake are in part responsible for blood pressure rises, psychophysiological stress processes may be involved as well (Poulter *et al.*, 1990).

An important issue in research on social relationships is whether social integration has direct effects on health, or whether social support acts as a buffer of other adverse experiences. Indeed, the notion of social support is used rather loosely in the literature, in that at least four general conceptualizations can be found. Social support is viewed in terms of *social integration, satisfying relationships* (intimacy and trust), *perceived support* or the appraisal that others can be relied upon in trying times, and in terms of actual or *received support* (Buunk and Hoorens, 1992). The measures used in cardiovascular epidemiology tend to have been global indices of networks and social integration which provide little information about the mechanisms through which vulnerability is increased. The fact that social isolation influences cardiovascular health in the absence of any known stressors suggests direct effects, but interactions with chronic life stress are also observed. For example, Dressler (1991) carried out a community study among black men in the Southern USA. A sub-group was characterized by 'lifestyle incongruity'. These were people whose lifestyle in terms of material goods and status displays exceeded their occupational position. It was argued that efforts to maintain such an ostentatious lifestyle in the absence of sufficient funds constituted a source of chronic life stress. Dressler found that blood pressures were indeed higher in the lifestyle incongruity subjects, but only when they enjoyed low levels of social support.

The inference behind these observations is that social isolation acts as a chronic stimulus to cardiovascular and neuroendocrine activation, thereby increasing risk of cardiovascular pathology in susceptible individuals. This phenomenon has not been examined extensively in experimental studies, although a buffering effect of social support on cardiovascular reactivity has been observed by Kamarck *et al.* (1990). They recorded blood pressure and heart rate responses from female students while they performed demanding mental tasks either alone or in the presence of a female friend. No differences in baseline levels were observed, so social support did not have an observable effect on cardiovascular function at rest. However, blood pressure and heart rate responses to tasks were smaller in the group who had a friend present, indicating that in the short-term at least, cardiovascular stress responses are modified by social support. More recent studies suggest that the impact of social contact is complicated by whether or not the friend is perceived as evaluative or supportive (Edens et al., 1992). As with so much research on social integration and support, both positive and negative effects of contact on cardiovascular reactivity can be found.

5.5 PERSONAL CONTROL AND THE PSYCHOSOCIAL ENVIRONMENT

A second major research theme to emerge over recent years has been the influence on stress responses and health of personal control over the psychosocial environment. The importance of the theme is that it helps to integrate findings from diverse research strategies (Steptoe and Appels, 1989). The basic result to emerge is that personal control over aversive events and experiences is positively related to reduced stress responsivity and favourable health outcomes. One important rider to this conclusion is that effects depend on the effort required to exert control, and that control which can only be maintained with high effort may itself generate heightened stress responses.

In the cardiovascular field, the role of personal control has been extensively studied in relation to the work environment. There is a well established literature on job stress which suggests that high job demands may have adverse consequences. Links with cardiovascular health have, however, been inconsistent. Karasek (1979) proposed that the influence of high job demand is moderated by personal control, so that even very pressurized occupations may be tolerated successfully if the person has some control over how the job is done and how skills can be developed (job decision latitude). High job strain will emerge when high job demands (perceived or actual) are accompanied by low job control.

A series of panel studies have been carried out in the USA and Sweden in which cardiovascular mortality and morbidity in occupations classified independently as high strain jobs were examined. Although not entirely consistent, these studies tend to show high rates of cardiovascular disease in high strain occupations (Karasek and Theorell, 1990). Cross-sectional studies corroborate the cardiovascular risk associated with job strain. For example, Schnall et al. (1990) found in a case control study that hypertensives report lower job decision latitude than normotensives. After controlling for other factors, high job strain was positively related to left ventricular mass (one of the forms of organ damage resulting from hypertension). Ambulatory blood pressure at work has been correlated with high job strain, particularly when the physical demands of the job are low (Theorell et al., 1991a; Light et al., 1992). Prospectively, high job strain has been associated with risk of reinfarction in young men returning to work following a myocardial infarction (Theorell et al., 1991b). In a study of variations in job strain over time, Theorell et al. (1988) found that self-monitored systolic pressure at work tended to be higher during periods of high demand/low control than at other times. In an important extension of this demand/control model, Johnson and Hall (1988) argued that social support makes a significant contribution. They provide evidence that people in high strain occupations may be protected from increased cardiovascular disease risk if they also enjoy high levels of co-worker support, but may be especially vulnerable if they are socially isolated. Marmot and Theorell (1988) have suggested that social support and psychosocial work conditions may together be responsible in part for social class variations in cardiovascular disease incidence. Studies of cardiovascular responses to mental stress in the laboratory add more pieces to the jigsaw (Steptoe et al., 1993). When subjects perform tasks during which they can either take breaks when they want (control) or else have the same number of breaks arbitrarily imposed (no control), blood pressure elevations are smaller in the control condition (Hokanson et al., 1971). Bohlin et al. (1986) conducted an experiment in which the cardiovascular responses of borderline hypertensives and normotensives were assessed during self-paced and externally-paced tasks. Normotensives showed the expected reduction of blood pressure responses in the self-paced (personal control) condition, but this was not true of hypertensives. It would appear that the hypertensives were not able to take advantage of personal control in regulating their cardiovascular stress responsivity.

Laboratory studies have also helped to define the limits of the beneficial effect of personal control. Obrist (1981) was perhaps the first systematically to investigate the parameters that might augment cardiovascular reactivity in potentially controllable situations. Subjects performed reaction time tasks in which the avoidance of aversive stimulation (electric shock) was contingent on fulfilling response speed criteria. Obrist observed that blood pressure and cardiovascular reactions tended to be larger or more sustained when subjects were given a difficult but potentially achievable criterion, than when the criterion was easy or virtually impossible. He surmised that effortful or active behavioural coping attempts were elicited under these circumstances, and that these were accompanied by sympathetically-mediated cardiovascular responses.

Evidence of this type suggests that, as far as cardiovascular responses are concerned, it is not only the controllability of events that is important, but also the effort required to maintain control. Individual differences in desire to control the personal environment and in efficacy expectancies are also crucial. Some people may strive to exert influence over situations in their lives that are outside their control, and may suffer as a result.

5.6 HYPERTENSION, CHRONIC STRESS AND EMOTIONAL COPING

The evidence presented thus far suggests that various aspects of the psychosocial environment, particularly levels of social isolation and the degree to which demands are controllable, may elicit heightened cardiovascular responses and increased risk of disease. However, only a minority of people exposed to these stressors actually succumb to cardiovascular disorders. The presence of risk cofactors is certainly important in accounting for individual differences in vulnerability. In addition, however, it is probable that psychological characteristics are significant, such that heightened cardiovascular stress responses are only observed in people with appropriate psychological predispositions (Johnson et al., 1992).

Two psychological characteristics have been the focus of much recent research.

One is hostility, which is linked to Type A behaviour and is discussed in the next section. The other is the suppression or inhibition of emotion when attempting to cope with threatening or aversive conditions.

Emotional suppression has been associated most consistently with risk of hypertension. A number of cross-sectional studies have shown that blood pressure levels are higher in people who inhibit anger and emotional expression (Harburg *et al.*, 1979; Gentry *et al.*, 1982). In a 12-year prospective survey, Julius *et al.* (1986) found that people who stated that they would suppress anger when being unjustly berated by the police or their spouse were at higher risk of premature death, after controlling for age, sex, weight and cardiovascular risk factors. The subjects at highest risk of all were those with elevated systolic blood pressure who also supressed their anger. Chronic occupational stress has been found in factory workers to predict the incidence of hypertension, but only among those who also suppress anger (Cottington *et al.*, 1986). A longitudinal study of young normotensives with a family history of hypertension followed up over 30 months showed that suppressed anger predicted the development of future borderline hypertension independently of other characteristics (Perini *et al.*, 1991).

Recognition of the importance of different forms of emotional coping may help reconcile the inconsistent literature relating cardiovascular stress responsivity with hypertension. There is a substantial body of research in which blood pressure, heart rate and other responses have been recorded during mental stress testing in normotensives and hypertensives. Most, although not all, such studies, tend to show heightened reactivity in hypertensives (Fredrikson and Matthews, 1990). Unfortunately, these studies only provide limited support for the causal argument that heightened stress responsivity leads to sustained hypertension (Steptoe, 1991). The reason is that elevated responsivity may be secondary to the presence of hypertension, and the alterations in central regulation and peripheral physiology that are characteristic of the disorder – it may, in other words, be an effect rather than a cause. Attempts have been made to circumvent this problem by studying normotensive people who are nevertheless at high risk for developing hypertension. It is reasoned that heightened reactivity in high risk individuals would indicate that stress responses are disturbed even before the establishment of fixed hypertension.

The most frequently employed index of risk has been family history, since the offspring of hypertensives are more likely than others to develop the condition themselves. Fredrikson and Matthews (1990) reported a meta-analysis (in which results from a large number of studies are aggregated) covering more than 25 comparisons of young people with and without a family history of hypertension, and concluded that there was greater blood pressure and heart rate reactivity in the positive history subjects, particularly when they were confronted with actively challenging tasks. However, it must be acknowledged that many of these studies relied on the young peoples' report of their parents' health. This might be inaccurate firstly through ignorance, and secondly because much hypertension is undiagnosed in the population. Some studies in which parental health information was systematically obtained have failed to show differences in cardiovascular stress reactivity. In particular, Ravogli *et al.* (1990) compared young subjects with two hypertensive parents, one hypertensive parent or no hypertensive parents, confirming parental status by blood pressure measurement or medication records.

Ambulatory blood pressure was higher in the positive history group, and left ventricular wall mass was greater. But the groups did not differ in their blood pressure or heart rate responses to laboratory stressors such as mental arithmetic, mirror drawing and the cold pressor test.

Recent work suggests that measurement of patterns of emotional coping may resolve this difficulty. Jorgensen and Houston (1986) found that subjects with a positive family history, who also showed denial or inhibition of hostile feelings, were the most responsive. In a recent study of low and high risk young men, we observed heightened blood pressure and heart rate reactions to mental arithmetic and mirror drawing tasks in high risk subjects, but only if they showed a pattern of neurotic inhibition of emotion (Vögele and Steptoe, 1992). Figure 5.1 illustrates results from another study conducted (Vögele and Steptoe, 1993) in which 14–16 year-old boys performed mental arithmetic and mirror drawing tasks in the laboratory. Blood pressures were measured from the parents of each youth. Twenty boys were classified as relatively high risk in that their father or mother had documented hypertension or a resting blood pressure greater than 140/85 mmHg, while the remaining 40 subjects were low risk. Participants also completed the Spielberger Anger Expression Inventory, and were classified according to whether they reported inhibition of anger. High anger inhibitors responded positively to questions such as 'When angry or furious I keep things in', or 'I tend to harbour grudges that I don't tell anyone about'. It can be seen in Figure 5.1 that there was little difference between the groups at rest. But the interaction between family risk and anger inhibition influenced blood pressure responses to mental stress tests. Systolic and diastolic blood pressure responses to mental arithmetic and mirror drawing were largest in subjects at high family risk who also showed high anger inhibition.

Investigations of this kind indicate that, when considering cardiovascular reactions relevant to hypertension, three factors need to be taken into account: the biological risk status of the person, situational factors such as the presence of demanding tasks, and patterns of inhibition of negative emotions. Ultimately, of course, the truth of these hypotheses must be tested in prospective longitudinal studies in which the psychological and cardiovascular reactions of young people are assessed. Follow-ups would determine whether those with large reactions who have appropriate psychological and biological characteristics and who also experience high levels of life stress are the ones who develop high blood pressure. A limited amount of longitudinal data suggests that people with larger than average cardiovascular reactions to mental stress tests are more likely to develop hypertension, particularly in the presence of other risk factors such as high body weight and disturbances of sodium metabolism (Steptoe, 1990). But no prospective studies have yet been conducted that include all three elements (biological, psychological and situational) specified here.

5.7 HOSTILITY AND CORONARY HEART DISEASE

Ten years ago, the status of Type A coronary-prone behaviour appeared secure (Dembroski et al., 1978). The complex pattern of competitiveness, hostility, time

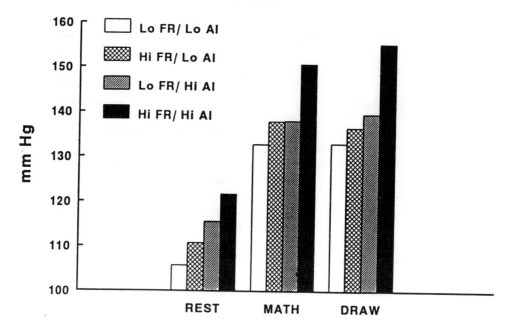

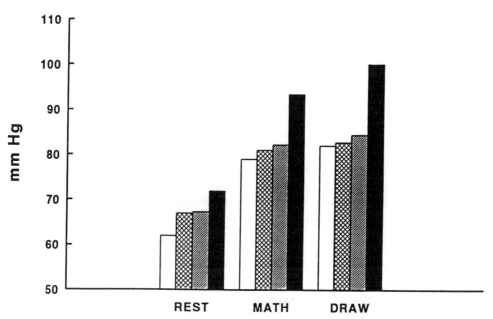

Figure 5.1 Mean levels of systolic blood pressure (top) and diastolic blood pressure (bottom) recorded during a 10 min initial rest period (REST), 5 min of non-verbal mental arithmetic (MATH), and 5 min of mirror drawing (DRAW) in 60 boys aged 14–16 yrs. Subjects were divided on the basis of parental blood pressure into high and low family risk of hypertension (Hi/Lo FR), and on the basis of anger expression measures into high and low anger inhibitors (Hi/Lo AI) (From Vögele and Steptoe, 1993).

urgency and job involvement had been documented as an independent risk factor for coronary heart disease in prospective studies, notably in the eight-year follow-ups of the Western Collaborative Group Study and the Framingham study (Matthews and Haynes, 1986). The National Heart, Lung and Blood Institute in the USA gave its seal of approval in 1981, when it acknowledged Type A as a major independent risk factor, standing alongside hypertension, serum cholesterol and smoking.

The situation has altered markedly in the last decade, although ironically it is psychologists who seem least aware of this, and who continue blithely to study Type A as if nothing had changed. The reason is that a number of prospective studies have been published in which Type A behaviour has failed to predict future coronary heart disease (Appels et al., 1987; Johnston et al., 1987; Mann and Brennan, 1987). Initially, this was ascribed to the use of questionnaires rather than the more robust Structured Interview method for assessing Type A behaviour, but this could not account for the failure to find an effect in the Multiple Risk Factor Intervention Trial (Shekelle et al., 1985). More disturbing still, extended follow-ups of the Western Collaborative Group and the Framingham studies cast doubt on the long-term association of Type A behaviour with myocardial infarction or coronary heart disease fatality (Ragland and Brand, 1988; Eaker et al., 1989).

Several explanations have been put forward for this change in pattern. The first is that there may be a reporting bias, so that Type A individuals might complain of cardiac symptoms earlier than others; in any prospective period, this would lead to an apparent difference in disease incidence. In support of this notion, scores on the Jenkins Activity Survey (a questionnaire measure of Type A behaviour) and the Framingham Type A scale are associated prospectively with angina incidence (Appels et al., 1987; Eaker et al., 1989). On the other hand, Chesney et al. (1981) discovered that Type A behaviour measured with the Structured Interview was not correlated with negative affect.

A second possibility is that Type A might be linked with risk behaviours that could themselves influence heart disease incidence. An association with alcohol consumption was shown in the Multiple Risk Factor Intervention Trial (Folsom et al., 1985), although it should be noted that earlier prospective analyses had shown Type A to be independent of conventional risk factors. A third suggestion is that the behaviour pattern has changed over time, with shifts either in the assessment system or in the socio-cultural context. Again, this does not appear to be a sufficient explanation, since a 27-year reassessment of participants in the Western Collaborative Group Study showed considerable stability in Type A assessments (Carmelli et al., 1991). Some investigators have argued that Type A behaviour may be a more reliable predictor when studied interactively with other psychosocial factors. Orth-Gomér and Undén (1990) followed up a cohort of Swedish men over ten years and found an interaction between Type A behaviour and social isolation. Mortality rates were significantly higher among socially isolated Type A men than all other groups. It was suggested that the less pleasant components of Type A behaviour may presage loosening of social networks and loss of social support.

The possibility that has attracted the greatest attention is that the original Type A concept was too broad, and that only certain components are responsible for the association with coronary heart disease. Several workers have argued that the hostility component is most crucial. Dembroski *et al.* (1989) showed that 'potential for hostility' was a predictor of myocardial infarction and sudden cardiac death in the Multiple Risk Factor Intervention Trial, and other prospective studies have found associations between components of hostility and heart disease incidence (e.g. Almada *et al.*, 1991; Hecker *et al.*, 1988). It must be acknowledged that negative prospective studies have been reported for hostility, so the matter remains inconclusive (Leon *et al.*, 1988; Hearn *et al.*, 1989).

Psychophysiological studies have also been published in which subjects scoring high in cynical or antagonistic hostility showed greater cardiovascular reactions to competitive or harassing situations than did low hostile people (Smith and Pope, 1991). This link between hostility and cardiovascular reactivity is rather different from that described in the last section, where suppression of hostile expression was implicated in hypertension. Engebretson, Matthews and Scheier (1989) suggest that the pattern depends on whether the subjects are able to use preferred modes of expression in the experimental context. Alternatively, it could be argued that inappropriate management of emotions is crucial, and that both restricted and excessive expression of hostility and other negative emotions will lead to disturbances in cardiovascular responses during stressful interactions. The possibility that sympathetically-mediated reactions are important in the development of atherosclerosis, independently of hypertension, has gained currency through animal and pharmacological studies (Åblad *et al.*, 1988).

5.8 STRESS, ISCHAEMIA AND ARRHYTHMIA

An area of cardiovascular research that has recently come to prominence concerns psychosocial influences on myocardial ischaemia and cardiac arrhythmia. This work has been stimulated by improved imagery techniques that provide more precise measures of changes in cardiac dimension and function, by the recognition of 'silent' and transient coronary ischaemia (reductions in coronary blood flow without symptoms), and by the increased attention paid clinically to arrhythmias and conduction disorders. The role of psychological factors has been thoroughly reviewed by Kamarck and Jennings (1991), so the literature described here is representative rather than comprehensive.

Depression of the ST segment of the electrocardiogram (ECG) complex is a well-established clinical marker of myocardial ischaemia. It is however a relatively insensitive index. Measures of regional perfusion abnormalities using positron emission tomography and cardiac wall motion abnormalities with radionuclide ventriculography now suggest that ischaemic responses to mental stress are more common than was assumed on the basis of ECG assessments (Deanfield *et al.*, 1984). For example, Rozanski *et al.* (1988) assessed 39 patients with coronary artery disease and 12 controls using radionuclide ventriculography during mental arithmetic, the Stroop colour-word interference task, a public speaking task and neutral reading. Of the patients, 29 showed cardiac wall motion abnormalities during exercise and, of these, 21 showed abnormalities during mental stress. Ischaemic responses were greatest in the speech task, even though this was not rated as any more stressful than the other conditions. The level of oxygen demand at which ischaemic responses occurred during mental stress was lower than that necessary to produce ischaemia during exercise. This indicates that mental stress testing was not simply mimicking exercise in placing higher work demands on the heart muscle.

In another study, both ischaemic and arrhythmic responses during mental stress were assessed in 165 post-infarction patients (Zotti et al., 1991). Mental arithmetic and a reaction time task were administered. The group was divided into high, medium and low cardiovascular responders on the basis of the magnitude of double product (heart rate × systolic blood pressure) reactions during mental stress. These three groups were indistinguishable in terms of age, infarction size and type, exercise tolerance, incidence of exercise-induced ischaemia or arrhythmia, and cardiac function. Nonetheless, as can be seen in Table 5.1, the occurrence of ECG-assessed ischaemia or arrhythmia was significantly more frequent in the high cardiovascular responder group than the medium and low responder groups. This suggests that individual differences in autonomic reactions during mental stress produced abnormalities of cardiac function. It can also be seen in Table 5.1 that there was a tendency for the higher responder group to have lower scores on measures of psychological distress than the low responders. This may reflect attempts to cope by denying or refusing to acknowledge threat and psychological disturbance.

Table 5.1 Cardiovascular stress responses in post-infarction patients. Characteristics of high, medium and low cardiovascular responders (from Zotti et al., 1991)

	High N = 52	Medium N = 56	Low N = 57	p
Age	53.5 ± 8.6	53.3 ± 10.2	51.4 ± 9.4	ns
Ischaemia/Arrhythmia during mental stress testing	26 (50%)	11 (19.6%)	4 (7.0%)	0.0001
Trait anxiety	36.4 ± 8.2	39.7 ± 10.1	40.6 ± 11.4	0.02
Psychophysiological symptoms	38.4 ± 5.6	39.7 ± 6.6	42.7 ± 11	0.02
Depression	3.5 ± 3.1	4.0 ± 3.6	4.9 ± 4.1	0.05

It is notable that ischaemic responses during mental stress testing are very rare in healthy subjects. They are characteristic of people who already suffer from coronary heart disease, and who already have narrowing of the coronary vessels. In this respect, the influence of cardiovascular stress responses on ischaemia and arrhythmia may be significant only in the short-term, triggering acute cardiac events in predisposed individuals. The clinical conditions to which this pattern is relevant are sudden cardiac death and ventricular arrhythmia. A limited amount of data from clinical studies has implicated acute psychological stress in these conditions, although much of the work on sudden cardiac death is rather weak methodologically. One well designed case study evaluated life events in the six months prior to sudden cardiac death in 81 women compared with neighbourhood controls (Cottington et al., 1980). Information about the cardiac death victims was provided by next of kin. A substantially elevated rate of major losses (such as death of spouse or family member) was reported for the cardiac death group compared with controls. In the case of ventricular arrhythmia, Reich et al. (1981) found that unusual and intense emotional states had been experienced by a substantial proportion of patients in the 24 hours previous to the episode. One prospective study has also provided telling data. Follick *et al.* (1988) assessed post-infarction patients who had been trained to transmit ECG readings by telephone to the hospital on a regular basis. Psychological distress assessed by self-report predicted the presence of cardiac arrhythmia over the one year study, even after controlling for clinical variables and medication.

The prospective significance of ischaemic and arrhythmic responses to mental stress has not yet been established. However, a small but provocative finding was recently reported by Manuck *et al.* (1992). They subjected 13 post-infarction patients to mental stress testing, recording cardiovascular and catecholamine responses to the Stroop colour-word interference test. Over a follow-up period of three to four and a half years, five patients experienced a new cardiac event. It was found that these people had displayed significantly higher systolic and diastolic pressure responses to mental stress than the remaining patients, and their plasma adrenaline levels were also elevated to a greater extent. They did not, however, differ in cardiovascular characteristics at rest or in response to exercise, or in factors such as age and in lipid levels. Although the sample size is rather small, this study does suggest that the magnitude of cardiovascular stress responses may be prognostic in patients with pre-existing heart disease.

5.9 PREVENTION AND MANAGEMENT OF CARDIOVASCULAR DISORDERS

It is widely recognized that efforts to reduce the high levels of cardiovascular disease prevalent in the Western world cannot be based simply on medical treatment, but that behaviour change is essential. Modifications in diet, smoking, physical activity, alcohol consumption and health care behaviours are central to the more successful programmes designed to reduce levels of cardiovascular mortality and morbidity on a population level (Winett *et al.*, 1989). Similarly, cognitive-behavioural techniques are valuable in the management of people with established heart disease, and can have beneficial effects on tertiary cardiovascular risk reduction. These interventions are not targeted on stress-related processes, but on other behavioural aspects of cardiovascular disorders. However, there has also been considerable interest in using stress management techniques such as relaxation and anxiety management in reversing the stress-related cardiovascular pathology discussed earlier in this chapter. The applications of stress management fall into three main categories.

5.9.1 HYPERTENSION

Relaxation, biofeedback, meditation and related procedures have been applied extensively to the management of hypertension. The literature tends to show few differences in the efficacy of these methods, so clinical applications typically involve relaxation training accompanied in some cases by biofeedback. A representative stress management programme has been detailed by Johnston and Steptoe (1989). It begins with an assessment and orientation phase in which the suitability of the client and his or her blood pressure level are determined, and education about the influence of psychological factors and the rationale for therapy are outlined. The training phase consists of about eight weekly sessions involving passive muscular relaxation and simple meditation, supplemented by home practice. The third phase (generalization) involves the transfer of relaxation skills into everyday life.

Training programmes of this type have been shown to be superior to attention-placebo control therapies, to reduce blood pressure outside as well as in the clinic, to lead to reduced medication consumption, and to be effective in general practice settings (Johnston and Steptoe, 1989). Follow-up studies of more than four years have been reported, with sustained differences between treatment and control groups (Patel et al., 1985). The persistence of blood pressure reductions has been associated with relaxation practice, and with the degree with which people attempt to integrate their coping skills into everyday life (Steptoe et al., 1987).

Despite this favourable profile, doubts concerning the efficacy of stress management in hypertension persist. The very sensitivity of cardiovascular activity to stressors makes it difficult to establish reliable baseline levels against which to assess treatment responses. Some well controlled trials have failed to demonstrate treatment effects, and there is an indication that methodologically superior trials produce less impressive responses (Irvine and Logan, 1991; Jacob et al., 1991). The conclusions to be drawn from this literature must therefore remain somewhat tentative.

5.9.2 CORONARY ARTERY DISEASE

An application of stress management that is less well explored concerns patients with established coronary heart disease who have not necessarily suffered a major event such as a heart attack. The most interesting results have been published by Ornish et al. (1990), who assessed the effect of a comprehensive lifestyle change programme on coronary atherosclerosis. The experimental group underwent a seven-day induction period followed by regular meetings over a year, and stress managment was accompanied by moderate exercise and a low fat vegetarian diet. Quantitative coronary angiography demonstrated a significant regression of coronary lesions in the experimental compared with control groups, together with decreases in chest pain, total cholesterol and body weight. Unfortunately, it is not clear which components of the complex intervention were responsible for the favourable response. A low fat diet has on its own been shown to produce coronary lesion regression (Watts et al., 1992), while regular exercise can itself lead to lower total cholesterol levels (Schlur et al., 1988). Nevertheless, these results are impressive in showing that behavioural interventions can not only influence parameters such as blood pressure, but also the underlying disease process.

5.9.3 POST-INFARCTION PATIENTS

The third area of stress management application is with patients following myocardial infarction. Acute myocardial infarction is a traumatic event which can be associated with severe emotional reactions and distress. Long-term adaptation to infarction is dependent less upon the clinical severity of the event than on early emotional responses (Philip, 1988).

Applications of stress management in this population has been focused on two distinct outcomes: effects on psychological state, and effects on subsequent cardiovascular morbidity. Interestingly however, the two may be related. For example, Van Dixhoorn et al. (1987) randomly assigned post-infarction patients to exercise training, or exercise training plus six relaxation sessions. Patients in the combined condition showed significantly greater improvements in subjective well-being and anxiety. At 2–3 year follow-up, new cardiac events (unstable angina requiring re-hospitalization, coronary bypass surgery, fresh infarction or death) were significantly more frequent in the exercise only (37%) compared with the exercise plus relaxation (17%) condition.

Another interesting programme was described by Frasure-Smith and Prince (1989). They conducted a longitudinal trial in which post-infarction patients were randomized to a 'life stress monitoring programme' or usual care. Subjects in the treatment programme were telephoned once a month and asked to complete the General Health Questionnaire (a standard measure of psychological disturbance). If the monthly call indicated high levels of distress, the patient received a visit from a research nurse, who helped the patient with counselling, teaching and support. On some occasions, referral to expert help was required. Over a seven year follow-up period, there was a significant difference in myocardial infarction recurrence, and this persisted after controlling for the initial characteristics of the two groups. Although this study has been questioned on methodological grounds, it is important in suggesting that relatively economical crisis orientated stress interventions may be valuable.

Certainly, the most spectacular results in terms of mortality have been reported from the Recurrent Coronary Prevention Program, a large trial of Type A behaviour modification for post-infarction patients (Friedman et al., 1986). This study involved 862 patients aged less than 65 who were randomized to group cardiac counselling or group cardiac counselling plus Type A behaviour counselling. The Type A intervention deployed a wide range of cognitive behavioural methods, including relaxation training, cognitive restructuring and belief, attitude and behaviour change. After four and a half years, Type A behaviour itself was reduced to a greater extent in the experimental group. In addition, cardiac recurrences occurred in only 12.9% of the experimental group compared with 21.2% of controls. As with other studies described in this section, it is difficult to pin down the precise role of cardiovascular stress response modifications in this treatment programme. Overall, however, the results of intervention studies suggest that psychosocial processes not only increase risk of cardiovascular disorder, but can be harnessed to good effect in prevention and management.

5.10 CONCLUDING REMARKS

Space has prevented the discussion of all aspects of stress research related to cardiovascular function. Important topics that have been omitted include the effects of stress on platelets and renal function, the influence of age and physical fitness, the psychological experiences surrounding an infarction, and the role of heart rate lability in panic (Appels, 1991; Steptoe and Johnston, 1991). My aim has been to demonstrate how salient themes such as social integration, personal control and emotional coping are addressed through a wide range of research strategies. This integrated approach helps to delineate more precisely the factors relating cardiovascular stress responses with disease risk. It can be seen that vigorous efforts have already been made to apply psychological methods in prevention and management. It is probable that hypotheses stimulated by the research themes discussed here will lead to further innovations in patient care, and to alterations of the psychosocial environment that may also help reduce the magnitude of cardiovascular stress responses.

5.11 REFERENCES

Åblad, B., Björkman, J-A., Gustafsson, D. *et al.* (1988) The role of sympathetic activity in atherogenesis: effects of β-blockade. *Am. Heart J.*, **116:** 322–7

Almada, S.J., Zonderman, A.B., Shekelle, R.B. *et al.* (1991) Neuroticism and cynicism and risk of death in middle-aged men in the Western Electric Study. *Psychosom. Med.*, **53:** 165–75

Anderson, E.A., Sinkey, C.A., Lawton, W.J. and Mark, A.L. (1989) Elevated sympathetic nervous system activity in borderline hypertensive humans: evidence from direct intraneural recordings. *Hypertension*, **13:** 177–83

Appels, A. (ed.) (1991) *Behavioral Observations in Cardiovascular Research.* Lisse, Amsterdam

Appels, A., Mulder, P., Van't Hof, M. *et al.* (1987) A prospective study of the Jenkins Activity Survey as a risk indicator for coronary heart disease in the Netherlands. *J. Chron. Dis.*, **40:** 959–65

Barlow, D.H., Beevers, D.G., Hawthorne, V.M. *et al.* (1977) Blood pressure measurement at screening and in general practice. *Br. Heart J.*, **39:** 7–12

Blanchard, E.B., Martin, J.E. and Dubbert, P. (1988) *Non-Drug Treatments for Essential Hypertension*, Pergamon, Oxford

Bohlin, G., Eliasson, K., Hjemdahl, P. *et al.* (1986) Personal control over work pace: circulatory, neuroendocrine and subjective responses in borderline hypertension. *J. Hypertension*, **4:** 295–305

Booth-Kewley, S. and Friedman, H.S. (1987) Psychological predictors of heart disease: a quantitative review. *Psychol. Bull.*, **101:** 343–62

Buunk, B.P. and Hoorens, V. (1992) Social support and stress: the role of social comparison and social exchange processes. *Br. J. Clin. Psychol.* **31:** 445–57

Byrne, D.G. and Rosenman, R.H. (eds.) (1990) *Anxiety and the Heart*, Hemisphere, Washington, DC

Carmelli, D., Dame, A., Swan, G. and Rosenman, R.H. (1991) Long-term changes in Type A behavior: a 27-year follow-up of the Western Collaborative Group Study. *J. Behav. Med.*, **14:** 593–606

Chesney, M.A. and Rosenman, R.H. (eds.) (1985) *Anger and Hostility in Cardiovascular and Behavioral Disorders*, Hemisphere, Washington, DC

Chesney, M.A., Black, G.W., Chadwick, J.H. and Rosenman, R.H. (1981) Psychological correlates of the Type A behavior pattern. *J. Behav. Med.*, **4:** 217–29

Cottington, E.M , Matthews, K.A., Talbott, E. and Kuller, L.H. (1980) Environmental events preceding sudden death in women. *Psychosom. Med.*, **42:** 567–74

Cottington, E.M., Matthews, K.A., Talbott, E. and Kuller, L.H. (1986) Occupational stress, suppressed anger, and hypertension. *Psychosom. Med.*, **48:** 249–60

Deanfield, J.E., Shea, M., Kensett, M. *et al.* (1984) Silent myocardial infarction due to mental stress. *Lancet*, **II:** 1001–4

DeFrank, R.S., Jenkins, C.D. and Rose, R.M. (1987) A longitudinal investigation of the relationships among alcohol consumption, psychosocial factors, and blood pressure. *Psychosom. Med.*, **49**: 236–49

Dembroski, T.M., Weiss, S., Shields, J.L. *et al.* (eds.) (1978) *Coronary-Prone Behavior*. Springer-Verlag, New York

Dembroski, T.M., MacDougall, J.M., Costa, P.T. and Grandits, G.A. (1989) Components of hostility as predictors of sudden death and myocardial infarction in the Multiple Risk Factor Intervention Trial. *Psychosom. Med.*, **51**: 514–22

Dressler, W.W. (1991) Social support, lifestyle incongruity, and arterial blood pressure in a Southern Black community. *Psychosom. Med.*, **53**: 607–20

Eaker, E.D., Abbott, R.D. and Kannel, W.B. (1989) Frequency of uncomplicated angina pectoris in Type A compared with Type B persons (the Framingham Study). *Am. J. Cardiol.*, **63**: 1042–5

Edens, J.L., Larkin, K.T. and Abel, J.L. (1992) The effect of social support and physical touch on cardiovascular reactions to mental stress. *J. Psychosom. Res.* **36**: 371–82

Engebretson, T.O., Matthews, K.A. and Scheier, M.F. (1989) Relations between anger expression and cardiovascular reactivity: reconciling inconsistent findings through a matching hypothesis. *J. Pers. Soc. Psychol.*, **57**: 513–21

Esler, M., Ferrier, C., Lambert, G. *et al.* (1991) Biochemical evidence of sympathetic hyperractivity in human hypertension. *Hypertension*, **17**(Suppl III): III29–35

Follick, M.J., Gorkin, L., Capone, R.J. *et al.* (1988) Psychological distress as a predictor of ventricular arrhythmias in a post-myocardial infarction population. *Am. Heart J.*, **116**: 32–6

Folsom, A.R., Hughes, J.R., Buehler, J.F. *et al.* (1985) Do Type A men drink more frequently than Type B men? Findings from the Multiple Risk Factor Intervention Trial (MRFIT). *J. Behav. Med.*, **8**: 227–35

Frasure-Smith, N. and Prince, R. (1989) Long-term follow-up of the Ischemic Heart Disease Life Stress Monitoring Program. *Psychosom. Med.*, **51**: 485–513

Fredrikson, M. and Matthews, K.A. (1990) Cardiovascular responses to behavioral stress and hypertension: a meta-analytic review. *Ann. Behav. Med.*, **12**: 30–9

Friedman, M., Thoresen, C.E., Gill, J.J. *et al.* (1986) Alteration of type A behavior and its effect on cardiac recurrences in post myocardial infarction patients: Summary results of the recurrent coronary prevention project. *Amer. Heart J.*, **112**: 653–65

Gentry, W.D., Chesney, A.P., Gary, H.E. *et al.* (1982) Habitual anger-coping styles 1. Effect on mean blood pressure and risk for essential hypertension. *Psychosom. Med.*, **44**: 195–203

Hanson, B.S., Isacsson, S-O., Janzon, L. *et al.* (1988) Social anchorage and blood pressure in elderly men – a population study. *J. Hypertension*, **6**: 503–10

Harburg, E., Blakelock, J.R. and Roeper, J. (1979) Resentful and reflective coping with arbitrary authority and blood pressure: Detroit. *Psychosom. Med.*, **41**: 189–202

Hearn, M.D., Murray, D.M. and Luepker, R.V. (1989) Hostility, coronary heart disease, and total mortality: a 33-year follow-up study of university students. *J. Behav. Med.*, **12**: 105–21

Hecker, M.H.L., Chesney, M.A., Black, G.W. and Frautschi, N. (1988) Coronary-prone behavior in the Western Collaborative Group Study. *Psychosom. Med.*, **50**: 153–64

Herd, J.A. (1991) Cardiovascular response to stress. *Physiol. Rev.*, **71**: 305–30

Hjemdahl, P. (1990) Physiology of the autonomic nervous system as related to cardiovascular function: Implications for stress research, in D.G. Byrne and R.H. Rosenman (eds.), *Anxiety and the Heart*, Hemisphere, Washington, DC, pp 95–158

Hokanson, J.F., DeGood, D.E., Forrest, M.S. and Brittain, T.M. (1971) Availability of avoidance behaviours in moderating vascular stress responses. *J. Pers. Soc. Psychol.*, **19**: 60–8

House, J.S., Robbins, C. and Metzner, H.L. (1982) The association of social relationships and activities with mortality: prospective evidence from the Tecumseh Community Health Study. *Am. J. Epidemiol.*, **116**: 123–40

House, J.S., Landis, A.R. and Umberson, D. (1988) Social relationships and health. *Science*, **241**: 540–4

Houston, B.K. and Snyder, C.R. (eds.) (1988) *Type A Behavior Pattern: Research, Theory, and Intervention*, Wiley-Interscience, New York

Irvine, M.J. and Logan, A.G. (1991) Relaxation behavior therapy as sole treatment for mild hypertension. *Psychosom. Med.*, **53**: 587–97

Jacob, R.G., Chesney, M.A., Williams, D.M. *et al.* (1991) Relaxation therapy for hypertension: design effects and treatment effects. *Ann. Behav. Med.*, **13**: 5–17

Johnson, E.H., Gentry, W.D. and Julius, S. (eds.) *Personality, Elevated Blood Pressure, and Essential Hypertension.* Hemisphere, Washington DC

Johnson, J.E. and Hall, F.M. (1988) Job strain, workplace social support, and cardiovascular disease: a cross-sectional study of a random sample of the Swedish working population. *Amer. J. Public Health*, **78**: 1336–41

Johnston, D.W. and Steptoe, A. (1989) Hypertension, in S. Pearce and J. Wardle (eds.), *The Practice of Behavioural Medicine*, Oxford University Press, Oxford, pp 1–25

Johnston, D.W., Cook, D.G. and Shaper, A.G. (1987) Type A behaviour and ischaemic heart disease in middle aged British men. *Br. Med. J.*, **295**: 86–9

Johnston, D.W., Anastasiades, P. and Wood, C. (1990) The relationship between cardiovascular responses in the laboratory and in the field. *Psychophysiol.*, **27**: 34–44

Jorgensen, R.S. and Houston, B.K. (1986) Family history of hypertension, personality patterns, and cardiovascular reactivity to stress. *Psychosom. Med.*, **48**: 102–17

Julius, M., Harburg, E., Cottington, E.M. and Johnson, E.H. (1986) Anger-coping types, blood pressure, and all-cause mortality: a follow-up in Tecumseh-Michigan (1971–1983). *Am. J. Epidemiol.*, **124**: 220–33

Kamarck, T.W. and Jennings, J.R. (1991) Biobehavioral factors in sudden cardiac death. *Psychol. Bull.*, **109**: 42–75

Kamarck, T.W., Manuck, S.B. and Jennings, J.R. (1990) Social support reduces cardiovascular reactivity to psychological challenge: a laboratory model. *Psychosom. Med.*, **52**: 42–58

Karasek, R.A. (1979) Job demands, job decision latitude and mental strain: Implications for job redesign. *Admin. Sci. Quart.*, **24**: 258–308

Karasek, R.A. and Theorell, T. (1990) *Healthy Work*, Basic Books, New York

Larsson, P.T., Hjemdahl, P., Olsson, G. *et al.* (1989) Altered platelet function during mental stress and adrenaline infusion in humans: evidence for an increased aggregability *in vivo* as measured by filtragometry. *Clin. Sci.*, **76**: 369–76

Leon, G.R., Finn, S.E., Murray, D. and Bailey, J.M. (1988) The inability to predict cardiovascular disease from hostility scores or MMPI items related to Type A behavior. *J. Cons. Clin. Psychol.*, **56**: 597–600

Light, K.C., Turner, J.R. and Hinderliter, A.L. (1992) Job strain and ambulatory work blood pressure in healthy young men and women. *Hypertension*, **20**: 214–18

Mann, A.H. and Brennan, P.J. (1987) Type A behaviour score and the incidence of cardiovascular disease: a failure to replicate the claimed associations. *J. Psychosom. Res.*, **31**: 685–92

Manuck, S.B., Olsson, G., Hjemdahl, P. and Rehnqvist, N. (1992) Does cardiovascular reactivity to mental stress have prognostic value in postinfarction patients? A pilot study. *Psychosom. Med.*, **54**: 102–8

Marmot, M.G. and Theorell, T. (1988) Social class and cardiovascular disease: the contribution of work. *Int. J. Health Serv.*, **18**: 659–74

Matthews, K.A. and Haynes, S.G. (1986) Type A behavior pattern and coronary disease risk. *Amer. J. Epidemiol.*, **123**: 923–60

McCann, B.S., Warnick, R. and Knopp, R.H. (1990) Changes in plasma lipids and dietary intake accompanying shifts in perceived workload and stress. *Psychosom. Med.*, **52**: 97–108

Obrist, P.A. (1981) *Cardiovascular Psychophysiology*, Plenum Press, New York

Ornish, D. (1990) *Dr. Dean Ornish's Program for Reversing Heart Disease*, Random House, New York

Ornish, D., Brown, S.E., Scherwitz, L.W. *et al.* (1990) Can lifestyle changes reverse coronary heart disease? *Lancet*, **336**: 129–33

Orth-Gomér, K. and Undén, A-L. (1990) Type A behavior, social support, and coronary risk: interaction and significance for mortality in cardiac patients. *Psychosom. Med.*, **52**: 59–72

Ostfeld, A.M., Kasl, S.V., D'Atri, D.A. and Fitzgerald, E.F. (1987) *Stress, Crowding, and Blood Pressure in Prison*, Lawrence Erlbaum, Hillsdale, NJ

Parati, G., Trazzi, A., Ravogli, A. *et al.* (1991) Methodological problems in evaluation of cardiovascular effects of stress in humans. *Hypertension*, **17**(Suppl III): III50–5

Patel, C., Marmot, M.G., Terry, D.J. *et al.* (1985) Trial of relaxation in reducing coronary risk: four year follow-up. *Br. Med. J.*, **290:** 1103–6

Perini, C., Müller, F.B. and Bühler, F.R. (1991) Suppressed aggression accelerates early development of essential hypertension. *J. Hypertension*, **9:** 499–503

Philip, A.E. (1988) Psychological predictors of outcome after myocardial infarction, in T. Elbert, W. Langosch, A. Steptoe and D. Vaitl (eds.), *Behavioural Medicine in Cardiovascular Disorders*, Wiley, Chichester, pp 193–204

Poulter, N.R., Khaw, K-T., Hopwood, B.E.C. *et al.* (1990) The Kenyan Luo migration study: observations on the initiation of a rise in blood pressure. *Br. Med. J.*, **300:** 967–72

Ragland, D.R. and Brand, R.J. (1988) Coronary heart disease mortality in the Western Collaborative Group Study: follow-up experience of 22 years. *Amer. J. Epidemiol.*, **127:** 462–75

Ravogli, A., Trazzi, S., Villani, A. *et al.* (1990) Early 24-hour blood pressure elevation in normotensive subjects with parental hypertension. *Hypertension*, **16:** 491–7

Reich, P., DeSilva, R.A., Lown, B. and Murawski, B.J. (1981) Acute psychological disturbances preceding life-threatening ventricular arrhythmias. *J. Amer. Med. Ass.*, **246:** 233–5

Rostrup, M. and Ekeberg, O. (1992) Awareness of high blood pressure influences on psychological and sympathetic responses. *J. Psychosom. Res.*, **35:** 117–23

Rostrup, M., Kjeldsen, S.E. and Eide, I. (1990) Awareness of hypertension increases blood pressure and sympathetic responses to cold pressor test. *Am. J. Hypertension*, **3:** 912–17

Rozanski, A., Bairey, N., Krantz, D.S. *et al.* (1988) Mental stress and the induction of silent myocardial ischemia in patients with coronary artery disease. *New Engl. J. Med.*, **318:** 1005–12

Ruberman, W., Weinblatt, E., Goldberg, J.D. and Chaudhary, B.S. (1984) Psychosocial influences on mortality after myocardial infarction. *New Engl. J. Med.*, **311:** 552–9

Salmond, C.E., Prior, I.A.M. and Wessen, A.F. (1989) Blood pressure patterns and migration – a 14 year cohort study of adult Tokelauans. *Am. J. Epidemiol.*, **130:** 37–52

Schlur, G., Schlierf, G., Wirth, A. *et al.* (1988) Low-fat diet and regular, supervised physical exercise in patients with symptomatic coronary artery disease: reduction of stress-induced myocardial ischemia. *Circulation*, **77:** 172–81

Schmidt, T., Dembroski, T.M. and Blümchen, G. (eds.) (1986) *Biological and Psychological Factors in Cardiovascular Disease*, Springer-Verlag, Heidelberg

Schnall, P.L., Pieper, C., Schwartz, J.E. *et al.* (1990) The relationship between 'job strain', work place diastolic blood pressure, and left ventricular mass index. *J. Amer. Med. Ass.*, **263:** 1929–35

Schneiderman, N., Weiss, S.M. and Kaufmann, P.G. (eds.) (1989) *Handbook of Research Methods in Cardiovascular Behavioral Medicine*, Plenum, New York

Shaper, A.G., Ashby, B. and Pocock, S.J. (1988) Blood pressure and hypertension in middle-aged British men. *J. Hypertension*, **6:** 367–74

Shekelle, R.B., Hulley, S.B., Neaton, J.D. *et al.* (1985) The MRFIT behavior pattern study. II. Type A behavior and incidence of coronary heart disease. *Am. J. Epidemiol.*, **122:** 559–70

Shekelle, R.B., Vernon, S.W. and Ostfeld, A.M. (1991) Personality and coronary heart disease. *Psychosom. Med.*, **53:** 176–84

Smith T.W. and Pope, M.K. (1991) Cynical thoughts as a health risk: current status and future directions, in M.J. Strube (ed.), *Type A Behavior*, Sage, Newbury Park, pp 77–88

Steptoe, A. (1981) *Psychological Factors in Cardiovascular Disorders*, Academic Press, London

Steptoe, A. (1990) The value of mental stress testing in the investigation of cardiovascular disorders, in L.R. Schmidt, P. Schwenkmezger, J. Weinman, and S. Maes (eds.), *Health Psychology: Theoretic and Applied Aspects*, Harwood, London, pp 309–29

Steptoe, A. (1991) Psychobiological processes in the etiology of disease, in P.R. Martin (ed.), *Handbook of Behavior Therapy and Psychological Science*, Pergamon, New York, pp 325–47

Steptoe, A. and Appels, A. (eds.) (1989) *Stress, Personal Control and Health*, Wiley, Chichester

Steptoe, A. and Johnston, D.W. (1991) Clinical applications of cardiovascular assessment. *Psychol. Assess.*, **3:** 337–49

Steptoe, A. and Vögele, C. (1991) The methodology of mental stress testing in cardiovascular research. *Circulation*, **83**(Suppl II): II14–24

Steptoe, A., Patel, C., Marmot, M. and Hunt, B. (1987) Frequency of relaxation practice, blood pressure reduction and the general effects of relaxation following a controlled trial of behaviour modification for reducing coronary risk. *Stress Med.*, **3**: 101–7

Steptoe, A., Fieldman, G., Evans, O. and Perry, L. (1993) Control over work pace, job strain and cardiovascular responses in middle-aged men. *J. Hypertension* **11**: 751–9

Stone, S.V. and Costa, P.T. (1990) Disease-prone personality or distress-prone personality? The role of neuroticism in coronary heart disease, in H.S. Friedman (ed.), *Personality and Disease*, Wiley-Interscience, New York, pp 178–200

Theorell, T., Perski, A., Åkerstedt, T. *et al.* (1988) Changes in job strain in relation to changes in physiological state. *Scand. J. Work Environ. Health*, **14**: 189–96

Theorell, T., de Faire, U., Johnson, J. *et al.* (1991a) Job strain and ambulatory blood pressure profiles. *Scand. J. Work Environ. Health*, **17**: 380–5

Theorell, T., Perski, A., Orth-Gomér, K., Hamsten, A. and de Faire, U. (1991b) The effects of the strain of returning to work on the risk of cardiac death after a first myocardial infarction before age 45. *Int. J. Cardiol.*, **30**: 61–7

Timio, M., Verdecchia, P., Venanzi, S. *et al.* (1988) Age and blood pressure changes: a 20-year follow-up in nuns in a secluded order. *Hypertension*, **12**: 457–61

Van Dixhoorn, J., Duivenvoorden, H.J., Staal, J.A. *et al.* (1987) Cardiac events after myocardial infarction: possible effect of relaxation therapy. *Euro. Heart J.*, **8**: 1210–14

Vögele, C., and Steptoe, A. (1992) Emotional coping and tonic blood pressure as determinants of cardiovascular reactions to mental stress. *J. Hypertension* **10**: 1079–87

Vögele, C. and Steptoe, A. (1993) Anger inhibition and family history as modulators of cardiovascular responses to mental stress in adolescent boys. *J. Psychosom. Res.* **37**: 503–14

Watson, D. and Pennebaker, J.W. (1989) Health complaints, stress, and distress: explaining the central role of negative affectivity. *Psychol. Rev.*, **96**: 234–54

Watts, G.F., Lewis, B., Brunt, J.N.H. *et al.* (1992) Effects on coronary artery disease of lipid-lowering diet, or diet plus cholestyramine, in the St. Thomas's Atherosclerosis Regression Study (STARS). *Lancet*, **339**: 563–9

Winett, R.A., King, A.C. and Altman, D.G. (1989) *Health Psychology and Public Health*, Pergamon, New York

Zotti, A.M., Bettinardi, O., Soffiantino, F. *et al.* (1991) Psychophysiological stress testing in post-infarction patients: Psychological correlates of cardiovascular arousal and abnormal cardiac responses. *Circulation*, **83**(Suppl II): II25–35

Part Three
Animal models of stress coping and resistance

6

Animal models of stress: An overview

PAUL WILLNER

DEPARTMENT OF PSYCHOLOGY
UNIVERSITY COLLEGE OF SWANSEA,
SWANSEA, UK

6.1. ANIMAL MODELS

Animal models of psychopathology have a number of distinct uses: they serve as screening tests within the framework of drug discovery and development programmes, as convenient methods for measuring the status of the underlying neurobiological mechanisms, and as simulations within which to investigate aspects of the human condition modelled (Willner, 1991). We are concerned here with the use of animal models as simulations of human stress. In this sense:

"Animal models represent experimental preparations developed in one species for the purpose of studying phenomena occurring in another species. In the case of animal models of human psychopathology one seeks to develop syndromes in animals which resemble those in humans in certain ways in order to study selected aspects of human psychopathology". (McKinney, 1984)

ISBN 0–12–663370–3

The development of valid animal models of psychiatric disorders can only proceed if procedures are in place by which their validity can be assessed. A general methodology exists for this purpose, under which models are assessed with respect to their predictive validity, face validity and construct validity: predictive validity refers to the accuracy of predictions made from the model, particularly in respect of drug actions; face validity refers to the phenomenological similarity between the model and the disorder modelled; and construct validity refers to the theoretical rationale for the model. These methods have been applied successfully across a range of psychiatric disorders, such as depression (Willner, 1984a), anxiety (Green and Hodges, 1991), schizophrenia (Ellenbroek and Cools, 1990), drug dependence (Goudie, 1991) and dementia (Sarter *et al.*, 1992).

Implementation of these validating procedures in relation to animal models of stress is more problematic than in many other areas, largely because stress is not a well-defined clinical syndrome. A substantial corpus of literature describes the treatment of specific stress-related disorders, such as anxiety or depression. However, because stress causes disorders rather than itself being a disorder, stress *per se* has not been viewed as a target for pharmacotherapy or for drug development. There is, therefore, little in the clinical literature against which to assess the predictive validity of animal models of stress. The assessment of face validity relies upon a comparison of symptoms displayed in the disorder and in the model, and again, this is far more straightforward in relation to disorders that can be described in terms of widely-agreed checklists of signs and symptoms such as those compiled in the American Psychiatric Association Diagnostic and Statistical Manuals. The assessment of construct validity requires a good understanding of the nature of the disorder. This precondition is always difficult to meet, but is particularly problematic in relation to the concept of stress, to which we now turn.

6.2 DEFINITIONS OF STRESS

6.2.1 THE STRESS RESPONSE

Selye (1952) defined stress as 'a non-specific response of the body to any demand made on it'. The stress response consists of a number of physiological adaptations, the most prominent of which is an activation of the hypothalamus-pituitary-adrenal (HPA) system. The endpoint of this element of the stress response, an elevation of plasma corticosteroid levels, is the most frequently studied stress indicator, and is usually considered to be definitive of a state of stress. The simplicity of this concept is deceptive.

Selye's studies were based upon the application of physical stressors, such as temperature change. Extremes of temperatures undoubtedly challenge the body: we know this because a thermal challenge activates a whole range of thermoregulatory mechanisms that attempt to restore and maintain homeostasis. In other words, the stressful nature of a thermal challenge is defined independently of the stress response, and this holds true also for other physical stressors, such as tissue damage or dehydration. This independent evidence that a situation is stressful is crucial: without it, the relationship between stressors and the putative 'stress response' is circular: corticosteroid release is assumed to be a stress response because it is elicited by stressors, but a stressor is defined by its ability to release corticosteroids.

Within the psychological domain, the definition of a stressor, independently of the stress response, is far less straightforward. At first sight, the concept of aversive motivation appears to provide a reasonable psychological analogue for the physiological processes that mediate bodily homeostasis. This position was adopted, for example by Miller (1953), who defined a stressor as 'any vigorous extreme or unusual stimulation which, being a threat, causes some significant change in behavior'. Indeed, in addition to their role in defining psychological stress, behavioural adaptations also form part of the integrated response to physical challenge such as thermal stress or tissue damage. From this perspective, a psychological stressor could be defined as a situation that elicits behaviours, such as escape or aggression, likely to terminate the situation: in other words, aversively motivated behaviours subserve homeostatic functions.

This integrated concept of stress, in which exposure to a stressor is assumed to evoke a short lived 'emergency reaction' characterized by fight or flight behaviours and activation of the sympathetic and adrenomedullary system (Cannon, 1936), accompanied by a longer-lasting activation of the adrenocortical 'stress response' (Selye, 1952), is widely accepted. However, when we examine more closely the relationship between aversion and the stress response, a number of problems emerge. In brief, the relationship between situations that evoke escape behaviours (psychological stressors) and activation of the HPA axis (the 'stress response') is far less straightforward than it at first appears.

One side of this coin is that, under some circumstances, putatively stressful situations can fail to activate the HPA axis. Repeated application of a stressor usually results in a decrease in the size of the corticosteroid response (e.g. Armario et al., 1984; Kant et al., 1985). However, after repeated administration of electric shock, under conditions that lead to habituation of the corticosteroid response, the shock remains aversive: the animal will terminate the exposure if given the opportunity to do so. Indeed, adrenalectomy, which abolishes the corticosteroid response, may actually increase the adverse behavioural consequences of stress (Edwards et al., 1990). The other side of the coin is that situations that cause corticosteroid release are not necessarily accompanied by behaviours directed towards their termination. For example, amphetamine stimulates corticosteroid release (Knych and Eisenberg, 1979), yet far from being aversive, this drug is avidly self-administered by animals (and people). Similarly, nicotine elevates plasma corticosteroids in habitual smokers (Pomerleau and Pomerleau, 1990). Despite the fact that they are self-administered, amphetamine and nicotine are usually considered to be stressors, not only because they activate the HPA axis, but also because, in many respects, these drugs are interchangeable with more conventional, undisputed stressors: for example, the locomotor stimulant effects of amphetamine and electric footshock show cross-sensitization (Antelman et al., 1980). However, this parallel goes even further and, under some circumstances which at present are poorly understood, animals choose to self-administer painful electric shocks (Stretch et al., 1968), which presumably activate the HPA axis. Corticosteroid release in the absence of behavioural evidence of aversion may also be seen in more mundane situations. The HPA axis is activated by novelty (Piazza *et al.*, 1990), yet rats will often display a behavioural preference for novel environments (Berlyne, 1960).

These observations, which may be analogous to human sensation-seeking behaviour (Zuckerman *et al.*, 1980), suggest that corticosteroid release may be related to a state of arousal, rather than to psychological stress as defined in terms of aversive behaviours. Such a relationship is suggested by the finding that the size of the corticosteroid response to novelty is predictive of the ease with which rats acquire amphetamine self-administration behaviour (Piazza *et al.*, 1990). The observation that acute administration of corticosteroids usually elevates mood in human subjects (Goodwin *et al.*, 1992) is also consistent with this notion and further underlines that, far from providing a simple, definitive measure of stress, corticosteroid levels are frequently unreliable, and can sometimes be frankly misleading.

Although the foregoing discussion has focused on corticosteroids as the most widely used marker of psychological stress, the same conclusion applies to other elements of the 'stress response': none bears an unequivocal relationship to behavioural adaptations adopted in the presence of a stressor. This leads to a far from satisfactory situation, with a choice of two equally unattractive options. One is to define stress in physiological terms (e.g. corticosteroid elevation) and to accept that there will be many situations in which events usually assumed to be stressors are not, and *vice versa*. In this case, the question of whether a particular experimental procedure provides an animal model of stress reduces to the trivial question of whether elevations in corticosteroids can be demonstrated. Alternatively, stress can be defined by the presence of events that are assumed to be psychological stressors. In this case, the question of the validity of animal models moves to centre stage.

6.2.2 THE STRESSFUL ENVIRONMENT

A good reason to move away from a physiological definition of stress is that it is rare for stress to be so defined in human studies. Indeed, in those rare studies in which physiological indices of stress have been recorded in human subjects, far from the physiological measures being taken as definitive, they have usually been the focus of the study. The problem is well illustrated by studies of preoperative stress in surgical patients (reviewed by Johnston, 1988). Increases in plasma cortisol levels are reliably observed prior to surgery, but these appear to be limited to a short period on the evening before surgery when preparations such as shaving were carried out, suggesting that cortisol elevations may be related to physical stress, rather than to the psychological stress of impending surgery (Czeisler *et al.*, 1976). Other studies in surgical patients have used palmar sweating as an index of sympathetic adrenomedullary activation. Again, consistent changes have been observed prior to (and following) surgery, but these appear to correlate with questionnaire measures of arousal rather than with perceived stress (Johnson *et al.*, 1970).

The absence of hard evidence that psychological stress in humans is closely correlated with increased activity in the HPA axis has not prevented the continued use of the 'stress response' as the standard for stress studies in animals. However, human studies have usually taken the very different approach of defining stress in stimulus terms. From this perspective, certain types of events are assumed to be stressful. For example, Lazarus and Cohen (1977) defined three categories of stressor, varying primarily in intensity: cataclysmic changes, such as war or natural disasters; major life events; and minor daily 'hassles': 'the little things than can irritate or distress people, such as one's dog being sick on the living room rug, dealing with an inconsiderate smoker, having too many responsibilities, feeling lonely, having an argument with a spouse, and so on' (Lazarus and Folkman, 1984). Another approach focuses on chronicity as a major dimension. For example, the Institute of Medicine report 'Stress and Human Health' defines four categories of stressor: '(1) *Acute, time-limited stressors*, such as parachute jumping, awaiting surgery, or encountering a rattlesnake; (2) *Stressor sequences*, or series of events that occur over an extended period of time, as the result of an initiating event such as job loss, divorce, or bereavement; (3) *Chronic intermittent stressors* such as conflict-filled visits to in-laws or sexual difficulties, which may occur once a day, once a week, once a month; and (4) *Chronic stressors* such as permanent disabilities, parental discord, or chronic job stress, which may or may not be initiated by a discrete event and which persist continuously for a long time' (Elliott and Eisdorfer, 1982). The most noteworthy aspect of this classification is that major life events (job loss, divorce, bereavement) are viewed not as acute, discrete (time-limited) events, but rather as initiating a chronic sequence of problems of living; we return to this point below.

6.2.3 MENTAL STRESS

Although superficially compelling, the situational approach to stress is also inadequate because the concept of psychological stress denotes a subjective state rather than a state of the external environment. Webster's Thesaurus, for example, defines stress as 'mental tension' and offers the following synonyms: 'tension, strain, pressure, burden, hardship, overexertion, agony, trial, affliction, anxiety, nervousness, fearfulness, apprehensiveness, apprehension, impatience, fear, ferment, disquiet, disquietude, tenseness, passion, intensity, fluster, expectancy, restlessness, trepidation, misgiving, mistrust, alarm, dread, flutter, trembling, pinch, urgency' (Laird, 1971). But mental stress bears no simple relationship to the presence of a putative stressor: when exposed to the same stressor, different people experience stress to different degrees, and the same person may experience stress to different degrees on different occasions.

This issue is not, in itself, a problem for human researchers. Several scales have been constructed to measure the intensity of perceived stress (Everly and Sobelman, 1987), and investigation of the factors giving rise to individual differences in stress cognition is relatively straightforward. A definition that takes account of the fact that events become psychological stressors only when they are perceived as such proposes that stress is: 'a particular relationship between the person and the environment that is appraised by the person as taxing or exceeding his or her resources and endangering his or her well-being' (Lazarus and Volkman, 1984). This definition introduces concepts of appraisal of, and coping with, stress as crucial dimensions of the experience, which mediate and determine the psychological impact of an environmental stressor.

While not problematic for research on humans (but see Chapter 2), accepting that stress is to be defined subjectively creates serious obstacles to the evaluation of animal models of stress. If we define stress in terms of either specific physiological responses, or particular types of environmental event, then these can be readily evaluated in animal models. However, while animals may well have subjective experiences, these are, for all practical purposes, outside the scope of scientific enquiry (Willner, 1984b). How, then, can we evaluate particular experimental procedures as animal models of human stress? And even more fundamentally, how can we ever know whether particular events are stressful in animals?

The problem of evaluating animal models in relation to the subjective nature of stress addresses the question of their construct validity (in contrast to issues such as severity and chronicity of stress, which contribute to the evaluation of face validity). In fact, the problem in this case does not differ greatly from that of evaluating the construct validity of other subjective states such as depression or anxiety. In all such cases, we can identify aspects of the disorder that, in principle, cannot be modelled in animals because they are knowable only through verbal report, and others that can in principle be operationalized (Willner, 1984b). According to Lazarus and Volkman (1984), stress is experienced when a cognitive appraisal of events gives rise to the perception of an inability to cope. Cognitive appraisal cannot, in principle, be studied in animal models. Loss of control, on the other hand, can readily be operationalized, and is a central feature of many animal models of stress.

The problem of recognizing stress in animals is also not as intractable as it appears; this is fortunate, since our ability to address this issue has important implications not only for experimental neuroscience, but also for animal welfare policies. Although exposure to a putative stressor does not guarantee the experience of stress in people, the likelihood of this response is greater in certain circumstances. Extreme conditions, such as natural disasters, torture or sudden loss of loved ones, generate stress in virtually all who experience them; it is only when we move to milder, more everyday stressors that individual differences assert themselves. Therefore, while an environmental definition of stress is theoretically flawed, it serves well in practice, under extreme conditions. By analogy, although we can have no direct access to the subjective experience of experimental animals, we may reasonably assume that they experience severe stressors as stressful, and that some of the behavioural and physiological changes associated with exposure to severe stressors are markers of the stressed state. If milder, more equivocal conditions evoke some of the same changes, then these might denote a state of stress also. In fact, as described in the following section, conditions that might be expected, on common sense grounds, to be mildly stressful do elicit effects similar to those of more extreme conditions and which cannot easily be attributed to causes other than stress.

6.3 EFFECTS OF STRESS

The consequences of stress in animals pervade virtually all aspects of adaptive functioning: indeed, one tongue-in-cheek definition of stress suggests that stress is the process that produces a change in your favourite physiological parameter (Murison and Ursin, 1982)! Effects of stress that have been widely reported include:

impairment of feeding, drinking, and sexual behaviours; decreased aggression; impaired acquisition of appetitively and aversively motivated behaviours; analgesia; changes in sleep architecture and biological rhythms; alterations in a variety of neuroendocrine parameters; weight loss; suppression of the immune system; and adverse cardiovascular and gastrointestinal effects. An exhaustive review of this literature would be neither practicable nor desirable. Instead, this section will focus on aspects of the behavioural effects of stress that are of particular theoretical significance.

6.3.1 'HELPLESSNESS'

The term 'learned helplessness' tends to be used rather indiscriminately to label procedures in which animals are exposed to inescapable stress. However, the term was originally coined by Seligman (1975) to describe one aspect of this procedure: that animals exposed to uncontrollable stress (usually electric shocks) show subsequent impairment in learning to escape shock. This effect is not seen in animals exposed to comparable, or indeed, identical, patterns of controllable shock. The term 'learned helplessness' carries a significant theoretical burden: it implies that the animals perform poorly because they have learned that their responses are ineffective in controlling their environment (Seligman, 1975).

However, inescapable (but not escapable) shock has a variety of other, simpler, effects that could also explain many of the behavioural impairments, such as decreased locomotor activity (Glazer and Weiss, 1976a, b; Anisman et al., 1979) and analgesia (Maier et al., 1982). To demonstrate that inescapable shock does, additionally, cause 'cognitive' impairments, Jackson and Minor (1988) assessed performance accuracy using a maze task in which performance would be independent of factors influencing motor speed. As predicted, accuracy was reduced in animals previously subjected to unavoidable shock, confirming the presence of a 'cognitive' component to the pattern of impairment. However, subsequent work showed that this 'cognitive' impairment arose from an increase in distractibility rather than from a learning disability; no learning impairment was evident in distraction-free conditions (Minor et al., 1984). So inescapable shock does cause 'cognitive' impairment, but at the level of attentional processes rather than 'helplessness'. As discussed below, the consequences of this reinterpretation for the status of learned helplessness as a model of human behaviour are considerable (see also Chapter 9).

6.3.2 RESPONSE SUPPRESSION

Although a cognitive impairment can be demonstrated following inescapable shock, under suitably distracting conditions, the design of the majority of experiments using the 'learned helplessness' paradigm is not adequate to demonstrate such an impairment. Most commonly, the effects of inescapable shock are tested using measures sensitive to changes to locomotor activity, such as running speed or the rate of lever-pressing. Therefore, in the majority of 'learned helplessness' studies, the impairments can be explained most parsimoniously as decreases in motor output. Decreased motor activity in the aftermath of severe stress has been widely reported, most commonly on the basis of open field studies (Anisman and Zacharko, 1982), or sometimes using a forced swimming test (Weiss et al., 1982). This behavioural suppression appears to differ from the behavioural inhibition that may also be observed in response to a stimulus previously paired with the stressor. The latter phenomenon, which is usually studied using the conditioned suppression ('conditioned fear') paradigm, is reversed by anxiolytic drugs (Geller and Seifter, 1960); in contrast, the motor consequences of inescapable shock are not reversed by anxiolytics (Sherman et al., 1982), and are independent of the presentation of specific stress-associated stimuli. However, as shown below, this distinction may be misleading.

The time course of stress-induced response suppression is an important issue in relation to the question of what types of human stress states are modelled by acute stress procedures in animals: as noted above, severe life events, in people, are stressful not only in themselves, but also as the initiators of periods of chronic stress (for example, bereavement leading to chronic loss of social support; job loss leading to chronic unemployment). Curiously, there appears to be an inverse relationship between the intensity of stress (shock) and the duration of the ensuing behavioural impairments (Glazer and Weiss, 1976a,b), such that the behavioural suppression following severe shock is at most 48 h in duration (Weiss et al., 1982), whereas the effects of mild shock can be detected for at least a week (Glazer and Weiss, 1976b); in both cases, the effects are observed only when shock is inescapable. This paradox appears to arise from a procedural difference: when mild shocks are used, their duration is usually longer, and this provides opportunities for inactivity to be adventitiously reinforced (Bracewell and Black, 1974; Glazer and Weiss, 1976b). Thus, the prolonged effects of mild stress may result from a conditioning process under which response suppression is induced by exposure to aspects of the experimental environment. If this interpretation is correct, then the long-term after-effects of inescapable shock may be more akin to conditioned fear than they at first appear.

However, not all long-term behavioural suppression is of this type. Locomotor activity has been reported to be suppressed for more than seven weeks, following a single shock session, provided activity was measured in the home cage (Desan et al., 1988). While this effect has proved difficult for other laboratories to replicate (e.g. Bauman and Kant, 1991), a related finding of a prolonged suppression of open field activity, which increases over time (van Dijken et al., 1992a,b), supports the concept of long-term behavioural suppression. What distinguishes this long-term suppression of locomotor behaviour from previously described after-effects of stress is that this effect is related to shock exposure per se rather than to shock controllability: it is seen equally in animals exposed to inescapable or to escapable shock (Woodmansee et al., 1991). In this respect, suppression of locomotor behaviour resembles stress-induced activation of the HPA axis, which is also largely independent of stress controllabilty (Maier et al., 1986; Prince and Anisman, 1990).

This example emphasizes that the interpretation of behavioural consequences of stress requires well-designed experiments. Although the comparison between controllable and uncontrollable shock was de rigeur in the early literature, it is now relatively uncommon for control groups to be tested with controllable shock: it is now more usual simply to compare inescapably shocked animals with unshocked controls. It is clear that the so-called learned helplessness paradigm should be considered more correctly as a mixture of paradigms, and care should be taken in generalizing conclusions between them.

6.3.3 ANHEDONIA

As indicated above, behavioural studies of stress have tended to focus largely on either gross measures of motor output or on performance in aversively motivated tasks. However, stress is also known to disrupt consummatory behaviours and performance in appetitively motivated tasks. A number of recent studies have addressed the question of whether these impairments result from a decrease in sensitivity to rewards (anhedonia).

A single session of inescapable (but not escapable) footshock has been shown to decrease responding for brain-stimulation reward (intracranial self-stimulation: ICSS) in mice. These studies have mainly used the rate of responding as the dependent variable, which is susceptible to a variety of non-specific influences. However, an impairment of sensitivity to reward is indicated by the observation that the effects of inescapable shock are anatomically specific: ICSS elicited from the ventral tegmental area (the origin of the mesolimbic dopamine projection), or from the nucleus accumbens or frontal cortex (two of its terminal fields) was suppressed by inescapable shock, but ICSS elicited from the substantia nigra (the origin of the nigrostriatal dopamine projection) was unaffected (Zacharko and Anisman, 1991). A similar decrease in sensitivity to sweet rewards, assessed by a decrease in preference for sweet solutions over plain water, has been reported in rats following a single session of restraint stress (Plaznik *et al.*, 1989) or social defeat (Koolhaas *et al.*, 1990). Another manipulation that has been reported to decrease responding for brain-stimulation reward, and to increase the threshold current necessary to sustain this ICSS behaviour, is withdrawal from chronic amphetamine treatment (Cassens *et al.*, 1981).

These effects tend to be relatively long-lasting: at least a week following inescapable shock (Zacharko and Anisman, 1991), at least ten weeks following defeat (Koolhaas *et al.*, 1990) and at least 18 days following amphetamine withdrawal (Leith and Barrett, 1976). However a prolonged decrease in ICSS responding was found only if the animals were tested for ICSS in the immediate aftermath of stress; otherwise the effect dissipated rapidly (Zacharko *et al.*, 1983). This suggests that the occurrence of prolonged anhedonia following brief stress exposure may depend to some extent on as yet undetermined conditioning processes.

A prolonged anhedonia is generated, and may be maintained for up to four months, by chronic sequential exposure to a variety of very mild stressors, such as overnight illumination, tilting of the cage, periods of food or water deprivation, or changes of cage mates. Over the first few weeks of exposure, rats subjected chronically to these low grade stressors reduced their consumption of, and preference for, weak solutions of sucrose and saccharin. Stressed animals were also subsensitive to reward in the place conditioning paradigm: the normal preference for environments paired with food, sucrose solutions, amphetamine or morphine was abolished or greatly attenuated in stressed animals (Willner *et al.*, 1987, 1992). Chronic mild stress has also been shown to increase the threshold for ICSS (Moreau *et al.*, 1991). Variety is essential to prevent habituation to these mild stressors (Muscat and Willner, 1992): of which more below.

6.3.4 VULNERABILITY TO STRESS

As noted above, responses to stress in humans depend to a large extent on the cognitive processing of stress-related information. The extent to which this issue can usefully be investigated in animal models is limited. However, the question of individual differences in vulnerability to stress has been addressed in a variety of ways.

One approach has used selective breeding techniques to develop inbred strains that differ in their responsiveness to stress. The best known examples are the Maudsley Reactive and Nonreactive and the Roman High and Low Avoidance rat strains (Broadhurst, 1975; Driscoll and Battig, 1982), which were selected on the basis of their responses to an acute stressor. The Flinders Sensitive Line (FSL) rat is the result of selective breeding for sensitivity to the hypothermic effect of cholinergic agonists. Relative to their control strain, FSL animals are more responsive to the motor suppressant effects of inescapable shock and show greater immobility in the forced swim test (Overstreet and Janowsky, 1991), and also are more vulnerable to the suppressive effect of chronic mild stress on responsiveness to sweet reward (Pucilowski *et al.*, 1992). However, FSL animals behave normally in the elevated plus maze, a putative animal model of anxiety (Schiller *et al.*, 1991).

A related approach has demonstrated that inescapable shock has variable behavioural effects in different inbred mouse strains. To take an extreme example, in the C57BL/67 strain, exposure to inescapable shock severely impaired subsequent learning to escape shock, but had no effect on ICSS responding, while the DBA/2J strain showed exactly the opposite pattern (Zacharko *et al.*, 1987; Shanks and Anisman, 1988). Clearly, studies of this kind have promise as a starting point from which to identify the physiological mechanisms underlying individual differences in responses to stress.

In addition to these genetic influences, vulnerability to stress also has environmental determinants. Social status is a significant example. A single defeat by a dominant male rat has been reported to cause a gradual increase over weeks in the motor suppressant effects of inescapable shock and in immobility in the forced swim test (Korte *et al.*, 1991b). In a chronic version of a similar procedure, mice were housed in social contact, but were physically separated except for once daily 3-min encounters; again, increased immobility in the forced swim test was observed in repeatedly defeated animals (Kudryatseva *et al.*, 1991). Dominant and submissive animals also differ in their neuroendocrine responses: within a colony of mice, dominant mice responded to repeated immobilization stress by increases in blood pressure and activation of the sympathetic-adrenomedullary system, while subordinate animals responded to the same stressor with an increase in corticosteroid secretion (Henry and Stevens, 1977).

The extent to which responsiveness to stress is influenced by stable housing conditions is an important question in relation to the extensive literature demonstrating that in people, the quality of available social support is an important determinant of vulnerability to stress (Brown and Harris, 1978). One study has reported that individually-housed rats were more sensitive than group-housed animals to the motor-suppressant effect of restraint stress (Dourish et al., 1989). In contrast, a number of studies have reported that housing conditions have little or no influence either on HPA responses to acute or chronic stressors (e.g. Giralt and Armario, 1989), or on the development of chronic mild stress-induced anhedonia (Muscat and Willner, 1992). The conclusion must be that the 'buffering' effect of social support cannot at present be modelled reliably in animals.

6.3.5 COPING STYLES

Responses to stress vary not only between individuals, but also between environments. As noted above, the response to stress is usually characterized as comprising two phases. The initial phase consists of attempts to cope with the stressor using active behavioural strategies (fight or flight), supported by activation of the sympathetic and adrenomedullary systems: if these coping attempts are unsuccessful, the body switches into a state of low behavioural activity, supported by the parasympathetic and adrenocortical systems; all of these later adaptations have the effect of conserving resources (Selye, 1952). Under some circumstances, however, the initial phase is bypassed. For example, if a cat is brought into the presence of a rat, the rat displays flight behaviour if flight is possible. But if there is no escape route, the rat immediately (and very sensibly) enters a state of total immobility and stays there. This response is innate: it does not require prior experience of cats (Blanchard and Blanchard, 1971).

A similar liability of behaviour in response to electric shock has been studied using the 'prod burying' technique. In this paradigm, an electrified probe, which delivers a shock whenever it is touched, is introduced into the animal's cage (Treit, 1985). After experiencing shock, rats attempted to bury the (now inactive) prod under sawdust (active coping); however, if no sawdust was present they displayed immobility (passive coping). Active coping in this model was accompanied by rises in plasma noradrenaline; passive coping was accompanied by rises in plasma corticosterone (De Boer et al., 1990b; Korte et al., 1991a). This may prove a useful research tool for investigating habitual active (Type A) or passive (Type B) modes of stress control, which have attracted considerable recent interest within health psychology (Matthews, 1982). In this context, it should be noted that active coping does not necessarily mean successful coping in the sense of escaping from the stressor. Electric shock elicits fighting in pairs of rats shocked together, and the occurrence of this non-adaptive response serves to minimize the development of gastric ulcers (Weiss et al., 1976).

6.4 DIMENSIONS OF STRESS

The discussion so far has focused on the effects of stress, with little attention to the nature of the stressor. We now examine some aspects of stressful situations that determine the form and magnitude of stress responses, or have been thought to do so.

6.4.1 MODALITY

A distinction is often drawn between physical and mental stress, with the implication that animal models that employ psychological stressors should be more relevant to the human condition. This position may be mistaken. Selye's (1952) account of the generalized stress response was based on the observation that a variety of noxious stressors, including heat and cold, had the common effect of increasing adrenocortical activity. However, in an important series of studies, Mason later showed that if the temperature is changed very gradually, heat actually decreases corticosteroid levels. Therefore the adrenocortical 'stress response' results, in this case, from the sudden introduction of a heat stimulus, rather than from temperature change *per se*. This strongly suggests that the emotional (or other psychological) concomitants of physical stressors are the important factor in the initiation of stress responses (Mason, 1968, 1971).

In fact, there are striking parallels between the form of stress responses following electric shock (a prototypical physical stressor) and social defeat (a prototypical psychological stressor). Perhaps the most dramatic parallels are in relation to stress-induced analgesia. Maier *et al.* (1982) reported that during exposure to inescapable shock, a biphasic pattern of analgesic responses is seen, with a brief early peak occurring within the first 20 shocks, followed by a more prolonged analgesia that appears after some 60 shocks have been received. Similarly, Rodgers and Randall (1988) have reported that analgesia is elicited by brief experience of defeat (occurring within one minute of the start of agonistic encounters, and in response to fewer than ten attack bites), and also by more prolonged defeat (resulting from exposure to extended attack). In both cases, the analgesia resulting from prolonged shock or defeat was opioid in nature (i.e. naloxone reversible and cross-tolerant to morphine), but the analgesia seen following brief exposure was not (Maier *et al.*, 1982; Rodgers and Randall, 1988). Parallels are also apparent in the behavioural suppressant effects of these two stressors. Both defeat and inescapable shock increase immobility, measured 24 h later in the forced swim test (Weiss *et al.*, 1982; Korte *et al.*, 1991b), and cause a prolonged increase in the freezing response elicited by presentation of a mild stressor (Koolhaas *et al.*, 1990; van Dijken *et al.*, 1992a). Finally, shock and defeat are reciprocally related: exposure to inescapable shock causes a decrease in social dominance (Williams, 1982), while experience of defeat impairs subsequent shock-avoidance learning (Scholtens and van de Poll, 1987). In view of these similarities, the often-expressed view that shock provides an 'unnatural' and therefore invalid model may be overstated; indeed, there may be some merit in the alternative view that shock may sometimes be preferable to more ethologically valid stressors. This is particularly the case in studies in which parametric control over the intensity of the stressor is an important consideration.

6.4.2 CHRONICITY

Repeated administration of a stressor is usually referred to as 'chronic stress'. This designation may sometimes be appropriate, but more frequently is not. When stressors are presented repeatedly, stress responses tend to habituate and an event that is acutely stressful may no longer evoke a stress response. In these circumstances, the term 'chronic stress' is inaccurate. Examples of habituation to the behavioural effects of acute stress, following repeated daily administration of the stressor are many and varied. Behavioural paradigms in which repeated application of the stressor leads to habituation of the response include: shock-induced hypomotility and hypophagia (Ottenweller et al., 1989); impairment of escape learning by exposure to cold water (Weiss et al., 1975); and hypomotility (Kennett et al., 1986) or immobility in the forced swim test (Platt and Stone, 1982) following restraint stress. Habituation of the HPA response to stress has also been observed following daily exposure to noise stress, forced swimming, forced running, footshock or restraint (Armario et al., 1984, 1985; Kant et al., 1985; Ottenweller et al., 1989; De Boer et al., 1990a).

Significantly, habituation in the HPA system is stressor specific: habituation of HPA responses is not observed in response to a stressor other than the one presented repeatedly (Kant et al., 1985). The specificity of adaptations to behavioural effects of stressors has not been systematically studied. However, it is clear that repeated presentation of single stressors results in more rapid habituation to stress-induced behavioural changes than presentation of a variety of stressors (Katz and Baldrighi, 1982; Muscat and Willner, 1992). Indeed, stress-induced corticosterone secretion may increase following chronic varied stress (Armario et al., 1985), and adverse behavioural effects of varied mild stressors can be observed over several months of presentation (Willner et al., 1992). Thus, chronic presentation of a variety of stressors does appear to cause chronic stress, unlike repeated presentation of the same stressor.

In contrast to the majority of studies, in which stress responses habituate on repeated presentation, some studies have reported no change in response (Ratner et al., 1989), or even an increase (Orr et al., 1990; Pitman et al., 1990). Indeed, while reports of sensitization to 'stress responses' are relatively uncommon, sensitization to the locomotor activating effects of stressors, and cross-sensitization between stressors and psychomotor stimulants, are well established phenomena (Robinson and Becker, 1986; Antelman et al., 1990). Habituation and sensitization to stress may conform to the dual process theory that successfully describes these phenomena in other domains (Groves and Thompson, 1970). According to this theory, habituation is more likely with frequent presentation of weak stimuli, whereas sensitization is more likely with infrequent presentation of strong stimuli. Some studies have reported data that accord with this dual process theory. Thus, greater habituation has been reported with more frequent stressors (De Boer et al., 1990a), and greater sensitization has been reported with a more intense stressor (Pitman et al., 1990). However, the inverse effect, greater habituation with a more intense stressor, has also been reported (Pitman et al., 1988).

6.4.3 PREDICTION AND CONTROL

The concept of control over stress has been of central importance in the development of both animal models and theoretical accounts of the relationship between stress and illness. In the human literature, the concept of helplessness has developed considerably, from a simple notion of failure to cope (Seligman, 1975), through various attributional formulations in which the crucial variables are the explanatory frameworks that people use to understand their failures to cope, and the domains in which those failures occur (Abramson *et al.*, 1978, 1989). In the animal literature the switch from active to passive patterns of stress response is reliably observed following exposure to uncontrollable stressors but not following exposure to controllable stressors; exceptions to this rule (e.g. Woodmansee *et al.*, 1991) are rare. The dimension of controllability thus appears to operationalize the psychological processes that, in humans, mediate between the stressor and the stress. However, this attractive and well-accepted concept may be mistaken.

In the typical experimental paradigm, inescapable shock is not only uncontrollable, but also unpredictable in onset and/or offset. There is some evidence that the important factor mediating the adverse effect of inescapable shock may be unpredictability, rather than uncontrollability. It was demonstrated many years ago that simply providing a feedback signal to accompany shock offset conferred protection against the ulcerogenic effect of inescapable (and therefore, uncontrollable) shock (Weiss, 1970; Overmeier *et al.*, 1985). More recently, providing signals at the onset or offset of uncontrollable shock has been shown to protect against subsequent behavioural suppression. It was hypothesized that unpredictability generates high levels of fear, which are maintained within manageable limits by signals denoting the presence of danger or safety (Jackson and Minor, 1988). This reconceptualization can be used to explain some hitherto puzzling observations, such as the prevention of shock-induced behavioural suppression by anxiolytic drugs (Drugan *et al.*, 1984). It also allows us to make sense of the data, discussed above, indicating that inescapably shocked animals perform poorly in 'cognitive' tasks primarily because they are easily distractable (Minor *et al.*, 1984). This is exactly as would be expected in a state of fearful hypervigilance, but is difficult to explain in terms of loss of control.

It is not at present clear whether the control dimension does, in fact, have any role in mediating the effects of stress in animal models, or whether predictability can explain all of the adverse effects of inescapable stress. Perhaps predictability determines the outcome on some response dimensions (e.g. behavioural suppression) while controllability determines the outcome on others. Further studies designed to investigate separately the influence of unpredictability and uncontrollability will be needed to address this important issue.

6.5 MODELS OF WHAT?

Exposure to putative stressors has a wide variety of behavioural consequences which are, to a large extent, independent of nature of the stressor. Although the present review has not been concerned with the question of the brain mechanisms underlying these effects, it is important to note that different behavioural effects of stress can be related to different physiological systems. Inescapable shock, for example, causes, *inter alia*, response suppression, analgesia and anhedonia. Shock-induced analgesia is abolished by removal of the pituitary gland, which does not, however, affect response suppression (MacLennan *et al.*, 1983). Some aspects of shock-induced response suppression are mediated by noradrenaline depletion in the forebrain (Minor *et al.*, 1988) which does not, however, cause anhedonia (Wise, 1978). Shock-induced anhedonia appears to depend on a decreased functional activity in the mesolimbic dopamine system (Zacharko and Anisman, 1991) which has little, if any, role in analgesia.

Given this diversity of both behavioural and neurobiological consequences, it is likely that aspects of the stress syndrome may be relevant to a variety of different psychopathologies. This chapter opened by pointing out the difficulty of assessing the validity of experimental procedures as 'animal models of stress', which at present is not feasible. However, in relation to specific psychopathologies the situation is potentially less recondite: in particular, some signposts are available for distinguishing between potential animal models of anxiety and depression. At the level of predictive validity, antidepressants are effective in both depression and anxiety, but benzodiazepine anxiolytics, at usual doses, are ineffective in depression; therefore, the way in which an animal model responds to benzodiazepine anxiolytics is more informative than its response to antidepressants. At the level of face validity, all symptoms of anxiety are observed, to a greater or lesser degree, in depression, but some symptoms of depression are not seen in anxiety disorders. In particular, the presence of anhedonia in an animal model points strongly towards depression rather than anxiety. At the level of construct validity, life events are implicated in the aetiology of both depression and anxiety, but a distinction has been drawn between threatening events, which tend to precipitate anxiety, and loss events, which tend to precipitate depression (see Chapter 2).

Applying these principles to the behavioural sequelae of stress, the most obvious conclusion is that anhedonia paradigms may be particularly relevant as animal models of depression. The inability to respond to pleasure is a core symptom of depression and resembles loss, in that both involve a decrease in positive reinforcement; by contrast, anhedonia is not seen in anxiety and has no obvious relationship to threat. The pharmacological data are consistent with these observations. Stress-induced anhedonia is reversible by antidepressants in all of the paradigms in which these effects have been tested (Koolhaas *et al.*, 1990; Zacharko and Anisman, 1991; Willner *et al.*, 1992); however, in the chronic mild stress paradigm, anhedonia was not reversed by a benzodiazepine (Muscat *et al.*, 1992).

Stress-induced response suppression, the behaviour most commonly studied under the rubric 'learned helplessness', has usually been considered to be an animal model of depression. This attribution assumed that the behavioural changes reflect a loss of motivation related to perceptions of loss of control (Seligman, 1975). However, as discussed above, recent data suggest that at least some of these effects reflect a state of hypervigilance resulting from excessive fear in an unpredictable environment (Jackson and Minor, 1988). From this perspective, behaviour in some stress-induced response suppression paradigms may be more appropriately considered to represent animal models of anxiety. However, while stress-induced response suppression may be prevented by benzodiazepines, they fail to reverse it once the effect is established (in contrast to antidepressants, which are effective both before and after stress; Drugan *et al.*, 1984). This suggests that the relevance of these paradigms may be to a form of anxiety that does not respond to benzodiazepines.

Phobia is one possibility, given the importance of conditioned fear in the prolongation of stress-induced response suppression: phobias do not respond to treatment with benzodiazepines, but there is evidence that some phobic states do respond to antidepressants (Liebowitz, 1992).

A different form of behavioural suppression is seen as a long-term decrease in locomotor activity, which apparently increases over time, following footshock or social defeat (Koolhaas et al., 1990; van Dijken et al., 1992a,b). It is difficult to see how these (apparently) spontaneous increases in behavioural suppression over time could be explained by conditioning; rather these paradigms may reflect a non-associative time-dependent sensitization, such as has been described following intermittent administration of psychostimulants (Paulson et al., 1991). These effects are probably relevant to anxiety rather than depression, since they are reversed by anxiolytics, and there are a number of interesting neurobiological parallels to posttraumatic stress disorder (van Dijken, 1992).

The object of presenting these examples is not to provide an exhaustive review of the relationship of stress phenomena to specific forms of psychopathology but rather, to indicate some of the approaches that can be used to address this question. Perhaps most important, the need to maintain a critical and open mind on animal models cannot be overemphasized: cynicism and propaganda are equally inappropriate as analytical tools for evaluating scientific procedures. There is an understandable, but perhaps counter-productive, tendency for the protagonists of particular experimental models to emphasize evidence supporting their relevance to particular disorders. Both sides in these debates should be alive to the possibility that an apparently invalid animal model might well be valid as a simulation of a different disorder. The further development and validation of animal models of specific stress-related disorders will also rely crucially on the availability of more and better clinical information concerning pharmacological profiles, differential diagnosis and psychological mechanisms. If stress-related disorders are themselves poorly understood, then animal models of those disorders must inevitably be subject to the same limitations.

6.6 REFERENCES

Abramson, L.Y., Seligman, M.E.P. and Teasdale, J.D. (1978) Learned helplessness in humans: Critique and reformulation. *J. Abnorm. Psychol.*, **87**: 49–74

Abramson, L.Y., Metalsky, G. and Alloy, L.B. (1989) Hopelessness depression: A theory-based subtype of depression. *Psychological Reviews*, **96**: 358–72

Anisman, H.A. and Zacharko, R.M. (1982) Depression: The predisposing influence of stress. *Behav. Brain Sci.*, **5**: 89–137

Anisman, H., Irwin, J. and Sklar, L.S. (1979) Deficits of escape performance following catecholamine depletion: implications for behavioural deficits induced by uncontrollable stress. *Psychopharmacology*, **64**: 163–70

Antelman, S.M., Eichler, A.J., Black, C.A. and Kocan, D. (1980) Interchangeability of stress and amphetamine sensitization. *Science*, **207**: 329–31

Armario, A., Castellanos, J.M. and Balasch, J. (1984) Adaptation of anterior pituitary hormones to chronic noise stress in male rats. *Behav. Neural Biol.*, **41**: 71–6

Armario, A., Restrepo, C., Castellanos, J. M. and Balasch, J. (1985) Dissociation between adrenocorticotrophin and corticosterone responses to restraint after previous chronic exposure to stress. *Life Sci.*, **36**: 2085–92

Bauman, R.A. and Kant, G.J. (1991) Circadian effects of escapable and inescapable shock on the food intake and wheelrunning of rats. *Physiol. Behav.*, **51**: 167–74

Berlyne, D.E. (1960) *Conflict. Arousal and Curiosity*, McGraw-Hill, New York

Blanchard, R.J. and Blanchard, D.C. (1971) Defensive reactions in the albino rat. *Learn. Motiv.*, **2**: 351–62

Bracewell, R.J. and Black, A.H. (1974) The effects of restraint and noncontingent preshock on subsequent escape learning in the rat. *Learn. Motiv.*, **5**: 53–69

Broadhurst, P.L. (1975) The Maudsley reactive and nonreactive strains of rats: A survey. *Behav. Genet.*, **5**: 299–319

Brown, G.W. and Harris, T. (1978) *Social Origins of Depression*, Tavistock, London

Cannon, W.B. (1936) *The Wisdom of the Body*, Norton, New York

Cassens, G.P., Actor, C., Kling, M. and Schildkraut, J.J. (1981) Amphetamine withdrawal affects threshold of intracranial self-stimulation. *Psychopharmacology*, **73**: 318–22

Czeisler, C.A., Ede, M.C. and Regestein, Q.R. (1976) Episodic 24-hour cortisol secretory patterns in patients awaiting elective cardiac surgery. *J. Clin. Endocrinol. Metab.*, **42**: 273–83

De Boer, S.F., Koopmans, S.J., Slangen, J.L. and van der Gugten, J. (1990a) Plasma catecholamine, corticosterone and glucose responses to repeated stress in rats: Effects of interstressor interval length. *Physiol. Behav.*, **47**: 1117–24

De Boer, S.F., Van der Gugten, J. and Slangen, J.L. (1990b) Plasma catecholamine and corticosterone levels during active and passive shock-prod avoidance behavior in rats: effects of chlordiazepoxide. *Physiol. Behav.*, **47**: 1089–98

Desan, P.H., Silbert, L.H. and Maier, S.F. (1988) Long-term effects of inescapable stress on daily running activity and antagonism by desipramine. *Pharmacol. Biochem. Behav.*, **30**: 21–9

Dourish, C.T., Gorka, Z., Williams, A.R. and Iversen, S.D. (1989) Potential influence of social support in a rodent model of depression. *J. Pyschopharmacol.*, **3**

Driscoll, P. and Battig, K. (1982) Behavioral, emotional and neurochemical profiles of rats selected for extreme differences in active, two-way avoidance performance, in I. Lieblich, (ed.), *Genetics of the Brain*, Elsevier, Amsterdam, pp 95–123

Drugan, R., Ryan, S.M., Minor, T.R. and Maier, S.F. (1984) Librium prevents the analgesia and shuttle-box escape deficits typically observed following inescapable shock. *Pharmacol. Biochem. Behav.*, **21**: 749–54

Edwards, E., Harkins, K., Wright, G. and Henn, F. (1990) Effects of bilateral adrenalectomy on the induction of learned helplessness behavior. *Neuropsychopharmacology* **3**: 109–14

Ellenbroek, B.A. and Cools, A.R. (1990) Animal models with construct validity for schizophrenia. *Behav. Pharmacol.* **1**: 469–90

Elliott, G. R. and Eisdorfer, C. (1982) *Stress and Human Health*, Springer-Verlag, New York

Everly, G.S. and Sobelman, S.A. (1987) *Assessment of the Human Stress Response*, AMS Press, New York

Geller, I. and Seifter, J. (1960) The effects of meprobamate, barbiturates, d-amphetamine and chlorpromazine on experimentally induced conflict in the rat. *Psychopharmacologia*, **9**: 482–92

Giralt, M. and Armario, A. (1988) Individual housing does not influence the adaptation of the pituitary-adrenal axis and other physiological variables to chronic stress in adult male rats. *Physiol. Behav*, **45**: 477–81

Glazer, H.I. and Weiss, J.M. (1976a) Long-term and transitory interference effects. *J. Exp. Psychol: Anim. Behav. Proc.*, **2**: 191–201

Glazer, H.I. and Weiss, J.M. (1976b) Long-term interference effect: an alternative to 'learned helplessness'. *J. Exp. Psychol: Anim. Behav. Proc.*, **2**: 202–13

Goodwin, G.M., Muir, W.J., Secki, J.R. *et al.* (1992) The effects of cortisol infusion upon hormone secretion from the anterior pituitary and subjective mood in depressive illness and in controls. *J. Affect. Disord.*, **26**: 73–84

Goudie, A. (1991) Animal models of drug abuse and dependence, in P. Willner, (ed.), *Behavioural Models in Psychopharmacology: Theoretical, Industrial and Clinical Perspectives*, Cambridge University Press, Cambridge, pp 453–84

Green, S. and Hodges, H. (1991) Animal models of anxiety, in P. Willner, (ed.), *Behavioural Models in Psychopharmacology: Theoretical, Industrial and Clinical Perspectives*, Cambridge University Press, Cambridge, pp 21–49

Groves, P.M. and Thompson, R.F. (1970) Habituation: A dual process theory. *Psychol. Rev.*, **77:** 419–50

Henry, J.P. and Stephens, P.M. (1977) The social environment and essential hypertension in mice: Possible role of the innervation of the adrenal cortex, in W. DeLong, A.P. Provoost, and A.P. Shapiro, (eds.), *Hypertension and Brain Mechanisms*, Elsevier, New York, pp 263–76

Jackson, R.L. and Minor, T.R. (1988) Effects of signalling inescapable shock on subsequent escape learning: Implications for theories of coping and 'learned helplessness'. *J. Exp. Psychol.: Anim. Behav. Proc.*, **14:** 390–400

Johnson, J.E., Dabbs, J.M. and Leventhal, J.H. (1970) Psychological factors in the welfare of surgical patients. *Nursing Res.*, **19:** 337–42

Johnston, M. (1988) Impending surgery, in Fisher, S. and Reason, J. (eds.), *Handbook of Life Stress, Cognition and Health*, Wiley, Chichester, pp 79–100

Kant, G.J., Eggleston, T., Landman-Roberts, L. *et al.* (1985) Habituation to stress is stressor specific. *Pharmacol. Biochem. Behav.*, **22:** 631–4

Katz, R.J. and Baldrighi, G. (1982) A further parametric study of imipramine in an animal model of depression. *Pharmacol. Biochem. Behav.*, **16:** 969–72

Kennett, G.A., Chaouloff, F., Marcou, M. and Curzon, G. (1986) Female rats are more vulnerable than males in an animal model of depression. *Eur. J. Pharmacol.*, **124:** 265–74

Knych, E.T. and Eisenberg, R.M. (1979) Effects of amphetamine on plasma corticosterone in the conscious rat. *Neuroendocrinology*, **29:** 110–18

Koolhaas, J.M., Hermann, P.M., Kemperman, C. *et al.* (1990) Single social defeat in male rats induces a gradual but long-lasting behavioural change: A model of depression? *Neurosci. Res. Commun.*, **7:** 35–41

Korte, S.M., Bouws, G.A.H., Koolhaas, J.M. and Bohus, B. (1991a) Neuroendocrine and behavioural responses during conditioned active and passive behaviour in the defensive burying/probe avoidance paradigm: effects of ipsapirone. *Physiol. Behav.*, **52:** 355–61

Korte, S.M., Smit, J., Bouws, G.A.H., Koolhaas, J.M. and Bohus, B. (1991b) Neuroendocrine evidence for hypersensitivity in serotonergic neuronal system after psychosocial stress of defeat, in B. Olivier, J. Mos, and J. Slangen, (eds.), *Animal Models in Psychopharmacology*, Birkhauser, Basel, pp 199–203

Kudryatseva, N.N., Bakshtanovskaya, I.V. and Koryakina, L.A. (1991) Social model of depression in mice of C57BL/6J strain. *Pharmacol. Biochem. Behav.*, **38:** 315–20

Lazarus, R.S. and Cohen, J.B. (1977) Environmental stress, in I. Altman, and J.F. Wohlwill, (eds), *Human Behavior and the Environment: Current Theory and Research*, Plenum, New York

Lazarus, R.S. and Folkman, S. (1984) *Stress, Appraisal and Coping*, Springer-Verlag, New York

Laird, C.G. (1971) *Webster's New World Thesaurus*, New American Library, New York

Leith, N.J. and Barrett, R.J. (1976) Amphetamine and the reward system: evidence for tolerance and post-drug depression. *Psychopharmacologia* **46:** 19–25

Liebowitz, M.R. (1992) Reversible MAO inhibitors in social phobia, bulimia, and other disorders. *Clin. Neuropharmacol.*, **15**(Suppl. 1A): 434–5

MacLennan, A.J., Drugan, R.C., Hyson, R.L. and Maier, S.F. (1983) Dissociation of long-term analgesia and the shuttle box escape deficit caused by inescapable shock. *J. Comp. Physiol. Psychol.*, **96:** 904–12

Maier, S.F., Drugan, R.C. and Grau, J.W. (1982) Controllability, coping behavior and stress-induced analgesia in the rat. *Pain*, **12:** 47–56

Maier, S.F., Ryan, S.M., Barksdale, C.M. and Kalin, N.H. (1986) Stressor controllability and the pituitary-adrenal system. *Behav. Neurosci.*, **100:** 669–74

Mason, J.W. (1968) A review of psychoendocrine research on the pituitary-adrenal cortical system. *Psychosom. Med.*, **30:** 576–607

Mason, J.W. (1971) A reevaluation of the concept of "non-specificity" in stress theory. *J. Psychiatr. Res.*, **8:** 323–33

Matthews, K.A. (1982) Psychological perspectives on the Type A behaviour pattern. *Psychol. Bull.*, **91:** 293–323

McKinney, W.T. (1984) Animal models of depression: An overview. *Psychiatr. Dev.*, **2:** 77–96

Miller, J.G. (1953) The development of experimental stress-sensitive tests for predicting performance in military tasks. *PRB Tech. Report 1079*, Psychological Research Associates, Washington DC

Minor, T.R., Jackson, R.L. and Maier, S.F. (1984) Effects of task-irrelevant cues and reinforcement delay on choice-escape learning following inescapable shock: Evidence for a deficit in selective attention. *J. Exp. Psychol.: Anim. Behav. Proc.*, **10:** 543–56

Minor, T.R., Pelleymounter, M.A. and Maier, S.F. (1988) Uncontrollable shock, forebrain norepinephrine, and stimulus selection during choice escape learning. *Psychobiology*, **16:** 135–45

Moreau, J.-L., Jenck, F., Martin, J.R. *et al.* (1992) Antidepressant treatment prevents chronic unpredictable mild stress-induced anhedonia as assessed by ventral tegmental self-stimulation behavior in rats. *Eur. Neuropsychopharmacol.*, **2:** 43–9

Murison, R. and Ursin, H. (1982) Stress and activation. *Behav. Brain Sci.*, **5:** 115–16.

Muscat, R and Willner, P (1992) Suppression of sucrose drinking by chronic mild unpredictable stress: A methodological analysis. *Neurosci. Biobehav, Rev.* **16:** 507–17

Muscat, R., Papp, M. and Willner, P. (1992) Reversal of stress-induced anhedonia by the atypical antidepressants fluoxetine and maprotiline. *Psychopharmacology* **109:** 433–8

Orr, T.E., Meyerhoff, J.L., Mougey, E.H. and Bunnell, B.N. (1990) Hyperresponsiveness of the rat neuroendocrine system due to repeated exposure to stress. *Psychoneuroendocrinology*, **15:** 317–28

Ottenweller, J.E., Natelson, B.H., Pitman, D.L. and Drastal, S.D. (1989) Adrenocortical and behavioral responses to repeated stressors: Toward an animal model of chronic stress and stress-related mental illness. *Biol. Psychiatr.*, **26:** 829–41

Overmeier, J.B., Murison, R., Skoglund, E.J. and Ursin, H. (1985) Safety signals can mimic responses in reducing the ulcerogenic effects of prior shock. *Physiol. Psychol.*, **13:** 243–7

Overstreet, D.H. and Janowsky, D.S. (1991) A cholinergic supersensitivity model of depression, in A. Boulton, G. Baker and M. Martin-Iverson, (eds.), *Neuromethods, Vol. 20: Animal Models in Psychiatry*, Birkhauser, Basel, pp 81–114

Paulson, P.E., Camp, D.M. and Robinson, T.E. (1991) Time course of transient behavioral depression and persistent behavioral sensitization in relation to regional monoamine concentrations during amphetamine withdrawal in rats. *Psychopharmacology*, **103:** 480–92

Piazza, P.-V., Deminiere, J.-M., Maccari, S. *et al.* (1990) Individual reactivity to novelty predicts probability of amphetamine self-administration. *Behav. Pharmacol.*, **1:** 339–45

Pitman, D.L., Ottenweller, J.E. and Natelson, B.H. (1988) Plasma corticosterone levels during repeated presentation of two intensities of restraint stress: Chronic stress and habituation. *Physiol. Behav.*, **43:** 47–55

Pitman, D.L., Ottenweller, J.E. and Natelson, B.H. (1990) Effect of stressor intensity on habituation and sensitization of glucocorticoid responses in rats. *Behav. Neurosci.*, **104:** 28–36

Platt, J.E. and Stone, E.A. (1982) Chronic restraint stress elicits a positive antidepressant response on the forced swim test. *Eur. J. Pharmacol.*, **82:** 179–81

Plaznik, A., Stefanski, R. and Kostowski, W. (1989) Restraint stress-induced changes in saccharin preference: The effect of antidepressive treatment and diazepam. *Pharmacol. Biochem. Behav.*, **33:** 755–9

Pomerleau, O.F. and Pomerleau, C.S. (1990) Cortisol response to a psychological stressor and/or nicotine. *Pharmacol. Biochem. Behav.*, **36:** 211–13

Prince, C.R. and Anisman, H. (1990) Situation specific effects of stressor controllability on plasma corticosterone changes in mice. *Pharmacol. Biochem. Behav.*, **37:** 613–21

Pucilowski, O., Overstreet, D.S., Rezvani, A. and Janowsky, D.S. (1992) Effects of acute and chronic stressors on saccharin preference in hypercholinergic rats. *Behav. Pharmacol.*, **3**(Suppl. 1): 50

Ratner, A., Yelvington, D.B. and Rosenthal, M. (1989) Prolactin and corticosterone response to repeated footshock stress in male rats. *Psychoneuroendocrinology*, **14:** 393–6

Robinson, T.E. and Becker, J.B. (1986) Enduring changes in brain and behavior produced by chronic amphetamine administration: A review and evaluation of animal models of amphetamine psychosis. *Brain Res. Rev.*, **11:** 157–98

Rodgers, R.J. and Randall, J.I. (1988) Environmentally induced analgesia: situational factors, mechanisms and significance. in R.J. Rodgers, and S.J. Cooper, (eds.), *Endorphins, Opiates and Behavioural Processes*, Wiley, Chichester, pp 107–42

Sarter, M., Hagan, J. and Dudchenko, P. (1992) Behavioral screening for cognition enhancers: From indiscriminate to valid testing. *Psychopharmacology*, 107: 144–59, 461–73

Schiller, G.D., Daws, L.C., Overstreet, D.H. and Orbach, J. (1991) Lack of anxiety in an animal model of depression with cholinergic supersensitivity. *Brain Res. Bull.*, 26: 433–5

Scholtens, J. and van de Poll, N.E. (1987) Behavioral consequences of agonistic experiences in the male S3 (Tyron Maze Dull) rat. *Aggress. Behav.*, 14: 371–87

Sherman, A.D., Sacquitne, and Petty, F. (1982) Specificity of the learned helplessness model of depression. *Pharmacol. Biochem. Behav.*, 16: 449–54

Seligman, M.E.P. (1975) *Helplessness: On Depression, Development and Death*, Freeman, San Francisco

Selye, H. (1952) *The Story of the Adaptation Syndrome*, Acta, Montreal

Shanks, N. and Anisman, H. (1988) Stressor-provoked behavioral changes in six strains of mice. *Behav. Neurosci.*, 102: 894–905

Stretch, R., Orlof, E.R. and Dalrymple, S.D. (1968) Maintenance of responding by fixed-interval schedule of electric shock presentation in squirrel monkeys. *Science*, 162: 583–6

Treit, D. (1985) Animal models for the study of anti-anxiety agents: A review. *Neurosci. Biobehav. Rev.*, 9: 203–33

van Dijken, H.H. (1992) Once is enough: An animal study on temporal aspects of stress-induced behavioural and neuroendocrine changes. *PhD thesis*, Free University of Amsterdam

van Dijken, H.H., Van der Heyden, J.A.M., Mos, J. and Tilders, F. (1992a) Inescapable footshocks induce progressive and long-lasting behavioural changes in male rats. *Physiol. Behav.*, 51: 787–94

van Dijken, H.H., Tilders, F.J.H., Olivier, B. and Mos, J. (1992b) Effects of anxiolytic and antidepressant drugs on long-lasting behavioural deficits resulting from one short stress experience in male rats. *Psychopharmacology*, 109: 395–402

Weiss, J.M. (1970) Somatic effects of predictable and unpredictable shock. *Psychosom. Med.*, 32: 397–408

Weiss, J.A., Glazer, L.A., Pohorecky, L.A. *et al.* (1975) Effects of chronic exposure to stressors on avoidance-escape behavior and on brain norepinephrine. *Psychosom. Med.*, 37: 522–33

Weiss, J.M., Pohorecky, L.A., Salman, S. and Gruenthal, M. (1976) Attenuation of gastric lesions by psychological aspects of aggression in rats. *J. Comp. Physiol. Psychol.*, 90: 252–9

Weiss, J.M., Bailey, W.H., Goodman, P.A. *et al.* (1982) A model for neurochemical study of depression, in M.Y.Spiegelstein and A Levy (eds.), *Behavioral Models and the Analysis of Drug Action*, Elsevier, Amsterdam, pp 195–223

Williams, J.L. (1982) Influence of shock controllability by dominant rats on subsequent attack and defensive behaviors towards colony intruders. *Anim. Learn. Behav.*, 10: 305–13

Willner, P. (1984a) The validity of animal models of depression. *Psychopharmacology*, 83: 1–16

Willner, P. (1984b) Drugs, biochemistry and subjective experience: Towards a theory of psychopharmacology. *Perspect. Biol. Med.*, 28: 49–64

Willner, P. (1991) Behavioural models in psychopharmacology, in P. Willner, (ed.). *Behavioural Models in Psychopharmacology: Theoretical, Industrial and Clinical Perspectives*, Cambridge University Press, Cambridge, pp 3–18

Willner, P., Towell, A., Sampson, D. *et al.* (1987) Reduction of sucrose preference by chronic mild unpredictable stress, and its restoration by a tricyclic antidepressant. *Psychopharmacology*, 93: 358–64

Willner, P., Muscat, R. and Papp, M. (1992) Chronic mild stress-induced anhedonia: A realistic animal model of depression. *Neurosci. Biobehav. Rev.* (in press)

Wise, R.A. (1978) Catecholamine theories of reward: A critical review. *Brain Res.*, 152: 215–47

Woodmansee, W.W., Silbert, L.H. and Maier, S.F. (1991) Stress-induced changes in daily activity in the rat are modulated by different factors than are stress-induced escape-learning deficits. *Soc. Neurosci. Abstr.*, 17: 146

Zacharko, R.M. and Anisman, H. (1991) Stressor-provoked alterations of intracranial self-stimulation in the mesocorticolimbic dopamine system: an animal model of depression, in

P. Willner, and J. Scheel-Kruger, (eds.), *The Mesolimbic Dopamine System, From Motivation to Action*, Wiley, Chichester, pp 411–42

Zacharko, R.M., Bowers, W.J., Kokkinidis, L. and Anisman, H. (1983) Region-specific reductions of intracranial self-stimulation after uncontrollable stress: Possible effects on reward processes. *Behav. Brain Res.* **9:** 129–41

Zacharko, R.M., Lalonde, G.T., Kasian, M. and Anisman, H. (1987) Strain specific effects of inescapable shock on intracranial self-stimulation from the nucleus accumbens. *Brain Res.,* **426:** 164–8

Zuckerman, M., Buchsbaum, M.T. and Murphy, L. (1980) Sensation seeking and its biological correlates. *Psychol. Bull.,* **88:** 187–214

7

Coping with stress

Robert Dantzer

INRA-INSERM U176
BORDEAUX CEDEX,
FRANCE

7.1 CONCEPTUAL ISSUES

The first formulations of theories of stress were built on the concepts of homeostasis and vital energy. Confronted with stimuli that threaten the constancy of their 'milieu interieur' (stressors), organisms were claimed to have little choice other than to initiate a counter-reaction to correct the disturbance and reinstate the original homeostasis (the stress response) (Selye, 1973). To do so, they needed energy. This was the job of pituitary-adrenal hormones which enabled the channelling of bodily resources in the adaptation process. When energy was insufficient or misused, the sanction was exhaustion and death; if the organism had the misfortune to be fragile, the strain to which it was submitted during this expenditure of energy could be responsible for pathology.

Although stress was at first seen as part of a catastrophe scenario (typical stressors were: haemorrhage, bone fractures, burns, explosions and poisons), Selye managed to generalize his stress theory to everyday life by claiming that, since homeostasis is constantly threatened, stress is part of life. Because stress is with us from birth to death, the only way to avoid it is to transform bad stress ('dystress',

ISBN 0–12–663370–3

responsible for exhaustion or pathology) into good stress ('eustress', a recipe for joy and pleasure). The popularity of Selye's theories was mainly due to their concordance with popular psychology. As everybody knows, there is no problem as long as we can manage to transform the bad things with which we are confronted into good things. But the trouble begins when we are unable to do so any longer. Selye was not a psychologist and his scientific approach to stress was based mainly on pathological and histological findings. He was therefore unable to do better than present his pseudo-philosophy in the form of cookbooks of recipes explaining why and how we need to live with stress.

It took some time for psychologists to foresee the potential of applying psychological tools to the concept of stress. By paying attention to the psychological dystress that people can experience when confronted with daily hassles, clinical psychologists were able to observe important differences in the way people deal with their problems. The ground was therefore set for relating psychological outcomes, not solely to the type of problem encountered, but also to the strategies used. This was an inportant move, since it allowed the replacement of the linear model of stress introduced by physiologists by a transactional model. In this model, the outcome was not simply a function of the eliciting situation, but a joint function of the way the situation is perceived and the efficacy of the strategies used by those exposed to it (Figure 7.1). The concept of coping refers to these intervening variables that the subject interposes between the situation and himself, in order to master the situation. Since these strategies are diverse and vary from one person to another or, for the same person, from one situation to another, clinical psychologists have been quite busy describing the infinite variety of coping strategies, classifying them, relating them to personality factors, finding out which ones are better adapted than others, depending on the situation, and developing tools for predicting which coping strategies will be selected in face of a given event, rather than merely describing them.

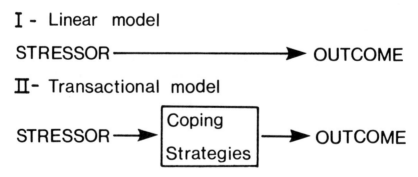

Figure 7.1 Linear and transactional models of stress

Psychobiologists have approached the issue of coping strategies from a different perspective. In the late 1960s, behaviourists, who were studying the behavioural and physiological consequences of exposure to diverse stressors, realized that the ability to control the situation and predict what is going to happen was actually more important than the physical characteristics of stressors. In these experiments, specific coping strategies were enforced by the experimental conditions to which animals were submitted and could be deduced only *post hoc*, from between-group comparisons (e.g. active attempts to cope *versus* resignation, corresponding respectively to the reactions of animals given the ability to control electric shocks *versus* those of animals denied this possibility). The issue of individual differences in coping strategies through the study of within-group variability took longer to materialize and became a prominent topic for research only recently.

In everyday language, the term 'coping' has strong implications of success. To cope with something is to do one's best to master it. When different coping strategies are available to deal with a problematic situation, the most successful strategy is the one which enables resolution of the situation at minimum cost. Even without taking into account socio-cultural constraints on what is or is not acceptable in terms of coping attempts, it is not always easy to define the cost of a given coping strategy, especially when the situation cannot be modified by the subject's actions. In subjective terms, it should be sufficient to question the subject, ask him/her to evaluate the effort which is needed to solve the problem and compare this effort to the dystress elicited by the problem. This would be much easier to do, however, if there was some objective indicator of the impact of the situation on the individual. For example, if the stressor elicits a response A and if, in individuals engaged in a specific coping attempt, the observed reaction is B, such that $B < A$, then it can be concluded that the subject's coping attempts have indeed been successful (Levine *et al.*, 1978).

To operationalize this way of assessing the efficacy of coping attempts, it is necessary to have appropriate indicators of the impact of the situation on the organism. In physiologically-oriented research, the relevant concept is that of arousal or activation. Although this concept has a long tradition in psychology and in neurophysiology, it is used in the coping literature to refer to a general, non-specific response to all situations in which there is a potential, or a real, threat to the organism (Ursin, 1985). Within this conceptual framework, coping responses refer to the means that are available to an individual for reducing activation induced by highly aversive situations. Several strategies are potentially relevant to this goal, but the important point is that a given strategy emerges and is retained if, and only if, it proves to be functional in the sense of reducing activation. This concept is important since it allows study of the organization and efficacy of coping strategies in non-human animals and the relationship of experimental findings in animals with clinical data.

The present review attempts to address the question of whether or not these findings support the basic premises of the coping theory.

7.2 METHODOLOGICAL ISSUES

Coping responses are conveniently categorized by clinical psychologists into two different classes, according to whether they aim to alter the stimulus situation ('problem-focused coping'), or whether they aim to control the emotion elicited by the stimulus situation ('emotion-focused coping'; Lazarus and Folkman, 1984). This classification has a counterpart in animal experiments, in the form of the distinction between behaviours that enable the subject to change the situation (e.g. escape/avoidance behaviour) and behaviours that are apparently irrelevant with respect to the eliciting conditions (e.g. displacement activities).

The physiological correlates of problem-focused coping in animals have mainly been studied by using negative reinforcers, i.e. stimuli whose removal or prevention increases responding. Electric shock has been the preferred stimulus because of its ease of application. When offset of electric shock is contingent upon responding, the negative reinforcement operation is designated 'escape'. When non-occurrence, or postponement, of electric shock is contingent upon responding, the negative reinforcement operation is designated 'avoidance'.

Experiments on coping are usually based on the comparison of animals which can escape or avoid electric shocks with others denied this possibility, but which receive the same electric shocks by being attached to the same electrodes (the yoked control design). In a typical experiment, rats in the escape/avoidance group are permitted to terminate electric shocks, or to delay their occurrence, by turning a wheel with their forepaws. Rats in the second group serve as yoked controls since their responses have no effect on the occurrence and duration of electric shocks; the shocks are in fact entirely dependent on the behaviour of their escape/avoidance partners. Electric shocks are delivered *via* electrodes fastened on the tail instead of an electrified grid floor to avoid possible position biases which decrease the effectiveness of shock. As a further control, a group of rats are placed in the same apparatus as the first two groups, but are not exposed to electric shocks (Figure 7.2).

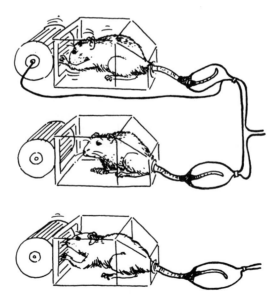

Figure 7.2 The yoked control design. The top and middle rats are connected to the same shock generator, whereas the bottom rat is not connected. Turning the wheel terminates electric shock for the first but not for the second rat (from Dantzer, 1989)

The yoked control design appears ideal for differentiating between contingencies and consequences, since one group receives response-consequences pairings, while the other receives only consequences. It therefore enjoys a wide popularity for demonstrating differential effects of predictable *versus* unpredictable reinforcers and controllable *versus* uncontrollable reinforcers. However, the inference of the importance of subjective states, such as predictability and controllability, in the mediation of the observed differences in such experiments can be questioned. As pointed out by Church (1964) in a largely ignored paper, the yoked control design cannot be used to infer the role of response-consequence contingency. The reason is that within-subjects and between-subjects random variability in the effectiveness of a consequence produces systematic differences between an experimental group and its yoked control design. These flaws are serious and cannot be corrected (Church, 1989).

Displacement activities are behavioural responses that typically occur in situations where a strongly motivated behaviour is interrupted or prevented. Responses produced in this way are thought to be energized by a surplus of motivational energy, the discharge of which through the normal paths is somehow prevented (Tinbergen, 1950). They usually consist of incomplete innate patterns of behaviour which are apparently irrelevant to the releasers present in the situation. Typical examples of displacement activities include courtship feeding and preening in birds and displacement digging and fanning in sticklebacks. There are a number of alternative explanations to the 'surplus energy' model proposed by early ethologists. In particular, it has been suggested that displacement activities are due to disinhibition and have the function of diverting attention away from the stimuli normally controlling ongoing behaviour (McFarland, 1985). They would therefore represent a form of emotion-focused coping.

Experimental analogues of displacement activies have been found in the laboratory, in the form of adjunctive or schedule-induced behaviours that typically occur when hungry animals are exposed to an intermittent schedule of food reinforcement (Falk, 1971). In a now classic experiment (Falk, 1961), food-deprived rats were trained to respond on a 1 min variable-interval (VI-1 min) schedule of food reinforcement in which a lever-press produced a small food pellet on an average of once each minute. Falk found that, after rats had eaten each pellet, they immediately began drinking water. Although post-prandial drinking is not remarkable, the quantity ingested was significantly above what occurs naturally. For example, during a 3 h test session, rats consumed more than three times their baseline 24 h water intake. Falk called this phenomenon 'schedule-induced polydipsia' (SIP) since it occurred only when food was delivered intermittently. He subsequently reported that it could be produced by a variety of different reinforcement schedules, including a fixed interval schedule of the same length (FI-60), or by delivery of a food pellet at 60 s intervals without requiring the animal to make an operant response (Falk, 1961, 1966) (Figure 7.3).

Falk (1966) suggested that SIP could be classified as an 'adjunctive behaviour' because it: (1) occurs as an adjunct to a reinforcement schedule; (2) is not directly involved in, or maintained by, the reinforcement contingency; and (3) has reinforcing properties. Many activities besides drinking can be elicited by intermittent schedules of food reward, including attack, pica, air-licking, wheel-running, escape and

Figure 7.3 Experimental set-up for studying schedule-induced polydipsia in rats. When water is freely available, rats submitted to an intermittent delivery of food engage in excessive drinking in the interval between successive presentations of food pellets (from Dantzer, 1989)

increased activity. They all share the following properties: (1) their rate of occurrence is an inverted U-function of the inter-reinforcement interval; (2) their rate of occurrence decreases as the food deprivation level decreases; (3) they are persistent and excessive; (4) they occur immediately after the delivery of the food reward; (5) the activity evoked depends on the stimuli that are available in the situation: whether a running-wheel or a drinking tube are present, for example.

All these characteristics have been interpreted as suggesting that adjunctive behaviours are adaptive since they permit animals to remain engaged in a situation that is favourable to their survival, but also contains an aversive component (Falk, 1971, 1977). Accordingly, a hungry rat that receives food on an intermittent schedule has its hunger partially appeased, but its feeding behaviour is also repeatedly thwarted. In this situation, drinking develops from two opposing tendencies: feeding behaviour as a response to the rewarding component of the situation and escape behaviour as a response to the aversive component.

7.3 PHYSIOLOGICAL CORRELATES OF PROBLEM-FOCUSED COPING

The coping hypothesis predicts that, in the yoked control design, experimental subjects (i.e. those receiving escapable stress) should show less arousal than yoked control subjects. To test this prediction, a number of physiological indices of arousal have been used.

7.3.1 PITUITARY-ADRENAL ACTIVITY

Plasma levels of adrenocorticotropic hormone (ACTH) and corticosterone or cortisol have been found to be lower in animals exposed to controllable negative reinforcers than in animals exposed to uncontrollable negative reinforcers. For example, rhesus monkeys, trained to terminate an intense noise by pressing a lever, responded to a 1 h session of noise presentation with a smaller increase in plasma cortisol than monkeys having no control over noise (Hanson et al., 1976). In addition, removal of the contingency between lever presses and noise termination resulted in a similar increase in plasma cortisol to that observed in animals having no control over noise. In the same way, dogs exposed to a series of electric shocks, which they could terminate or prevent, showed a smaller increase in plasma cortisol than yoked control subjects (Dess et al., 1983). The beneficial effect of controllability extends to predictability since the corticosterone response to white noise decreased from the first to the twentieth noise presentation in animals which were exposed to predictable stimulation but not in those which were exposed to an unpredictable stimulation (de Boer et al., 1989) (Figure 7.4).

Several studies, however, failed to find different pituitary-adrenal responses in experimental and yoked control subjects. For example, rats exposed to a free-operant procedure in which they had to pull a disk to avoid or escape tailshock had elevated levels of plasma corticosterone at the end of a shock session lasting 3, 6 or 21 h. These hormonal levels did not differ from those of yoked rats (Tsuda and Tanaka, 1985).

Some authors have suggested that the main difference between experimental and yoked control subjects is not in the peak level of the pituitary-adrenal response but in the time taken to return to baseline. Since the negative feedback of corticosterone on the stress response is mediated by hippocampal influences on the hypothalamus (McEwen et al., 1986), this would imply that exposure to uncontrollable stressors can affect limbic functions. For example, Swenson and Vogel (1983) reported that plasma corticosterone levels remained elevated significantly longer in rats exposed to 60 mins of inescapable foot-shock than in rats exposed to escapable shock. However, this finding could not be replicated by Maier et al. (1986) in an experiment where rats were submitted to 80 trials of shock-escape training in wheel-turn boxes. Under these conditions, escapable and inescapable electric shocks led not only to equal rises of both plasma corticosterone and ACTH, but also to a similar rate of decline in plasma levels of these two hormones after termination of the shock session.

It could be argued that the failure to find differences between experimental and yoked control animals is due to the use of a single test session; there would be little opportunity to develop a dissociation between the possible de-arousal effects of

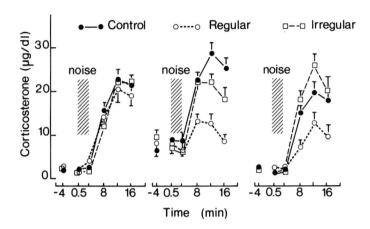

Figure 7.4 Time course of the plasma corticosterone response to predictable or unpredictable white noise stimulation in rats. Rats bearing a permanent jugular catheter were exposed to a series of 20 successive noise stimuli, 4 min duration, spaced regularly or irregularly. Control animals were exposed only to the first and the last stimuli of the series. The left part of the figure represents changes in plasma corticosterone in response to the first signal; the middle part the response to the last signal; the right part the response to a single noise stimulus presented 24 h later. Note that the plasma corticosterone response habituates more rapidly in rats exposed to regular than to irregular noise stress (adapted from de Boer *et al.*, 1989)

coping and the arousal-enhancing effects of the effort needed to gain control over the stressor. If this were the case, attenuation of the pituitary-adrenal response in experimental animals should be easier to demonstrate in chronic than in acute stress conditions. In accordance with this prediction, the plasma corticosterone levels of rats exposed to five daily sessions of a free operant schedule of avoidance responding did not differ from those of non-shocked rats, whereas the corticosterone response was still maximal in yoked controls (Tsuda and Tanaka, 1985). Similarly, rats with an extensive history of avoidance behaviour in a Skinner box displayed lower plasma levels of corticosterone at the end of a one-hour avoidance session than yoked animals (Herrmann *et al.*, 1984). In this latter case, however, pre-session corticosterone concentrations in avoidance animals were significantly higher than those in yoked and control subjects. A potential problem with chronic stress exposure is the rapid development of tolerance of the pituitary-adrenal response to the test situation. This tolerance was observed to occur even in rats which did not learn to avoid electric shocks (Coover *et al.*, 1973).

In social conflict situations, studies in tree-shrews, rats and mice suggest that pituitary-adrenocortical activity is elevated more in animals which behave defensively or passively and become submissive than in offensive animals which remain active and become dominant. These differences may, however, be situation specific, as shown by systematic observations of changes in plasma cortisol concentrations in guinea-pigs tested in different environments (Haemisch, 1990). Offensive males displayed a smaller increase in cortisol than defensive males when tested in their home cage, but not when tested in the cage of their opponent.

In summary, it is clear that coping strategies have a profound influence on the intensity and duration of the pituitary-adrenal response. However, it would be premature to conclude from these findings that coping is always accompanied by lower physiological activation.

7.3.2 SYMPATHETIC-ADRENOMEDULLARY RESPONSES

Acute exposure to stressors evokes significant increases in plasma noradrenaline and adrenaline; these increases are positively correlated with intensity of the stressor. Since plasma noradrenaline comes mainly from sympathetic nerve terminals and plasma adrenaline from the adrenal medulla, changes in plasma catecholamines give an estimate of the functional activity of the sympathetic-adrenomedullary system.

Swenson and Vogel (1983) reported that chronically catheterized rats exposed to a single session of inescapable footshock displayed higher peak plasma concentrations of adrenaline and noradrenaline than rats receiving escapable shocks. Yoked rats also displayed prolonged elevations of plasma catecholamines after termination of footshock. Since these changes were observed in conditions where peak plasma corticosterone levels did not differ between the two groups, it is possible that plasma catecholamine concentrations are more sensitive to the influence of control than plasma corticosterone concentrations, but the generality of this finding remains to be tested.

Since repeated exposure to stressors is accompanied by tolerance of plasma catecholamine responses to the familiar stressors and sensitization of the response to a novel stressor, Konarska et al. (1990) hypothesized that predictability of chronic intermittent stress should lead to more rapid habituation and decreased sensitization of the hormonal response. However, rats presented with a relatively unpredictable stress regimen on each of 26 consecutive days did not differ in their response to restraint stress on the 28th day from rats chronically exposed to highly predictable restraint stress. Body weight gain was more sensitive than plasma catecholamines since unpredictable chronic intermittent stress was attended by a greater attenuation of body weight gain than in predictable stress conditions. The conclusion that the sympathetic-adrenomedullary system is relatively insensitive to stress predictability must be tempered, however, by the finding that plasma noradrenaline and adrenaline responses differ in rats exposed to either regularly or irregularly scheduled white noise pulses in a single day (de Boer et al., 1989) (Figure 7.5).

7.3.3 STRESS-INDUCED ANALGESIA

Another physiological concomitant of stress is the increase in pain threshold that occurs in individuals exposed to a wide variety of stressful situations. This

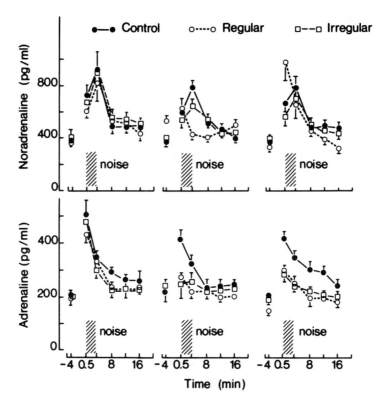

Figure 7.5 Time course of the plasma catecholamine response to predictable or unpredictable white noise stimulation in rats. The protocol is the same as for Figure 7.4. Note that variations in plasma noradrenaline but not adrenaline discriminate the predictable and unpredictable stress conditions (de Boer *et al.*, 1989)

phenomenon is known as stress-induced analgesia (SIA). It can be demonstrated easily by measuring changes in the latency of threshold responses to pain, such as tail-flick in rats and mice when radiant heat is focused on their tail, or paw-licking in animals placed on a hot-plate. After exposure to stress, latencies of these responses to pain increase in a manner which resembles the effect of injection of an analgesic dose of morphine.

Based on ethological considerations, Bolles and Fanselow (1980) proposed that activation of the endogenous analgesic systems that mediate SIA might play a critical role in modulating defensive responses to stressors which threaten the organism's physical integrity. To respond quickly and efficiently to innate or acquired danger signals, animals must ignore injurious pain and concentrate their attention on the best way to escape danger. Obviously, licking an injured limb is better carried out in a safe and remote place than in front of the predator which is responsible for the injury. Endogenous analgesic systems are activated by the same stimuli that activate defensive behaviour, so that analgesia prevents the disruptive influences of pain on coordinated defence activities. Stated in psychological terms, the fear-motivated defensive system inhibits the pain-motivated system (Figure 7.6).

SIA is a heterogeneous phenomenon since different stressors activate different analgesic substrates, including opioid/non-opioid and hormonal/non-hormonal types. Different forms of SIA appear to be associated with different coping strategies. Exposure of rats to a series of inescapable electric shocks produced a

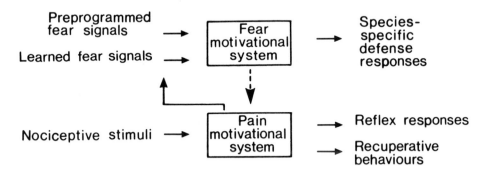

Figure 7.6 Reciprocal interactions between the fear and pain motivational systems. Solid lines represent facilitatory influences whereas dashed lines represent inhibitory influences (adapted from Bolles and Fanselow, 1980)

non-opioid form of SIA at the beginning of the session, which was replaced by an opioid SIA at the end of the shock session (Grau *et al.*, 1981; see also Chapter 6). In the same manner, mice exposed to a highly aggressive conspecific initially exhibited a non-opioid form of analgesia that was blocked by benzodiazepines. Opioid SIA occurred later in the encounter when the attacked animals no longer tried to fight back or flee and when these active responses were replaced by passive defence and submissive behaviour. Based on these observations and other data from the literature relating SIA to the uncontrollability of stressors, Rodgers and Randall (1987) proposed that non-opioid SIA is typically associated with active defensive behaviour initiated in the face of uncertain and remote dangers. The function of analgesia in this context would be to facilitate the shift between strategies, depending on their respective outcomes. In contrast, prolonged contact with the threatening stimulus induces an opioid form of analgesia. The function of this analgesia would be to strengthen the immobility and loss of reflexive movements, so that the attacker loses interest in its prey.

The influence of controllability on SIA is not limited to negative reinforcers, since rats that could obtain food by pressing a lever (controllable food) were found to display less analgesia at the end of the session than rats that received food independently of their behaviour (uncontrollable food) (Tazi *et al.*, 1987) (Figure 7.7). In the same manner, predictable food evokes less SIA than unpredictable food (Tazi *et al.*, 1987).

In summary, predictability and controllability of appetitive and aversive events appear to be critical determinants of the intensity and nature of SIA.

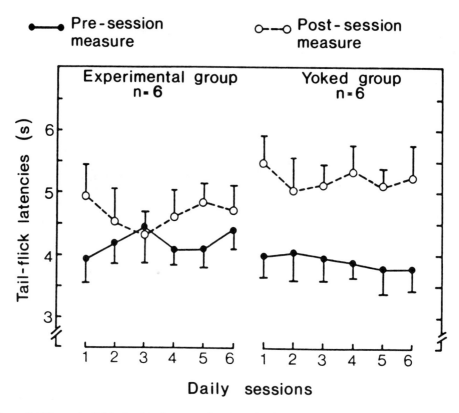

Figure 7.7 Uncontrollable food induces analgesia in food-deprived rats. Experimental animals were able to get food pellets by pressing a lever according to a continuous schedule of food reinforcement during six consecutive daily sessions whereas yoked animals received the same amount and temporal distribution of food as their active partners. Each point represents tail flick latencies (mean + SEM) measured before and after each session (from Tazi *et al.*, 1987)

7.3.4 CIRCADIAN RHYTHMS

Considerable evidence indicates that alterations in circadian rhythms occur during stress, but there has been little systematic investigation of the nature and mechanisms of these effects. Using a within-subject design, Stewart *et al.* (1990a) observed that exposure to four sessions of inescapable electric shock over an 8-day period lengthened the free-running period of locomotion in a running wheel; there was a positive correlation between period lengthening and escape performance tested 1 week later, i.e. those animals which performed poorly in the escape test showed little or no change in period. In a subsequent study, they replicated these findings in both free-running and entrained animals. By comparing groups of animals presenting long or short free-running periods, it was confirmed that there was no influence of baseline period on the response to inescapable shock (Stewart *et al.*, 1990b).

   In another study, chronic stress, in the form of around-the-clock intermittent footshock, greatly decreased the amplitude of the daily body temperature rhythm and this effect was more marked in yoked than in experimental rats (Kant *et al.*, 1991). In addition, reinstatement of the body temperature rhythm was less rapid in yoked than in experimental rats.

These findings on the relationships between circadian rhythms and coping are important since they suggest that plasticity of circadian rhythmicity is associated with the ability to adapt to environmental challenges.

7.3.5 IMMUNE FUNCTION

Exposure to a variety of environmental and psychosocial stressors can alter the functioning of the immune system. However, the old notion that stress exacerbates the progression of disease *via* its corticosteroid-mediated immunosuppressive effects must be revised. Experimental and clinical studies demonstrate that both laboratory and natural stressors alter lymphocyte and macrophage functions in a complex way. This depends on the type of immune response, the physical and psychological characteristics of stressors and the timing of stress relative to the induction and expression of the immune event (Dantzer and Kelley, 1989). In addition, the influences of stress on immunity are mediated not only by glucocorticoids, but also by catecholamines, endogenous opioids and pituitary hormones.

Among the many studies showing an effect of stress on immunity, some have specifically looked at the possibility of a differential effect of controllability and predictability. In mice exposed to a single session of inescapable shock 24 h following tumour cell transplantation, the rate of tumour growth was found to be accelerated, whereas mice exposed to escapable shock did not differ from non-shocked mice either in tumour size or time of tumour appearance (Sklar and Anisman, 1979). However this detrimental effect of uncontrollability on tumour development was no longer apparent in animals exposed chronically to the stressor (Sklar *et al.*, 1981). In rats, a single session of inescapable shock reduced the incidence of rejection of transplanted non-syngeneic tumour cells whereas escapable shock had no effect (Visintainer *et al.*, 1982). The possibility that such changes are mediated by suppression of the immune system is suggested by studies showing that inescapable, but not escapable, shocks decrease cellular immune responses and natural killer cell activity. Rats submitted to inescapable electric shocks on the first day and re-exposed to a few 'reminder' shocks on the second day displayed a lower proliferative response of blood lymphocytes to T-cell mitogens than that in rats submitted to escapable electric shock or in apparatus-control animals (Laudenslager *et al.*, 1983). This impairment of lymphoproliferative responses due to inescapable electric shock appears to carry over to chronic conditions, since it was also observed in yoked animals paired with rats exposed to ten sessions of free-operant avoidance learning in a shuttle-box (Mormède *et al.*, 1988). In this last experiment, however, uncontrollability did not have generalized immunosuppressive effects since, in contrast to the changes observed in lymphoproliferative responses, the primary antibody response against sheep erythrocytes was decreased in the avoidance rats, compared with yoked animals and apparatus-control animals (Figure 7.8). The factors responsible for these differential changes in humoral and cellular immune responses are not known.

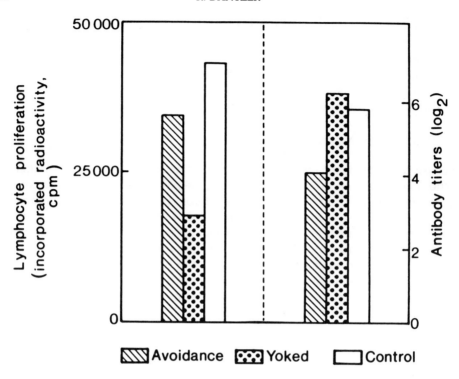

Figure 7.8 Behavioural control over electric shock has differential influences on lymphocyte proliferation and antibody titers in rats. Rats were submitted to 10 daily sessions of continuous avoidance in a shuttle-box, using the yoked control design. All rats were injected with sheep erythrocytes on the fifth day and killed after the last test session. Splenocytes were stimulated by concanavalin A and the maximum lymphocyte proliferation is expressed as amount of ³H-thymidine incorporated (left). Hemagglutinin titers are presented in the right portion of the figure (from Mormède *et al.*, 1988)

Behavioural control also modulates stress-induced immune changes in humans. This was demonstrated by comparison of lymphocyte proliferation and leukocyte percentages in male subjects subjected to mild electric shock and loud white noise, administered in an unpredictable, intermittent fashion (Weisse *et al.*, 1990). Subjects exposed to the controllable stressors had a lower post-stress lymphocyte proliferative response to concanavalin A and a lower percentage of monocytes. These alterations were not observed in subjects who could not control the stressor.

In summary, the available evidence points to controllability and predictability of stressors as critical factors in modulation of immune function.

7.3.6 BRAIN NEUROCHEMISTRY

A vast body of literature indicates that uncontrollable stressors produce a significant depletion of brain noradrenaline in different brain areas, particularly the hypothalamus and brain stem. Depletion of brain noradrenaline results from the inability of increased synthesis to compensate for the high utilization rate of neurotransmitter. In contrast, synthesis of noradrenaline keeps up with increased utilization during exposure to controllable stressors (see Chapter 11). An important charcteristic of the decreased noradrenaline levels observed in animals exposed to uncontrollable shock is that it is a long-lasting phenomenon which can be reinstated after exposure to only a small number of inescapable shocks. This suggests that the lack of control over stressors can lead to sensitization of the neuronal structures involved.

The possibility that changes in noradrenaline metabolism may help to differentiate coping attempts from successful coping is suggested by the results of two independent studies. Swenson and Vogel (1983) attributed the larger depletion of hypothalamic noradrenaline observed in non-coping rats compared with coping animals to the aversive nature of the stimulus. In contrast, a fall in hippocampal noradrenaline was seen only in coping rats and was attributed to the ability to cope with electric shocks. In rats exposed to a free operant schedule of avoidance responding, Tsuda and Tanaka (1985) found a higher noradrenaline turnover in the hypothalamus, amygdala and thalamus of avoidance rats compared to yoked controls after a 3 h or 6 h experimental session, but not after a 21 h session. In this last case, only yoked rats displayed an increased noradrenaline turnover, whereas turnover rates were back to normal in avoidance rats. Successful coping, in the form of significant avoidance performance, was only achieved after a minimum of 6 h exposure to the experimental conditions.

Brain concentrations of 5-hydroxytryptamine (5-HT) appear to be less sensitive to stressful conditions than noradrenaline. Following uncontrollable shock, 5-HT remained unchanged in the brain stem and hypothalamus, but decreased in the locus coeruleus, frontal cortex and septum. These changes were not observed after controllable shock (Weiss et al., 1981).

The mesolimbic and mesocortical dopamine systems, but not the nigrostriatal system are very sensitive to stress. Inescapable electric shock increases dopamine turnover in the first two systems but not the nigrostriatal pathway (see Chapter 11). Under certain conditions, repeated exposure to stress can cause sensitization of these neural structures. This has been demonstrated indirectly, by scoring the intensity of stereotyped activities induced by dopamine agonists, such as cocaine and amphetamine. In an influential paper, McLennan and Maier (1983) demonstrated that electric footshocks enhanced the subsequent stereotyped response to amphetamine and cocaine only if the shocks were uncontrollable. In this experiment, rats that received the same number of shocks as yoked control animals, but were allowed to escape from them, showed levels of stereotypy similar to those of unshocked rats. In comparison, the group that had received uncontrollable shocks showed significantly higher levels of stereotypy than the other two groups.

In summary, the data available on the effects of inescapable electric shocks on brain neurochemistry suggest that uncontrollability has profound and long-lasting influences on the metabolism of central neurotransmitters, especially noradrenaline. Such changes involve processes of neuronal sensitization.

7.4 DISPLACEMENT ACTIVITIES

7.4.1 PHYSIOLOGICAL CORRELATES

The possibility that displacement activities serve a coping function has been tested mainly in the schedule-induced drinking model. Brett and Levine (1979) were the first to report that exposure to intermittent schedules of food delivery elevated plasma corticosterone levels, but that drinking suppressed this elevation. The low levels of plasma corticosterone associated with schedule-induced polydipsia (SIP) could not be attributed to simple haemodilution because significant reductions in corticosterone were observed when the amount of water consumed was still relatively small (Brett and Levine, 1981). These corticosterone-reducing effects of SIP have been replicated (Tazi et al., 1986) and it has further been shown that, if SIP is prevented by removal of the drinking tube, then plasma corticosterone levels are significantly elevated in comparison to those seen when SIP is permitted. Enhanced plasma corticosterone levels are also displayed by those rats which do not drink, despite having access to water (Dantzer et al., 1988a) (Figure 7.9). The consistency of the relationship between adjunctive behaviour and pituitary-adrenal activity has also been investigated using other paradigms. Pigs which are exposed to an intermittent distribution of food with a chain in front of them develop chain chewing behaviour in the interval between food deliveries and display lower plasma cortisol

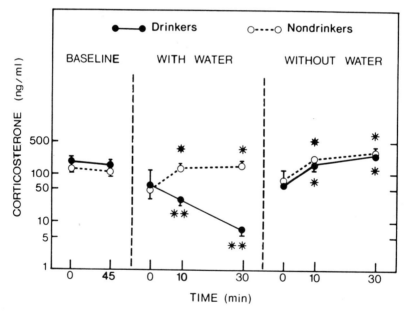

Figure 7.9 Effects of schedule-induced polydipsia on plasma corticosterone levels in rats classified as drinkers or nondrinkers according to their water consumption. Animals had an implanted catheter in their jugular vein. Note that when water was removed, drinkers displayed a significant increase in plasma corticosterone levels. Each point represents the mean (+S.E.M.) of 4 rats. * $p < 0.05$, ** $p < 0.01$ compared to values measured at the start of the session (Time 0) (from Dantzer et al., 1988a)

levels than when the chain is removed (Dantzer and Mormède, 1981). This significant drop in plasma cortisol occurs only when the animals are fed intermittently and not when they receive all their food at the beginning of the test session (Dantzer *et al.*, 1986).

However, the view that schedule-induced activities function as coping responses that reduce corticosterone levels has become controversial as results inconsistent with this hypothesis have also been reported. Rather than decreasing corticosterone levels, it has been demonstrated that SIP can significantly increase plasma corticosterone in comparison to baseline levels (Wallace *et al.*, 1983), or compared with levels exhibited by animals exposed only to the schedule of food presentation but not permitted to drink (Mittleman *et al.*, 1988). In addition to these puzzling data, other physiological indices of stress or activation have been investigated with similarly inconsistent results. In catheterized rats, plasma corticosterone levels declined within the session in drinking animals, but plasma concentrations of catecholamines and prolactin failed to show similar reductions and instead remained constant during the same sampling period (Dantzer *et al.*, 1988a). In rats allowed to develop schedule-induced wheel running, an increase instead of a decrease in plasma levels of corticosterone was observed, and this change was in direct relation to the amount of running performed (Tazi *et al.*, 1986). In pigs which had developed schedule-induced chain chewing, the decreased pituitary-adrenal activity displayed at the end of the session of intermittent feeding was actually significant only when compared with elevated pre-session levels. In contrast, plasma levels of cortisol remained elevated throughout the session in pigs which did not have access to a chain during the session (Dantzer *et al.*, 1986).

If it is assumed that adjunctive behaviours do not fulfil a coping or stress-reducing function, a logical next question is: what mechanism can account for the reduction in plasma corticosterone levels? To answer this question, it is necessary to refer to studies on the effects of feeding schedules on pituitary-adrenal activity (Heybach and Vernikos-Daniellis, 1979a,b; Honma *et al.*, 1986). In rats, and perhaps in other species, restriction of feeding to a single daily meal alters the circadian rhythmicity of pituitary-adrenal activity. There is a shift of the normal afternoon peak in plasma corticosterone to the time just before the daily feeding session. Plasma levels of corticosterone as well as ACTH fall rapidly upon presentation of food and water. This rapid fall can also occur in response to the cues that have been previously associated with the daily presentation of food and water. However, sustained decreases in pituitary-adrenal activity are dependent on actual food or water consumption. Consummatory activities appear to have potent suppressive influences on the pituitary-adrenal axis, not only directly at the level of the hypothalamus and the pituitary, but also indirectly, *via* some unknown mechanism controlling the clearance of circulating corticosteroids. Two sets of data support this last possibility. First, plasma corticosterone decreases significantly during schedule-induced drinking in adrenalectomized rats implanted with a corticosterone pellet (Levine and Levine, 1989). Secondly, the decline in plasma corticosterone concentrations that results from drinking occurs too rapidly (2–3 min) to be accounted for by pituitary or hypothalamic influences (Wilkinson *et al.*, 1982). Taken together, these results suggest that the decline in plasma corticosterone levels may be attributed, at least in part, to metabolic mechanisms rather than psychological or coping processes.

Other physiological indices of stress also provide ambiguous data concerning the coping function of SIP. In rats submitted to an intermittent schedule of food distribution, body temperature monitored by implanted transmitters increased by about 1°C in the course of 30 min sessions in well-trained rats. Although there was a slight tendency for drinkers to show a smaller elevation than non drinkers, this difference did not reach significance. This increase in body temperature was not a function of general activity since wheel-running was not accompanied by a higher hyperthermic response (Seguy, 1990).

In a systematic study of physiological correlates of SIP in rats submitted to an intermittent schedule of food distribution, Mittleman et al. (1990) observed that oxygen consumption was higher in drinkers than in nondrinkers. However, heart rate was lower in the drinkers, which also tended to be less active than the nondrinkers. It would therefore appear that schedule-induced drinking is associated with inhibition of sympathetic nervous system activity, which should normally decrease energy mobilization. However, ingesting and then heating large quantities of room temperature water to body temperature before excreting it results in excessive energy expenditure which cannot be counterbalanced by the decreased sympathetic activity.

There has been little interest in the direct investigation of sympathetic nervous system activity in individuals displaying displacement activities, in spite of the fact that this is relatively easy to do. For example, tethering of adult male cynomolgus macaques has been shown to result in persistent elevations in heart rate compared to other stressful conditions. Administration of propranolol, a β-adrenergic antagonist, induced an abrupt, sustained decrease in heart rate, indicating that the increase in heart rate associated with tethering was due to persistent stimulation of the sympathetic nervous system (Adams et al., 1988; see also Chapter 4).

The possibility that displacement activities induce a shift from sympathetic nervous system arousal to parasympathetic nervous system activation is intriguing and would be worth testing more systematically. It is supported by data from both infants and children. In infants, oral activities involved in sucking a pacifier activate vagal mechanisms, as deduced from the variation in the plasma pattern of gastro-intestinal hormones (Uvnas-Moberg, 1987). In children moving from kindergarten to primary school, there is an increase in leg-swinging stereotypes during the time children have to remain seated at their desks and learn writing, reading and arithmetic. These stereotypies are accompanied by small but significant reductions in heart rate (Soussignan and Koch, 1985) (Figure 7.10).

Concerning endogenous analgesic systems, there is evidence both in rats (Tazi et al., 1987) and pigs (Rushen et al., 1990) that schedule-induced activities are accompanied by an attenuation of the analgesic response to food distribution.

7.4.2 NEUROCHEMICAL CORRELATES

Building on the observation that SIP does not develop with the same probability in different individuals exposed to an intermittent schedule of food distribution,

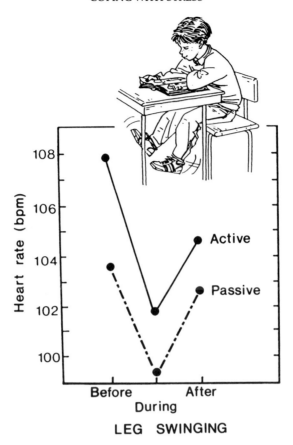

Figure 7.10 Influence of displacement activities on heart rate. Leg-swinging stereotypies in children at school are accompanied by small but significant decreases in heart rate whether children participate actively (active) or not (passive) in school activities (from Soussignan and Koch, as modified by Dantzer, 1989)

Mittleman and Valenstein (1984) initiated a series of systematic studies on the neurobiological basis of these individual differences. Rats that display spontaneous excessive drinking (SIP-positive animals) when exposed to an intermittent distribution of food and those that do not (SIP-negative animals) were compared both within SIP sessions and in other experimental situations. The main result of these studies was the demonstration of a consistent relationship between the predisposition to develop SIP and the propensity to respond by eating or drinking in response to electrical stimulation of the lateral hypothalamus (LH). As brain catecholamines, and especially dopamine, have been implicated in the regulation of oral activities elicited by exposure to a wide variety of mild stressors, they suggested that differences in predisposition to display SIP and LH stimulation-induced drinking might be related to individual differences in the responsiveness of forebrain dopaminergic systems. They went on to test this hypothesis by comparing the behavioural response of drinkers to amphetamine and their dopaminergic response to footshock stress. In both cases, SIP-positive rats showed a higher response, indicating that their central dopaminergic neurones were sensitized (Mittleman and Valenstein, 1984; Mittleman et al., 1986) (Figure 7.11). This is in contrast to the decreased response to amphetamine observed in SIP-positive rats immediately after the end of the SIP session (Tazi et al., 1988). Repeated injections of amphetamine were found to transform a negative response to LH-stimulation into a positive response (Mittleman and Valenstein, 1984). On the behavioural side, SIP-positive animals displayed more rapid acquisition of active avoidance in a shuttle-box and less freezing when confronted with an aggressive resident male in a defeat test than SIP-negative rats (Dantzer et al., 1988b).

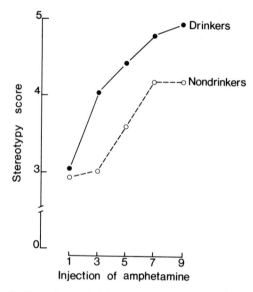

Figure 7.11 Influence of adjunctive activities on response to amphetamine. Rats classified as drinkers or nondrinkers on the basis of their water consumption in the schedule-induced polydipsia experiment were repeatedly injected with amphetamine and their stereotypy score was assessed by an observer blind to the drinking category. Note that drinkers developed a higher response to amphetamine than nondrinkers (from Mittleman et al., 1986)

Taken together, these findings can be interpreted as suggesting that the predisposition to develop SIP is another facet of a more general profile of behavioural and neurochemical reactivity to aversive situations. The behavioural characteristic of this profile could be a reduced ability to shift motor programs, i.e. behavioural rigidity. SIP-positive animals may be individuals who easily develop routines and become more stereotyped in their way of responding to a given stimulus situation. This could be due to a greater sensitivity to sensitization processes of the neuronal structures underlying the behavioural response.

7.5 CONCLUDING REMARKS

A cursory view of the present chapter would appear to provide some support for the fact that different coping strategies are associated with different patterns of physiological and neurochemical activation and for the qualification that this rather vague statement precludes any specific prediction concerning the specific consequences of a given coping strategy.

This pessimistic conclusion is unwarranted for a number of reasons. First, many interesting findings have emerged, from the observation of distinct psychobiological profiles of reactivity to stressors to the importance of neuronal sensitization processes. Secondly, the accumulated data clearly demonstrate that unidimensional constructs such as stress, arousal and activation are of limited use in describing coping strategies. Complexity has to be accepted as such, rather than being masked by simple constructs. We have now entered a generation of studies on coping in which the emphasis is on inter-individual differences in the propensity to respond to a given situation (coping styles). The objective is directly to study these inter-individual differences not only as a state variable, in response to a given situation, but also as a trait variable, i.e., as capable of influencing responding to different situations. Such research cannot be carried out superficially, with the different perspectives isolated from each other because of the specialized disciplines to which they belong. It requires in depth studies and integration and therefore represents a challenge for future studies (Dantzer, 1989).

7.6 REFERENCES

Adams, M.R., Kaplan, J.R., Manuck, S.B. *et al.* (1988) Persistent sympathetic nervous system arousal associated with tethering in cynomolgus macaques. *Lab. Anim. Science,* **38**: 279–81

de Boer, S.F., van der Gugten, J. and Slangen, J.L. (1989) Plasma catecholamine and corticosterone responses to predictable and unpredictable noise stress in rats. *Physiol. Behav.,* **45**: 789–95

Bolles, R.C. and Fanselow, M.S. (1980) A perceptual defensive recuperative model of fear and pain. *Behav. Brain Sci.,* **3**: 291–323

Brett, L. and Levine, S. (1979) Schedule-induced polydipsia suppresses pituitary-adrenal activity in rats. *J. Comp. Physiol. Psychol.,* **93**: 946–56

Brett, L. and Levine, S. (1981) The pituitary-adrenal response to 'minimized' schedule-induced drinking. *Physiol. Behav.,* **16**: 153–8

Church, R.M. (1964) Systematic effect of random error in the yoked control design. *Psychol. Bull.,* **62**: 122–31

Church, R.M. (1989) The yoked control design, in T. Archer and L.G. Nilsson (eds.) *Aversion, Avoidance and Anxiety,* Lawrence Erlbaum, Hillsdale, NJ, pp 403–15

Coover, G.D., Ursin, H. and Levine, S. (1973) Plasma corticosterone levels during active avoidance learning in rats. *J. Comp. Physiol. Psychol.,* **82**: 170–4

Dantzer, R. (1989) *L'illusion Psychosomatique,* Odile Jacob, Paris

Dantzer, R. and Kelley, K.W. (1989) Stress and immunity: An integrated view of relationships between the brain and the immune system. *Life Sci.,* **44**: 1995–2008

Dantzer, R. and Mormède, P. (1981) Pituitary-adrenal correlates of adjunctive activities in pigs. *Horm. Behav.,* **16**: 78–92

Dantzer, R., Gonyou, H.W., Curtis, S.E. and Kelley, K.W. (1986) Changes in serum cortisol reveal functional differences in frustration-induced chain chewing in pigs. *Physiol. Behav.,* **39**: 775–7

Dantzer, R., Terlouw, C., Mormède, P. and Le Moal, M. (1988a) Schedule-induced polydipsia decreases plasma corticosterone levels but increases plasma prolactin levels. *Physiol. Behav.,* **43**: 275–9

Dantzer, R., Terlouw, C., Tazi, A. *et al.* (1988b) The propensity for schedule-induced polydipsia is related to differences in conditioned avoidance behaviour and in defense reaction in a defeat test. *Physiol. Behav.,* **43**: 269–73

Dess, N.K., Linwick, D., Patterson, J. *et al.* (1983) Immediate and proactive effects of controllability and predictability on plasma cortisol responses to shock in dogs. *Behav. Neurosci.*, **97**: 1005–16

Falk, J.L. (1961) Production of polydipsia in normal rats by an intermittent food schedule. *Science*, **133**: 195–6

Falk, J.L. (1966) The motivational properties of schedule-induced polydipsia. *J. Exp. Anal. Behav.*, **9**: 19–25

Falk, J.L. (1971) The nature and determinants of adjunctive behaviour. *Physiol. Behav.*, **6**: 577–88

Falk, J.L. (1977) The origin and functions of adjunctive behavior. *Anim. Learn. Behav.*, **5**: 325–35

Grau, J.S., Hyson, R.L., Maier, S.F. *et al.* (1981) Long-term stress-induced analgesia and activation of the opiate system. *Science*, **203**: 1409–12

Haemisch, A. (1990) Coping with social conflict, and short-term changes of plasma cortisol titers in familiar and unfamiliar environments. *Physiol. Behav.*, **47**: 1265–70

Hanson, J.D., Larson, M.E. and Snowdon, C.T. (1976) The effect of control over high intensity noise on plasma cortisol levels in rhesus monkeys. *Behav. Biol.*, **16**: 333–40

Herrmann, T.F., Hurwitz, H.M.B. and Levine, S. (1984) Behavioral control, aversive stimulus frequency and pituitary-adrenal response. *Behav. Neurosci.*, **98**: 1094–9

Heybach, J.P. and Vernikos-Danellis, J. (1979a) Inhibition of adrenocorticotrophin secretion during deprivation-induced eating and drinking in rats. *Neuroendocrinology*, **28**: 329–38

Heybach, J.P. and Vernikos-Danellis, J. (1979b) Inhibition of pituitary-adrenal response to stress during deprivation-induced feeding. *Endocrinology*, **104**: 967–73

Honma, K., Honma, S., Hirai, T. *et al.* (1986) Food ingestion is more important to plasma corticosterone dynamics than water intake in rats under restricted daily feeding. *Physiol. Behav.*, **37**: 791–5

Kant, G.J., Bauman, R.A., Pastel, R.H. *et al.* (1991) Effects on controllable vs. uncontrollable stress on circadian temperature rhythms. *Physiol. Behav.*, **49**: 625–30

Konarska, M., Stewart, R.E. and McCarty, R. (1990) Predictability of chronic intermittent stress: Effects on sympathetic-adrenal medullary responses of laboratory rats. *Behav. Neural Biol.*, **53**: 231–43

Leudenslager, M.L., Ryan, S.M., Drugan, R.C. *et al.* (1983) Coping and immunosuppression: inescapable but not escapable shock suppresses lymphocyte proliferation. *Science*, **221**: 568–70

Lazarus, R.S. and Folkman, S. (1984) Coping and adaptation, in W.D. Gentry (ed.) *Handbook of Behavioral Medicine*, Guilford Press, New York, pp 282–325

Levine, R. and Levine, S. (1989) Role of the pituitary-adrenal hormones in the acquisition of schedule-induced polydipsia. *Behav. Neurosci.*, **103**: 621–37

Levine, S., Weinberg, J. and Ursin, H. (1978) Definition of the coping process and statement of the problem, in H. Ursin, E. Baade and S. Levine (eds.) *Psychobiology of Stress: A Study of Coping Men*, Academic Press, New York, pp 3–21

MacLennan, A.J. and Maier, S.F. (1983) Coping and the stress-induced potentiation of stimulant stereotypy in the rat. *Science*, **219**: 1091–3

Maier, S.F., Ryan, S.M., Barksdale, C.M. and Kalin, N.H. (1986) Stressor controllability and the pituitary-adrenal system. *Behavioral Neuroscience*, **100**: 669–78

McEwen, B.S., De Kloet, E.R. and Rostene, W. (1986) Adrenal steroid receptors and actions in the nervous system. *Physiol. Rev.*, **66**: 1121–88

McFarland, D.J. (1985) *Animal Behaviour*, Pitman, London

Mittleman, G. and Valenstein, E.S. (1984) Ingestive behaviour evoked by hypothalamic stimulation and schedule-induced polydipsia are related. *Science*, **224**: 415–17

Mittleman, G., Castaneda, E., Robinson, T.E. and Valenstein, E.S. (1986) The propensity for nonregulatory ingestive behaviour is related to differences in dopamine systems: Behavioural and biochemical evidence. *Behav. Neurosci.*, **100**: 213–20

Mittleman, G., Jones, G.H. and Robbins, T.W. (1988) The relationship between schedule-induced polydipsia and pituitary-adrenal activity: Pharmacological manipulations and individual differences. *Behav. Brain Res.*, **28**: 315–24

Mittleman, G., Brener, J. and Robbins, T.W. (1990) Physiological correlates of schedule-induced activities in rats. *Am. J. Physiol. (Regul. Integr. Comp. Physiol.)*, **259**: R485–91

Mormède, P., Dantzer, R., Michaud, B. *et al.* (1988) Influence of stressor predictability and behavioral control on lymphocyte reactivity, antibody responses and neuroendocrine activation in rats. *Physiol. Behav.*, **43**: 577–83

Rodgers, R.J. and Randall, J.I. (1987) On the mechanisms and adaptive significance on intrinsic analgesia systems. *Rev. Neurosci.*, **1**: 185–200

Rushen, J., de Passillé, A.M.B. and Schouten, W. (1990) Stereotypic behaviour, endogenous opioids and postfeeding hypoanalgesia in pigs. *Physiol. Behav.*, **48**: 91–6

Seguy, F. (1990) Polymorphisme de l'hyperthermie émotionnelle chez le rat: Etude psychobiologique et pharmacologique. *DEA Neurosciences*, University of Bordeaux II

Selye, H. (1973) The evolution of the stress concept. *Am. Sci.*, **61**: 692–9

Sklar, L.S. and Anisman, H. (1979) Stress and coping factors influence tumor growth. *Science*, **205**: 513–15

Sklar, L.S., Bruto, V. and Anisman, H. (1981) Adaptation to the tumor-enhancing effects of stress. *Psychosom. Med.*, **43**: 331–42

Soussignan, P. and Koch, P. (1985) Rhythmical stereotypies (leg-swinging) associated with reductions in heart rate in normal school children. *Biol. Psychol.*, **21**: 161–7

Stewart, K.T., Rosenwasser, A.M., Hauser, H. *et al.* (1990a) Circadian rhythmicity and behavioral depression: I. Effects of stress. *Physiol. Behav.*, **48**: 149–55

Stewart, K.T., Rosenwasser, A.M., Levine, J.D. *et al.* (1990b) Circadian rhythmicity and behavioral depression: Effects of lighting schedules. *Physiol. Behav.*, **48**: 157–64

Swenson, R.M. and Vogel, W.H. (1983) Plasma catecholamine and corticosterone as well as brain catecholamine changes during coping in rats exposed to stressful footshock. *Pharmacol. Biochem. Behav.*, **18**: 689–93

Tazi, A., Dantzer, R., Mormède, P. and Le Moal, M. (1986) Pituitary-adrenal correlates of schedule-induced polydipsia and wheel-running in rats. *Behav. Brain Res.*, **19**: 249–56

Tazi, A., Dantzer, R. and Le Moal, M. (1987) Prediction and control of food rewards modulate endogenous pain inhibitory systems. *Behav. Brain Res.*, **23**: 197–204

Tazi, A., Dantzer, R. and Le Moal, M. (1988) Schedule-induced polydipsia experience decreases locomotor response to amphetamine. *Brain Res.*, **445**: 211–15

Tinbergen, N. (1950) Derived activities: Their causation, biological significance, origin and emancipation during evolution. *Quart. Rev. Biol.*, **27**: 1–32

Tsuda, A. and Tanaka, M. (1985) Differential changes in noradrenaline turnover in specific regions of rat brain produced by controllable and uncontrollable shocks. *Behav. Neurosci.*, **99**: 802–17

Ursin, H. (1985) The instrumental effects of emotional behavior, in P.P.G. Bateson and P.H. Klopfer (eds.), *Perspectives in Ethology vol. 6*, Plenum, New York, pp 45–62

Uvnas-Moberg, K. (1987) Gastro-intestinal hormones and pathophysiology of functional gastrointestinal disorders. *Scand. J. Gastroenterol.*, **22**(Suppl. 128): 138–46

Visintainer, M.A., Volpicelli, J.R. and Seligman, M.E.P. (1982) Tumor rejection after inescapable or escapable electric shock. *Science*, **216**: 437–9

Wallace, M., Singer, G., Finlay, J. and Gibson, S. (1983) The effect of 6-OHDA lesions of the nucleus accumbens system on schedule-induced drinking, wheel running and corticosterone levels in the rat. *Pharmacol. Biochem. Behav.*, **18**: 129–36

Weiss, J.M., Goodman, P., Losito, B. *et al.* (1981) Behavioral depression produced by an uncontrollable stressor. Relationship to norepinephrine, dopamine and serotonin levels in various regions of rat brain. *Brain Res. Rev.*, **3**: 167–205

Weisse, C.S., Pato, C.N., McAllister, C.G. *et al.* (1990) Differential effects of controllable and uncontrollable acute stress on lymphocyte proliferation and leukocyte percentages in humans. *Brain Behav. Immun.*, **4**: 339–51

Wilkinson, C.W., Shimsako, J. and Dallman, M.F. (1982) Rapid decreases in adrenal and plasma corticosterone concentrations are not mediated by changes in plasma adrenocorticotropin concentrations. *Endocrinology*, **110**: 1599–606

8

Stress and behavioural inhibition

Neil McNaughton

Department of Psychology and Centre for Neuroscience
University of Otago
Dunedin
New Zealand

8.1 INTRODUCTION

Behavioural inhibition is the measured variable in most popular animal models of anxiety. It is the source of both the perceived psychological and the perceived pharmacological validation of the models. The psychological validation derives from the intuition that anxiety will inhibit behaviour which leads to a threatening situation. The pharmacological validation derives from the fact that many of the behavioural effects of the 'classical' (sedative/hypnotic) anxiolytic drugs (e.g. barbiturates and benzodiazepines) can be characterized as a loss of behavioural inhibition. That is, whereas there are many cases where an anxiolytic drug has no effect on the learning of a response, a characteristic effect is to impair the 'unlearning' of the same response when it is no longer appropriate (e.g. when it is punished or no longer rewarded).

This led Gray (1982) to view anxiety as consisting of activity in a 'Behavioural

191

ISBN 0–12–663370–3

Inhibition System', critical parts of which he attributed to the hippocampus. The main problem to be addressed in this chapter is that 'recently developed anxioselective compounds have been reported to release . . . behaviour . . . to a smaller extent . . . then benzodiazepines' and so, according to Graeff (1988), 'existing animal models . . . have little claim to represent the pathophysiological mechanisms'.

I will argue that many existing animal models do in fact capture significant aspects of human psychopathology. In particular, I will suggest that the 'smaller extent' of the effects of the antianxiety drug, buspirone, is not due to any difference between it and the classical anxiolytics in primary neural effect. Rather, as originally suggested by Johnston and File (1988), it appears to be the result of the release of stress hormones by buspirone which, I speculate, could be a significant factor in its clinical actions also. However, I will also argue that it is the 'hippocampal' rather that the 'behavioural inhibition' features of animal models that make these models successful.

Since this is a book about stress, and since the release of stress hormones is common to both anxiety and depression, I should emphasize that I shall not only treat anxiety and depression as categorically distinct from each other, but also treat as categorically distinct: generalized anxiety, panic, atypical depression and unipolar depression. I will argue that it is only generalized anxiety that should be seen as responding to the specifically anxiolytic actions of the drugs we will be considering. Likewise, stress should be seen as accompanying each of these categorically different conditions, rather than defining any one of them. This last point should become clearer as we consider the relationship between stress and behavioural inhibition.

8.2 THE RELATION BETWEEN STRESS, BEHAVIOURAL INHIBITION AND CORTICOSTERONE RELEASE

If we take stress to be indicated by activity in the hypothalamic-pituitary-adrenal system, it is clear that we can discriminate at least two general types of stressors. There are physical challenges, such as low temperature or infection, where the release of stress hormones can be attributed directly to effects of the physical stimulus on the body. Then there are psychological challenges. These are effective by virtue of the *interpretation* put on them by the organism rather than by virtue of their physical characteristics.

In dealing with animal models of human psychological stressors it is necessary to distinguish between at least two types: those related to anxiety and those related to depression. The present chapter concentrates on animal models of anxiety (or, more specifically, of anxiolytic action).

Many animal models of anxiety involve conditioning: this is often seen as the province of particularly behaviourist approaches to psychology. It is important to emphasize, therefore, that anxiety and stress depend on the subject's interpretation of stimuli. For example, take a neutral stimulus (one which does not produce an observable behavioural response). Present it repeatedly, always followed by an inescapable shock. The stimulus will no longer be neutral: it will have acquired the capacity to inhibit ongoing behaviour. As this 'conditioned suppression' develops, the stimulus also develops the capacity to release corticosterone (Brady, 1975a,b). Neither the neutral stimulus nor the shock when presented alone elicit the same behavioural or hormonal response as the conditioned stimulus (Brady, 1975b). So, these experiments not only show a correlational relationship between the development of behavioural inhibition and hormonal release, but they are also evidence for development of a novel mental state, anxiety, and its stressful nature. A radical behaviourist (e.g. Watson, 1924, or Blackman, 1983), would not accept that one can infer mental states in this way (but see Dickenson, 1980).

In conditioned suppression, the neutral stimulus is rendered stressful by pairing with an inescapable shock. More complex results are obtained with escapable shock. Again, the crucial factor is the subject's interpretation of the stimuli. If an animal is trained to make a response to avoid a shock ('active avoidance'), the early stages of training are accompanied by increasing corticosterone release. However, as the animal becomes successful in avoiding the shock, corticosterone release decreases again and returns to pretraining levels as performance becomes perfect (Coover et al., 1973). This is difficult to explain in terms of simple association. The avoidance response and hormonal release initially increase together and so could be explained by the same type of process. However, the loss of corticosterone release as conditioning proceeds cannot be explained in the same way as the continued increase in avoidance responding.

One could argue that corticosterone release is a consequence of anticipation of shock (as in the conditioned suppression case); and that, since perfect avoidance eliminates shock, this anticipation extinguishes. However, it is clear that the decrease in corticosterone release is the result of the animals' knowledge that responding makes it safe rather than because it has forgotten about the shock. If it is prevented from making the avoidance response, then corticosterone is released even if no more shocks are delivered (Coover et al., 1973).

It is interesting to compare conditioned suppression and active avoidance with the results of training with a food reinforcer rather than shock. Learning to respond in order to obtain food (as opposed to when the response is made to avoid shock) is not accompanied by release of corticosterone. Instead, once the response is learned, corticosterone is released if the expected food is omitted ('extinction'). This shows that active learning of a response (which also occurs in active avoidance, but not conditioned suppression) does not release corticosterone. It also shows that corticosterone release is not simply the result of pairing a neutral stimulus with a harmful one such as shock. Remember too, that, as discussed above, shock by itself does not release corticosterone.

It is tempting at this point to conclude that in each of conditioned suppression, active avoidance and extinction of a rewarded response, a state of anxiety is generated and that this results in the release of corticosterone. However, we have already distinguished physical stressors from psychological ones and, albeit without much argument, have separated anxiety from depression. The release of corticosterone in all three types of experiment does not, therefore, guarantee that this release is due to the same psychological state in each case. Indeed, experiments with anxiolytic drugs suggest that we should distinguish 'fear' in the active avoidance situation from 'anxiety' in conditioned suppression and extinction of reward.

8.3 DRUGS WHICH RELIEVE ANXIETY

Even now, despite the availability of diagnostic systems such as DSM–III, it is difficult to get agreement among psychologists and psychiatrists as to the nature of human anxiety. To attempt to identify its homologue in animals might seem somewhat ambitious, therefore.

There are two reasons for taking the contrary position. First, anxiety, as an adaptive response to anticipated threat (Blanchard and Blanchard, 1990), is likely to involve fundamental mechanisms common to rats and humans, in the same way that there are similarities in their digestive systems. Of course, the specific stimuli which give rise to anxiety will be different in the two species. Secondly, there is much better agreement about which drugs are effective in treating anxiety than there is about what anxiety itself is. Therefore, these drugs can be administered to animals and, provided there has not been some strong adaptive force producing divergence, they can be expected to affect homologous systems. The viability of this approach is illustrated by the fact that anxiolytic drugs of widely different generic groups can be shown to have common effects in animals. We will also see that differences in the action of the drugs in animals can be related to differences in their clinical actions. But which drugs are anxiolytic?

In the clinic, anxiety often coexists with depression. As a result, differential diagnosis of the two types of disorder can be difficult. Not suprisingly, therefore, centrally-acting drugs which have been claimed to be effective in treating anxious sympoms include 'antidepressants'. These include: tricyclic antidepressants, such as imipramine; monoamine oxidase inhibitors, such as phenelzine; and the 'novel' anxiolytic, buspirone. To start with, I will treat all of these drugs as potentially anxiolytic. However, the data I will review suggest that phenelzine should not be classified as being anxiolytic in the same sense as the others. Drugs with a mainly peripheral action on anxiety symptoms, such as β-adrenoceptor blockers, will not be discussed.

Before I analyse similarities and differences between these classes of drug, I will first look at what classical (sedative/hypnotic) anxiolytics can tell us about animal models of anxiety. The classical anxiolytics are of particular interest here because they are not generally effective in treating depression or panic and so, in one sense, can be used as the paradigm example of an anxiolytic drug. On the other hand, they are all to some extent muscle relaxant, anticonvulsant and CNS depressants: this problem will be dealt with later.

8.4 ANIMAL MODELS OF ANXIETY AND BEHAVIOURAL INHIBITION

There are at least two desiderata which most people would seek in an animal model of anxiety (File, 1980; see Chapter 6). The first is face validity, i.e. the behavioural properties of the test and, ideally, the psychological theory underlying it, should fit with our current notions of the nature of anxiety. The second is pharmacological validity.

"It is necessary to show: (a) that most of the clinically useful drugs do have an anxiolytic profile in the animal test; and (b) that other classes of drugs do *not* have an anxiolytic profile . . .

[However,] from a practical point of view a low rate of false positives would be acceptable in a screening procedure, whereas missing potentially useful drugs would be more serious" (File, 1980).

A range of apparently quite different tests appear to fulfil these criteria to at least some extent, for example: the Geller–Seifter conflict test; the conditioned emotional response; punished locomotor activity; punished drinking; the social interaction test and the elevated plus-maze (File, 1980; Pellow et al., 1985 for an overview of these tests). In each of these tests a stimulus (a novel conspecific, elevated open maze arms, a signal which predicts shock) inhibits behaviour and is presumed to generate anxiety. Also, in each of these tests, classical anxiolytics of quite different generic groups have similar effects.

The dependent measure of anxiolytic drug action in each of these animal models of anxiety is the release of behaviour. For example, in the social interaction test anxiolytics increase social interaction in an unfamiliar environment; in the elevated plus-maze they increase exploration of open elevated arms relative to 'closed' arms (i.e. arms with side walls); and, in conflict schedules such as the Geller–Seifter test, they increase responding which has been suppressed by previous pairing with shock. Gray (1977, 1982) has therefore argued that anxiolytic drugs act specifically by reducing activity in a 'Behavioural Inhibition System'.

To illustrate this point of view, let us consider the three 'stressful' schedules we discussed above: conditioned suppression, active avoidance and extinction of reward. All three, as we noted, involve release of corticosterone. But only conditioned suppression and extinction of reward schedules produce behavioural inhibition. Anxiolytic drugs reduce behavioural inhibition in these two schedules but do not reduce active avoidance. The active avoidance test, lacking behavioural inhibition, fails the pharmacological challenge as a model of anxiety (Gray, 1977).

This last result may seem surprising to those who treat anxiety and fear as synonymous. However, the face validity of active avoidance as a model of anxiety can also be questioned on purely behavourial grounds. Blanchard and Blanchard (1990) have shown that the behaviour of wild rats is quite different when they are certain that a cat is present, compared with when they are uncertain. The former can be equated with active avoidance and, to some extent, the latter with behavioural inhibition, although Blanchard and Blanchard (1990) emphasize that a battery of measures is required to draw valid conclusions. There are, therefore, good behavioural reasons for distinguishing fear from anxiety (as these are defined by Blanchard and Blanchard). As we might expect, anxiolytics are more effective in tests for anxiety than in those for fear.

8.5 TYPES OF BEHAVIOURAL INHIBITION

When considering a term such as 'anxiety', there is a strong temptation to believe that it refers to some single state of the mind or brain. I have argued elsewhere that it is a mistake to approach any emotion from this point of view (McNaughton, 1989). As we shall see, there are also good reasons not to equate anxiety with behavioural inhibition in any simplistic manner. First, whereas many models of anxiety use behavioural inhibition as a measure, others do not: for example, fear-potentiated startle. Secondly, behavioural inhibition is itself complex: there are instances where it should not be equated with anxiety and, even in those where it can, it may not always reflect the same underlying state.

Let us first consider behavioural inhibition in the absence of anxiety. In a conflict schedule such as the Geller–Seifter paradigm, the animal receives a shock only if it makes a response during the warning stimulus. At other times, it will receive a food reward and no shock. It soon learns to inhibit responding when the warning signal is on. As a result, it ceases to receive shock. As with the active avoidance schedule, in which fear is present and corticosterone is released only during acquisition, we would think that anxiety should be present only during acquisition. Consistent with this view, a benzodiazepine such as chlordiazepoxide given at 5 mg/kg to rats throughout acquisition impairs behavioural inhibition. Furthermore, if the animals are first trained to suppress and only subsequently given chlordiazepoxide at 2.5 mg/kg on one day and 5 mg/kg on subsequent days, there is no change in behavioural inhibition. By contrast, in conditioned suppression, which is comparable to conflict except that the shock is unavoidable, chlordiazepoxide remains effective even after many days of training (McNaughton, 1985). It is notable that anxiolytics have often been reported to reduce behavioural inhibition when given after training is complete. However, the drug is invariably given in a pattern across days such that the results may be due to state-dependency, rather than a truly anxiolytic action.

Let us next consider the possibility of different types of behavioural inhibition and hence of anxiety. To do this I will compare three closely related operant schedules: successive discrimination of reward omission; differential reinforcement of low rates of response and fixed interval. All of these are sensitive to anxiolytics such as chlordiazepoxide given during acquisition and all are theoretically similar. In successive discrimination, a stimulus such as a light signals that reward will be unavailable. This is comparable to the Geller-Seifter schedule where the stimulus signals that shock will be presented. In differential reinforcement of low rates, the animal receives a reward only if it delays responding by some fixed time interval since its last response. Here the animal's own response signals that reward will be unavailable, therefore. In a fixed interval schedule, delivery of a reward signals that reward will be unavailable for an interval (e.g. 1 min).

In all three schedules, behavioural inhibition occurs as the result of an explicit or implicit signal which predicts that reward is unavailable. In all three schedules, anxiolytics, such as chlordiazepoxide, reduce behavioural inhibition. However, the mechanisms of behavioural inhibition are not identical. In the case of differential reinforcement of low rates, the effects of chlordiazepoxide are blocked by the opiate receptor antagonist, naloxone. This shows that chlordiazepoxide is working through an endogenous opiate system (Tripp et al., 1987). In the case of successive discrimination, the effects of chlordiazepoxide are unaffected by naloxone (Tripp and McNaughton, 1987). In the case of the fixed interval schedule, naloxone blocks the effects of chlordiazepoxide at the beginning of the interval, but does not change the effects of chlordiazepoxide in the later parts of the interval (Tripp and McNaughton, 1991).

We therefore have evidence for at least two types of anxiety-related behavioural inhibition: an opioid and a non-opioid type. It is particularly interesting that these two types are equally sensitive to classical anxiolytic drugs, and can both operate within a single schedule (such as fixed interval). Moreover, the gradual transfer from one type to another is indistinguishable on behavioural measures.

The possibility of multiple neurochemical types of behavioural inhibition might seem daunting. However, a degree of coherence can be brought back by the realization that these particular effects of the anxiolytic drugs may all be achieved through one brain structure: the hippocampus.

8.6 THE HIPPOCAMPUS AND ANXIOLYTIC DRUGS

Gray's (1982) theory of the 'Behavioural Inhibition System' includes a complex neuropsychological architecture. A key component of this system, with respect to behavioural inhibition itself, is the hippocampus.

The hippocampus is a temporal lobe structure more usually viewed as being involved in memory than emotion. However, Gray (1977, 1982; Gray and McNaughton, 1983) noted that the majority of the behavioural effects of classical anxiolytic drugs in animals are reproduced by hippocampal lesions. For example, hippocampal lesions have very similar effects to anxiolytic drugs on the three tests we have just been considering (successive discrimination, differential reinforcement of low rates and fixed interval).

Gray (Gray and Ball, 1970; McNaughton et al., 1977) also discovered that anxiolytic drugs impair the control of slow waves by the medial septal nucleus. This appears to be an important feature of hippocampal information processing (Winson, 1990). This effect of anxiolytic drugs is reproduced by impairment of noradrenergic input to the hippocampus from the locus coeruleus. The locus coeruleus is probably the target for action of the *classical* anxiolytics in producing this particular electrophysiological change. Lesion of the noradrenergic input to the hippocampus reproduces about half of the common behavioural effects of anxiolytics and hippocampal lesions (McNaughton and Mason, 1980).

In addition to affecting septal control of hippocampal slow waves, anxiolytic drugs also reduce the effectiveness with which ascending information from the pons and midbrain can produce slow waves (Stumpf, 1965; McNaughton and Sedgwick, 1977; McNaughton et al., 1986). This effect is neurophysiologically and pharmacologically distinct from the 'septal' effect of the anxiolytics and almost certainly accounts for common effects of anxiolytics and hippocampal lesions which are not reproduced by loss of hippocampal noradrenaline.

All these data suggest that anxiolytic drugs have at least two separate actions which impair hippocampal function and hence reduce anxiety. However, the classical anxiolytics (ethanol, barbiturates, meprobamate and benzodiazepines) all act, in different ways, to increase GABA neurotransmission. As a result, all are to some degree depressant, anticonvulsant, muscle relaxant, hypnotic and addictive. The extensive parallels between anxiolytics and hippocampal lesions (and by implication virtually all the animal models of anxiolytic action) could be the result of one of these other side effects, therefore. The anticonvulsant action of the drugs is particularly relevant since the hippocampus is an extremely common focus for epileptic seizures.

We recently set out, therefore, to test other drugs which had been reported to be effective in treating clinical symptoms of anxiety: buspirone, imipramine and phenelzine. These drugs share none of the side effects of the sedative/hypnotic anxiolytics. We found that buspirone and imipramine had strikingly similar effects to those of the classical anxiolytics; this was apparent in tests of *both* septal *and* midbrain control of hippocampal slow waves (McNaughton and Coop, 1991; Zhu and McNaughton, 1991). Phenelzine was entirely different. I will argue later that this is because it is not an anxiolytic drug in the same sense as the others.

So, all drugs known to be centrally-acting anxiolytics, but with no other actions in common, have two effects on the control of hippocampal electrophysiology. This seems strong evidence for the view that the clinical effects of these drugs are mediated by the hippocampus. It also suggests that the two electrophysiological tests, *taken together*, are a good model of anxiolytic action. Can the same be said of behavioural models?

8.7 STRESS AND MODELS OF ANXIOLYTIC ACTION

Most conventional behavioural models of anxiolytic action are sensitive to hippocampal lesions. We have seen that all anxiolytic drugs impair hippocampal function. Therefore, it might be expected that conventional behavioural models should also be sensitive to the novel anxiolytic, buspirone. I shall argue that, contrary to the views of Graeff (1988) and others, this is true in principle, but can be complicated by stress hormones in practice.

Given the intuitive plausibility of most animal models of anxiolytic action and the conventional linking of the hippocampus with memory, rather than anxiety, the behavioural effects of buspirone appear positively mischievous. In tests such as the elevated plus-maze, which has been developed as an 'anxiolytic' rather than 'hippocampal' screening test, buspirone has inconsistent 'anxiolytic' actions (Pellow *et al.*, 1985). In conflict tests which are used both as anxiolytic screens and as probes for hippocampal damage, buspirone has weak effects and has an inverted U-shaped dose-response curve (Sanger *et al.*, 1985; Meert and Colpaert, 1986; Pich and Samanin, 1986; see also Graeff, 1988). Finally, in the Morris water maze and with rearing in a low stress open field test, buspirone has a linear dose-response curve (Panickar and McNaughton, 1991; McNaughton and Morris, 1993).

The water maze is a key test of hippocampal function (Morris *et al.*, 1982) which is sensitive to classical anxiolytics (McNaughton and Morris, 1987). This is particularly interesting because, being an active avoidance task, it has no obvious relevance to anxiety. Further, the effect of anxiolytic drugs is specific to the spatial navigation component of the task: simple rule-based improvements in performance are unaffected by these drugs. The psychological status of rearing is less clear and it is not a conventional test of anxiolytic action. Nevertheless, rearing is reduced by classical anxiolytics and hippocampal lesions, as well as by buspirone. Thus buspirone has the clearest effects in tests of hippocampal function and least effect in tests of anxiolysis. Moreover, when compared with chlordiazepoxide, the relative potencies of buspirone are similar for effects on hippocampal slow waves and for therapeutic anxiolytic effects.

A number of lines of evidence suggest that release of stress hormones is the cause of the anomalous effects of buspirone on 'anxiolytic' tests and that, in particular, it gives rise to buspirone's inverted U-shaped dose-response curve. Johnston and File (1988) noted that, in the social interaction test, stress hormones had opposite effects to anxiolytic drugs. This suggested that the effects of buspirone were confounded by activation of the hypothalamic-pituitary-adrenal axis. Classical anxiolytic drugs also release stress hormones at high doses but, at the doses normally used in behavioural experiments, they inhibit this stress-induced release. Buspirone is different: 'it fails to inhibit stress-induced release and, instead, releases corticosterone at doses (1 mg/kg) close to the optimum dose in many behavioural tests for anxiolytics (de Boer *et al.*, 1990).

Whereas the effects of buspirone on hippocampal slow waves resemble those of classical anxiolytics, they also share some features of the effects of hypothalamic-pituitary-adrenal hormones. We have shown that these effects of buspirone can be reproduced by administering corticosterone together with chlordiazepoxide (McNaughton and Coop, 1991). Similarly, in preliminary experiments, we have shown that, if endogenous release of corticosterone is prevented with a synthesis inhibitor, the effects of buspirone on hippocampal electrophysiology are identical to those of chlordiazepoxide. The anxiolytic, but not the depressant, effects of buspirone on a fixed interval schedule also appear to be sensitive to corticosterone levels.

On the basis of present evidence, then, anxiolytic drugs affect behavioural inhibition by at least two actions mediated by the hippocampus. Their reduction of behavioural inhibition can, at least in the case of buspirone, be antagonized by stress hormones. In the case of classical anxiolytics, by contrast, the endogenous release of corticosterone is inhibited and drug-induced release occurs only at relatively high doses.

8.8 CLINICAL IMPLICATIONS

The data we have reviewed show that classical anxiolytics, the novel anxiolytic buspirone and the tricyclic antidepressant imipramine all impair hippocampal function. Since we started with the assumption that these drugs are all anxiolytic this might appear satisfactory. In fact, it raises two questions: first, how do we account for the apparently different time course of clinical action of the classical and novel anxiolytics? Second, what is the common 'anxiety' that these three different types of drug change and how do they change it? In particular, how do we account for the fact that phenelzine does not impair hippocampal function and yet we started with the assumption that this monoamine oxidase inhibitor antidepressant might also be anxiolytic.

The first question arises from the frequent observation that the clinical effects of classical anxiolytics appear to be immediate in onset whereas buspirone and imipramine improve anxiety only over about two weeks of treatment. We have shown that the hippocampal effects of all of these drugs are immediate and do not change markedly with repeated administration over several weeks (Zhu and McNaughton, 1991). Careful inspection of the literature shows that, if anxiety is assessed using the Hamilton rating scale, benzodiazepines do in fact take about two weeks to reduce anxiety (Wheatley, 1982). It appears that, on initial administration, classical anxiolytics produce muscle relaxant and euphoriant effects. These cause the patients to say that they feel 'better'. Although these effects undergo rapid tolerance, by the time tolerance has developed, the anxiolytic effects of the drugs are apparent. On this view, classical anxiolytics, buspirone and imipramine all produce immediate effects on the hippocampus which eliminate cognitive or 'reinforcement' mechanisms supporting anxiety. In the absence of reinforcement, the anxiety effectively undergoes extinction and this takes some time to become complete.

The question of the types of anxiety affected by the drugs is best approached *via* Table 8.1. In the top two rows we note that classical anxiolytics, buspirone and imipramine all impair hippocampal function. This is not true of phenelzine. In the next row, we note that the same pattern is true for generalized anxiety disorder. In this context, it is probably important to note that the anxiolytic effects of imipramine, conventionally regarded as an antidepressant, are independent of any concurrent depression or changes in depressive symptoms (Kahn *et al.*, 1986). However, panic (which is often viewed as part of an anxiety disorder 'panic/agoraphobia') does not fit this pattern. Buspirone is ineffective in treating panic. By contrast, buspirone *is* effective in treating unipolar depression – but this is clearly separate from its anxiolytic action since it is not convincingly shared by classical anxiolytics (except alprazolam).

In considering, next, 'atypical depression' (depression with concurrent panic and somatic symptoms of anxiety), it is notable that this disorder is particularly responsive to treatment with phenelzine; the efficacy of this drug is appreciably greater than imipramine. Where phenelzine has been reported to improve anxiety, however, it seems that depression may have been the underlying disorder. For example, Sheehan *et al.* (1981) reported that phenelzine reduced anxiety. However, virtually all the patients in this study had been treated with classical anxiolytics (in some cases, for years) with no relief of symptoms. In addition to anxiety, the reported symptoms included depression and panic, suggesting that patients had atypical depression. On this basis, although the symptoms of atypical depression resemble anxiety, it should properly be classified as a purely depressive disorder, or placed in a class of its own.

The overlap of drug effects in Table 8.1 indicates that drugs are never specific for any one neuronal system. For example, we have shown in recent experiments that the effects of buspirone and imipramine on the hippocampus (and, presumably, generalized anxiety) are both achieved through an agonist action on $5-HT_{1A}$ receptors. Phenelzine does not share imipramine's affinity for $5-HT_{1A}$ receptors (Kahn *et al.*, 1986) and so does not have the same effects on hippocampal function. Phenelzine does share with imipramine an agonist action at other $5-HT$ receptors, whereas buspirone does not. These differences in the actions of buspirone and other antidepressants probably account for its lack of anti-panic effects.

Of course, another possibility is that panic is more affected by stress hormones than is generalized anxiety; in this way buspirone's effect would be nullified, as in the animal models. In this context, one might speculate that, even with generalized anxiety, an inverted U dose-response curve, coupled with differences in prescription practices, could account for the greater preference shown for buspirone by American as opposed to British psychiatrists (cf. Goldberg, 1990 and Marks, 1990).

Table 8.1 Effects of 'anxiolytic' drugs

	Classical Anxiolytics	Buspirone	Imipramine	Phenelzine
Impaired hippocampal function ('septal')	+	+	+	0/–
Impaired hippocampal function ('midbrain')	+	+	+	0
Relief of:				
Generalized anxiety	+	+	+	0(?)
Panic disorder	(+)	0	+	+
Unipolar depression	(+)	+	+	+
Atypical depression	0	?	(+)	+
5–HT$_{1A}$ receptor activation	0	+	+	0
Effect on other 5–HT systems	–	–	+	+

+: active; 0: inactive; –: reduced activity

8.9 AN OVERVIEW OF ANXIETY AND DEPRESSION

In this final section I will attempt to provide an overview of anxiety and depression. This summary (Figure 8.1) must be treated as highly tentative – particularly since the role I have assigned to stress hormones in the effects of buspirone has little experimental support at present. Equally, it is not intended to be original but, rather, combines ideas which have appeared separately and together in many different sources.

On this view both anxiety and psychological stress are the result of the perception of *potential* danger. An existing danger ('present'), by contrast, activates the fear system (Blanchard and Blanchard, 1990). Panic attacks result from activity in the fight/flight/freeze aspects of the fear system. Simple phobias also operate through this system which is probably located in or associated with 5-hydroxytryptamine releasing systems in the dorsal periaqueductal grey (Graeff, 1991).

Agoraphobia is viewed as being a consequence of panic attacks and is *anxiety* (not phobia) conditioned by the unexpected occurrence of panic. One can speculate that the peripheral symptoms of agoraphobic anxiety increase the incidence of panic attacks (in the same way as a lactate infusion can) and that the muscle relaxant and euphoriant effects of the benzodiazepines achieve some apparent anti-panic potency through this route – whereas buspirone does not. There seems to be little evidence that agoraphobia, as opposed to specific place phobia, occurs in the absence of concurrent panic attacks (Friend and Andrews, 1990).

Primary anxiety disorder, prototypically generalized anxiety disorder, reflects overactivity in a risk assessment system which is normally activated by potential danger; many stimuli of potential danger will be conditioned, but innate anxiety stimuli, e.g. novelty, are also possible. One common output of this risk assessment system is behavioural inhibition, but active risk analysis may be even more characteristic (Blanchard and Blanchard, 1990).

The view that the hippocampus is the key to the action of anxiolytic drugs is consistent with a more 'cognitive' and less 'emotional' interpretation of their action. On this view, the effects of the drugs on central anxiety are due to an immediate

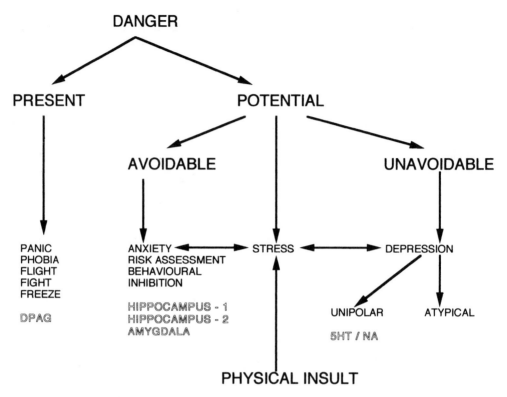

Figure 8.1 Suggested relationship between different classes of danger or insult (large capitals) and different types of human and animal response (small capitals). Reciprocal interactions are suggested between anxiety, stress and depression. Neuronal systems proposed to underlie the types of response are given in outline capitals. For details of derivation and explanation, see text. (DPAG: dorsal periaqueductal grey)

action on cognitive/reinforcement mechanisms. This is reflected in their effects on spatial navigation and is, perhaps, related to their high dose amnestic effects. Their effectiveness in anxiety would, then, be a result of the fact that risk assessment requires a large input from hippocampal cognitive mechanisms. There would be no direct effect of the drugs on anxiety itself (see also Green 1991). Given their action on the amygdala, some additional direct effects on more 'emotional' mechanisms cannot be ruled out, however.

Given this suggested neurology, it is an open question how far 'the anxiety system' can be viewed as homogenous. We have already seen that all anxiolytics have at least two actions on the hippocampus, each of which contributes to their behavioural profile. In addition they almost certainly all act on the amygdala as well. Thus, 'anxiolytic action' probably has at least three components. It remains to be seen whether these components are essentially independent and additive (each site of drug action contributing some specific aspect of the overall effect), or whether they are more integrated (action at one site combining with actions at others to produce the overall effects).

While accepting the need to subdivide both normal and pathological 'anxiety', it may also be useful to see pathological anxiety as part of a more global category of neurosis. We have made a categorical distinction between fear (panic), anxiety and depression in terms of the brain systems immediately involved in their control (there may also be a case for viewing obsessive-compulsive disorder as distinct from them, possibly located in the basal ganglia, e.g. Rapoport, 1989). How can we reconcile this with the fact that a patient presenting with any one of the 'neurotic' disorders on one occasion is likely to present with a different one on a second occasion (e.g. Andrews, 1990) suggesting that all the disorders share a common predisposing factor?

A possible explanation results from the fact that our distinction between the systems affected by the anxiolytic drugs appears to depend on receptor subtype and anatomical location more than on the neurotransmitters involved. Thus, buspirone and imipramine appear to be more selective for 5-HT_{1A} than is phenelzine; clomipramine (which is particularly effective in obsessive-compulsive disorder) may be a more potent 5-HT uptake inhibitor than other antidepressant drugs (Modigh, 1990) and so on. Thus, because of variations in receptor subtype and in the importance of reuptake as a means of inactivation, some drugs appear to have greater effects than others on anatomically distinct 5-HT systems.

It may be, then, that a predisposition to all of the 'neurotic' disorders can arise from a common maladjustment of 5-HT systems; that combines with additional precipitating factors to manifest itself as a local disorder (of the hippocampus in the case of generalized anxiety). This would account for the strong genetic linkage between these different 'neuroses' (Andrews, 1990) despite their diagnostic and apparent pharmacological differentiation.

Figure 8.1 also includes some speculations about stress hormones. I have assumed that their release is, in principle, independent of the generation of anxiety or depression. Benzodiazepines, by mimicking a putative endogenous anxiolytic hormone (or by blocking the actions of an endogenous anxiogenic one), both reduce anxiety and reduce the release of stress hormones. The purpose of this might be akin to the temporary analgesia produced by endogenous opiates, and it may be no accident that some of the effects of the benzodiazepines are achieved *via* systems with opioid links. Stress hormones, in turn, appear capable of inhibiting 5-HT_{1A} agonists, at least as exemplified by buspirone. The benzodiazepine and 5-HT_{1A} routes to anxiolytic action appear to be neurally and pharmacologically independent, sharing only a hippocampal final common path. The potential interactions between stress and other endogenous modulators of anxiety are very complex, therefore. A challenge for the future is to determine and understand these interactions.

8.10 CONCLUDING REMARKS

The main conclusion to be drawn is that anxiety, equated with generalized anxiety disorder, involves quite discrete neural systems which are acted on by all anxiolytic drugs; these, in principle, reduce risk analysis and behavioural inhibition. The failure of many animal models of anxiety to show clear results with drugs like buspirone appears attributable to an antagonistic action by stress hormones. Some tests, such as the elevated plus-maze, may be sensitive to one or several of the non-anxiolytic effects of these drugs. The clinical effects of the drugs, and hence interpretation of animal models, require the categorical distinction of panic, generalized anxiety, atypical depression and unipolar depression. Nonetheless, it remains possible that these aetiologically different disorders share a common predisposing factor which interacts with physiological or environmental factors to precipitate the different specific pathologies. There is good reason to suppose that these pathologies all represent inappropriate activation of mechanisms which are adaptive in normal circumstances. Despite the apparent pharmacological failure of some animal models, it appears that their psychological validity is good. However, care will be needed in the future to differentiate the various aspects of the response to danger which result in fear, anxiety and depression, respectively.

ACKNOWLEDGEMENT

The work reported in this paper was supported by grants from the HRC of New Zealand.

8.11 REFERENCES

Andrews, G. (1990) Neurosis, personality and cognitive behaviour therapy, in N. McNaughton and G. Andrews (eds.), *Anxiety*, University of Otago Press, Dunedin, New Zealand, pp 3–14

Blackman, D.E. (1983) On cognitive theories of animal learning: extrapolation from humans to animals, in G.C.L. Davey (ed.), *Animal Models of Human Behaviour*, Wiley, New York, pp 37–50

Blanchard, R.J. and Blanchard, D.C. (1990) An ethoexperimental analysis of defense, fear and anxiety, in N. McNaughton and G. Andrews (eds.), *Anxiety*, Univesity of Otago Press, Dunedin, New Zealand, pp 124–33

Brady, J.V. (1975a) Conditioning and emotion, in L. Levi (ed.), *Emotions: Their Parameters and Measurement*, Raven Press, New York, pp 309–40

Brady, J.V. (1975b) Toward a behavioural biology of emotion, in L. Levi (ed.), *Emotions: Their Parameters and Measurement*, Raven Press, New York, 17–46

Coover, G.D., Ursin, H. and Levine, S. (1973) Plasma corticosterone levels during active avoidance learning in rats. *J. Comp Physiol. Psychol*, **82**: 170–4

de Boer, S.F., Slangen, J.L. and van der Gugten, J. (1990) Effects of chlordiazepoxide and buspirone on plasma catecholamine and corticosterone levels in rats under basal and stress conditions. *Endocrin. Experimentalis*, **24**: 229–39

Dickenson, A. (1980) *Contemporary Animal Learning Theory*, Cambridge University Press, Cambridge

File, S.E. (1980) The use of social interaction as a method for detecting anxiolytic activity of chlordiazepoxide-like drugs. *J. Neurosci. Methods*, **2**: 219–38

Friend, P. and Andrews, G. (1990) Agoraphobia without panic attacks, in N. McNaughton and G. Andrews (eds.), *Anxiety*, University of Otago Press, Dunedin, New Zealand, pp 67–73

Goldberg, H.L. (1990) Modern treatment modalities in anxiety, in N. McNaughton and G. Andrews (eds.), *Anxiety*, University of Otago Press, Dunedin, New Zealand, pp 102–7

Graeff, F.G. (1988) Animal models of aversion, in P. Simon, P. Soubrie and D. Wiedlocher (eds.), *Selected Models of Anxiety, Depression and Psychosis*, Karger, Basle, pp 115–41

Graeff, F.G. (1991) Neurotransmitters in the dorsal periaqueductal grey and animal models of panic anxiety, in M. Briley and S.E. File (eds.), *New Concepts in Anxiety*, Macmillan Press, Basingstoke, pp 288–312

Gray, J.A. (1977) Drug effects on fear and frustration: possible limbic site of action of minor tranquillisers, in L.L. Iversen, S.D. Iversen and S.H. Snyder (eds.), *Handbook of Psychopharmacology: volume 8*, Plenum Press, New York

Gray, J.A. (1982) *The Neuropsychology of Anxiety: an Enquiry into the Functions of the Septo-Hippocampal System*, Oxford University Press, Oxford

Gray, J.A. and Ball, G.G. (1970) Frequency-specific relation between hippocampal theta rhythm, behaviour and amobarbital action. *Science N.Y.*, **168**: 1246–8

Gray, J.A. and McNaughton, N. (1983) Comparison between the behavioural effects of septal and hippocampal lesions. *Neurosci. Biobeh. Rev.*, **7**: 119–88

Green, S. (1991) Benzodiazepines, putative anxiolytics and animal models of anxiety. *TINS*, **14**: 101–4

Johnston, A.L. and File, S.E. (1988) Effects of ligands for specific 5-HT receptor sub-types in two animal tests of anxiety, in M. Lader (ed.), *Buspirone: a New Introduction to the Treatment of Anxiety*, Royal Society of Medicine Services, pp 31–41

Kahn, R.J., McNair, D., Lipman, R.S. *et al.* (1986) Imipramine and chlordiazepoxide in depressive and anxiety disorders: II. Efficacy in outpatients. *Arch. Gen. Psychiatr.*, **43**: 79–85

Marks, J. (1990) The current role of benzodiazepines in anxiety, in N. McNaughton and G. Andrews (eds.), *Anxiety*, University of Otago Press, Dunedin, New Zealand, pp 109–13

McNaughton, N. (1985) Chlordiazepoxide and successive discrimination – different effects on acquisition and performance. *Pharmacol. Biochem. Behav.*, **23**: 487–94

McNaughton, N. (1989) *Biology and Emotion*, Cambridge University Press, Cambridge

McNaugton, N. and Coop, C.F. (1991) Neurochemically dissimilar anxiolytic drugs have common effects on hippocampal rhythmic slow activity. *Neuropharmacol.*, **30**: 855–63

McNaughton, N. and Mason, S.T. (1980) The neuropsychology and neuropharmacology of the dorsal ascending noradrenergic bundle: a review. *Prog. Neurobiol.*, **14**: 157–219

McNaughton, N. and Morris, R.G.M. (1987) Chlordiazepoxide, an anxiolytic benzodiazepine, impairs place navigation in rats. *Behav. Brain Res.*, **24**: 39–46

McNaughton, N. and Morris, R. G. M. (1993) Buspirone produces a dose-related impairment in spatial navigation. *Pharmacol. Biochem. Behav.*, **43**: 167–71

McNaughton, N., James, D.T.D., Stewart, J., Gray, J.A., Valero, I. and Drewnarski, A. (1977) Septal driving of hippocampal theta rhythm as a function of frequency in the male rat: effects of drugs. *Neurosci.*, **2**: 1019–27

McNaughton, N., Richardson, J. and Gore, C. (1986) Reticular elicitation of hippocampal slow waves: common effects of some anxiolytic drugs. *Neurosci.*, **19**: 899–903

McNaughton, N. and Sedgwick, E.M. (1978) Reticular stimulation and hippocampal theta rhythm in rats: effects of drugs. *Neurosci.*, **3**: 629–32

Meert, T.F. and Colpaert, F.C. (1986) The shock probe conflict procedure: a new assay responsive to benzodiazepines, barbiturates and related compounds. *Psychopharmacol.*, **88**: 445–50

Modigh, K, (1990) The pharmacology of clomipramine in anxiety disorders, in N. McNaughton and G. Andrews (eds.), *Anxiety*, University of Otago Press, Dunedin, New Zealand, pp 93–100

Morris, R.G.M., Garrud, P., Rawlins, J.N.P. and O'Keefe, J. (1982) Place navigation impaired in rats with hippocampal lesions. *Nature*, **297**: 681–3

Panickar, K.S. and McNaughton, N. (1991) Dose-response analysis of the effects of buspirone on rearing in rats. *J. Psychopharmacol.*, **5**: 72–6

Pellow, S., Chopin, P., File, S.E. and Briley, M. (1985) Validation of open:closed arm entries in an elevated plus-maze as a measure of anxiety in the rat. *J. Neurosci. Methods*, **14**: 149–67

Pich, E.M. and Samanin, R. (1986) Disinhibitory effects of buspirone and low doses of sulpiride and haloperidol in two experimental anxiety models in rats: possible role of dopamine. *Psychopharmacol.*, **89**: 125–30

Rapoport, J.L. (1989) The biology of obsessions and compulsions. *Scientific Am.* (March), 63–9

Sanger, D.J., Joly, D. and Zivkovic, B. (1985) Behavioural effects of non-benzodiazepine anxiolytic drugs: a comparison of CGS 9896 and zopiclone with chlordiazepoxide. *J. Pharmacol. Exp. Ther.*, **232**: 831–7

Sheehan, D.V., Ballenger, J. and Jacobson, G. (1982) Relative efficacy of monoamine oxidase inhibitors and tricyclic antidepressants in treatment of endogenous anxiety, in D.F. Klein and J. Rabking (eds.), *Anxiety: New Research and Challenging Concepts*, Raven Press, New York, pp 47–60

Stumpf, C. (1965) Drug action on the electrical activity of the hippocampus. *Int. J. Neurobiol.*, **8:** 77–138

Tripp, E.G. and McNaughton, N. (1987) Naloxone fails to block the effects of chlordiazepoxide on acquisition and performance of successive discrimination. *Psychopharmacol.*, **91:** 119–23

Tripp, E.G. and McNaughton, N. (1991) Naloxone and chlordiazepoxide: effects on acquisition and performance of signalled punishment. *Pharmacol. Biochem. Behav.*, **41:** 475–81

Tripp, E.G., McNaughton, N. and Oei, T.P.S. (1987) Naloxone blocks the effects of chlordiazepoxide on acquisition but not performance of differential reinforcement of low rates of response (DRL). *Psychopharmacol.*, **91:** 112–19

Watson, J.B. (1924) *Behaviourism*, Norton, New York

Wheatley, D. (1982) Buspirone, multicentre efficacy study. *J. Clin. Psychiatr.*, **43:** 92–4

Winson, J. (1990) The meaning of dreams. *Scientific Am.* (November), 42–8

Zhu, X. and McNaughton, N. (1991) Effects of long-term administration of imipramine on reticular-elicited hippocampal rhythmical slow activity. *Psychopharmacol.*, **105:** 433–8

9

Learned helplessness: Relationships with fear and anxiety

STEVEN F. MAIER

DEPARTMENT OF PSYCHOLOGY
UNIVERSITY OF COLORADO
BOULDER
COLORADO, USA

9.1 INTRODUCTION

The topic of learned helplessness cannot be comprehensively reviewed in a single chapter. It is both large and complex, ranging from issues concerning changes in specific neurotransmitters in particular brain regions to the attribution of causality by humans. Furthermore, there have been many recent reviews of various aspects of learned helplessness (Maier, 1990; Anisman *et al.*, 1991). Instead, this chapter will focus on a small set of issues that have not been extensively discussed. However, first a bit of history and terminology to orient the reader so that the issues to be addressed can be placed into context.

The term 'learned helplessness' has been used to refer to a particular behavioural phenomenon, to a general category of experimental outcome, and to a theoretical explanation. It grew out of a set of experiments conducted in the mid and late 1960s, the goal of which was to understand better the processes involved in escape and avoidance learning. In a typical escape/avoidance procedure a signal such as a light or a tone precedes an aversive stimulus, usually footshock. if the subject makes a designated response (jumping over a hurdle in a shuttlebox, pressing a lever, etc.) in the interval between the onset of the signal and the onset of the shock, the signal is terminated and the shock does not occur. The failure of an avoidance response leads to the onset of shock, and the shock remains until the designated response occurs, now called an escape response. Trials are terminated if an escape response does not occur after some fixed period of time, say 30 s.

A number of investigators working in the laboratory of Richard L. Solomon wished to determine whether a stimulus that had been paired with shock in a Pavlovian fashion would produce avoidance responding if it were presented in the escape/avoidance situation after the subject had been trained to escape and avoid. The obvious way to do this experiment was to first expose the subject to the Pavlovian conditioning procedure in which a tone, for example, was paired with shock while the subject was restrained. The subject would then be trained in a shuttlebox to escape and avoid using a light, for example, as the warning signal. After the subject was well-trained to avoid (jump over the hurdle) upon presentation of the light, the tone would then be presented. Would the subject jump the hurdle? This experiment was attempted, but it failed (Overmier and Leaf, 1965). The reason for the difficulty was that the subjects who had been first given Pavlovian conditioning later failed to learn to escape and avoid, so the tone could not be tested. This phenomenon was dramatic. The subjects not only failed to learn to avoid the shock, they did not even learn to escape it.

What aspect of the Pavlovian conditioning procedure was critical in producing the later failure to learn to escape? Was it exposure to the tone, the contingency between the tone and shock, the shock, etc.? The tone and the contingency between the tone and the shock proved to be irrelevant: prior exposure to shock was the key. However, the shock or any other unconditioned stimulus (UCS) used in Pavlovian conditioning has a particular characteristic: it is inescapable and unavoidable. This led to a series of studies (Overmier and Seligman, 1967; Seligman and Maier, 1967) in which the escapability of the initial shocks was manipulated. Subjects were given initial shocks in a particular environment, say while restrained in a hammock or small enclosure. One set of subjects could terminate each shock by performing some response, say pressing a panel or turning a wheel. A second set was yoked to the first set. Each subject was paired with one of the subjects from the group that could escape and was simply given a series of inescapable shocks, each shock identical in duration to that produced by the escape subject. Thus each escape-yoke pair received identical durations and distributions of shocks, but one could escape and one could not. A third set of subjects received no shock at all and was simply placed in the apparatus. All subjects were then tested for escape learning in a different environment, say a shuttlebox. The results were that the subjects who were originally exposed to escapable shocks learned the new escape response at a rate indistinguishable from the non-shocked controls, while the animals who had received identical inescapable shocks learned to escape very poorly.

So, the critical feature was initial exposure to shock that was *inescapable*. Why should this lead to later failure to learn to escape? What is important about the presence *versus* absence of an escape response? Maier *et al.* (1969) argued that the escape response *per se* was not the crucial feature, but that the key was the degree of *behavioural control* over the aversive event that the organism was able to exert. We argued that organisms are sensitive to their degree of control over events, and that when events are not under their control they learn this. Thus it was argued that the subject that does not have an escape response had learned that it had no control over the termination of the shocks. We further argued that this learning would produce later impairments in learning to escape because it would (a) reduce the organisms' motivation to persist in escape attempts, a motivational or response-initiation deficit, and (b) interfere with the actual associative process, a cognitive deficit (see: Maier and Seligman, 1976; Maier, 1990). This theory was called 'learned helplessness theory' because of this explanatory emphasis on learning that the shock is uncontrollable by the yoked subjects. The failure to learn to escape after exposure to inescapable shock came to be called the 'learned helplessness effect'. The term is now more generally used to refer to any experimental outcome, behavioural or physiological, that is demonstrated to result from uncontrollable events.

At roughly the same time Jay M. Weiss, working in the laboratory of Neal E. Miller, was also conducting experiments exploring whether an organism's ability to terminate shocks to which it was exposed would make a difference. However, Weiss was not interested in whether escapability would influence later learning but in whether it would modulate the 'stressfulness' of the shocks as measured by ulcer formation, adrenocortical responses, neurotransmitter changes and the like. He found that the escapability of the shock had a profound impact on these measures (Weiss, 1968; Weiss *et al.*, 1970). If the animal could terminate the shocks it showed very little ulcer formation and brain noradrenaline changes, and resembled an animal that had not been exposed to shock at all. On the other hand, animals that could not terminate the shocks showed the usual changes associated with stress.

These two sets of studies each began a research tradition which has persisted to the present day. One has explored the behavioural consequences of variations in controllability and offered behavioural/psychological explanations for the phenomena uncovered. The other has focused on physiological consequences of controllability and has offered neurochemical/molecular explanations. The interaction between these two traditions has typically involved disputes concerning whether a particular behavioural consequence of controllability is best explained at the psychological or physiological level. Obviously, behavioural phenomena can be appropriately explained at either level, and the goal ought to be an integrated explanation (Gray, 1982; Willner, 1985), not an 'either-or'. The purpose of this chapter is to make a small step in this direction. A final preliminary matter concerns two restrictions. This chapter will consider only the consequences of exposure to a single session of uncontrollable stressors. Different processes come into play when exposure becomes chronic or repeated (Weiss *et al.*, 1975), and both circumstances cannot be discussed in a single chapter. Second, the controllability of non-stressful and even appetitive events has also been studied, but this chapter will be restricted to a consideration of aversive events.

9.2 BEHAVIOUR

Although the shuttlebox shock-escape deficit is the most intensively studied consequence of exposure to uncontrollable stressors, it is far from being the only behavioural change. The first step in an integration is to determine just what sorts of subsequent behaviours are modulated by the controllability of stressors. To count as a learned helplessness effect, a change must not only follow exposure to uncontrollable shock or some other stressor, but must be demonstrated to occur *only* if the stressor is uncontrollable. Thus only behavioural sequelae that have been shown to meet this criterion will be reviewed here. A complete catalogue will not be provided, but instead the goal will be to give the reader an appreciation of the kinds of behaviour affected (see also Chapter 6). Any complete explanation of the impact of shock or stressor controllability will have to account for the full range of outcomes, not just the shuttlebox escape deficit.

9.2.1 BEHAVIOURS

The first thing to note is that interference with escape learning is not limited to the shuttlebox or to shock. Inescapable shock exposure (but not escapable shock) also interferes with later lever press (Seligman and Beagley, 1975) and runway (Elmes *et al.*, 1975) escape from shock, and escape from water in a swim task if the water is cold (Irwin *et al.*, 1980). Inescapable shock even interferes with escape from frustration (Rosellini and Seligman, 1975). The failure of an expected reward to occur is aversive, and organisms will learn and perform responses to escape or leave a situation in which this occurs. Rats that have been first exposed to inescapable shock show a reduced tendency to escape from a situation in which an expected reward does not occur (Rosellini and Seligman, 1975).

Escape learning is not the only behaviour motivated by aversive stimuli that is altered by prior inescapable shock. Shock elicits a burst of vigorous movement as an unconditioned response (UCR), and this burst of running and jumping is reduced in animals that have been previously exposed to inescapable shock in a different situation (Anisman *et al.*, 1978; Drugan and Maier, 1982). In similar fashion, inescapably shocked animals are quick to give up struggling and swimming when placed in water that cannot be escaped, and adopt a 'floating' posture (Weiss *et al.*, 1981). Although motor activity is the most obvious UCR to events such as shock, there are others. These UCRs are also reduced by prior uncontrollable shock. For example, shock elicits 'aggressive' behaviour if two rats are shocked together. After the occurrence of a shock the two rats face each other and adopt upright 'boxing' postures and occasionally direct bites at each other. Prior exposure to inescapable but not to escapable shock reduces these shock-elicited behaviours (Maier *et al.*, 1972). A behaviour that is probably related, 'defensive burying', is also reduced by prior exposure to inescapable shock. Here a rat is placed into an enclosure that contains an electrified 'prod'. The floor of the enclosure is covered with bedding material. The rat will eventually contact the prod, and the normal response of the rat is to use its paws to spray bedding material at the prod, thereby burying the prod. This is considered to be defensive behaviour, and it is reduced by prior exposure to inescapable, but not to escapable, shock administered in a different situation (Williams, 1987).

Although shock-elicited aggression is best characterized as defensive behaviour (Blanchard and Blanchard, 1984) rather than offensive aggression, behaviour more properly characterized as aggression is similarly affected. Aggression that resembles the behaviour that occurs in the wild can be produced by introducing a stranger into an established colony, usually two males and a female. The dominant resident male rat will attack the intruder, and all the components of aggressive attack observed under naturalistic conditions are present in the behaviour of the attacking male (lateral attack, biting, etc.). The intruder typically attempts to escape, shows defensive aggressive behaviours, and eventually adopts a species-typical 'defeat posture' characterized by standing up and holding forepaws limp, or turning over onto the back and exposing the ventral surface. The behaviour of both the resident and the intruder is altered by prior exposure to inescapable shock. If the dominant male is removed and exposed to inescapable shock, it becomes less aggressive and shows a reduced tendency to attack if an intruder is later introduced to the colony (Williams, 1982). An intruder that has been inescapably shocked shows a reduced tendency to escape and engage in defensive aggression, and 'gives up' and adopts defeat postures sooner than normal (Williams and Lierle, 1986).

Moreover, inescapable shock reduces the animals' tendency to be dominant in situations requiring competition for resources. Rapaport and Maier (1978) began by assessing the dominance hierarchy in a group of rats. Pairs of hungry rats were placed in an apparatus containing a food cup. The apparatus was constructed in such a way that only one rat could have access to the food cup at any moment, and the rats were allowed to compete for control of the food cup. The measure of interest was the percentage of time that a given animal was in control of the food cup. All possible pairs of rats were repeatedly tested, and a stable hierarchy emerged such that one of the rats won against all of the others, another won against all but the first rat, etc. The rats were then exposed to a single session of escapable shock, yoked inescapable shock, or no shock. Dominance was then again measured 24 h later. The result was that animals that had received inescapable shock became less dominant: they now lost in competition to animals that they had previously beaten. Rats that received escapable shock actually became more dominant than they were before. This result could have been confounded with a reduction in the inescapably shocked subjects' motivation to eat. However, food intake was shown to be normal 24 h after inescapable shocks of the number and intensity used by Rapaport and Maier. The inescapably shocked rats ate as much as controls if they did not have to compete, but lost control of the food cup if competition was necessary.

However, there are conditions under which food and water intake is more disrupted by uncontrollable than by controllable shock. Weiss (1968) found a greater disruption in food and water intake after uncontrollable shock, with this effect persisting for roughly 24 h after stressor exposure. Among the differences between the procedures used by Rapaport and Maier (1978) and Weiss (1968), was shock intensity which was much greater in the Weiss study. Using shocks in the 1.0–1.6 mA range, we have found that disruptions in food and water intake attributable to the controllability of the stressor are quite brief. However, even when overall intake has returned to normal, inescapably shocked rats may still be more easily disputed by aversive tastes added to their diet (Dess *et al.*, 1988).

The behavioural changes described so far are all ones in which a behaviour that normally occurs is inhibited in some way by prior exposure to an uncontrollable stressor. There are also behaviours that are facilitated. The most notable are fear-related behaviours. Organisms engage in species characteristic behaviours when exposed to innate signals for danger (e.g. a cat for a rat) or to previously neutral stimuli that have become signals for danger by virtue of having been previously followed by an aversive event such as shock. The rats' dominant reaction to signals for danger is to freeze (Blanchard and Blanchard, 1969). Freezing is not merely a lack of motion. It involves muscular tension, with the rat inhibiting all movement except that required for respiration. It is usually directed in some fashion, for example into a corner or near a wall. It has been argued that this response would aid the rat in avoiding predators that are detected by the rat, since predators typically respond to movement (Fanselow and Lester, 1988). Rats that have received inescapable shock show more freezing when exposed to the environment in which shock has occurred than do rats that have received equal amounts of escapable shock (Mineka *et al.*, 1984). They also show greater reductions in food intake in the shock environment (Desiderato and Newman, 1971), another measure of fear. In addition, rats that have received inescapable shock show enhanced freezing in a new environment if shock occurs in that environment (Maier, 1990). That is, the conditioning of fear in the new environment is facilitated.

A final set of changes relate to pain sensitivity/reactivity. Stressors produce analgesia (see Amir and Galena, 1986 for a review), a decrease in pain reactivity. Both inescapable and escapable shock produce analgesia (Drugan *et al.*, 1985a), but the characteristics of the analgesia are quite different (see Maier, 1986 for a review). The number of inescapable shocks required to produce the behavioural changes described above (80–100 inescapable shocks) produces an 'opioid' analgesia, while escapable shock leads to a 'nonopioid' analgesia (Grau *et al.*, 1981; Hyson *et al.*, 1982). The term opioid analgesia is applied to decreases in pain reactivity that are reversed by traditional opiate antagonists such as naloxone and naltrexone and cross-tolerant with morphine. Non-opiate analgesia refers to analgesia which is not altered by these treatments. Further, analgesia can be later reinstated in inescapably shocked subjects with a small amount of shock insufficent to produce analgesia by itself. This is not the case for escapably shocked subjects (Jackson *et al.*, 1979). Relatedly, inescapably shocked subjects are also later hyper-reactive to morphine and become more analgesic than normal (Grau *et al.*, 1981). This sort of exaggerated reaction to pharmacological agents is not limited to morphine. Stimulants such as amphetamine and cocaine produce high levels of stereotyped behaviour in that the organism repeatedly engages in motor patterns such as 'hand washing'. Prior exposure to inescapable but not to escapable shock enhances this stimulant-induced stereotypy (MacLennan and Maier, 1983).

9.2.2 TIME COURSE

The time interval between uncontrollable stressor exposure and behavioural testing was not specified in the above discussion. Surely, the behavioural sequelae of inescapable shock would not be expected to be permanent and should dissipate with time? In most of the studies above, the interval between inescapable shock and testing was 24 h. How long would the behavioural changes persist if the subjects were not exposed to stressors during the interval? The answer is not known for all of the outcomes. The time interval between inescapable shock and testing has been manipulated in studies of shuttlebox escape learning (Jackson et al., 1978), activity in response to shock (Jackson et al., 1978), freezing after exposure to shock in a new environment (Maier, 1990), analgesia upon re-exposure to shock (Jackson et al., 1979), shock-elicited aggression (Maier, unpublished), and struggling when placed in water (Weiss et al., 1981). In all cases the behaviour was present 48 h after inescapable shock, but not after 72 h. Thus these effects dissipated between 48 and 72 h after exposure to inescapable shock.

9.2.3 ANALYSIS

This *partial* list of behavioural outcomes of exposure to uncontrollable stressors is bewildering. Why are the effects of a single session of 60–100 relatively moderate inescapable shocks so pervasive? A simple answer that is sometimes proposed is that a session of inescapable shock is a severe 'stressor', more 'stressful' than escapable shock, and so naturally many systems will be altered. First, this hardly explains anything. Why should a severe stressor have all of these consequences? Secondly, it is not clear that the term 'stressful' has a clear enough definition to be useful. For example, there are measures according to which inescapable shocks with the parameters used in my laboratory are *not* more stressful than escapable shocks. The degree of activation of the pituitary-adrenal system is often taken as an index of the degree of stress. Maier et al. (1986) measured levels of both adrenocorticotropic hormone (ACTH) and corticosterone immediately after escapable and yoked inescapable shock, and at various times thereafter. The entire timecourse of ACTH and corticosterone response to the shock session was assessed. Escapable and inescapable shock led to exactly equal rises in both hormones, and the timecourse of recovery to baseline did not differ. The rats from both groups were re-exposed to a small amount of shock 24 h later to determine whether there might be differential sensitivity to later stressors, and there was not. It might seem that the failure of an effect of controllability might be attributable to a 'ceiling effect'. For example, the adrenal cortex is not able to 'keep up' with very high levels of ACTH (Urquhart, 1974). However, there is no such ceiling effect for release of pituitary ACTH, and the groups did not differ on this either. Nor was there a difference 24 h later when the animals were exposed to only a very small amount of shock in a situation different from that in which the original shocks had occurred. This is not to argue that there are no conditions under which controllability can modulate pituitary-adrenal activity, nor that controllability might not exert subtle influences on central processes involved in pituitary-adrenal regulation. There clearly are such circumstances (Swenson and Vogel, 1983). The point is that escapable and inescapable shocks identical to those that produce many of the behavioural differences noted above do not produce any obvious differences in pituitary-adrenal activity, and so are not differentially stressful by this criterion.

Another approach would be to examine whether the large number of behavioural changes might fit into a small number of categories. Putting the learning-cognitive changes and shifts in drug reactivity aside for the moment, many of the behaviours can be captured by suggesting that fight/flight behaviour is inhibited and fear related behaviour is facilitated. Another way to say this is that it would seem the behaviours directly elicited by an aversive UCS or innate signal are reduced, and responses to learned signals for danger are increased. Motor activity in reaction to shock or water, escape from frustration, shock-elicited aggression, and defensive burying, can all be considered UCRs to aversive events. Aggressive attack by a resident, defensive aggression by an intruder, and competition with a conspecific for food, can all be described as UCRs to an innate signal (the other rat). Freezing or reduced appetitive behaviour when placed in the presence of stimuli that had been present during shock are reactions to learned danger signals: conditioned responses (CR) to the context in which the shocks had occurred. Indeed, it has been shown that freezing is never a UCR to shock, but is always a conditioned fear response (Fanselow and Lester, 1988). Further, opioid analgesia has been argued to be part of the reaction to signals for danger (Fanselow, 1986).

9.3 FEAR AND ANXIETY

This behavioural categorization of learned helplessness phenomena does not lead to an explanation by itself. However, a number of investigators have suggested that fear might be a key consideration. Inescapable shock does lead to the conditioning of more fear to stimuli present during shock than does an equal amount of escapable shock. There are a number of possible mechanisms whereby this could happen (Jackson and Minor, 1988; Warren et al., 1990), but this issue is beyond the scope of this chapter. How could enhanced fear be responsible for the later behaviours? Perhaps there are stimuli present in the test environment that are sufficiently similar to those that had been present in the original shock situation so that conditioned fear transfers or generalizes to the test situation. Since more fear was conditioned in inescapably than in escapably shocked subjects, there would be more fear to transfer. High levels of conditioned fear in the test environment might then lead to the behaviours observed. For example, it has been argued that this fear both produces freezing which can interfere with other behaviours such as fight/flight, and also produces a motivational state which can interact with the normal motivation produced in the task to interfere with the normal behavioural reactions that occur (Williams and Lierle, 1986).

This is an attractive hypothesis and there are data to support it. Both behavioural (Volpicelli et al., 1984) and pharmacological (Drugan et al., 1984) treatments that reduce the amount of fear that is produced by inescapable shock, or that is conditioned to cues present during the inescapable shock, reduce or eliminate the later behaviours characteristic of inescapable shock. However, there are two major difficulties for the hypothesis.

The first difficulty is the time course of learned helplessness effects. Why should these effects be absent if 48–72 h or more intervene between inescapable shock and testing? Conditioned fear is not a transient phenomenon. There is only one known mechanism whereby a Pavlovian CR diminishes without further training of some kind (counterconditioning, extinction, etc.). This mechanism is forgetting. However, Pavlovian CRs in general, and conditioned fear in particular, are known not to be forgotten in as little time as 48–72 h (Hoffman *et al.*, 1963). There is often some short-term forgetting soon after training, but this is typically complete by 24 h after training (Gleitman and Holmes, 1967). I know of no instance in which a Pavlovian conditioned response or conditioned fear was present 24 h after training, but had been forgotten by 48–72 h after training. If fear is conditioned to a stimulus, the conditioned fear response will occur if the stimulus is presented weeks or months after the original experience (Hoffman *et al.*, 1963; Gleitman and Holmes, 1967). There is little or no forgetting of conditioned fear over substantial amounts of time, even after only a very small number of shocks (Gleitman and Holmes, 1967). Certainly, one would not expect the forgetting of fear, to cues that had been present during 100 inescapable shocks, to occur over a 72 h period. Consistent with this argument, Irwin *et al.* (1986) placed mice back into the shock environment two weeks after a single session of inescapable shock and found that neurochemical changes characteristic of fear were produced. It might be argued that in the learned helplessness studies the exact conditioned stimuli are not presented during testing, and so perhaps it is the tendency to generalize the conditioned fear that disappears over 72 h. However, the tendency to generalize to stimuli different from those that had been present during training increases rather than decreases with time since conditioning (Thomas *et al.*, 1985).

The second difficulty is that, although reducing fear at the time of inescapable shock blocks learned helplessness effects, reducing fear at the time of later behavioural testing does not. For example, the administration of the benzodiazepines (BDZ), chlordiazepoxide and diazepam, before inescapable shock eliminates the subsequent shuttlebox escape deficit and reinstates analgesia (Drugan *et al.*, 1984), but the same drugs have no effect on these behaviours if administered before the test rather than before the inescapable shock (Drugan *et al.*, 1984). Indeed, Maier (1990) measured both fear, as indexed by freezing, and shuttlebox escape in the same animals. Diazepam administered before the shuttlebox test reduced fear to control levels, but did not even reduce the escape deficit. This was not attributable to possible interference with escape produced by the diazepam itself. It had no effect at all on escape behaviour in control subjects. Further, a pharmacological agent (naltrexone) that eliminated the escape deficit when given before testing actually increased fear/freezing.

In summary, it seems to be important that inescapable shock arouses fear, or a state related to fear. However, it may not be that this intense fear leads to learned helplessness effects because there are environmental stimuli in the test situation that elicit this fear as a CR. Or, perhaps this mechanism is responsible for some sequelae of inescapable shock, but there are additional fear-related mechanisms. What could such mechanisms be? The time course suggests an *unconditioned* rather than a conditioned change. It suggests that uncontrollable stressors *sensitize* a fear-related process, a change which dissipates over 48–72 h. Thus stimuli that would normally result in fear might lead to an exaggerated level of fear in previously inescapably shocked subjects for a period of 48–72 h. To test this idea Maier (1990) exposed rats to either one or two footshocks 24 h after either escapable or inescapable tailshock in a different apparatus. The one or two shocks produced freezing in all subjects, but this was greatly enhanced in the inescapably shocked subjects. However, this exaggerated freezing did not occur if 72 h had intervened between original shock exposure and testing.

What is a sensitized fear process? Perhaps it is what is normally meant by 'anxiety', as distinct from fear or conditioned fear. Anxiety is generally defined as a fear of what might happen, or a fear response for which there is no clear precipitating event in the environment, or a fear response that is out of proportion to the real threat. Fear is an emotional response that has an identifiable target (a predator, an aggressive conspecific, a tone that signals the occurrence of shock, an environmental context in which shock has previously occurred, etc.) and is 'appropriate' to the degree of danger. Here, there is an event that can produce directed defensive behaviour. Anxiety, on the other hand, does not have these characteristics. Thus anxiety is not conditioned to a stimulus and, once aroused, may simply dissipate.

This distinction between fear and anxiety is similar to distinctions made between different stages of defensive behaviours produced by a threat such as a predator. Several different classification schemes have been offered (Blanchard and Blanchard, 1988; Fanselow and Lester, 1988), but they have in common the assertion that the kind of behaviour that occurs depends on the spatial and temporal proximity of the threat to the organism. For example, Fanselow and Lester argue that whenever an organism such as a rat enters a situation with increased risk of predation it engages in 'pre-encounter defence' in which 'vigilance' and 'wariness' predominate: the rat will minimize exposure and time in the open. If a predator is detected, behaviour switches to 'post-encounter' defence, the predominant behaviour being freezing. If the predator comes very close and/or attack ensues, then defence shifts to 'circa-strike', with running, jumping, and biting now predominating. Blanchard and Blanchard (1988) have argued for a somewhat similar scheme, and call their proposed first stage of defence, during which a predator has not yet been detected, 'risk assessment'. Blanchard and Blanchard noted that this first stage of defence might correspond to what is typically labelled anxiety.

Does inescapable shock produce anxiety that persists for a period of time and then disappears? Investigators interested in anxiety have developed a variety of behavioural tests that are said to be sensitive to anxiety (see File, 1985 for a review). There are two major categories. In one type a conflict is produced, usually by rewarding the same response with food or water and punishing it with shock. Thus responding yields an item the rat wants, but also produces something it does not want. A variety of evidence indicates that the degree of response suppression produced by the punishment covaries with level of anxiety induced. A second type of test places the rat in a somewhat novel environment, and examines whether this alters behaviour in which the rat might normally engage. The greater the degree of suppression of the behaviour, the more anxiety. This is called anxiety because there are no stimuli that clearly signal a danger.

We (Short and Maier, 1990; 1991) have used these sorts of tests to pursue the anxiety question. We did not want to use a test that involved shock or food and water, because any result obtained could be confounded by alterations in shock sensitivity and hunger or thirst. The 'social interaction' test (File, 1985) seemed to be the best documented non-conflict measure, so this was chosen. In the social interaction test two rats that have not interacted with each other previously are placed into an open arena. The behaviour of the pair is scored for the amount of time spent in social interaction, defined as sniffing, following, grooming, kicking, pushing, standing on, wrestling, crawling over or under, or mounting the partner. Both environmental and pharmacological treatments known to increase anxiety (anxiogenic) decrease social interaction, and treatments that are known to decrease anxiety (anxiolytic) increase interaction (File, 1980).

Rats were given escapable shock, yoked inescapable shock, or were only restrained. The animals were tested for social interaction 24h later, with both members of a pair having been similarly treated the previous day. Great care was taken to eliminate any possible cues during testing that had present during inescapable shock. The testing arena was even kept in a room in which no animal had ever been shocked, and the arena was carefully cleaned after each pair to eliminate odours. Locomotion as well as interaction was measured, so that we could correct for any effect of inescapable shock on locomotion during testing. The data are shown in Figure 9.1. There was little if any effect of inescapable shock on locomotion in the arena. Nevertheless, since it is usual to correct for differences in locomotion, the amount of time in social interaction was divided by the amount of locomotor activity. Clearly, inescapable but not escapable shock reduced the amount of time spent interacting. This is also true for the uncorrected interaction data. A subsequent experiment found the decrement in social interaction to dissipate with time.

Thus there does appear to be a state that persists for some period of time after inescapable shock that is akin to sensitized fear or anxiety. Perhaps this state is at the heart of many of the other behavioural changes. This hypothesis circumvents the

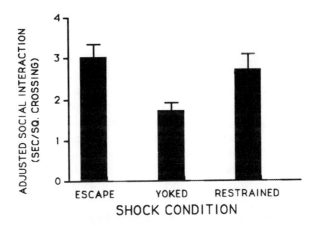

Figure 9.1 The time spent in social interaction, divided by the number of squares crossed, for rats that had received escapable, yoked inescapable, or no shock 24h earlier. Error bars show S.E.M.

first difficulty, with respect to the time course, noted with the conditioned fear hypothesis. However, it does not circumvent the second. Agents such as diazepam and chlordiazepoxide should reduce anxiety if given before testing, so should they not eliminate any behavioural effects caused by anxiety? I will return to this later.

9.4 NEUROCHEMISTRY OF FEAR, ANXIETY AND LEARNED HELPLESSNESS

Stressor controllability modulates many of the neurochemical consequences of stressors. Space does not permit a summary and there are many excellent reviews (e.g. Anisman et al., 1991). The present discussion will focus on whether there are commonalities between what is known about the neurochemistry of fear/anxiety and learned helplessness, with a view to gaining further understanding of the multiple behavioural consequences of inescapable shock.

9.4.1 ANXIETY AND GABA

Much of the recent research concerning the neurochemical basis of anxiety has revolved around the neurotransmitter γ-aminobutyric acid (GABA) and one class of GABA receptor, the GABA$_A$ receptor (see also Chapter 12). This is because many of the agents that reduce anxiety in humans (e.g. benzodiazepines, barbiturates and alcohol) facilitate the actions of GABA at the GABA$_A$ receptor, whereas agents that interfere with GABA transmission increase anxiety. GABA is the major inhibitory transmitter in the brain. It inhibits neuronal activity by opening chloride channels, thereby causing hyperpolarization. The GABA$_A$ receptor is composed of a number of polypeptide subunits (Martin, 1987): GABA binds to the β-subunit, and this leads to the opening of the chloride channel associated with the receptor. Another subunit, the α-subunit, is a modulatory site: that is, activity at this site alters the β-subunit so that the ability of GABA to bind to the site is changed, as is the amount of chloride ion admitted into the neurone by the binding of a given amount of GABA. BDZs such as chlordiazepoxide and diazepam bind to the α-site, and so it has been called a BDZ binding site or receptor. The binding of a BDZ to this site alters the β-subunit in such a way that GABA binding is facilitated, and the entry of chloride augmented. These two effects have the outcome of increasing the neuronal inhibition exerted by a fixed amount of GABA. Barbiturates and alcohol work by a slightly different mechanism, but they also increase the inhibition induced by a given amount of GABA. Interestingly, there is a class of compounds that bind to the BDZ site in such a way that they interfere with the ability of GABA to bind to the β-subunit and *reduce* chloride entry. These compounds, called inverse agonists because they exert effects opposite to those of BDZs, increase anxiety. They lead to behaviours and physiological changes characteristic of anxiety (Ninan et al., 1982), and in humans feelings of 'uneasiness and impending doom' (Dorow, 1982).

The question of whether GABA changes constitute 'the mechanism of anxiety' remains controversial (see Gray, 1987 for a discussion) and is not resolved. However, it is clear that facilitation of GABA transmission is often anxiolytic, and interference with GABA action is often anxiogenic. It thus becomes sensible to inquire whether there is a relation between GABA and learned helplessness effects.

9.4.2 LEARNED HELPLESSNESS AND GABA

It has already been noted that administration of BDZs before inescapable shock blocks the shuttlebox escape learning deficit and reinstated opioid analgesia that normally follows inescapable shock. To this it can be added that chlordiazepoxide administered before inescapable shock also blocks the reduced social interaction that would normally follow 24 h later (see Figure 9.2). Conversely, the anxiogenic partial inverse agonist (N-methyl-β)-carboline-3-carboxamide (FG 7142) produces the shuttlebox deficit (Drugan *et al.*, 1985b) and analgesia (Maier, unpublished) and the reduction in social interaction (Figure 9.3). These were all tested 24 h after the administration of the drug, long after the drug had cleared and its acute effects (e.g. increased autonomic activity) had disappeared. Thus a single administration of a BDZ partial inverse agonist produced a change that persisted for at least 24 h. Figure 9.3 also shows that the impact of FG 7142 on social interaction 24 h later is dose dependent and can be blocked by the BDZ receptor antagonist flumazenil (Ro 15-1788). This latter finding suggests that the effect of FG 7142 is mediated by the BDZ receptor. Consistent with these findings, Petty and Sherman (1981) found that injecting a GABA$_A$ antagonist (bicuculline) into the hippocampus produced an escape learning deficit. Thus agents which facilitate GABA action can prevent learned helplessness effects and agents which interfere with GABA action can produce them.

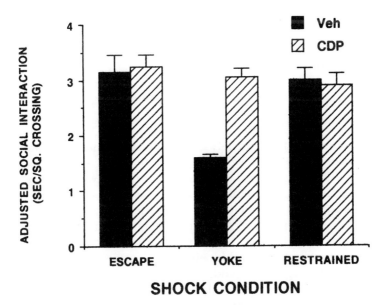

Figure 9.2 The time spent in social interaction, divided by the number of squares crossed. Animals received either 10 mg/kg (i.p.) chlordiazepoxide (CDP) or vehicle (Veh) before escapable, yoked inescapable, or no shock on Day 1. Social interaction testing was carried out 24 h later. Error bars show S.E.M.

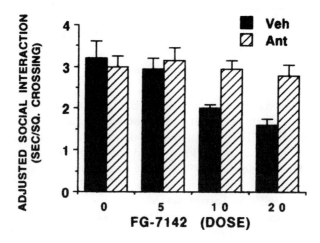

Figure 9.3 The time spent in social interaction, divided by the number of squares crossed. Subjects had received 0, 5, 10 or 20 mg/kg FG-7142 (i.p.) 24 h earlier, in combination with 10 mg/kg (i.p.) of the BD antagonist (Ant) flumazenil, or the vehicle (Veh) for flumazenil. Error bars show S.E.M.

The findings that BDZ agonists mimic 'having control' and that BDZ inverse agonists mimic 'lack of control' raise two possibilities with regard to stressor controllability. Aversive events, such as shock, increase release of GABA (Soubrie *et al.*, 1980) and perhaps (a) learning that the stressor is controllable, or 'having control', releases an *endogenous anxiolytic* substance, a naturally occurring BDZ-like substance in the brain which facilitates GABA action, and/or (b) 'having no control', or learning that there is no control, releases an *endogenous anxiogenic* substance, a naturally occurring BDZ inverse agonist which interferes with GABA action. The existence of a BDZ receptor suggests that there may be endogenous substances or ligands that act at the receptor and claims have been made for the existence of both anxiolytic (De Blas *et al.*, 1987) and anxiogenic (De Robertis *et al.*, 1988) ligands.

Both positions predict that the administration of a BDZ before shock would eliminate the later difference in anxiety between animals that did and did not have control, and lead to anxiety levels similar to those in non-shocked controls. However, they differ with regard to what they predict would happen if a BDZ receptor *antagonist* were to be administered before the shock experience. A pure BDZ receptor antagonist would exert neither an anxiolytic nor an anxiogenic action on its own, but would occupy the BDZ receptor and prevent either an endogenous agonist or inverse agonist from binding to the receptor. If lack of control releases an anxiogenic ligand that is responsible for effects such as decreased social interaction, then a BDZ receptor antagonist should reduce later anxiety in inescapably shocked subjects and have no effect on escapably shocked subjects. It should reduce the increased anxiety in the inescapably shocked subjects because the putative endogenous anxiogenic cannot now have an effect on the BDZ receptor, since the receptors are occupied by the antagonist. In contrast, if having control releases an anxiolytic ligand that is responsible for the difference in anxiety, then a BDZ receptor antagonist administered before shock should increase anxiety after escapable shock and have no effect on the anxiety measured after inescapable shock.

There is some debate concerning whether substances proposed to be BDZ receptor antagonists are actually pure antagonists without either agonist or inverse agonist properties (File *et al.*, 1982). Nevertheless, Short and Maier (1991) chose to explore one of these, flumazenil (Ro15-1788). Rats received either escapable shock, yoked inescapable shock, or restraint, followed by a social interaction test 24 h later. Separate groups received flumazenil at a dose (10 mg/kg) known to have no effect on social interaction (File and Pellow, 1986) either before the Day 1 treatment or the Day 2 interaction test (they were injected with only the vehicle on the other day). Controls received vehicle on both days. The data are shown in Figure 9.4. The vehicle controls (VEH-VEH) show that, as usual, inescapable but not escapable shock decreased social interaction. Now examine the groups given flumazenil before the Day 1 treatment (ANT-VEH). Flumazenil completely eliminated the difference in social interaction between groups, and did so by blocking the decrease in interaction produced by the inescapable shock! It might be noted that, as expected, flumazenil at the dose used had no effect on social interaction by itself, confirming the relatively pure antagonist nature of the drug in this test. Thus these results would seem to suggest that lack of control releases an endogenous anxiogenic substance which acts at the BDZ receptor. Interestingly, flumazenil administered before the social interaction test had no effect at all, a finding to which I will return.

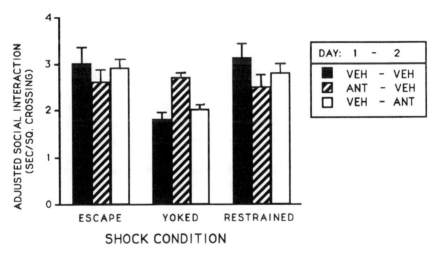

Figure 9.4 The time spent in social interaction, divided by the number of squares crossed. Subjects received escapable, yoked inescapable, or no shock on Day 1 and social interaction testing on Day 2. Subjects were given either flumazenil (Ant) or vehicle (Veh) before the Day 1 treatment or Day 2 test. Error bars show S.E.M.

There is further evidence on this issue. Interference with GABA action can produce seizure activity as well as anxiety. For example, a large dose of the GABA receptor antagonist bicuculline produces seizures. As might be expected, BDZ agonists moderate such seizures (since they facilitate GABA action) and inverse agonists potentiate them (since they further interfere with GABA). As noted previously, stressors release GABA or augment GABA activity. It might then be expected that a stressor would alleviate seizures produced by GABA antagonists, and this is indeed the case (Soubrie *et al.*, 1980). But what if uncontrollable stressors in amounts sufficient to produce learned helplessness effects release an endogenous anxiogenic inverse agonist? What should prior exposure to 60–100 inescapable shocks do to seizure activity produced by a GABA antagonist such as bicuculline? It should enhance seizures and make the subject more susceptible to seizure activity. This is exactly what was reported by Drugan *et al.* (1985c): inescapable but not escapable shock increased bicuculline-induced seizures. Finally, recall that inverse agonists interfere with GABA action by ultimately decreasing the amount of chloride ion that enters the neurone when GABA interacts with the $GABA_A$ receptor. Drugan *et al.* (1989b) measured the amount of chloride ion uptake by cortical neurones stimulated with muscimol, a GABA agonist. Chloride uptake was reduced in inescapably shocked animals, just as would be produced by an anxiogenic inverse agonist. Moreover, there was reduced binding of flumazenil to neuronal tissue, suggesting that the BDZ receptors in inescapably shocked subjects were already occupied with an endogenous substance.

9.4.3 ANXIETY AND 5-HYDROXYTRYPTAMINE

An alteration in GABAergic inhibition is not by itself sufficient to explain anxiety or to provide an understanding of the multiple effects of inescapable shock. GABA and GABA receptors are widely distributed throughout the brain and spinal cord, and most of the regions containing GABA and GABA receptors are presumed to have nothing to do with anxiety. Thus one crucial issue concerns the sites in the brain where changes in GABAergic inhibition are important for anxiety and learned helplessness. Moreover, the large majority of GABA neurones in the brain are interneurones that function in local recurrent inhibitory feedback loops (Haefely, 1984). In these loops, activation of a neurone which might release dopamine, noradrenaline, or 5-hydroxytryptamine (5-HT), for example, also activates a GABA interneurone which synapses back onto the original neurone, inhibiting its action. Thus the GABA loop operates as a negative feedback loop, restraining the firing of the principal neurone. Thus it would seem likely that any anxiolytic effect of GABAergic facilitation would occur because GABAergic restraint on *some other system* is increased. Conversely, anxiogenic effects of interference with GABAergic inhibition would likely be produced by decreased GABA restraint *on some other critical system* or systems.

GABAergic inhibition has been reported to exist in many of the brain's major systems and pathways. It would be naive to expect that only one of these would be involved in producing and regulating anxiety. Anxiety is a label that we give to a constellation of behavioural and physiological events. For example, Gray (1982) has argued that anxiety is characterized by behavioural inhibition, increased vigilance and arousal. These are unlikely to all be produced by a single discrete system. Moreover, any particular task used to assess anxiety will involve different mixes of these, and so the exact mediating mechanisms can be expected to differ from measure to measure. Since we have found stressor controllability to modify behaviour on the social interaction test, I will focus on systems that might modulate behaviour in this situation.

Two different neurotransmitter systems have frequently been argued to be important in anxiety (see also Chapter 11). One is the noradrenergic locus coeruleus-dorsal bundle system. The locus coeruleus is a compact nucleus located in the pons, and its small number of widely bifurcating neurones contain the majority of noradrenaline in the brain. The noradrenergic cells in the locus coeruleus contain GABA receptors, and GABA inhibits the spontaneous firing rate of these neurones (Cederbaum and Aghajanian, 1978). Activation of this system has been argued to be a component of anxiety (Redmond, 1987) and has also been argued to play a key role in learned helplessness (Weiss et al., 1986). However, this system will not be discussed here. This is because there is good reason to question whether the noradrenergic system arising from the locus coeruleus plays a key role in anxiety (see Johnston, 1991 for a recent review), particularly for the social interaction measure. Neither lesion of the locus coeruleus nor administration of noradrenergic receptor antagonists alters anxiety-related behaviour in the social interaction test (File et al., 1979; File, 1980). In addition, the role of noradrenaline in learned helplessness has been frequently reviewed (e.g. Weiss and Simson, 1986).

A clearer case can be made for the involvement of 5-HT in anxiety. Neurones containing 5-HT are largely confined to the raphe nuclei in the brain stem. One of these nuclei, the dorsal raphe nucleus (DRN) has been clearly shown to contain GABA (Steinbusch and Nieuwenhuys, 1983), and GABA or BDZs microinjected into the DRN reduce DRN electrical activity. Systemic injection of BDZs also reduces DRN firing rates (Laurent et al., 1983). The DRN projects to a variety of structures including the nigro-striatal system, limbic, and cortical regions.

An understanding of 5-HT systems requires a brief discussion of 5-HT receptors. There are a large number of different types of 5-HT receptors, with general agreement concerning 5-HT_{1A}, 5-HT_{1B}, 5-HT_{1C}, 5-HT_{1D}, 5-HT_2 and 5-HT_3 receptors (see Peroutka, 1988 for a review). 5-HT_{1A} receptors are present both on the soma and dendrites of raphe neurones, as well as postsynaptically. The somatodendritic 5-HT_{1A} receptors function as 'autoreceptors'. Activity of raphe neurones releases 5-HT onto these receptors, with activation of these receptors inhibiting firing of the neurone (Hamon et al., 1988). Thus the autoreceptors function in a negative feedback fashion to restrain activation of raphe cells. The postsynaptic 5-HT_{1A} receptors are concentrated in the cortex, septum and hippocampus and inhibit activity at these sites (Bockaert et al., 1987; Sprouse and Aghajanian, 1988). In the rat, 5-HT_{1B} receptors are also a kind of autoreceptor and are present on axon terminals. Activation of these receptors by 5-HT or other agonists inhibits release of 5-HT from the terminals. 5-HT_2 receptors are generally agreed to be entirely postsynaptic. The role of 5-HT_3 receptors is less clear, but they appear to be postsynaptic as well.

The relationship between 5-HT and anxiety is not a simple one and has been the subject of numerous reviews (e.g. Briley et al., 1991; Iversen, 1984; Soubrie, 1986). The general trend is that manipulations which decrease 5-HT activity are anxiolytic and manipulations which increase 5-HT activity are anxiogenic in behavioural tests. However, there are numerous exceptions and the results are not nearly as consistent as those obtained with BDZ agonists and inverse agonists. This need not concern us here. The issue addressed is not whether 5-HT alterations mediate anxiety, but whether the effects of controllability on anxiety might be caused by alterations in 5-HT. This requires only that 5-HT be *involved* in anxiety or behaviour that we label as anxiety, and this is demonstrably the case.

9.4.4 LEARNED HELPLESSNESS AND 5-HYDROXYTRYPTAMINE

Recall that inescapable shock reduces social interaction measured 24 h later. A plausible hypothesis was that uncontrollability or learning that the stressor is uncontrollable releases an endogenous anxiogenic ligand. Could 5-HT and the DRN be involved? Here, discussion is limited to the DRN because the focus is on interactions between GABA and 5-HT; whereas GABA is clearly present in the DRN, its presence in other raphe nuclei, such as the median raphe nucleus, is not clear. However, it is recognized that the median raphe nucleus can alter fear and related processes (Graeff and Silviera Filho, 1978) and undoubtedly plays an important role in defensive behaviour.

Lesion of the DRN by microinjection of the neurotoxin 5,7 dihydroxytryptamine (5,7-DHT) directly into the nucleus increases social interaction, just as do BDZs (File *et al.*, 1979). This anxiolytic effect is specific to the DRN and was not produced by destruction of the median raphe (File *et al.*, 1979). It has already been noted that BDZ inverse agonists reduce social interaction. This anxiogenic effect of peripheral injections of inverse agonists can be completely blocked by microinjecting the BDZ antagonist flumazenil directly into the DRN (Hindley *et al.*, 1985), suggesting that the DRN is the site of action of the inverse agonist. Consistent with this suggestion, microinjection of the inverse agonist into the DRN decreases social interaction, but not microinjection outside the DRN in nearby sites (Hindley *et al.*, 1985).

Thus if inescapable shock releases an endogenous BDZ inverse agonist, perhaps it decreases social interaction by acting at the DRN to interfere with normal GABAergic inhibition of DRN cells, thereby producing exaggerated responding by these cells. To begin to test this idea Short and Maier (1991) microinjected 10 ng of the BDZ receptor antagonist, flumazenil, or vehicle into the region of the DRN before rats received either escapable shock, yoked inescapable shock, or restraint. The next day they were tested for social interaction. If inescapable shock reduces social interaction because it releases an endogenous BDZ inverse agonist that acts at the DRN, then flumazenil administered into the DRN should block the usual effect. The results are shown in Figure 9.5. Social interaction was reduced in the animals that had received yoked inescapable shock, and this effect was completely blocked by flumazenil administered into the DRN. We conducted an identical experiment, but administered the flumazenil in the region of the locus coeruleus. This had no effect on the reduction in social interaction produced by inescapable shock.

However, a little thought reveals a difficulty with the present hypothesis. The putative endogenous anxiogenic would be released at the time of inescapable shock, and would interfere with GABAergic inhibition at the DRN and increase 5-HT activity *at that time*. But, the social interaction test was 24–72 h later. Why would 5-HT activity be augmented 24–72 h after the supposedly critical event? It is possible, although very unlikely, that the inverse agonist stays bound to the BDZ receptor for that period of time. However, recall that flumazenil did not reduce the effect of inescapable shock on social interaction when it was administered before the social interaction test, but only when it was given before the shock experience. Flumazenil competes with known inverse agonists for the BDZ receptor, and should have displaced any inverse agonist still bound.

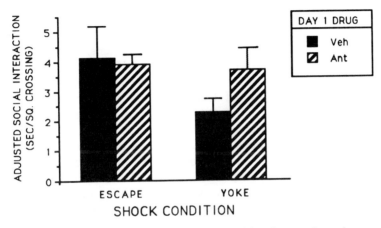

Figure 9.5 The time spent in social interaction divided by the number of squares crossed. Subjects received escapable or yoked inescapable shock on Day 1, preceded by either flumazenil (Ant) or vehicle (Veh) injected into the DRN. Error bars show S.E.M.

An admittedly very speculative possibility can be offered. The foregoing suggests that inescapable shock might produce an unusually high degree of activation of the DRN because it simultaneously activates the DRN (as do stressors in general, Jacobs, 1987) and interferes with GABAergic feedback circuits which normally restrain the activity of the cells. The 5-HT_{1A} somatodendritic autoreceptors also participate in this self-regulatory process, but they appear to have the peculiar characteristic that they are very easy to desensitize or 'downregulate'. Many kinds of receptors decrease in sensitivity when stimulated by large amounts, or by repeated applications, of agonists. This happens because receptor number is reduced, affinity of the receptor for its ligand decreases, or both. The desensitizing process usually requires multiple exposures to an agonist, but 5-HT_{1A} receptors have been reported to desensitize after a single application of 8-hydroxy-2-(di-n-propylamino)tetralin (8-OH-DPAT), a relatively pure 5-HT_{1A} agonist (Kennett et al., 1987). This appears to be peculiar to the somatodendritic 5-HT_{1A} receptors and does not hold for the post-synapatic 5-HT_{1A} receptors (Kennett et al., 1987). Thus it is possible that inescapable shock leads to a sufficiently large accumulation of 5-HT in the DRN (because GABAergic inhibition has been reduced) that the 5-HT_{1A} autoreceptors are downregulated, and that this desensitized state persists for some period of time. Inescapable shock has indeed been shown to lead to an increase in the level of 5-HT and of the 5-HT metabolite 5-hydroxyindoleacetic acid (5-HIAA) in the region of the raphe nuclei (Edwards et al., 1986), and it is not uncommon for receptor changes to persist for a number of days. The outcome would be a *hyper-responsive 5-HT system*, since 5-HT released during the course of activation would not now produce the normal amount of feedback inhibition. Thus inescapably shocked animals might later display an exaggerated amount of 5-HT activity when exposed to an event that stimulates some 5-HT activity, the social interaction test for example.

Although there has been much less research concerning the role of 5-HT in learned helplessness than there has been with regard to the catecholamines, there is some evidence that excessive 5-HT activity is indeed involved in mediating learned helplessness effects such as poor escape learning. Inescapable shock does lead to greater activation of 5-HT systems than does escapable shock, as indicated by the accumulation of the 5-HT metabolite 5-HIAA (Helhammer *et al.*, 1984). As would be expected by the present argument, increasing 5-HT levels above normal by administering large amounts of trytophan (the precursor of 5-HT) interferes with escape learning, and amounts of trytophan and inescapable shock that are insufficient to produce escape deficits when given alone summate to produce deficits (Brown *et al.*, 1982). Consistent with these findings, Drugan *et al.* (1989a) found that the inescapable shock-produced escape deficit was prevented by the administration of the partial 5-HT$_{1A}$ agonist, buspirone, before the inescapable shock. 5-HT$_{1A}$ agonists such as buspirone reduce 5-HT activity by acting at the 5-HT$_{1A}$ autoreceptors. It should be noted, however, that this effect of buspirone could have been mediated instead by action at postsynaptic 5-HT receptors (Martin *et al.*, 1990). Furthermore, rats selectively bred to show escape failure revealed a number of changes in 5-HT receptor function (Edwards *et al.*, 1991).

There is, however, conflicting evidence. Petty *et al.* (1991) used *in vivo* microdialysis to measure 5-HT release in frontal cortex and hippocampus. Inescapably shocked animals did reveal enhanced basal levels of 5-HT after the shock experience and before escape testing, but there was no correlation between 5-HT levels after escape testing and whether a given subject failed to learn to escape. Moreover, stimulating 5-HT release in the frontal cortex with potassium did not produce escape failure. It might be noted that frontal cortex and hippocampus receive major 5-HT innervation from the median raphe, and so these results may not speak directly to the hypothesis entertained here. Brain regions other than the frontal cortex and hippocampus will be discussed below.

A number of studies have manipulated 5-HT activity in the time period between the inescapable shock and escape testing. For example, Martin *et al.* (1989) gave rats inescapable shock followed by three daily sessions of escape/avoidance testing, beginning 48 h after the inescapable shock. Re-uptake blockers for 5-HT (citalopram, fluvoxamine, idalpine, and zimelidine) were administered twice a day, beginning 6 h after the inescapable shock treatment. The result was that poor escape behaviour was reversed by the third day of escape/avoidance testing. This would also seem to be inconsistent with the present 5-HT hypothesis because re-uptake blockers increase 5-HT levels, and this should have exacerbated the learned helplessness effect, not diminished it. However, a number of factors should be noted. First, there is evidence to suggest that different processes are involved in the production and the reversal of learned helplessness effects (Petty, 1986). Second, Martin *et al.* (1989) did not include non-inescapably shocked controls. Thus it is possible that the drug treatment facilitated escape responding *per se*, rather than reversing learned helplessness. That is, the drug regimen might have improved the escape performance of controls if they had been included, and so might not be operating on the neurochemical changes produced by inescapable shock. Third, there was no comparison between controllable and uncontrollable shock, so it is not known whether the poor escape which followed inescapable shock reflects a learned helplessness effect. Fourth, the use of repeated administrations of the re-uptake blockers may have been sufficient to downregulate postsynaptic 5-HT receptors. Thus 5-HT released during the escape testing might have produced decreased levels of postsynaptic activity.

In sum, there is sufficient evidence to encourage the notion that 5-HT is involved in the anxiety and escape learning deficits that are modulated by stressor controllability. However, the available evidence is far too fragmentary and inconsistent to justify strong conclusions. Clearly, this is an area in which more research is needed. Interestingly, many of the other behaviours modified by uncontrollable stressors are altered by manipulations of 5-HT activity. In general terms, increasing 5-HT activity produces results similar to inescapable shock, and decreasing 5-HT activity produces effects in the opposite direction. Increasing 5-HT activity by administering either tryptophan, 5-HT re-uptake blockers, or post-synaptic 5-HT receptor agonists has been reported to decrease offensive aggression (see Miczek and Donat, 1989, for a review), shock elicited aggression (Miczek and Donat, 1989), defensive burying (Broekamp et al., 1986), non-opioid analgesia (Curzon et al., 1986), competition for food at doses that do not alter food intake in the absence of competition (Gentsch et al., 1989), and food intake at higher doses (Gentsch et al., 1989). These manipulations also increase stimulant induced stereotypy (Baldessarini et al., 1975) and reactivity to morphine (Roberts, 1985). Although I was not able to find a study of intruder behaviour after such manipulations, the decreased aggressiveness and decreased tendency towards active escape that is produced might be expected to lead to the rapid adoption of defeat postures. In contrast, decreases in 5-HT activity produced by lesion, depletion of 5-HT, 5-HT$_{1A}$ agonists, or 5-HT$_2$ and 5-HT$_3$ antagonists, have produced the opposite outcomes wherever data are available.

9.5 THE AMYGDALA, DORSAL RAPHE AND PERIAQUEDUCTAL GRAY

How could all these effects of 5-HT come about? To proceed, it will be necessary to briefly consider what is know about neural systems involved in the regulation of fear, anxiety, and defensive behaviour. The role of the septo-hippocampal system will not be considered since its involvement in anxiety and learned helplessness has been discussed at length (Gray, 1982). Instead, this presentation will focus on the amygdala, the DRN, and the periaqueductal gray (PAG).

A considerable amount of research has been directed at understanding the role of the amygdala and the PAG in mediating defensive behaviour, conditioned fear, and anxiety. The PAG plays a key role in organizing defensive behaviour (see Bandler, 1988 for a review). Electrical or chemical stimulation in the PAG is able to produce fully organized flight/fight behaviours such as running, jumping, freezing and biting, as well as autonomic nervous system changes characteristic of flight or fight such as increases in heart rate and blood pressure (Bandler, 1988). The key role of the PAG is highlighted by the fact that stimulation of the PAG produces many of these changes even if other structures important to defensive behaviour such as the amygdala and medial hypothalamus have been destroyed (De Molina and Hunsperger, 1962). Of importance here, the PAG is functionally divided. Stimulation in the dorsal regions of the PAG (dPAG) produces behaviours such as running, jumping and biting, that are similar to reactions to an aversive stimulus itself (UCRs), while stimulation in the ventral PAG (vPAG) produces freezing, a behaviour that is thought to be a conditioned fear response. Lesions in the dPAG eliminate running and jumping in reaction to shock, but do not reduce the freezing response conditioned to the cues present during shock (Fanselow, 1991). On the other hand, vPAG lesions have no effect on the motor UCR to shock, but eliminate the development of conditioned freezing (Fanselow, 1991). Further, stimulation in the vPAG produces opioid analgesia, while stimulation in the dPAG produces non-opioid analgesia. These studies indicate that UCSs such as shock have access to the dPAG, and that the dPAG is involved in the organization of UCRs to stimuli such as shock. Signals for shock, on the other hand, seem to operate to produce behaviour through the vPAG.

The amygdala is generally assigned a pivotal role in the organization of conditioned fear (see review by Davis et al., 1991). It seems to organize information concerning aspects of the environment that are dangerous, and to integrate the various components of the fear response. Afferent input from auditory and visual channels converge in nuclei in the basolateral portions of the amygdala with input from aversive stimuli such as shock (Iwata et al., 1986), and associations may form between them in this location. This information is conveyed to the central nucleus of the amygdala (CEA), which organizes the fear behaviour and provides output to autonomic nuclei in the brainstem, to the medial hypothalamus, and to the vPAG via a monosynaptic connection (Rizvi et al., 1991). Lesions of the CEA eliminate fear responses (freezing, heart rate changes, potentiated startle, etc.) to both learned signals for danger such as a tone that has been paired with shock (Hitchcock and Davis, 1987), and unlearned signals such as the sight of a cat (Blanchard and Blanchard, 1972). Conditioned opioid analgesia is also eliminated (Helmstetter, 1992). Lesions of the CEA do not, however, exert major effects on defensive behaviour in reaction to aversive UCSs such as shock. Defensive burying soon after a shock (Roozendaal et al., 1991), the unconditioned analgesia which occurs immediately after shock (Fox and Sorenson, 1991), and the activity burst after a shock (Helmstetter, personal communication), are all unaffected. Further, offensive aggression is not reduced (Blanchard and Takahashi, 1988). It should be noted, however, that CEA lesions do reduce the peak amplitude of activity during a series of rapidly occurring shocks by 10–20% (Hitchcock et al., 1989). Stimulation of the amygdala produces freezing and an opioid analgesia (Helmstetter et al., 1991), and only produces behaviours such as running and jumping at very high current intensities (Applegate et al., 1983).

In sum, conditioned fear responses involve convergence of CS and UCS input at the amygdala, output from the CEA to a variety of structures including the vPAG, and freezing and opioid analgesia as consequences of the vPAG activation. Aversive stimuli such as shock directly activate structures in the dPAG, and the amygdala can communicate with the dPAG as well by an indirect route through the medial hypothalamus. Activation of dPAG sites leads to fight/flight and nonopioid analgesia (see Fanselow, 1991 for a more elaborate model that includes these mechanisms).

But how are the DRN and 5-HT involved? The DRN sends projections to both the amygdala and dPAG (Steinbusch and Nieuwenhuys, 1983). Importantly, stimulation of the DRN *inhibits the output of the dPAG*. Recall that electrical or chemical stimulation in the dPAG elicits running and jumping. Stimulation of the dPAG is also aversive (Bandler, 1988), and rats will perform responses such as pressing a lever to terminate stimulation. However, simultaneous stimulation in the DRN eliminates the running and jumping produced by dPAG stimulation, and reduces the aversiveness of the stimulation (Kiser *et al.*, 1980). The inhibition of the consequences of dPAG activation produced by DRN stimulation is mediated by 5-HT$_2$ receptors in the dPAG, since the inhibition is blocked by 5-HT$_2$ antagonists (Schutz *et al.*, 1985). In addition to inhibition of dPAG function, DRN stimulation produces freezing, autonomic nervous system activation, and opioid analgesia (Cannon *et al.*, 1982; Graeff and Silveira Filho, 1978), behaviours characteristic of amygdala activation. There is no evidence to indicate that these outcomes of DRN stimulation are mediated by the DRN projection to the amygdala, but conflict-induced anxiety can produce accumulation of 5-HT metabolites in the amygdala (Sakurai-Yamashita *et al.*, 1989). Moreover, 5-HT$_2$ antagonists microinjected into the amygdala have an anxiolytic effect (Hodges *et al.*, 1987). This is not to argue that conditioned fear is mediated by the DRN projection to the amygdala, since the DRN is not necessary for the production of conditioned fear (Srebo and Lorens, 1975; Davis *et al.*, 1991). The only suggestion being made is that DRN activity might enhance conditioned fear that is produced by other circuitry, by releasing 5-HT at the amygdala. As expected from the above, DRN lesions increase the fight/flight behaviour produced by aversive UCSs (Jacobs and Cohen, 1976), and inhibition of DRN activity by microinjection of GABA or BDZs decrease conditioned fear (Thiebot *et al.*, 1980).

9.5.1 SUMMARY

The overall hypothesis can now be stated succinctly. Perhaps, when organisms are exposed to inescapable shock, learning that the shock is uncontrollable (or some other aspect of the procedure) leads to the release of an endogenous anxiogenic BDZ inverse agonist which acts at the DRN (and perhaps elsewhere). This interferes with GABA inhibition of DRN activity and, in combination with the strong activation of DRN activity by shock, leads to exaggerated 5-HT activity and an accumulation of 5-HT in the DRN itself. This might downregulate somatodendritic autoreceptors in the DRN, leaving the DRN neurones in a hyper-sensitive state for some period of time. If the subject encounters a subsequent event during this period which leads to 5-HT activity, the activity would be greater than that normally produced by that event. The DRN activation would facilitate fear-related amygdala processes and inhibit fight/flight-related dPAG processes. In addition, the DRN projects to the nigro-striatal system and is capable of producing alterations in dopamine related behaviours such as stimulant-induced stereotypy (Baldessarini *et al.*, 1975). Finally, 5-HT from the DRN modulates the analgesic effect of morphine by interacting with opioid processes at the nucleus raphe magnus (Roberts, 1985). Indeed, the administration of re-uptake blockers for 5-HT into the nucleus raphe magnus produces analgesia to subthreshold amounts of morphine (Roberts, 1985). Thus a hyper-sensitive 5-HT system would be capable of exaggerating stereotypic responding to stimulants and analgesic responses to morphine, both of which have been demonstrated to follow 24 h after exposure to inescapable but not escapable shock. This could also lead to the appearance of reinstated opioid analgesia to small amounts of shock. An otherwise inadequate amount of endogenous opioid release might summate with a larger than normal quantity of 5-HT.

It is now possible to return to the peculiar finding that BDZs block learned helplessness effects if administered before inescapable shock, but not when given before testing. Recall that a BDZ given before testing reduced the exaggerated fear resulting from prior inescapable shock, but not the shuttlebox escape deficit or analgesia. It might seem that these latter findings are inconsistent with the present hypothesis since BDZs should enhance GABAergic inhibition of the DRN, thereby countering DRN hyper-reactivity. This should reduce fear, as should the direct action of BDZs at the amygdala (Nagy *et al.*, 1979). This was indeed the result. But, it should also reduce inhibition of dPAG output, so why are defensive behaviours or shock UCRs still reduced in inescapably shocked animals? It can be noted that BDZs and GABA also play a role in the PAG itself. The dPAG contains BDZ receptors and intrinsic GABA neurones, and GABA or BDZs reduce dPAG activity and behavioural output (Graeff, 1991). Thus, BDZs would be expected to both reduce DRN mediated inhibition of dPAG output and dPAG activity itself. These effects are in opposed directions in terms of dPAG output, thus cancelling and leaving no effect of the BDZ at the time of testing.

Although highly speculative, this hypothesis is able to account for the myriad of behavioural changes produced by uncontrollable stressors and the general pattern that aversive CRs or conditioned fear are increased, but many UCRs or defensive behaviours are decreased. Regardless of the merits of this particular hypothesis, however, it is clear that some sort of sensitized state or process persists after inescapable shock and dissipates over a matter of a few days. The foregoing represents one speculation about what that might be, how it might come about, and how it could account for the pattern of behavioural changes. It should be re-emphasized that other systems such as noradrenaline pathways are also involved in mediating effects of uncontrollability, and that other brain circuitry such as the septo-hippocampal system also surely plays a role. Indeed, the DRN projects to the hippocampus, although the hippocampus receives its largest 5-HT input from the median raphe.

9.6 ATTENTION

Is it this simple? Can all learned helplessness effects be behaviourally interpreted as reflecting anxiety-produced potentiation of fear and inhibition of defensive behaviour? And what of the assertions concerning alterations in the learning process with which learned helplessness began? Is there a cognitive change, what is its nature, and can it result from increased anxiety or fear?

It has not been easy to demonstrate that inescapable shock actually alters the learning process rather than only the behavioural tendencies that the organism exhibits during escape tasks. This is because the escape tasks typically used, such as shuttling or pressing levers, confound poor learning with increases in fear behaviour or decreases in defensive behaviour. If an animal stops shuttling, it is because it has failed to learn the contingency between its behaviour and shuttling, or because flight tendencies are being inhibited? We have used the strategy of examining escape learning in situations in which there is no direct relation between successful learning and speed of responding. The first of these was a Y-maze which contained a centre region and three arms going off at equal angles (Jackson et al., 1980). The maze was very small so little movement was required to go from one arm into another. Each trial began with shock onset, and shock terminated whenever the rat moved to an arm in a particular direction from its location or last arm occupancy, say a left turn. So, the rat had to turn left to turn off shock. Any number of errors were allowed on a given trial, trials simply did not terminate until a left turn was made. If the rat began with a right turn, it then had to go back and enter the arm that it had occupied at the beginning of the trial, etc. There were no external cues to indicate to the rat which arm would be correct; the rat had to learn to make a response relative to its own body or internal cues, proprioceptive or otherwise. Here, learning consists of choosing correctly, not responding quickly or slowly. Inescapably shocked rats learned to choose only poorly: they continued to make errors trial after trial. They did respond more slowly than controls, but continued to respond and to choose; they just learned slowly (Jackson et al., 1980). Furthermore, there was no correlation between speed of responding and accuracy. Thus the poor choice accuracy of inescapably shocked subjects could not be explained as resulting from reductions in movement or inhibition of flight. Moreover, speed and accuracy were dissociated experimentally (also see below). Increasing the shock intensity in the Y-maze was able to overcome the reduced speed of responding in the previously inescapably shocked subjects, but it did not improve their choice accuracy (Jackson et al., 1980).

These and other findings which have been reviewed elsewhere (Maier, 1990) suggest that exposure to uncontrollable aversive events alters the learning process itself, not just the organism's behavioural tendencies (see also Chapter 6). However, the learning process involves many components (registration of stimuli, attention, association formation, rehearsal, memory formation, retrieval, etc.), and these data do not by themselves indicate the locus of the effect. A series of studies reported by Minor et al. (1984) are instructive in this regard. They demonstrated that the poor Y-maze choice escape learning by inescapably shocked rats only occurred if external irrelevant stimuli were present in the Y-maze situation. By an irrelevant stimulus is meant one which cannot lead to problem solution. For example, a light could be illuminated at the end of each arm of the Y-maze. Inescapably shocked subjects are poor to learn to choose to go left if one of the lights was illuminated at the beginning of each trial, with the arm being chosen randomly. The light was an irrelevant cue because it was in the correct arm 50% of the time and the incorrect arm 50% of the time. Animals that had received escapable shock, or no prior shock, quickly learned to ignore the cue and to learn to use position (left) to solve the problem, whereas inescapably shocked animals continued to direct their behaviour toward the light and used it to guide their choices (some rats consistently approached the light and some consistently avoided the light). If there were no external irrelevant stimuli, then inescapably shocked animals did not show a choice escape learning deficit. For example, if the lights behind each of the three arms were illuminated at the beginning of the trial, the inescapably shocked animals learned at normal rates, since now the light could not be used to guide behaviour (approaching or avoiding light would not lead to a choice of any particular arm).

There are a variety of ways to interpret these results. One is that exposure to inescapable shock shifts attention away from internal response-related cues, such as proprioception, towards external cues in the environment. Thus inescapably shocked animals would be expected to learn poorly if external irrelevant cues are present since they should be biased towards attending to them, but should learn normally if no such cues are present to draw attention. Notice that this implies that the altered learning in inescapably shocked subjects does not represent a deficit, as poor choice learning should only result when there are irrelevant external cues. Indeed, this position predicts that inescapable shock ought to *enhance* choice learning if external cues are made *relevant* and response-related cues made *irrelevant*!

Lee and Maier (1988) tested this prediction in a water-maze choice escape task. Rats were placed into water in a start area, and could get out of the water by swimming into one of two goal areas located next to each other, one on the right and one on the left. A platform that allowed the rat to climb out of the water was located in one of the two compartments. A large black or white stimulus was located at the entrance to the goal compartments. In the first experiment the correct solution was to swim to a specific side, with no external stimuli designating the side. Inescapably shocked animals were poor at learning this task if black-white was an irrelevant stimulus (one side black and one white, randomly alternated), but learned normally if the irrelevant black-white cue was removed (both sides black, or both sides white). In the critical experiment black-white was made the relevant cue. Here black and white alternated randomly from left to right, but the escape platform was always behind one of the cards (black for some subjects and white for others) rather than always being on the left or right. Animals that had received inescapable shock 24 h earlier learned faster than escapably shocked or non-shocked subjects. For further experiments along these lines see Jackson et al. (1978).

Why do inescapably shocked animals show an attentional bias towards external cues? Is this another manifestation of heightened fear or anxiety? Is this outcome produced by the same processes as described earlier in this chapter? After all, increased vigilance is a frequently cited component of anxiety (Gray, 1982), and hypervigilance has been argued to be crucial to human anxiety (Eysenck, 1991). Alternatively, the explanation may be closer to the original learned helplessness argument. Perhaps a natural consequence of learning that one's own behaviour is ineffective in controlling shock or other aversive events is a shift in attention away from cues relating to behaviour towards cues in the environment. It might be that the shift in attention is not directly driven by anxiety and anxiety-related processes, but rather by what the subject has learned.

How can these possibilities be separated? Recall that the anxiety-related changes produced by inescapable shock dissipated in 48–72 h, were blocked by the

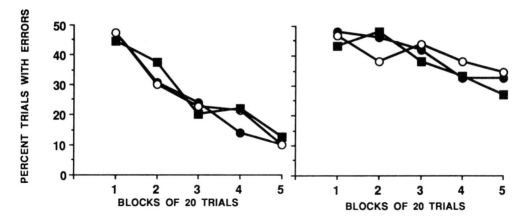

Figure 9.6 Mean percent trials with an error across blocks of 20 Y-maze trials. Subjects had been restrained (left panel) or given inescapable shock (right panel) 24 h earlier. Day 1 treatment was preceded by 0, 5 or 10 mg/kg (i.p.) diazepam. ●: 0; ○: 5; ■: 10

administration of BDZs before inescapable shock, and were produced by administration of the BDZ inverse agonist, FG 7142. The Y-maze choice escape learning deficit fails on each of these criteria. It does not dissipate in 48–72 h, and is undiminished even seven days after inescapable shock (Jackson *et al.*, 1980). Figures 9.6 and 9.7 show the results of an experiment in which diazepam was administered

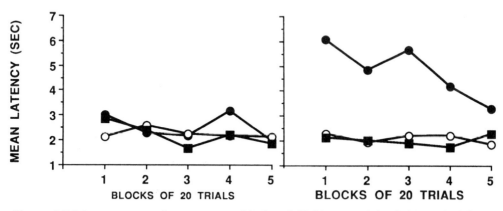

Figure 9.7 Mean response latency across blocks of 20 Y-maze trials. Subjects had been restrained (left panel) or given inescapable shock (right panel) 24 h earlier. Day 1 treatment was preceded by 0, 5 or 10 mg/kg (i.p.) diazepam. ●: 0; ○: 5; ■: 10

before inescapable shock or restraint at doses of either 0, 5 or 10 mg/kg. Y-maze testing with an irrelevant light cue present was conducted 24 h later. Figure 9.6 contains the choice accuracy data. The left panel shows the results for restrained subjects, and the right panel the results for the inescapably shocked rats. Clearly, diazepam did not even reduce the effect of inescapable shock on choice escape learning, even though these doses are more than sufficient to block the shuttle escape deficit, reinstate analgesia, and increase freezing. Figure 9.7 shows the latencies to make the first choice response in the Y-maze. As usual, inescapable shock produced slow responding (right panel), but this effect was blocked by diazepam. That is,

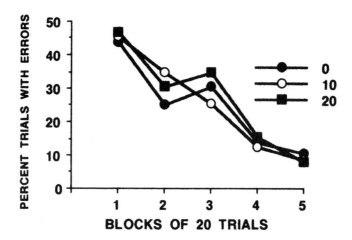

Figure 9.8 Mean percent trials with an error across blocks of 20 Y-maze trials. Subjects had received 0, 5 or 20 mg/kg FG-7142 (i.p.) 24 h earlier

diazepam did block slow responding to shock, an outcome here argued to be related to fear-anxiety, but had no effect on choice accuracy, all in the same task in the same subjects. Figures 9.8 and 9.9 depict the outcome of an experiment in which either 0, 10, or 20 mg/kg of FG 7142 was administered before 1½ h of restraint, the procedure used by Drugan et al. (1985b) that produced a shuttle escape deficit and reinstatable analgesia. Y-maze testing was 24 h later. As can be seen in Figure 9.8, FG 7142 had no

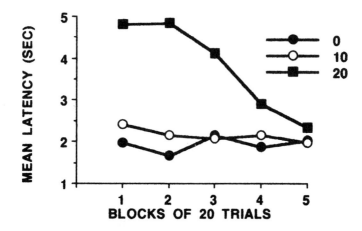

Figure 9.9 Mean response latency across blocks of 20 Y-maze trials. Subjects had received 0, 5 or 20 mg/kg FG-7142 (i.p.) 24 h earlier

effect on choice accuracy. However, FG 7142 did produce slow responding (Figure 9.9).

These results offer no encouragement for the view that the poor choice escape learning and, by implication, shift in attention are attributable to the persisting anxiety produced by inescapable shock. An explanation needs to be sought elsewhere. Many neurochemical systems are potential candidates, and here the locus coeruleus-dorsal bundle noradrenergic system may be pivotal. Consistent with this possibility, Minor et al. (1988) found that rats with 6-hydroxydopamine lesions of the dorsal bundle showed a pattern of choice escape behaviour in the Y-maze similar to that produced by inescapable shock. Lesioned animals learned to choose at a rate similar to non-lesioned controls when there were no irrelevant external cues present, but learned very poorly relative to controls if an irrelevant light cue was added. However, the dorsal bundle lesioned animals were not slow to respond, escape latencies were unaffected by the lesion. Thus this system might play a special role in the attentional changes produced by inescapable shock.

An additional speculation might be that the septo-hippocampal system is crucially involved. Support for this idea must wait on the appropriate experimental evidence. However, it can be noted that lesion of the septal input to the hippocampus prevents inescapable shock from producing shuttlebox and lever press escape deficits (Beagley and Beagley, 1978; Kelsey and Baker, 1983). Shuttlebox and lever press escape can be argued to be 'mixed tasks', with the inescapable shock produced deficit reflecting *both* decreased flight and attentional shifts. The evidence for the flight or fear component has already been reviewed. It can be added here that manipulations designed to draw attention to the organism's responses, or to internal feedback for responses in the shuttlebox and lever press tasks, *eliminate* the deficit (Maier and Testa, 1975; Maier et al., 1987). For example, caudate-putamen lesions have been shown to interfere with learning tasks, such as lever pressing, because they disrupt the processing of proprioceptive feedback (Potegal, 1982). Provision of a brief external stimulus, such as a tone after a response, eliminates the learning deficit in caudate-putamen lesioned animals, the argument being that the stimulus feedback brings attention to the response or to internal feedback from the response. That is, attention to response-related cues is restored. In similar fashion, providing a brief external stimulus after the escape response eliminates the shuttlebox and lever press escape deficits produced by inescapable shock (Maier et al., 1987). There is no obvious reason why a 0.25 s tone should facilitate flight or reduce fear. It might reduce fear if it became a safety signal, but that requires many trials. The impact of the feedback stimulus was almost immediate. Indeed, providing feedback actually *interfered* with escape learning in animals that had previously received escapable shock or no shock. It is thus not surprising that both noradrenaline and 5-HT are involved in the mediation of the shuttlebox and lever press escape deficits.

9.7 CONCLUDING REMARKS

The purpose of this chapter has been to focus on a number of less well known and less often reviewed aspects of learned helplessness. The research in this area has concentrated on shuttlebox and lever press escape acquisition as the behavioural endpoint, and noradrenaline as a mediator of the effects obtained. However, uncontrollable, as opposed to controllable, stressors produce a much larger constellation of behavioural changes, and this aspect of uncontrollability has been highlighted here. It seemed that many of these changes could be captured as decreased fight or flight and increased fear. This is how anxiety is often characterized, and so a goal of the chapter was to explore the possibility that exposure to uncontrollable shock produces a state that can be characterized as heightened anxiety and which persists for a period of time after the organism is removed from the uncontrollable shock situation. This is distinct from increased fear conditioned to cues present during shock, which should terminate shortly after the organism is removed from the shock environment and should be re-aroused whenever the subject is re-exposed to the appropriate cues. The evidence indicated that some process is sensitized by inescapable shock, which dissipates over a 48–72 h period. It leads to behaviour characterized as anxious in the social interaction test, and augments conditioned fear.

The GABA/BDZ complex seemed to be important for this process. The behaviours in question were blocked by administration of BDZs before inescapable shock and, in the one case tested, by the BDZ receptor antagonist flumazenil. These behaviours also were produced by the BDZ partial inverse agonist FG 7142, without the administration of inescapable shock. However, BDZs did not reduce the behavioural effects of inescapable shock when administered before later testing, suggesting that the GABA/BDZ manipulations must be exerting their ultimate behavioural effect by altering some other system in which the changes persist. The available literature suggests 5-HT as the target system, with the DRN being a likely important structure. A hyper-sensitive 5-HT system would explain the pattern of behavioural results by facilitation of processes in the amygdala and inhibition of processes in the dPAG. Other projections of the DRN and functions of 5-HT provided a potential avenue to explain some of the alterations in reactivity to stimulants and opiates produced by inescapable shock.

However, not all of the behavioural sequelae of inescapable shock are amenable to this sort of explanation. Inescapable shock leads to a shift in attention towards external events in the environment, and this is not affected by manipulation of the GABA/BDZ complex. This cognitive alteration may result from the learning of uncontrollability itself, and perhaps involve mediation by the noradrenergic, locus-coeruleus dorsal bundle system with the septo-hippocampal system playing a key role.

It might be noted, in concluding, that these sorts of considerations may help in understanding the complexity and often apparently contradictory nature of results in the literature. Performance in different behavioural test tasks would be expected to involve different mixes of these processes, and so results could easily shift with seemingly minor modifications in the behavioural parameters. There is no single 'learned helplessness effect'; there are a number of them, and they have different bases. Two (heightened anxiety and attentional changes) have been highlighted here, but there are undoubtedly others.

9.8 REFERENCES

Amir, Z. and Galena, H. (1986) Stress-induced analgesia: Adaptive pain suppression. *Physiol. Rev.*, **66:** 1091–119

Anisman, H., de Catanzaro, D. and Remington, G. (1978) Escape performance deficits following exposure to inescapable shock: Deficits in motor response maintenance. *J. Exper. Psychol. Anim. Behav. Proc.*, **4:** 197–218

Anisman, H., Zakman, S., Shanks, N. and Zacharko, R.M. (1991) Multisystem regulation of performance deficits induced by stressors. An animal model of depression. in A. Boulton, G. Baker and M.T. Martin Iverson (eds.), *Animal Models in Psychiatry II*, Humana Press, Clifton, NJ, pp 1–61

Applegate, C.D., Kapp, B.S., Underwood, M.D. and McNoll, C.C. (1983) Autonomic and somatomotor effects of amygdala central n. stimulation in awake rabbits. *Physiol. Behav.*, **31:** 353–60

Baldessarini, R.J., Amatruda, T.T., Griffith, F.F. and Gerson, S. (1975) Differential effects of serotonin on turning and stereotypy induced by apomophine. *Brain Res.*, **93:** 158–63

Bandler, R. (1988) Brain mechanisms of aggression as revealed by electrical and chemical stimulation. Suggestion of a central role for the midbrain periaqueductal grey region. *Prog. Physiol. Psychol. Psychobiol.*, **13:** 67–155

Beagley, G.H. and Beagley, W.K. (1978) Alleviation of learned helplessness following septal lesions in rats. *Physiol. Psychol.*, **6:** 241–4

Blanchard, D.C. and Blanchard, R.J. (1969) Crouching as an index of fear. *J. Compar. Physiol. Psychol.*, **67:** 370–5

Blanchard, D.C. and Blanchard, R.J. (1972) Innate and conditioned reactions to threat in rats with amygdaloid lesions. *J. Compar. Physiol. Psychol.*, **81:** 281–90

Blanchard, D.C. and Blanchard, R.J. (1984) Inadequacy of pain-aggression hypothesis revealed in naturalistic settings. *Aggr. Behav.*, **10:** 33–46

Blanchard, D.C. and Blanchard, R.J. (1988) Ethoexperimental approaches to the biology of emotion, in M.R. Rosenzweig and L.W. Porter (eds.), *Annual Review of Psychology, Vol. 39*, Annual Review, Inc., Palo Alto, pp 43–69

Blanchard, D.C. and Takahaski, S.N. (1988) No change in intermale aggression after amygdala lesions which reduce freezing. *Physiol. Behav.*, **42:** 613–6

Bockaert, J., Dumius, A., Bouhelal, R., Sebben, M. and Cory, R.N. (1987) Piperazine derivatives including the putative anxiolytic drugs, buspirone and ipsapirone are agonists at 5-HT1A receptors negatively coupled with adenylate cyclase in hippocampal neurons. *Arch. Pharm.*, **335:** 588–92

Briley, M., Chopin, P. and Moret, C. (1991) The role of serotonin in anxiety: Behavioural approaches, in M. Briley and S.E. File (eds.), *New Concepts in Anxiety*, CRC Press, Boca Raton, pp 56–73

Broekkamp, C.L., Rÿk, H.W. Joly-Gelsim, D. and Lloyd, K.L. (1986) Major tranquillizers can be distinguished from minor tranquillizers on the basis of effects on marble burying and swim-induced grooming. *Eur. J. Pharm.*, **126:** 223–9

Brown, L.R., Rosellini, A., Samuels, O.B. and Riley, E.P. (1982) Evidence for a serotonergic mechanism of the learned helplessness phenomenon. *Pharm. Biochem. Behav.*, **17:** 877–83

Cannon, J.T., Prieti, G.J., Lal, Al and Liebeskind, J.C. (1982) Evidence for opioid and nonopioid forms of stimulation-produced analgesia in the rat. *Brain Res.*, **243:** 315–21

Cederbaum, J.M. and Aghajanian, G.K. (1978) Afferent projections to the rat locus coeruleus as determined by a retrograde tracing technique. *J. Compar. Neurol.*, **178:** 1–16

Curzon, G., Hutson, P.H., Kennett, G.A. *et al.* (1986) Characteristics of analgesias induced by brief or prolonged stress. *Ann. New York Acad. Sci.*, **467:** 93–104

Davis, M., Hitchcock, J.M. and Rosen, J.B. (1991) Neural mechanisms of fear conditioning measured with the acoustic startle reflex, in J. Madden IV (ed.), *Neurobiology of Learning, Emotion, and Affect*, New York, Raven Press, pp 67–97

De Blas, A.L., Park, D. and Friedrich, P. (1987) Endogenous benzodiazepine-like molecules in the human, rat, and bovine brains studied with a monoclonal antibody to benzodiazepines. *Brain Res.*, **413:** 275–84

De Molina, F.A. and Hunsperger, R.W. (1962) Organization of the subcortical system governing defense and flight in the cat. *J. Physiol.*, **160**: 200–13

De Robertis, E., Pena, C., Paladini, A.C. and Malina, J.H. (1988) New developments on the search for the endogenous ligand(s) of central benzodiazepine receptors. *Neurochem. Int.*, **13**: 1–11

Desiderato, O. and Newman, A. (1971) Conditioned suppression produced in rats by tones paired with escapable or inescapable shock. *J. Compar. Physiol. Psychol.*, **77**: 427–43

Dess, N.K., Raizer, J., Chapman, C.D. and Garcia, J. (1988) Stressors in the learned helplessness paradigm: Effects on body weight and conditioned taste aversion in rats. *Physiol. Behav.*, **44**: 483–90

Dorow, R. (1982) β-Carboline monomethylamide causes anxiety in man. *CINP Congr. Jerusalem*, **13**: 176

Drugan, R.C. and Maier, S.F. (1982) The nature of the activity deficit produced by inescapable shock. *Anim. Learning & Behav.*, **10**: 401–6

Drugan, R.C., Ryan, S.M., Minor, T.R. and Maier, S.F. (1984) Librium prevents the analgesia and shuttlebox escape deficit typically observed following inescapable shock. *Pharmacol. Biochem. Behav.*, **21**: 749–54

Drugan, R.C., Ader, D.N. and Maier, S.F. (1985a) Shock controllability and the nature of stress-induced analgesia. *Behavioral Neurosci.*, **99**: 791–801

Drugan, R.C., Maier, S.F., Skolnick P. *et al.* (1985b) A benzodiazepine receptor antagonist induces learned helplessness, *Eur. J. Pharmacol.*, **113**: 453–7

Drugan, R.C., McIntyre, T.D., Alpern, H.P. and Maier, S.F. (1985c) Coping and seizure susceptibility. Control over shock protects against bicuculline-induced seizures. *Brain Res.*, **342**: 9–17

Drugan, R.C., Deutsch, S.I., Weizman, A. *et al.* (1989a) Molecular mechanisms of stress and anxiety: Alterations in the benzodiazepine/GABA receptor complex, in H. Weiner, I. Florin, R. Murison and D. Hellhammer (eds.), *Frontiers in Stress Research, Vol. 3*, Hans Huber, Toronto, pp 148–59

Drugan, R.C., Morrow, A.L., Weizman, R. *et al.* (1989b) Stress-induced behavioral depression in the rat is associated with a decrease in GABA receptor-mediated chloride ion flux and brain benzodiazepine receptor occupancy. *Brain Res.*, **487**: 45–51

Edwards, E., Johnson, J., Anderson, D. *et al.* (1986) Neurochemical and behavioral consequences of mild, uncontrollable shock. Effects of PCPA. *Pharmacol. Biochem. Behav.*, **25**: 415–521

Edwards, E., Harkins, K., Wright, G. and Henn, F. (1991) Modulation of [³H]paroxetine binding to the 5-hydroxytryptamine uptake site in an animal model of depression. *J. Neurochem.*, **56**: 1581–6

Elmes, D.G., Jarrard, L.E. and Swart, P.D. (1975) Helplessness in hippocampectomized rats. Response perseveration? *Physiol. Psychol.*, **3**: 51–5

Eysenck, M.W. (1991) Cognitive factors in clinical anxiety: Potential relevance to therapy, in P. Bevan, A.R. Cools and T. Archer (eds.), *Behavioural Pharmacology of 5-HT*, Lawrence Erlbaum, Hillsdale, NJ, pp 418–434

Fanselow, M.S. (1986) Conditioned fear induced opiate analgesia: A competing motivational state theory of stress-analgesia. *Ann. New York Acad. Sci.*, **467**: 40–54

Fanselow, M.S. (1991) The midbrain periaqueductal gray as a coordinator of action in response to fear and anxiety, in A. Depaulis and R. Bandler (eds.), *The Midbrain PAG*, Plenum Press, New York, pp 151–73

Fanselow, M.S. and Lester, L.S. (1988) A functional behavioristic approach to aversively motivated behavior: Predatory imminence as a determinant of the topography of defensive behavior, in R.C. Bolles and M.D. Beecher (eds.), *Evolution and Learning*, Lawrence Erlbaum, Hillsdale, pp 185–212

File, S.E. (1980) The use of social interaction as a method for detecting anxiolytic activity of chlordiazepoxide-like drugs. *J. Neurosci. Methods*, **2**: 219–38

File, S.E. (1985) Animal models for predicting clinical efficacy of anxiolytic drugs: Social behavior. *Neuropsychobiology*, **13**: 55–62

File, S.E. and Pellow, S. (1986) Intrinsic actions of the benzodiazepine receptor antagonist Ro15-1788. *Psychopharmacology*, 1–11

File, S.E., Hyde, J.R.G. and Macleod, N.K. (1979) 5,7-Dihydroxytryptamine lesions of dorsal and median raphe nuclei and performance in the social interaction test of anxiety and in a home-cage aggression test. *J. Affec. Disord.*, **1**: 115–22

File, S.E., Lister, R.G. and Nutt, D.J. (1982) The anxiogenic actions of benzodiazepine antagonists. *Neuropharmacology*, **21**: 1033–87

Fox, R.J. and Sorenson, C.A. (1991) Bilateral lesions of the central nucleus of the amygdala attenuate analgesia induced by environmental challenges. *Soc. for Neurosci. Abstracts*, **17**: 108

Gentsch, C., Lichtsteiner, M. and Feer, H. (1989) Competition for sucrose pellets in triads of male Wistar rats: The effects of eight 5-HT agonists, in P. Bevan, A.R. Cools and T. Archer (eds.), *Behavioral Pharmacology of 5-HT*, Lawrence Erlbaum, Hillsdale, NJ, pp 149–54

Gleitman, H. and Holmes, P.A. (1967) Retention of incompletely learned CER in rats. *Psychonomic Sci.*, **7**: 19–20

Graeff, F.G. (1991) Neurotransmitters in the dorsal periaqueductal grey and animal models of panic anxiety, in M. Briley and S.E. File (eds.), *New Concepts in Anxiety*, CRC Press, Boca Raton, pp 288–313

Graeff, F.C. and Silviera Filho, N.G. (1978) Behavioral inhibition induced by electrical stimulation of the raphe nucleus of the rat. *Physiol. Behav.*, **21**: 477–84

Grau, J.W., Hyson, R.L., Maier, S.F. *et al.* (1981) Long-term stress-induced analgesia and activation of the opiate system. *Science*, **213**: 1409–11

Gray, J.A. (1982) *The Neuropsychology of Anxiety: An inquiry into the function of the septo-hippocampal system*. Oxford University Press, London

Gray, J.A. (1987) *The Psychology of Fear and Stress*, Cambridge University Press, Cambridge

Haefely, W. (1984) Actions and interactions of benzodiazepine agonists and antagonists at GABAergic synapses, in N.G. Bowery (ed.), *Actions and Interactions of Benzodiazepines*, Raven Press, New York, pp 263–85

Hamon, M., Fattaccini, C.M., Adrein, J. *et al.* (1988) Alterations of central 5-HT and DA turnover in rats treated with ipsaprone and other 5-HT1A agonists. *J. Pharmacol. Exper. Therap.*, **246**: 745–52

Hellhammer, D.H., Rea, M.A., Bell, M. *et al.* (1984) Learned helplessness: Effects on brain monoamines and the pituitary-gonadal axis. *Pharmacol. Biochem. Behav.*, **21**: 481–5

Helmstetter, F.A. (1992) The amygdala is essential for the expression of conditional hypoalgesia. *Behav. Neurosci.* (in press)

Helmstetter, F.A., Brozoski, E.L. and Frost, J.A. (1991) Kappa opioid agonists produce hypoalgesia on the tailflick test after application to the amygdala. *Soc. Neurosci. Abstracts*, **17**: 296

Hindley, S.W., Hobbs, A., Paterson, I.A. and Roberts, M.H.T. (1985) The effects of methyl b-carboline-3-carboxylate on social interaction and locomotor activity when microinjected into the nucleus raphé dorsalis of the rat. *Br. J. Pharma.*, **86**: 753–61

Hitchcock, J.M. and Davis, M. (1987) Fear-potentiated startle using an auditory conditioned stimulus: Effect of lesions of the amygdala. *Physiol. Behav.*, **39**: 403–9

Hitchcock, J.M., Sananes, B. and Davis, M. (1989) Sensitization of the startle reflex by footshock: Blocked by central nucleus of the amygdala or its efferent pathway to the brainstem. *Behav. Neurosci.*, **103**: 509–19

Hodges, H., Green, S. and Glenn, B. (1987) Evidence that the amygdala is involved in benzodiazepine and serotonergic effects on punished responding but not on discrimination. *Psychopharmacology*, **92**: 491–504

Hoffman, H.S., Fleshler, M. and Jensen, R. (1963) Stimulus aspects of aversive controls: The retention of conditioned suppression. *J. Exper. Anal. Behav.* **6**: 575–83

Hyson, R.L., Ashcraft, L.J., Drugan, R.C. *et al.* (1982) Extent and control of shock affects naltrexone sensitivity of stress-induced analgesia and reactivity to morphine. *Pharmacol. Biochem. Behav.*, **17**: 1019–25

Irwin, J., Suissa, A. and Anisman H. (1980) Differential effects of inescapable shock on escape performance and discrimination learning in a water maze escape task. *J. Exp. Psychol: Animal Behavior Processes*, **6**: 21–40

Irwin, J., Ahluwalia, P. and Anisman, H. (1986) Sensitization of norepinephrine activity following acute and chronic footshock. *Brain Res.*, **379**: 98–103

Iversen, S.D. (1984) 5-HT and anxiety. *Neuropharmacol*, **23**: 1553–60

Iwata, J., Ledown, J.E., Meeley, M.P. *et al.* (1986) Intrinsic neurons in the amygdaloid field projected to by the medial geniculate body mediate emotional responses conditioned to acoustic stimuli. *Brain Res.*, **371**: 395–9

Jackson, R.L. and Minor, T.R. (1988) Effects of signaling inescapable shock on subsequent escape learning implications for theories of coping and 'learned helplessness'. *J. Exp. Psychol: Animal Behavior Processes*, **14**: 390–400

Jackson, R.L., Maier, S.F. and Rapaport, P.M. (1978) Exposure to inescapable shock produces both activity and associative deficits in the rat. *Learning and Motivation*, **9**: 69–98

Jackson, R.L., Maier, S.F. and Coon, D.J. (1979) Long-term analgesic effects of inescapable shock and learned helplessness. *Science*, **206**: 91–4

Jackson, R.L., Alexander, J.H. and Maier, S.F. (1980) Learned helplessness, inactivity, and associative deficits: Effects of inescapable shock on response choice escape learning. *J. Exp. Psychol: Animal Behavior Processes*, **6**: 1–20

Jacobs, B.L. (1987) Central monoaminergic neurons: Single unit studies in behaving animals, in H.E. Meltzer (ed.), *Psychopharmacology, the Third Generation of Progress*, Raven Press, New York, pp 159–71

Jacobs, B.L. and Cohen, A. (1976) Differential behavioral effects of lesions of the medial or dorsal raphe nuclei in rats: Open field and pain-elicited aggression. *J. Compar. Physiol. Psychol.*, **90**: 102–8

Johnston, A.L. (1991) The implication of noradrenaline in anxiety, in M. Briley and S.E. File (eds.), *New Concepts in Anxiety* CRC Press, Boca Raton, pp 347–66

Kelsey, J.E. and Baker, M.D. (1983) Ventromedial septal lesions in rats reduce the effects of inescapable shock on escape performance and analgesia. *Behav. Neurosci.*, **97**: 945–61

Kennett, G.A., Marcou, M., Dourish, C.T. and Curzon, G. (1987) Single administration of 5-HT$_{1A}$ agonists decrease 5-HT$_{1A}$ presynaptic, but not postsynaptic receptor mediated responses: Relationship to antidepressant-like actions. *Eur. J. Pharmacol.*, **138**: 53–60

Kiser, R.S., Brown, C.A., Sanghera, M.K. and German, D.C. (1980) Dorsal raphe nucleus stimulation reduces centrally-elicited fearlike behavior. *Brain Res.*, **191**: 265–72

Laurent, J.P., Margold, M., Hunkel, V. and Haefly, W. (1983) Reduction by two benzodiazepines and pentobarbitone of the multiunit activity in substantia nigra, hippocampus, nucleus, locus coeruleus, and dorsal raphe nucleus of 'encephale isole' rats. *Neuropharmacology*, **22**: 501–12

Lee, R.K.K. and Maier, S.F. (1988) Inescapable shock and attention to internal versus external cues in a water escape discrimination task. *J. Exp. Psychol: Animal Behavior Processes*, **14**: 302–11

MacLennan, A.J. and Maier, S.F. (1983) Coping and the stress-induced potentiation of stimulant stereotypy in the rat. *Science*, **219**: 1091–3

Maier, S.F. (1986) Stressor controllability and stress-induced analgesia, in D.D. Kelly (ed.), *Stress-induced Analgesia*, New York Academy of Sciences, New York, pp 55–78

Maier, S.F. (1990) The role of fear in mediating the shuttle escape learning deficit produced by inescapable shock. *J. Exp. Psychol.: Animal Behavior Processes*, **16**: 137–50

Maier, S.F. and Seligman, M.E.P. (1976) Learned helplessness: Theory and evidence. *J. Exp. Psychol: General*, **105**: 3–46

Maier, S.F. and Testa, T.J. (1975) Failure to learn to escape in rats previously exposed to inescapable shock is partly produced by associative interferences. *J. Compar. Physiol. Psychol.*, **88**: 554–64

Maier, S.F., Seligman, M.E.P. and Solomon, R.L. (1969) Pavlovian fear conditioning and learned helplessness: Effects on escape and avoidance behavior of (a) the CS-US contingency, and (b) the independence of the US and voluntary responding, in B.A. Campbell and R.M. Church (eds.), *Punishment*, Appleton-Century-Crofts, New York

Maier, S.F., Anderson, C. and Lieberman, D. (1972) The influence of control of shock on subsequent shock-elicited aggression. *J. Compar. Physiol. Psychol.*, **81**: 94–101

Maier, S.F., Ryan, S.M., Barksdale, C.M. and Kalin, N.H. (1986) Stressor controllability and the pituitary-adrenal system. *Behav. Neurosci.*, **100**: 669–78

Maier, S.F., Jackson, R.L. and Tomie, A. (1987) Potentiation, overshadowing, and prior exposure to inescapable shock. *J. Exp. Psychol: Animal Behavior Processes*, **13**: 260–72

Martin, I.L. (1987) The benzodiazepines and their receptors: 25 years of progress. *Neuropharmacol.*, **26**: 957–70

Martin, P., Laporte, A.M., Soubrie, P. *et al.* (1989) Reversal of helpless behavior in rats by serotonin uptake inhibitors, in P. Bevan, A.R. Cools and T. Archer (eds.), *Behavioural Pharmacology of 5-HT*, Lawrence Erlbaum, Hillsdale, NJ, pp 231–35

Martin, P., Beninger, R.J., Hamon, M. and Puech, A.J. (1990) Antidepressant-like action of 8-OH-DPAT, a 5-HT$_{1A}$ agonist, in the learned helplessness paradigm: Evidence for a postsynaptic mechanism. *Behav. Brain Res.*, **38**: 135–44

Miczek, K.A. and Donat, P. (1989) Brain 5-HT system and inhibition of aggressive behavior, in P. Bevan, A.R. Cools and T. Archer (eds.), *Behavioural Pharmacology of 5-HT*, Lawrence Erlbaum, Hillsdale, NJ, pp 117–45

Mineka, S., Cook, and Miller, S. (1984) Fear conditioned with escapable and inescapable shock: The effects of a feedback stimulus. *J. Exp. Psychol.: Animal Behavior Processes*, **10**: 307–23

Minor, T.R., Jackson, R.L. and Maier, S.F. (1984) Effects of task irrelevant cues and reinforcement delay on choice escape learning following inescapable shock: Evidence for a deficit in selective attention. *J. Exp. Psychol.: Animal Behavior Processes*, **10**: 543–56

Minor, T.R., Pelleymounter, M.A. and Maier, S.F. (1988) Uncontrollable shock, forebrain NE, and stimulus selection during choice escape learning. *Psychobiology*, **16**: 135–46

Nagy, J., Zambi, K. and Desci, L. (1979) Anti-anxiety action of diazepam after intra-amygdaloid application in the rat. *Neuropharmacology*, **18**: 573–6

Ninan, P., Insel, T., Cohen, R. *et al.* (1982) Benzodiazepine receptor-mediated experimental 'anxiety' in primates. *Science*, **218**: 1332–4

Overmier, J.B. and Leaf, R.C. (1965) Effects of discriminative Pavlovian fear conditioning upon previously or subsequently acquired avoidance responding. *J. Compar. Physiol. Psychol.*, **60**: 213–18

Overmier, J.B. and Seligman, M.E.P. (1967) Effects of inescapable shock on subsequent escape and avoidance behavior. *J. Compar. Physiol. Psychol.*, **63**: 28–33

Peroutka, S.J. (1988) 5-HT receptor subtypes: Molecular, biochemical and physiological characterization. *Trends in Neurosci.*, **11**: 496–500

Petty, F. (1986) GABA mechanisms in learned helplessness, in G. Bartholini, K.G. Lloyd and R.L. Morselli (eds.), *GABA and Mood Disorders*, Raven Press, New York, pp 61–7

Petty, F. and Sherman, A.D. (1981) GABAergic modulation of learned helplessness. *Pharmacol. Biochem. Behav.*, **15**: 567–70

Petty, F., Kramer, T.R., Philips, T.R. *et al.* (1991) Learned helplessness and serotonin: *In vivo* microdialysis. *Soc. Neurosci. Abstracts*, **16**: 752

Potegal, M. (1982) Vestibular and neostriatal contributions to spatial orientation, in M. Potegal (ed.) *Spatial Abilities: Development and physiological foundations*, Academic Press, New York, pp 361–87

Rapaport, P.M. and Maier, S.F. (1978) Inescapable shock and food competition dominance in rats. *Anim. Learn. Behav.*, **6**: 160–5

Redmond, D.E. (1987) Studies of the nucleus locus coeruleus in monkeys and hypotheses for neuropsychopharmacology, in H.E. Meltzer (ed.), *Psychopharmacology, The Third Generation of Progress*, Raven Press, New York, pp 967–77

Rizvi, T.A., Ennis, M., Behbehani, M. and Shipley, M.T. (1991) Connections between the central nucleus of the amygdala and the midbrain periaqueductal gray: Topography and reciprocity, *J. Compar. Neurol.*, **303**: 121

Roberts, M.H.T. (1985) 5-Hydroxytryptamine and antinociception. *Neuropharmacology*, **23**: 1529–36

Roozendaal, B., Koolhaas, J.M. and Bohus, G. (1991) Central amygdala lesions affect behavioral and autonomic balance during stress in rats. *Physiol. Behav.*, **50**: 77–81

Rosellini, R.A. and Seligman, M.E.P. (1975) Frustration and learned helplessness. *J. Exper. Psychol.: Animal Behavior Processes*, **104**: 149–57

Sakurai-Yamoshita, Y., Kataoka, Y., Yamashita, K. *et al.* (1989) Conflict behavior and dynamics of monoamines of various brain nuclei in rats. *Neuropharmacology*, **28**: 1067–73

Schutz, M.T.B., De Aguiar, J.C. and Graeff, F.G. (1985) Antiaversive role of serotonin in the dorsal periaqueductal gray matter. *Psychopharmacology*, **85**: 340–5

Seligman, M.E.P. and Beagley, G. (1975) Learned helplessness in the rat. *J. Compar. Physiol. Psychol.*, **88**: 534–41

Seligman, M.E.P. and Maier, S.F. (1967) Failure to escape traumatic shock. *J. Exp. Psychol.*, **74**: 1–9

Short, K.R. and Maier, S.F. (1990) Uncontrollable but not controllable stress produces enduring anxiety in rats despite only transient benzodiazepine receptor involvement. *Soc. Neurosci. Abstracts*, **16**: 388

Short, K.R. and Maier, S.F. (1991) Localization of a benzodiazepine receptor mediated control dependent increase in anxiety following stress in the rat. *Soc. Neurosci. Abstracts*, **17**: 283

Soubrie, P. (1986) Reconciling the role of central serotonin neurons in human and animal behavior. *Behav. Brain Sci.*, **9**: 319–64

Soubrie, P., Thiebot, M.J., Jobert, A. et al. (1980) Decreased convulsant potency of picrotoxin and pentetrazol and enhanced [^{3}H] flunitrazepam cortical binding following stressful manipulations in rats. *Brain Res.*, **189**: 505–17

Sprouse, J.S. and Aghajanian, G.K. (1988) Responses of hippocampal pyramidal cells to putative serotonin 5-HT$_{1A}$ and 5-HT$_{1B}$ agonists: A comparative study with dorsal raphe neurons. *Neuropharmacology*, **7**: 707–15

Srebro, B. and Lorens, S.A. (1975) Behavioral affects of selective midbrain raphe lesions in the rat. *Brain Res.*, **89**: 303–25

Steinbusch, H.M.W. and Nieuwenhuys, R. (1983) The raphe nuclei of the rat brainstem: A cytoarchitectonic and immunohistochemical study, in P.C. Emson (ed.), *Chemical Neuroanatomy*, Raven Press, New York, pp 131–209

Swenson, R.M. and Vogel, W.H. (1983) Plasma catecholamine and corticosterone as well as brain catecholamine changes during coping in rats exposed to stressful footshock. *Pharm. Biochem. Behav.*, **18**: 689–93

Thiebot, M.H., Jobert, A. and Soubrie, P. (1980) Conditioned suppression of behavior: Its reversal by intra-raphe microinjection of chlordiazepoxide and GABA. *Neurosci. Lett.*, **16**: 213–17

Thomas, D.R., Windell, B.T., Bakke, I. et al. (1985) Long-term memory in pigeons: I. The role of discrimination problem difficulty assessed by reacquisition measures. II. The role of stimulus modality assessed by generalization slope. *Learning and Motivation*, **16**: 464–77

Urquhart, J. (1974) Physiological actions of ACTH, in R.O. Grepp and E.B. Astwood (eds.), *Handbook of Physiology: Section 7. Endocrinology: Vol. 4. The pituitary gland and its endocrine control: Pt 2*, American Physiological Society, Washington, DC, pp 133–58

Volpicelli, J.R., Ulm, R.R. and Altenor, A. (1984) Feedback during exposure to inescapable shocks and subsequent shock-escape performance. *Learning and Motivation*, **15**: 279–86

Warren, D.A., Rosellini, R.A. and Maier, S.F. (1990) Fear, stimulus feedback, and stressor controllability, in G.H. Bower (ed.), *The Psychology of Learning and Motivation, Vol 22*, Academic Press, New York, pp 167–207

Weiss, J.M. (1968) Effects of coping responses on stress. *J. Compar. Physiol. Psychol.*, **65**: 251–60

Weiss, J.M. and Simson, P.G. (1986) Depression in an animal model: Focus on the locus coeruleus, in R. Porter and G. Bock (eds.), *Antidepressants and Receptor Function*, Wiley, New York, pp 191–216

Weiss, J.M., Stone, E.A. and Harrell, N. (1970) Coping behavior and brain norepinephrine level in rats. *J. Compar. Physiol. Psychol. 3*, **72**: 153–60

Weiss, J.M., Glazer, H.I., Pohorecky, L.A. et al. (1975) Effects of chronic exposure to stressors on avoidance-escape behavior and on brain norepinephrine. *Psychosom. Med.*, **37**: 522–34

Weiss, J.M., Goodman, P.A., Losito, B.G. et al. (1981) Behavioral depression produced by an uncontrollable stressor: Relationship to norepinephrine, dopamine and serotonin levels in various regions of rat brain. *Brain Res. Rev.*, **3**: 167–205

Weiss, J.M., Simson, P.G., Hoffman, L.J. et al. (1986) Infusion of adrenergic receptor agonists and antagonists into the locus coeruleus and ventricular system of the brain. Effects on swim-motivated and spontaneous activity. *Neuropharmacol.*, **25**: 367–84

Williams, J.L. (1982) Influence of shock controllability by dominant rats on subsequent attack and defensive behaviors toward colony intruders. *Animal Learning & Behavior.*, **10**: 305–13

Williams, J.L. (1987) Influence of conspecific stress odors and shock controllability on conditioned defensive burying. *Animal Learning & Behavior.*, **15**: 333–41

Williams, J.L. and Lierle, D.M. (1986) Effects of stress controllability, immunization, and therapy on the subsequent defeat of colony intruders. *Animal Learning & Behavior*, **14**: 305–14

Willner, P. (1985) *Depression: A Psychobiological Synthesis*, Wiley, New York

Part Four
Neurochemistry of stress: introduction to techniques

10

Neurochemistry of stress: Introduction to techniques

MARIANNE FILLENZ

UNIVERSITY LABORATORY OF PHYSIOLOGY
UNIVERSITY OF OXFORD
OXFORD, UK

ISBN 0–12–663370–3

10.1 INTRODUCTION

External stimuli elicit specific responses, appropriate to a particular stimulus; nonspecific responses are elicited by stimuli described as stressors. The response to stress consists of physiological and behavioural changes which are attributable to the activation of a complex neuroendocrine system involving neurones in the peripheral and central nervous system, as well as the secretion of a number of hormones from the pituitary and the adrenal glands.

The nervous system consists of chains and networks of neurones which communicate by means of chemical mediators, and there are a number of phenomena which signal activity of a neuronal pathway (Figure 10.1). Arrival of the nerve impulse at the nerve terminal leads to the release of transmitter which interacts with receptors. These are both postsynaptic, for the onward transmission of the signal, and presynaptic autoreceptors, which result in self-regulation of the activity of the neurone. The released transmitter is replaced by accelerated synthesis. Neurochemical changes associated with neuronal activity occur with a wide range of time courses: some are rapid and brief, measured in seconds or minutes, others are delayed and long-lasting, measured in days or weeks.

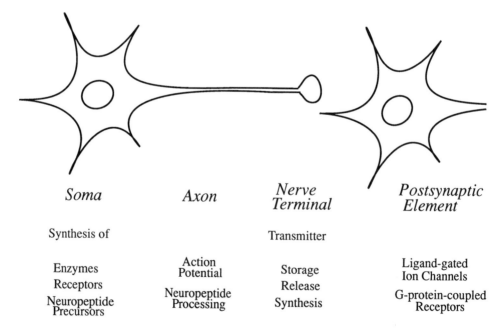

Soma	Axon	Nerve Terminal	Postsynaptic Element
Synthesis of		Transmitter	
Enzymes	Action Potential	Storage	Ligand-gated Ion Channels
Receptors		Release	
Neuropeptide Precursors	Neuropeptide Processing	Synthesis	G-protein-coupled Receptors

Figure 10.1 Schematic drawing of the main events which occur in different parts of the neurone

The *effects* of neurotransmitters depend on interaction with specific receptors, which are of two kinds: receptors which are ligand-gated ion channels, and receptors coupled to so-called G-proteins which are, in turn, coupled to a wide variety of effector mechanisms *via* second messengers. Corresponding to these two classes of receptors there are two kinds of interactions between neurones. The first is the transmission of action potentials across specialized synapses, mediated by rapid excitatory or inhibitory synaptic potentials, produced by opening of ligand-gated ion channels; this mechanism subserves the specialized responses and is properly called neurotransmission. The second, which uses the G-protein coupled receptors, rather than initiating transmission across synapses modulates this transmission by actions on both the presynaptic terminal and the postsynaptic element. This process, called neuromodulation, accounts for many of the so-called non-specific responses to stress.

The receptors that are ligand-gated ion channels are activated predominantly by amino acid transmitters; the numerous G-protein coupled receptors are activated by amines and peptides, as well as some amino acids. The G-protein associated receptors have effects which are more diffuse, both spatially and temporally, than the ligand-gated ion channels.

Specific responses have been studied by electrophysiological recording of impulse discharge from single neurones organized into specific pathways; this technique is much less appropriate for studying neuromodulation. Monitoring neurochemical changes has proved a more fruitful approach to the nonspecific stress response.

Organisms exposed to stress show either coping responses or, when these fail, develop pathological conditions, which may be either behavioural, structural, or both. These various aspects are dealt with in other chapters of this volume. This chapter examines techniques that have been used to study the neurochemical changes associated with acute and chronic stress in animals and humans, and the mechanism of action of drugs used in the clinical treatment of what are regarded as diseases resulting from stress. There are severe limitations to the study of these changes in humans, although advances in techniques have greatly extended the range of measurements that are now possible. The interpretation of these measurements, however, relies heavily on a wide range of animal experiments.

10.2 PERIPHERAL CHANGES

10.2.1 MEASUREMENT OF COMPOUNDS IN PLASMA

10.2.1.1 Catecholamines

Stress stimulates central and peripheral monoaminergic neurones. The concentration of the catecholamines, adrenaline and noradrenaline, in plasma is very low, and measurement of plasma catecholamines had to wait for the development of sufficiently sensitive assays. The earliest of these were radioenzymatic assays; these are sensitive but expensive, labour-intensive and not absolutely specific. They have been replaced by high performance liquid chromatography with electrochemical detection (HPLC-ECD; Hjemdahl, 1986).

The concentration of noradrenaline in plasma is widely used as an index of peripheral noradrenaline release, despite the lack of a linear relationship between the activity of the sympatho-adrenal system and plasma noradrenaline concentration. This disparity arises because there are active uptake mechanisms for the removal of the released catecholamines. These include a high affinity neuronal, and a lower affinity extraneuronal, uptake mechanism (Iversen, 1967). Released noradrenaline which escapes these uptake mechanisms appears in the plasma and constitutes the noradrenaline 'overflow'. A number of factors determine the extent of the noradrenaline overflow from a particular organ: these include the density of innervation, the width of the synaptic cleft and the degree of vascularization. Thus in organs such as the vas deferens, or the iris, with a rich sympathetic innervation, where sites of transmitter release are 200 nm away from the smooth muscle cells and where there is a relatively sparse vascularization, only a small proportion of the released noradrenaline will overflow from the synapse and reach the general circulation. This is in contrast to the heart, where the innervation consists of bundles of sympathetic fibres which lie at a considerable distance from the cardiac muscle cells, and where there is a rich supply of capillaries. As a result, a much greater proportion of released noradrenaline will overflow to the plasma. Such complications explain the marked variation in concentration of plasma noradrenaline depending on the site of sampling.

The contributions of noradrenaline by particular organs have been estimated by measuring both adrenaline and noradrenaline arterio-venous differences (Brown et al., 1981). Assuming similar clearances for the two catecholamines, the adrenaline arterio-venous difference can be used as a measure of the removal of noradrenaline, because adrenaline is not secreted by peripheral neurones. Subtraction of this difference from the arterio-venous difference for noradrenaline in any given organ gives an estimate of noradrenaline release into the circulation. However, estimates of the contribution to total plasma noradrenaline have to take into account the fraction of cardiac output that goes to a particular organ.

The sympatho-adrenal system does not act as a single unit, but shows considerable regional differentiation in its responses. Evidence for this comes from measurements of noradrenaline turnover in various organs (Landsberg and Young, 1986: see below). This has shown that there is considerable variation in the pattern of activation of the sympatho-adrenal system in response to various stressors. For this reason, differences in plasma noradrenaline concentration may signal changes in the amount of noradrenaline released from a particular set, or combination, of sympathetic nerve terminals which contribute to the noradrenaline overflow. In addition, a contribution to plasma noradrenaline may be derived from the adrenal medulla.

The time course of any changes must also be taken into account, because there is evidence that a constant stimulus, such as cold exposure, produces phasic changes in plasma noradrenaline. In rats with an indwelling intra-atrial cannula, assays of plasma noradrenaline carried out at hourly intervals show an initial steep rise which reaches a peak after 4 h of cold exposure, after which plasma noradrenaline begins to decline (Benedict et al., 1979). Measurement of the stores of vesicular noradrenaline in the nerve terminals of the heart suggest that the decline is not due to adaptation to the cold stimulus, but rather to a depletion of the releasable store of noradrenaline (Fillenz et al., 1979).

The most serious drawback to the use of measurements of plasma noradrenaline in response to stress is the fact that obtaining the sample is itself stressful; in humans at least, there are marked individual differences in the perception of such a stimulus as a stressor, which depend to a large extent on the previous history of the individual. This means that not only are comparisons between individuals unreliable, but even changes in the same individual are difficult to interpret. Additional sources of variation are diet, sodium intake, age, time of day, alcohol consumption and tobacco smoking (Lake *et al.*, 1984).

In addition to the parent catecholamines, plasma contains both their deaminated and methylated metabolites (Figure 10.2). These are also frequently used as indices of transmitter release. Monoamine oxidase (MAO), the enzyme which deaminates catecholamines, is found mainly in nerve terminals, and the deaminated metabolites arise both from deamination following reuptake of released transmitter and from intraneuronal deamination of newly synthesized noradrenaline, or its precursor dopamine, which has failed to be taken up into storage vesicles. Catechol-*O*-methyltransferase (COMT) is extraneuronal and methylates catecholamines after their release; methylation can be preceded, or followed, by deamination, giving rise to metabolites which are both methylated and deaminated. Both methylated and/or deaminated metabolites are freely diffusable across the neuronal membrane. Because of their origin, changes in metabolite concentration may signal changes in either synthesis rate or release rate, and these are not always closely coupled. Furthermore, plasma catecholamine metabolites are derived both from peripheral and central catecholaminergic neurones; although the blood brain barrier is impermeable to catecholamines, the metabolites are removed from the extracellular compartment of the brain by an active transport mechanism located at the blood brain barrier. The same considerations apply to the other two central monoamines, dopamine and 5-hydroxytryptamine, and their metabolites: dihydroxyphenylacetic acid (DOPAC) and 5-hydroxyindoleacetic acid (5-HIAA).

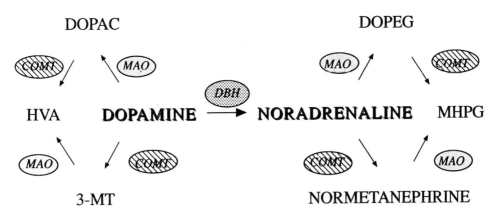

Figure 10.2 Metabolites of dopamine and noradrenaline together with the appropriate enzymes. DOPAC = 3,4-dihydroxyphenylacetic acid; HVA = homovanillic acid; 3-MT = 3-methoxytyramine; DOPEG = 3,4-dihydroxyphenylglycol; MHPG = 3-methoxy-4-hydroxyphenylglycol. DBH = dopamine-β-hydroxylase (intravesicular); MAO = monoamine oxidase (intraneuronal); COMT = catechol-*O*-methyl transferase (extraneuronal)

There has been considerable debate concerning the extent to which changes in plasma catecholamine metabolites can be used as an index of the turnover of central catecholamines (Blomberry *et al.*, 1980) or release of peripheral catecholamines.

10.2.1.2 Neurone-derived proteins

Catecholamines in noradrenergic nerve terminals are stored in two populations of storage particles: the small and large vesicles. Small vesicles co-store noradrenaline with adenosine triphosphate (ATP) (Lagercrantz and Fried, 1982); large vesicles and adrenomedullary vesicles also contain soluble proteins and peptides (Klein, 1982). Vesicular exocytosis will therefore result in the release of a number of compounds in addition to catecholamines.

The soluble proteins co-stored with catecholamines are dopamine-β-hydroxylase (DBH, the enzyme that catalyses the conversion of dopamine to noradrenaline) and chromogranin. The fact that DBH is not taken up after release raised the possibility that measurements of plasma DBH may be a more accurate measure of the activity of the sympatho-adrenal system than measurement of catecholamines. Accordingly, a number of attempts have been made to monitor changes in plasma DBH in response to stress.

One of the first observations was that the difference in basal levels of DBH between rats and human subjects was much greater than the difference in plasma catecholamine concentration. This is explained by the very low percentage of large catecholamine storage vesicles in rat as compared to human sympathetic nerve terminals. There is also evidence that there is independent release from small and large vesicles. Thus, whereas low frequency stimulation causes release from small vesicles, high frequency bursts are a much more effective stimulus for release from large vesicles (Lundberg *et al.*, 1986); release from small and large vesicles of sympathetic nerve terminals is mediated by different voltage-gated calcium channels (Lipscombe *et al.*, 1989), and there is evidence for a difference in the site of exocytosis for small and large vesicles (Thureson-Klein, 1983). In addition, because of its large molecular size, DBH is slow to enter the circulation. All these factors may account for the lack of parallelism between changes in plasma noradrenaline and DBH in response to stress (Weinshilboum and Axelrod, 1971).

Similar problems confront the use of measurements of chromogranin in the peripheral circulation as an index of sympatho-adrenal activity (O'Connor *et al.*, 1991).

10.2.1.3 Glucocorticoids

Stress causes a marked increase in the secretion of adrenal glucocorticoid hormones, of which cortisol is the main hormone in humans and corticosterone in rodents. Following secretion, approximately 90% of the endogenous hormone is bound to the blood-borne carriers, corticosteroid-binding globulin and albumin, while only 5–10% circulates unbound. Only free cortisol is believed to be biologically active. The presence of the bound hormone complicates the measurement of the biologically relevant cortisol because there is considerable controversy concerning the relationship between bound and free cortisol (Ekins, 1990). There are a number of other problems associated with the measurement of plasma cortisol. Like the majority of hormones, the secretion of cortisol is not continuous but occurs in a pulsatile fashion. The pulse amplitude in humans is highest early in the morning and lowest in the evening, resulting in a marked circadian rhythmicity of cortisol output. In rats, which are nocturnal, the rhythm is reversed. In parallel with the variation in pulse amplitude, there is an inverse relation between basal release of cortisol and the size of the stress-induced response. As with plasma catecholamines, measurement of changes in plasma cortisol is complicated by the variable stress of sample collection.

The demonstration that corticosteroids are present in saliva (Katz and Shannon, 1964) and the development of sensitive radioimmuno assay (RIA) techniques in the late 1970s have led to the measurement of salivary cortisol for the investigation of the hypothalamic-pituitary-adrenal axis activity. Salivary samples are collected using the salivette, which involves chewing a sterilised cotton swab, which is then centrifuged. The sample can be stored at room temperature for up to two weeks and is then assayed by RIA. A number of alternative biochemical assays have been developed which, however, are not yet commercially available. The ease of this technique means that it can be used to investigate stress under a wide variety of conditions.

A number of questions have been raised concerning the relationship between salivary cortisol and plasma cortisol; however salivary cortisol shows circadian variation and values normally parallel those of the free fraction in plasma. A variety of physical and psychological stressors have been shown to produce elevations in salivary cortisol. Among these is the demonstration of the large individual variation in the salivary cortisol response to venipuncture. This has important implications for techniques which rely on plasma sampling techniques (Kirschbaum and Hellhammer, 1989).

10.2.1.4 Peptides

There are a number of plasma peptides which show stress-related changes. Plasma peptides are derived from four sources: (1) the sympatho-adrenal system, (2) the anterior pituitary, (3) the hypothalamus, and (4) the enteric nervous system.

The peptides that are released from the sympatho-adrenal system are co-stored with catecholamines in adrenomedullary vesicles and the large vesicles in sympathetic nerve terminals; they include the opioids, met- and leu-enkephalin, substance P and neuropeptide Y (Lundberg and Hokfelt, 1986). All these peptides are also found in the alimentary tract. Stress-related changes in circulating peptides have been reported, although frequently there is no close parallelism with changes in plasma catecholamines (Lundberg et al., 1985; Zukowska-Grojek et al., 1988).

The peptides derived from the anterior pituitary which show stress-related changes are the hormones adrenocorticotropic hormone (ACTH), prolactin, growth hormone, as well as the opioid β-endorphin which, together with ACTH, is derived from the large precursor molecule pro-opiomelanocortin (POMC).

Finally, plasma contains a number of releasing factors; these are derived from nerve terminals of hypothalamic neurones in the median eminence, where they enter the hypophyseal portal system to reach the anterior pituitary. The most important of these is corticotropin releasing factor (CRF: see Chapter 13). CRF stimulates the release of ACTH which, in turn, stimulates the release of adrenal glucocorticoids; there is a negative feedback mechanism whereby glucocorticoids inhibit the release of both CRF and ACTH (Figure 10.3). All three compounds show a diurnal variation. There have been a number of attempts to measure peripheral plasma CRF immunoreactivity, but these studies have yielded conflicting results concerning the diurnal variation of plasma CRF. Since it is known that there are peripheral sources of CRF (pancreas, testes, adrenal medulla, placenta) and independent mechanisms for regulating release from these various sources, the absence of diurnal variation has been interpreted as indicating a non-hypothalamic source for the plasma CRF. The presence of a recently described CRF binding protein (Orth and Mount, 1987) may give rise to spurious results with unextracted plasma, where free CRF has not been separated from bound CRF.

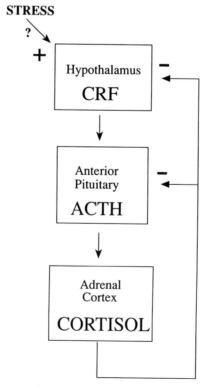

Figure 10.3 Regulation of the release of cortisol. CRF = corticotropin releasing factor; ACTH = adrenocorticotropic hormone

10.2.2 MEASUREMENT OF URINARY COMPOUNDS

The demonstration that stress caused the activation of the sympatho-adrenal system (Cannon, 1915) and the pituitary-adrenal axis (Selye, 1936) led to early attempts to monitor the changes in urinary catecholamines and adrenocortical hormones. The range of compounds found in urine has expanded as new assay methods have been developed, and such studies have yielded some useful results. Thus the substitution of high performance liquid chromatography with electrochemical detection (HPLC-ECD) for the earlier techniques means that various metabolites of adrenaline, noradrenaline, dopamine and 5-hydroxytryptamine (5-HT) can now be assayed in a single sample (Peaston, 1988). Radioimmunoassay techniques are still used for the measurement of steroids, such as cortisol and various peptides including anterior pituitary hormones, although HPLC can also be used for some of these compounds. Recently, CRF immunoreactivity has been described in urine (Maser-Gluth *et al.*, 1989); if this observation is confirmed, the measurement of 24 h excretion patterns of CRF may prove more useful than isolated plasma concentrations.

The main limitation associated with measurements of urine is the considerable time-lag between the appearance of a substance in the blood and its excretion in the urine. Samples are often collected overnight to give a mean concentration over a long period; when random samples are collected, allowance has to be made for the rate of urine production. This is done by normalizing against the concentration of creatinine. A further complication is that catecholamine metabolites can also be derived from the diet (Odink *et al.*, 1988).

10.3 MONITORING THE ACTIVITY OF NEURONES IN THE CNS

Monitoring the activity of neurones in the central nervous system (CNS) of humans is difficult, and the interpretation of such measurements relies heavily on animal experiments. Release from neurones in the CNS is into a narrow synaptic cleft, and direct measurement of such release is technically not possible at present. Indirect measurements of release are used, therefore; these include *ex vivo* and *in vivo* techniques. An advantage of *ex vivo* techniques is that one can look at numerous brain regions and a number of transmitters in the same animal; the disadvantages are poor time resolution and the use of large numbers of animals because of the need for separate controls. In contrast, with *in vivo* techniques, each animal can act as its own control but, in general, only one brain region can be studied in any animal.

10.3.1 *EX VIVO* MEASUREMENTS OF MONOAMINES

Early studies of neuronal systems involved in the response to stress were concerned first with noradrenergic, later with 5-hydroxytryptaminergic and most recently with dopaminergic pathways. In experiments using noxious stress, such as inescapable footshock, a decrease in the concentration of noradrenaline stores was found in certain brain areas (Maynert and Levi, 1964). However, with non-noxious stress there is no such change in concentration, presumably because synthesis rate is adjusted to match release rate in such a way that there is little or no depletion in spite of considerable variation in the rate of transmitter release (see Chapter 12). If such a steady state exists, then the rate of turnover of the transmitter can be calculated and used as an index of release rate. This can be estimated from: the rate of decline of transmitter concentration when synthesis is inhibited (Brodie et al., 1966), the decline of the specific activity of radiolabelled noradrenaline after prelabelling stores with a tracer dose of [^{3}H] noradrenaline (Neff et al., 1968), or the rate of formation of [^{3}H] noradrenaline from [^{3}H] tyrosine (Gordon et al., 1966). These methods apply equally to noradrenaline, dopamine and 5-HT.

Another measure of neuronal activity uses changes in the concentration of metabolites. In general, released transmitter is metabolized after reuptake into the nerve terminal. This suggests that increased release should be accompanied by an increase in the concentration of transmitter metabolites. Some of the difficulties in the interpretation of changes in the concentration of metabolites have already been discussed (see Section 10.2.1). However, there are a number of other factors which specifically affect their concentration in the brain: metabolites are readily diffusable across the neuronal membrane into the extracellular space from which they are removed by a transport mechanism across the blood brain barrier and enter the systemic circulation. Metabolites also arise from intraneuronal deamination of the products of the first, rate-limiting step of the synthetic pathway which have failed to be taken up into the storage vesicles, and hence are exposed to the action of MAO localized on the outer membrane of the intraneuronal mitochondria. Metabolite concentration therefore depends on synthesis rate and availability of storage vesicles, as well as release rate. Rarely are there statistically significant increases in metabolite concentration on exposure to stress; however, the ratio of metabolite to transmitter concentration has been found to increase in response to stress in selected brain areas. In 5-HT releasing neurones a high proportion of the basal 5-HT synthesis is not destined for release but is deaminated intraneuronally to 5-HIAA; this synthesis rate is unaffected by a reduction in the basal impulse traffic. This means that although tissue levels of 5-HIAA are a good index of overall 5-HT turnover, they do not reveal whether turnover is related to functional release of transmitter.

There are also a number of possible sources of error in the *measurement* of metabolites. Comparison of concentrations of the methylated metabolites, 3-methoxytyramine and normetanephrine, following microwave fixation, (which rapidly denatures degradative enzymes), with those obtained from unfixed tissue have indicated rapid and very large *post-mortem* increases (Ikarashi et al., 1985). There may also be some oxidation of 3-methoxytyramine to homovanillic acid, explaining why changes in homovanillic acid are smaller than changes in DOPAC. In mice, but not in rats and humans, 3-methoxy-4-hydroxyphenylglycol (MHPG) and other catabolites are unsulphated (Kuchel and Buu, 1984) and hence represent the total catabolite. In rats and humans the sulphated MHPG must also be measured; this requires different methodological procedures.

A further possible complicating factor arises from the demonstration that stress in animals can cause a reduction in MAO activity. The discovery of two endogenous inhibitors of MAO, whose concentration is increased by stress, may provide the explanation of this decrease in MAO activity. There is a large molecular weight polypeptide found in plasma (Giambalvo, 1984), cerebrospinal fluid (CSF; Becker *et al.*, 1983) and brain (Isaac *et al.*, 1986) and a low molecular weight non-peptide, which has been called tribulin (Elsworth *et al.*, 1986) and is now identified as isatin (Glover *et al.*, 1988). The urinary output of tribulin is increased by various forms of stress and there is evidence for an increase of a MAO inhibitor in brain as a result of stress (Glover *et al.*, 1989). Such changes would tend to reduce the concentration of the deaminated metabolites derived either from increased synthesis or transmitter reuptake.

10.3.2 *EX VIVO* MEASUREMENTS OF PEPTIDES

A large number of peptides have now been identified in the brain; to determine which of these play a role in the stress response, one needs to show an alteration in release in response to a wide variety of stressors. The various indirect techniques described for monitoring the activity of aminergic neurones are not applicable to peptidergic neurones. For instance, the replacement of released peptides is not nearly as rapid and efficient as that of the small molecular weight transmitters which are synthesized in the nerve terminal. This is because peptides are synthesized in the cell body and reach the nerve terminal by means of axoplasmic transport. Enhanced release is therefore likely to lead to a depletion of the nerve terminal store of peptides. Advantage has been taken of this; a neuropeptide with consistently altered concentrations in discrete brain regions in response to a variety of stressors can be considered to play a role in the stress response.

Brain regions dissected by micropunch are assayed for changes in neuropeptide concentration. There are a number of problems associated with these measurements. The earliest technique used was RIA. This technique is open to error, because of cross-reactivity between related peptides, or the various breakdown products of a given peptide, by the various intracellular and extracellular peptidases. The problem of cross-reactivity has been tackled by combining separation techniques with sensitive assay techniques. The most commonly used separation technique is HPLC, which can be combined with either absorbance, or derivatization followed by fluorescence detection, or electrochemical detection or RIA. A more recent advance is the use of capillary electrophoresis, where separation results from the differential speed of migration of molecules in an electric field, followed by mass spectrophotometric methods for their detection. These techniques are still in the stage of development (Davis, 1991). An alternative test for hypothalamic neuropeptide release is measurement of peptides in the hypophyseal portal system.

A recently developed technique which maps the spatial distribution of released neuropeptides is the use of antibody microprobes. Glass micropipettes bearing a uniform layer of immobilized antibodies on their outer surface are introduced into the CNS; released endogenous ligand for the antibody binds at a localized area of the microprobe. Probes are then withdrawn from the brain and incubated with a radiolabelled form of the ligand before autoradiographs of the probe are taken. Binding of endogenous ligand is detected on microprobe autoradiographs as a deficit in the binding of radiolabelled ligand (Duggan, 1990). So far this technique has been used only to study the release of neuropeptides in the spinal cord in response to nociceptive stimulation (Duggan *et al.*, 1990).

Using some of these assay techniques it has been possible to show that CRF is unevenly distributed in micropunch dissected brain regions, the highest concentration being found in the median eminence. A significant decrease in median eminence CRF is found after a single 3 h restraint stress or a two week regimen of daily exposure to a variety of stressors which include cold swim, cold restraint, tail pinch (Chappell et al., 1986). That the reduction in CRF content in the median eminence in response to stress probably represents depletion of nerve terminals following release is supported by the increase in CRF found in the hypophyseal portal plasma (Plotsky and Vale, 1984).

Since replacement of released neuropeptide does not occur in the nerve terminal, changes in the cell bodies of neuropeptide releasing neurones have also been studied. Here increased activation is likely to lead to an increase in the concentration of the precursor protein. Both acute and chronic stress result in an increase in CRF concentration in the locus coeruleus (Chappell et al., 1986), a brain region which contains cell bodies and which shows immunocytochemical staining for CRF (Swanson et al., 1983). Chronic but not acute stress leads to an increase in CRF in the periventricular nucleus and the anterior hypothalamic nucleus (Chappell et al., 1986). An increase in median eminence CRF is seen 24 h after a 5 min restraint stress with no change at earlier time points; after colchicine administration (which blocks axoplasmic transport) and anisomycin inhibition of protein synthesis during the restraint, there is a reduction in the median eminence CRF concentration (Haas and George, 1988).

These results have been interpreted as showing that some, but not all, neuronal pathways in the brain containing CRF are relevant to the stress response and that activation of these neurones leads to an initial depletion of CRF stores in the nerve terminals. This is replaced by newly synthesized CRF transported from the cell body; stimulation of synthesis leads to an increase in cell body CRF concentration. However, measurement of concentration alone cannot distinguish betwen synthesis, release or degradation; no evidence is at present available concerning CRF synthesis rates as assessed by in situ hybridization. This technique uses short strands of deoxyribonucleic acid (DNA) as probes ('cDNA') for the messenger ribonucleic acid (mRNA) encoding CRF protein. Similarly, virtually nothing is known about the regulation of the processing of CRF from its precursor molecule, or about the peptidase/s responsible for its metabolic degradation.

Another peptide of considerable interest is neuropeptide Y which, besides being found in intrinsic neurones in the alimentary tract, is co-stored with catecholamines in the adrenal medulla, in sympathetic ganglia and in various brain regions. The paraventricular nucleus, which is the source of CRF, has a dense innervation of neuropeptide Y immunopositive fibres. In spite of the co-localization of noradrenaline and neuropeptide Y, various forms of stress which produce a signficant decrease in noradrenaline concentration have failed to produce any changes in neuropeptide Y concentration (Schon et al., 1986; Mormede et al., 1990). Hence neurotransmitter co-localization does not necessarily imply parallel changes in concentration after stress. The regulation of neuropeptide Y release in the brain is as yet incompletely understood.

Another brain peptide whose concentration is changed by stressors is thyrotropin releasing hormone (TRH). Stressors such as immobilisation or haemorrhage, which do not present a thermoregulatory challenge, lead to a reduction in the concentration of TRH in only some of the brain regions which contain this peptide (Takayama *et al.*, 1986; Ono *et al.*, 1989).

10.3.3 *IN VIVO* TECHNIQUES

The *ex vivo* measurements described above give no indication of the temporal relation between the applied stress and the neurochemical changes. Such a correlation became possible with the development of *in vivo* monitoring techniques. Two techniques have become available in the last decade for monitoring *in vivo* neurochemical changes: voltammetry and microdialysis. The two *in vivo* techniques are complementary. Voltammetry has a higher time resolution and electrodes (6 μm) are much smaller than dialysis probes (300 μm). However the life-time of the electrodes is short and the technique monitors only single compounds. The dialysis probe can be used both for sampling the chemical composition of the extracellular compartment and for local drug application. The number of compounds that can be monitored is limited only by the available assays. With both techniques, animals act as their own controls.

10.3.3.1 *Voltammetry*

Kissinger and his colleagues in the 1970s developed *in vivo* electrochemistry. This technique uses a variety of carbon electrodes, implanted into selected brain areas; oxidation, at the surface of the electrode, of compounds found in the extracellular compartment of the brain gives rise to a current whose size is a measure of the concentration of the compound; the identity of the compound is characterized by the potential at which oxidation occurs.

The technique was originally developed to monitor changes in extracellular dopamine as an index of its release. However, there are two main problems associated with the technique: selectivity and sensitivity. First, not only is there overlap in the oxidation potentials of dopamine and its deaminated metabolite DOPAC, but the concentration of dopamine is in the nM range, whereas the concentration of the metabolite is in the μM range. Secondly, ascorbate, whose concentration in the extracellular compartment is in the range of 250–500 μM, has an oxidation potential close to that of catecholamines. A similar situation applies to 5-HT, its metabolite 5-HIAA, and uric acid, a metabolite of adenosine which is present in μM concentrations.

A number of approaches have been used to improve the selectivity and sensitivity of the technique. These include modifications of the electrode and of the configuration of the applied potential. Although these modifications have resulted in improved selectivity and sensitivity, no technique at present enables measurement of changes in the basal concentrations of noradrenaline, dopamine or 5-HT, or changes in response to physiological stimuli (Marsden *et al.*, 1988).

The selectivity of carbon fibre electrodes is greatly enhanced by electrical pretreatment, which consists of subjecting them to a number of rapid potential changes *in vitro*. Although the sensitivity of carbon fibre electrodes pretreated in this way is still not sufficient for monitoring either basal or physiologically-stimulated dopamine release, changes in dopamine concentration can be monitored under two conditions. The first requires that animals be pretreated with a MAO inhibitor, usually pargyline, which prevents formation of DOPAC and unmasks the presence of dopamine (Gonon et al., 1985). The drawback is that the systemic administration of a MAO inhibitor has widespread effects. With the second approach, which involves electrical stimulation of dopaminergic fibres resulting in their synchronous activation, increases in dopamine concentration can be recorded (Gonon, 1988). This approach has been used successfully to demonstrate the close parallelism between impulse traffic and variations in extracellular dopamine, and has shown that, at a given mean impulse frequency, phasic burst discharges are much more effective than tonic frequencies in producing release (Gonon et al., 1991). The improved selectivity achieved by electrical pretreatment is shortlived (about 5 h), and the electrodes have therefore been used most widely for pharmacological experiments in anaesthetized animals.

Since metabolites are present in much higher concentrations than the transmitter, voltammetry has been used for in vivo monitoring of metabolite changes in response to stress. In brain regions containing noradrenergic neurones, there is a substantial concentration of extracellular DOPAC, which is increased in response to stressful stimuli (Curet et al., 1985; Gonon et al., 1985). DOPAC in noradrenergic neurones is derived from the intraneuronal deamination of dopamine by MAO, and there is some evidence that the concentration of DOPAC in noradrenergic neurones, although not a measure of noradrenaline release, provides a measure of the rate of synthesis. This is because tyrosine hydroxylation is the rate-limiting step in catecholamine synthesis and its product, l-dihydroxyphenylalanine (l-DOPA), is converted to dopamine by dopa decarboxylase, followed by deamination to DOPAC.

Voltammetry has also been used to measure changes in extracellular DOPAC in brain areas with a rich dopaminergic innervation; changes in DOPAC in these neurones are more difficult to interpret, since DOPAC is derived from both intraneuronal deamination of newly synthesized as well as released dopamine after its reuptake into the nerve terminal. Nevertheless, increases in DOPAC concentration in response to some forms of stress in some brain areas have been reported (Bertolucci-D'Angio et al., 1990).

10.3.3.2 Microdialysis

The other technique for in vivo monitoring of brain neurochemical changes is microdialysis. This is a technique for sampling the extracellular compartment with minimal interference, compared to the push-pull cannula technique in which the pumping of liquid into the brain inflicts considerable damage. In microdialysis, artificial CSF is perfused through an implanted dialysis probe. Chemical compounds, which have diffused into the perfusion fluid from the extracellular compartment of the brain, are then assayed by a variety of techniques. The damaged blood brain barrier reforms within 30–60 min after which sampling is from the brain extracellular compartment, not blood (Benveniste, 1989; Dykstra et al., 1992). The limiting factors are the molecular cut-off size of the dialysis membrane and the assay systems available. Although microdialysis was first used largely to study the effects of drugs in anaesthetized animals, the technique is increasingly used in unanaesthetized animals. This enables study of neurochemical changes in freely-moving animals under nearly normal conditions. The period over which such experiments can be carried out is limited; this is attributed to a progressive development of diffusion barriers comprised of glial cells around the dialysis probe. The ideal period for experimentation appears to be from 24–48 h after the implantation of the probe (Westerink and De Vries, 1988). Prolonged monitoring of changes resulting from repeated or prolonged stress requires the implantation of guide cannulae through which new dialysis probes can be introduced; the risk is that each new insertion involves damage to the blood-brain barrier.

Initially the HPLC techniques, already developed for catecholamines and indoleamines, were used to measure the concentration of these compounds in the dialysate. Both the transmitters and their metabolites can be measured in the same sample; the usefulness of this will be discussed in the section on synthesis. Since the level of these transmitters in the extracellular compartment is very low, samples had to be collected for minimum periods of 10 min, although 20 and 30 min were more common, to have enough material for assay. This meant a better selectivity, but a much lower time resolution, than that obtained with voltammetry. Recently, the scope of microdialysis has been greatly expanded with the development of new assay techniques. The amino acid transmitters can be assayed by electrochemical detection following precolumn derivatization and separation by reverse phase HPLC (Kehr and Ungerstedt, 1988). There are also a variety of enzyme based assay methods; immobilized enzymes interact with substrates in the dialysate and the products of the reaction are measured by electrochemical detection (Boutelle et al., 1992). The enzyme-based assays have a high sensitivity and selectivity and obviate the need for chromatographic separation.

The use of the acetylcholine enzyme electrode has revealed that even the mild stress of handling increases the release of acetylcholine in the rat hippocampus (Kalen et al., 1990). Previous attempts to show the participation of acetylcholine in the stress response using measurements of acetylcholine concentration, turnover rate, activity of choline acetyltransferase had all failed to show consistent effects, presumably because of the very efficient coupling of release and synthesis in cholinergic neurones (Tucek et al., 1989). Recently developed enzyme electrodes for glucose and lactate provide ways of monitoring changes in metabolic activity in response to stress (Schasfoort et al., 1988; De Bruin et al., 1990; van der Kuil and Korf, 1991).

Recently, neuropeptides have been added to the list of compounds whose in vivo release is monitored using microdialysis. Currently the technique is best suited to neuropeptides with a molecular weight less than 3000. However, peptides are assayed by RIA, which imposes a serious limitation on the technique (Kendrick, 1990).

10.3.4 MEASUREMENTS IN HUMANS

Although some measurements using microdialysis have been carried out in humans during neurosurgical operations, such measurements raise serious ethical problems. However, attempts have been made to deduce the function of CNS neurones from measurements carried out on human CSF which contains monoamines and their metabolites, as well as peptides.

Measurements of the concentrations of noradrenaline and its metabolites, as well as DBH, in CSF suggest that these compounds provide some indication of the activity of central noradrenergic neurones under a variety of conditions (Major *et al.*, 1984). Similar measurements have been made of CSF concentrations of the 5-HT metabolite, 5-HIAA. Lumber CSF is more likely to reflect spinal, rather than brain, 5-HT metabolism; also, although 5-HIAA may reflect 5-HT metabolism, it does not necessarily reflect the functional state of 5-hydroxytryptaminergic neurones (see above). These factors may explain the inconclusive results in patient studies.

A large proportion of patients with major depression or endogenous depression exhibit hyperactivity of the hypothalamo-pituitary-adrenal axis as assessed by the concentration of plasma cortisol, urinary free cortisol excretion and non-suppression of the secretion of both cortisol and ACTH by the synthetic glucocorticoid, dexamethasone. To discover whether the abnormality arises in the pituitary or the central nervous system, CRF concentrations in CSF have been measured using a sensitive and specific radioimmunoassay (Nemeroff *et al.*, 1984: see Chapter 13). These studies show an increase in CRF concentrations in the CSF of depressed patients (Banki *et al.*, 1987). A similar elevation has been found in CSF from suicide victims: a comparison of cisternal and lumbar CSF demonstrated a rostro-caudal gradient, indicating a supraspinal source for the peptide (Arato *et al.*, 1989).

10.4 MEASUREMENT OF SYNTHESIS RATE

The chemical compounds released from nerve terminals have to be replaced by synthesis. Fluctuations in release are therefore accompanied by fluctuations in synthesis rate. Although synthesis of the transmitter occurs throughout the neurone, synthesis relevant to release of amino acids and monoamines occurs in the nerve terminal. Peptides, on the other hand, are derived from precursor proteins, propeptides, synthesized in the cell body, where they are packaged into vesicles. These vesicles travel to the nerve terminal by axoplasmic transport, during which propeptides undergo cleavage by intravesicular enzymes. This enzymic cleavage produces the active peptides released from the nerve terminal. Different mechanisms regulate the synthesis of the various neurotransmitters and the way in which changes in synthesis rate are coupled to impulse traffic is as yet incompletely understood. Nevertheless, changes in synthesis rate or in the characteristics of the rate-limiting enzyme have been used extensively as an index of neuronal activity.

10.4.1 SYNTHESIS OF MONOAMINES

10.4.1.1 *Catecholamines*

The synthesis of catecholamines (noradrenaline, dopamine and adrenaline) involves a number of enzymes; of these tyrosine hydroxylase, which catalyzes the first step in the synthesis, is the rate-limiting step (Figure 10.4). Tyrosine hydroxylase is a cytoplasmic enzyme; its substrate is tyrosine. In addition, it has an absolute requirement for oxygen and a cofactor, tetrahydrobiopterin, which is a reduced pteridine. Whereas the concentration of tyrosine is normally sufficient to saturate the enzyme, the concentration of reduced cofactor is not. Supply of cofactor therefore limits enzyme activity. The product of tyrosine hydroxylation is l-DOPA, which is converted to dopamine by the enzyme, dopa decarboxylase. In noradrenergic neurones the dopamine is taken up into vesicles to be converted to noradrenaline by the enzyme DBH which is confined to the inner aspect of the vesicle membrane. Tyrosine hydroxylase is subject to endproduct inhibition by l-DOPA, dopamine and noradrenaline.

During increased activity of catecholaminergic neurones, the increase in the rate of transmitter release is paralleled by an increase in the rate of tyrosine hydroxylation. There have been numerous efforts to discover how neuronal activity is linked to the activity of tyrosine hydroxylase. Although the enzyme is subject to endproduct inhibition, the acceleration of synthesis is not attributable to a decline in levels of the end product resulting from transmitter release. This is because released transmitter is inactivated by reuptake into the nerve terminal, a process likely to increase rather than decrease its concentration in the cytoplasm. Instead, the acceleration of synthesis is attributable to an activation of the enzyme. This results from phosphorylation of the enzyme, leading to a conformational change which alters some of its kinetic parameters. The most important of these changes is an *increase* in the affinity of the enzyme for the reduced pteridine cofactor and a *reduction* in its affinity for the catechols responsible for feedback inhibition. These changes result in an increase in the activity of the enzyme.

Tyrosine hydroxylase is phosphorylated by a number of protein kinases which include: Ca^{2+}/calmodulin dependent protein kinase, adenosine 3',5'-cyclic monophosphate (cAMP)-dependent protein kinase and Ca^{2+}-phospholipid protein kinase. Each of these protein kinases phosphorylates the enzyme at a different site, and produces characteristic changes in kinetic parameters. The activation of protein kinases in turn depends on either the influx of Ca^{2+} or the stimulation of receptors (Zigmond et al., 1989). The changes in kinetic parameters of the enzyme produced by phosphorylation, which in turn is mediated by activation of receptors, constitutes rapid regulation of synthesis.

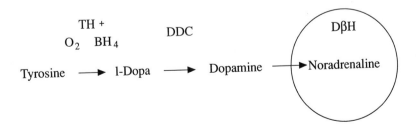

Figure 10.4 Synthesis of catecholamines. TH = tyrosine hydroxylase, which requires oxygen and the reduced pteridine co-factor, tetrahydrobiopterin (BH_4); DDC = dopa decarboxylase; l-Dopa = l-dihydroxyphenylalanine; DβH = dopamine-β-hydroxylase. TH and DDC are cytoplasmic, DβH is intravesicular

Catecholaminergic neurones have receptors for their own transmitter, which are called autoreceptors; these are situated on the cell body and dendrites (somato-dendritic autoreceptors) and on the nerve terminals (presynaptic autoreceptors). The somato-dendritic autoreceptors regulate impulse discharge while the presynaptic autoreceptors regulate both the release and synthesis of transmitter. Both the somato-dendritic membrane and the nerve terminal also have receptors that respond to transmitters of other neurones, called heteroreceptors (or more commonly, postsynaptic receptors in the case of the somato-dendritic receptors).

Dopaminergic and noradrenergic neurones have both somato-dendritic and presynaptic autoreceptors which inhibit impulse discharge, transmitter release and transmitter synthesis. These are dopamine D_2 receptors in the case of dopaminergic neurones and α_2 receptors in the case of noradrenergic neurones. The noradrenergic, but not the dopaminergic, neurones also have presynaptic autoreceptors which facilitate release and synthesis of noradrenaline; these are β-adrenoceptor, presynaptic autoreceptors. The inhibitory autoreceptors on both dopaminergic and noradrenergic neurones appear to be tonically active *in vivo* since administration of selective receptor antagonists causes an increase in transmitter release and synthesis (Figure 10.5).

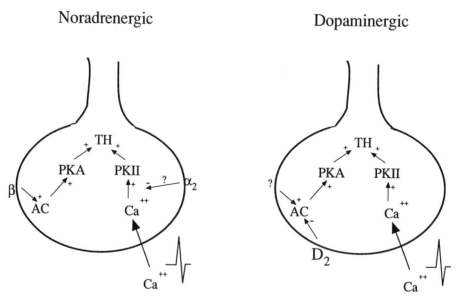

Figure 10.5 Schematic representation of regulation of tyrosine hydroxylase activity in noradrenergic and dopaminergic nerve terminals by impulse-dependent influx of Ca^{2+} and stimulation of presynaptic autoreceptors

In noradrenergic neurones, repeated or prolonged stimulation results in synthesis of additional enzyme protein: this process is called enzymic induction. Such induction affects both the transmitter-specific enzymes, tyrosine hydroxylase and DBH. Induction of enzyme involves DNA transcription and translation; it is a much slower process than activation and the newly-synthesized enzyme appears in adrenomedullary cells and in the cell bodies of catecholaminergic neurones with a delay of 12–18 h. In neurones, this additional enzyme is transported to the nerve terminals, the site of release-relevant synthesis, by axoplasmic transport. The increase in enzyme in the nerve terminal appears with a delay which is proportional to its distance from the cell body; it is measured in days rather than hours. The increase in enzyme protein may persist for periods of up to 1–3 weeks (Zigmond, 1980).

10.4.1.2 5-Hydroxytryptamine synthesis

The rate-limiting enzyme for synthesis of 5-HT is tryptophan hydroxylase, which shares many of the characteristics of tyrosine hydroxylase: it has the same requirement for oxygen and the reduced pteridine cofactor, tetrahydrobiopterin; its substrate is tryptophan. In contrast to tyrosine hydroxylase, the concentration of cofactor is normally saturating but the concentration of tryptophan is subsaturating. This means that supply of tryptophan can influence 5-HT synthesis. The evidence for this is that synthesis is increased by the addition of tryptophan but not by the addition of cofactor. Furthermore, unlike tyrosine hydroxylase, the enzyme does not appear to be subject to endproduct inhibition.

Purified tryptophan hydroxylase is phosphorylated by Ca^{2+}/calmodulin-dependent protein kinase and the increased rate of 5-HT synthesis, resulting from electrical stimulation of 5-hydroxytryptaminergic pathways, is attributed to altered kinetic properties resulting from phosphorylation of tryptophan hydroxylase by Ca^{2+}/calmodulin-dependent protein kinase (Boadle-Biber et al., 1986); it is thus linked to the influx of Ca^{2+}, which accompanies the arrival of the nerve impulse at the nerve terminal.

Like the catecholaminergic neurones, 5-HT releasing neurones have both somato-dendritic and presynaptic inhibitory autoreceptors: the somato-dendritic auto-receptors are $5-HT_{1A}$ receptors and reduce impulse discharge; presynaptic autoreceptors are $5-HT_{1B}$ in rats and $5-HT_{1D}$ in humans and reduce the influx of Ca^{2+}. This causes a depression of both release and synthesis of 5-HT.

Chronic stimulation of 5-hydroxytryptaminergic neurones leads to an activation of tryptophan hydroxylase which outlasts the stimulation and, since it is not reversed by incubation with alkaline phosphatase, may represent induction of new enzyme. There is as yet no direct evidence for this, however (Boadle-Biber et al., 1989).

10.4.2 TECHNIQUES USED TO MEASURE MONOAMINE SYNTHESIS RATE

A number of techniques have been used to measure changes in monoamine synthesis rate.

10.4.2.1 Accumulation of l-DOPA and 5-HTP

Changes in synthesis in vivo are reflected by changes in the accumulation of l-DOPA, the product of tyrosine hydroxylation, or 5-HTP, the product of tryptophan hydroxylation, when the enzyme dopa decarboxylase is inhibited. Changes in l-DOPA accumulation in response to stressors such as drug-induced hypotension, insulin-induced hypoglycaemia and footshock have been shown to occur in selected brain regions. Measurement of l-DOPA accumulation is a widely used technique and shares the advantages and limitations of other *ex vivo* techniques. However the procedure itself interferes with synthesis; since release depends on newly synthesized transmitter, inhibition of synthesis will inhibit release and so eliminate the regulatory effect on synthesis of the stimulation of the presynaptic auto-receptors. In the case of tyrosine hydroxylase an accumulation of l-DOPA may also exert feedback inhibition on the enzyme.

10.4.2.2 Synaptosomal synthesis rate

When brain tissue is homogenized the nerve terminals, torn off their axons, reseal. These isolated nerve terminals are called synaptosomes and there is extensive evidence that they retain many of their normal functions such as metabolism, transmitter synthesis, transmitter release and receptor-mediated regulation of these functions. To identify the effect of stress on specific pathways, synthesis rate has been measured in synaptosomes prepared *ex vivo* from selected brain regions of animals subjected to *in vivo* stress. Synthesis rate has been calculated either from the rate of accumulation of noradrenaline or dopamine in the presence of the MAO inhibitor pargyline, or from the rate of formation of $[^3H] H_2O$ from $[^3H]$ tyrosine. Such studies have shown that, depending on the nature of the stress and the neuronal pathway, increases in synthesis rate may persist in synaptosomes prepared from animals killed at varying times after the stress. Thus a single, mild footshock results in an increase in the rate of tyrosine hydroxylation in hippocampal synaptosomes prepared from animals killed 10 s, but not 30 min, after the shock (Graham-Jones *et al.*, 1983); after 14 i.p. saline injections, given over a period of either 3 days or 14 days, noradrenaline synthesis in hippocampal synaptosomes is increased at both 24 and 48 h after the last injection (Birch *et al.*, 1986). The response of the dopaminergic mesocortical pathway to stress differs from the noradrenergic pathway: a single i.p. saline injection results in an increase in dopamine synthesis in synaptosomes from frontal cortex, which is seen 24 h after the stress (Anderson and Fillenz, 1984). After 14 i.p. saline injections, the increased synthesis is still present seven days after the last stress (Anderson, 1986).

Such results imply that changes in tyrosine hydroxylase greatly outlast the immediate effect of the stress.

10.4.2.3 Changes in solubilized enzyme

To discover the mechanism responsible for the increase in synthesis rate, the characteristics of solubilized enzyme have been studied. This is prepared from organs or brain regions of stressed animals and compared with that from unstressed animals. Such experiments have shown an increase in the affinity of the enzyme for the reduced cofactor induced by stress. Treatments such as electroconvulsive shock (Masserano *et al.*, 1989), hypotension and hypoglycaemia (Fluharty *et al.*, 1985) have all been shown to result in an increase in affinity for cofactor of tyrosine hydroxylase prepared from the adrenal medulla of the stressed rats. Similar changes in soluble tyrosine hydroxylase extracted from brain tissue have been reported following electrical stimulation of central catecholaminergic pathways (Roth *et al.*, 1975; Haycock and Haycock, 1991) but, so far, not after activation by stress.

To demonstrate induction of tyrosine hydroxylase by chronic stress, the activity of the solubilized enzyme is measured in the presence of saturating concentrations of cofactor. More direct evidence for induction is provided by immunoprecipitation of the enzyme or by measurement of the levels tyrosine hydroxylase mRNA, whose increase precedes that of the enzyme protein. It is now even possible to measure tyrosine hydroxylase mRNA, enzyme protein and enzyme activity in a single locus coeruleus. Such measurements have shown that after prolonged cold stress, there is a greater increase in levels of tyrosine hydroxylase mRNA than in enzyme protein, but there is little or no increase in enzyme activity (Richard *et al.*, 1988). When noradrenaline release *in vivo* is measured in these animals, using microdialysis, no increase in basal release or synthesis is found, in spite of the increase in enzyme protein. However, when challenged by an acute stress, the release and synthesis of noradrenaline in previously stressed animals is greater than in unstressed rats. Thus exposure to chronic stress produces an increase in the *capacity* to respond to a novel stressor (Nisenbaum *et al.*, 1991).

It has been proposed that the development of resistance following repeated exposure to stress is mimicked by the chronic administration of antidepressant drugs (Stone, 1979b); this hypothesis is based on the similarity of receptor changes found in the two conditions (see below). Doubt is cast on this hypothesis by the recent finding that, in contrast to chronic stress, repeated administration of either antidepressant drugs or electroconvulsive shocks, decreases levels of both tyrosine hydroxylase immunoreactivity and mRNA (Nestler *et al.*, 1990).

Stress can affect 5-HT synthesis in a number of ways. Some stressors produce changes in the concentration of plasma tryptophan, which is one possible mechanism for the regulation of synthesis. However a recent study suggests that sound stress, consisting of 100 dB tones of 2 s duration presented at random over a 2 h period, produces a phosphorylation-dependent activation of the enzyme. The increase in activity seen in soluble tryptophan hydroxylase prepared from the cortex of the stressed rats and which persists *in vivo* for up to 1 h after the end of the stress, is reversed by incubation with alkaline phophatase and cannot be further increased by *in vitro* phosphorylation (Boadle-Biber *et al.*, 1989). In contrast, when sound stress is applied intermittently for periods of up to six weeks, there is a much more enduring increase in the activity of the enzyme. This increase in enzyme activity following chronic stress has characteristics which suggest that it might represent enzyme induction (Boadle-Biber *et al.*, 1989).

10.4.3 SYNTHESIS OF PEPTIDES

Released peptides are derived from large precursor molecules, which are synthesized in the cell body and then packaged into vesicles in the Golgi apparatus. During their transport along the axon they undergo cleavage whereby inactive precursor propeptides are broken down into their active peptide subunits.

Replacement of released peptides involves processes which affect both the synthesis of propeptide precursor and its cleavage into peptide subunits, i.e. both pre- and posttranslational mechanisms are involved. For instance, the two anterior pituitary peptides, ACTH and β-endorphin, which show marked changes in response to stress, are derived from the precursor molecule pro-opiomelanocortin (POMC). Both synthesis of the precursor protein and its subsequent processing can be measured by incubating anterior pituitary cells from control and from stressed rats with [^{3}H]lysine. The synthesis rate of POMC is calculated from measurements of the rate of incorporation of [^{3}H]lysine into protein; the subsequent processing is monitored in pulse-chase experiments in which incubation with [^{3}H]lysine for 15 min is followed by incubation with unlabelled lysine for periods of 15, 30, 60 and 90 min. The rate of appearance of the label in the active peptide cleavage products at the various time intervals is a measure of the rate of processing of the precursor. Such experiments have shown that, immediately after an acute stress, both the rate of synthesis (translation) of POMC and the rate of conversion of precursor into product are doubled. There is no change, at this stage, in POMC mRNA levels. However, after repeated stress, there is an increase in the rate of transcription, as reflected by an increase in mRNA, but the posttranslational processing is no longer accelerated (Shioma *et al.*, 1986).

10.5 RECEPTOR CHANGES

10.5.1 EFFECTS OF RECEPTOR ACTIVATION

There are three main classes of receptors: (1) ionophore-coupled receptors, (2) G-protein coupled receptors, and (3) intracellular steroid receptors.

Whereas neurotransmitters which interact with receptors coupled to ion channels lead to the opening of ion channels, binding of neurotransmitters or hormones to G-protein coupled receptors gives rise to a wide variety of effects. G-protein associated receptors can either interact with an ion channel or regulate the concentration of intracellular second messengers which, in turn, activate protein kinases. The main second messengers are cAMP and Ca^{2+} (Figure 10.6).

For receptors coupled to the G_s protein, binding of agonists stimulates the membrane-bound enzyme adenylate cyclase which converts ATP to cAMP; cAMP stimulates the cAMP-dependent protein kinase, protein kinase A, which, by phosphorylating proteins, produces a variety of biological effects. A second group of receptors, coupled to a G_i protein, produces the opposite effects by inhibiting adenylate cyclase. A third group of receptors, coupled to a different G-protein which is variously called Go or Gp, stimulates the membrane-bound enzyme phospholipase C. This causes the breakdown of polyphosphatidylinositols, an important class of membrane proteins; among their breakdown products are a number of inositol phosphates and diacyl glycerol. Inositol triphosphate releases Ca^{2+} from intracellular stores; Ca^{2+} can produce effects either on its own, or by binding to calmodulin, or by activating Ca^{2+}/calmodulin protein kinase. Diacyl glycerol activates the Ca^{2+}, phospholipid-dependent protein kinase C. These second messengers produce a wide variety of biological effects.

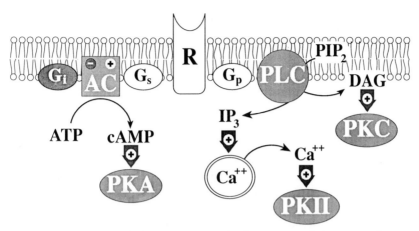

Figure 10.6 G-protein-coupled receptors and their intracellular second messengers. R = receptor; AC = adenylate cyclase; PLC = phospolipase C; PIP_2 = phosphatidylinositol diphosphate; DAG = diacyl glycerol; IP_3 = inositol triphosphate; PKA = cAMP-dependent protein kinase (protein kinase A); PKC = Ca^{2+}-phospholipid-dependent protein kinase (protein kinase C); PKII = Ca^{2+}-calmodulin-dependent protein kinase (protein kinasae II)

Steroid receptors are intracellular and, when bound by an agonist, are transferred to the nucleus where they control gene expression. Recent work has shown that both cAMP and Ca^{2+} are also involved in the expression of the so-called early genes which, in turn, regulate DNA transcription (Morgan and Curran, 1991).

The interaction of the transmitters with their receptors, as well as activating the appropriate effector mechanisms, also produces changes in the receptors themselves. When exposed to their agonist, receptors undergo desensitization and when deprived of their agonist, show sensitization. The receptors whose mechanism of desensitization is most clearly understood are the G-protein coupled receptors which regulate adenylate cyclase.

Receptors, when bound to their agonist, are phosphorylated by a special protein kinase, originally called β-adrenoreceptor kinase, but now known to phosphorylate other agonist-bound receptors such as α_2-adrenoceptors; this phosphorylation first results in an uncoupling of the receptor from the G-protein, which inactivates it, followed by the internalization of the receptor. This latter process results in a reduction in the density of the cell surface receptors. This is *homologous* desensitization, and occurs when a receptor is exposed to its own agonist. In *in vitro* experiments it has been shown that homologous desensitization occurs within minutes of continuous exposure to the agonist, and is equally rapidly reversible on washing. This is due to the action of an intracellular phosphatase, which dephosphorylates the receptor, after which it is re-inserted into the membrane (Sibley *et al.*, 1987).

G-proteins themselves can also be phosphorylated by other protein kinases such as protein kinase A and C, which will interfere with their binding to the receptors. This is *heterologous* desensitization, where exposure to one agonist leads to the desensitization of other receptors which share the same effector mechanism.

Finally, adenylate cyclase can also be phosphorylated and this may enhance its activity resulting in heterologous sensitization. It is becoming increasingly clear that both the immediate and the delayed effects of receptor activation often involve complex receptor interactions.

In addition to these relatively rapid mechanisms, there are much slower changes in receptor number which involve breakdown or new synthesis of receptors.

10.5.2 MEASUREMENT OF RECEPTOR CHANGES

10.5.2.1 Receptor binding studies

Receptor binding studies have been carried out with membrane preparations from various brain regions after acute stress, chronic stress and after chronic administration of antidepressant drugs. There are relatively few reports of changes in receptor affinity or density resulting from acute stress. Among these are the increase in binding of [^{3}H] ketanserin, a 5-HT$_2$ receptor ligand, in the frontal cortex following a single 2.5 h immobilization (Torda et al., 1989) and the changes in GABA receptors following acute stress (Biggio et al., 1987; Skerritt et al., 1987).

After chronic stress a number of receptors have been shown to be decreased in density. A reduction in the number of cortical β-adrenoceptor binding sites was the earliest report of a stress-induced receptor change (U'Prichard and Kvetnansky, 1980; Stone and Platt, 1982). Although there is considerable variation in the reports of the effects of repeated stress on cortical α$_2$-adrenoceptors, decreases in B$_{max}$ of cortical α$_2$-adrenoceptors after chronic stress have been reported by some workers (Stone, 1981; Stanford et al., 1984). The reduction in receptor number is by no means uniform throughout the brain, but shows the same regional specificity as the changes in transmitter release or turnover. Receptor changes also vary with the form and severity of the stress.

In general, increased exposure to the agonist causes a reduction in the number of binding sites. The mechanism which underlies the increase in the number of 5-HT$_2$ binding sites reported after acute stress as well as chronic stress, which causes an increase in 5-HT turnover, suggests that the regulation of 5-HT$_2$ receptors is different from the generally accepted model (Frazer et al., 1988).

The scope for receptor binding studies in humans is limited. There have been a number of receptor binding studies carried out on brains of suicides post-mortem which found increases in the density of muscarinic receptors (Meyerson et al., 1982) and of 5-HT$_2$ receptors (Mann et al., 1989); others failed to find a change in 5-HT$_2$ receptors (Cheetham et al., 1988). Measurements of the density of imipramine binding, which represents a site which may be an allosteric regulator of the serotonin uptake mechanism (Paul and Skolnick, 1984), have also yielded conflicting results. The reasons for these discrepancies may lie in the selectivity of the radioligand used, the presence of drugs, or the diagnosis. A recent development is the use of positron emission tomography (PET) and single photon emission tomography (SPET) for the in vivo measurement of receptor binding sites (Ell, 1990). Studies in human subjects on a variety of neuroreceptors have already been carried out.

There have also been receptor binding studies in humans using elements in the blood; these are platelets, (which have α$_2$- and α$_1$-adrenoceptors, 5-HT$_2$ receptors, high affinity benzodiazepine receptors and imipramine-binding sites) and lymphocytes (which have β-adrenoceptors). The significance of abnormalities in binding sites in such studies is not clear at present. Antidepressant drugs that increase the concentration of amines to which receptors are exposed could have an effect on all the receptors, both central and peripheral. For stress-related changes in transmitter release, however, which show a high degree of regional specificity, peripheral blood cells are unlikely to serve as a model for central neurones. Also, receptors on blood cells are not protected by the blood brain barrier and could even be affected by diet. If, on the other hand, the differences in binding sites are determined by trait rather than state, then the elements in blood could serve as a model for abnormalities in central nervous receptors.

10.5.2.2 Measurement of receptor function

Receptors consist of a ligand binding site and an effector mechanism; agonist binding and receptor function are therefore by no means synonymous. In order to understand the functional implications of changes in receptor binding it is essential to test receptor *function* as well as *binding*. Such tests of receptor function can address either the second messenger mechanism in *in vitro* preparations, or the physiological function of receptors in the intact animal.

10.5.2.2.1 In vitro preparations. The earliest demonstration of a parallel change in receptor binding and function was the demonstration that repeated immobilization stress resulted in both a reduction in noradrenaline stimulated cAMP production (Stone, 1979a) and downregulation of β-adrenoceptor binding sites (Stone and Platt, 1982). However, further analysis of these phenomena revealed that noradrenaline-stimulated cAMP formation involves both β-adrenoceptor and non-β-receptor stimulation: the maximal cAMP accumulation in response to noradrenaline is three times greater than that in response to the β-selective agonist, isoprenaline. The non-β-receptor has pharmacological characteristics of both α_1- and α_2-adrenoceptors. Since agonists for neither of these two receptors cause an increase in cAMP, the effect is a potentiation of the effect of the β-adrenoceptor, therefore. The underlying mechanism is not known, but it may involve phosphorylation of the catalytic subunit of adenylate cyclase (Stone et al., 1985).

The reduction in β-adrenoceptor-stimulated cAMP formation after repeated immobilization is transient and is no longer present 24 h after the last stress; at this stage only the noradrenaline-stimulated, but not the isoprenaline stimulated, cAMP formation is reduced, implying that it is the non-β-receptor potentiation of β-adrenoceptor stimulation of adenylate cyclase that has been lost.

The effector system of β-adrenoceptors is the most firmly established. For other receptor types, functional tests which look at processes distal to the second messenger system have been used. For instance, α_2-adrenoceptors act as inhibitory somato-dendritic and presynaptic autoreceptors and presynaptic heteroreceptors; the autoreceptors regulate impulse discharge, release and synthesis of noradrenaline (see Section 10.4.1). Changes in the responsiveness of α_2-adrenoceptors have been measured by changes in firing frequency of locus coeruleus neurones in response to the α_2-adrenoceptor agonist, clonidine (McMillen et al., 1980), the effect of clonidine on K^+-evoked release of noradrenaline in cortical slices (Schoffelmeer and Mulder, 1982), or the effect of clonidine on K^+-dependent enhancement of noradrenaline synthesis in cortical synaptosomes (Birch *et al.*, 1986). The same principles apply to dopaminergic and 5-hydroxytryptaminergic autoreceptors which are both somato-dendritic and presynaptic.

K^+-evoked release and acceleration of synthesis in brain slices is a widely used preparation, although it is vulnerable to serious errors; in contrast to synaptosomes, slices retain much of the structural complexity of the intact brain. Application of K^+ causes release of transmitters from all the nerve terminals in the slice which allows extensive interaction between them; in view of the important role of presynaptic receptors in the regulation of both release and synthesis, this introduces serious complications. In a synaptosomal suspension, which also consists of a heterogeneous collection of nerve terminal derived structures, the dilution of the released transmitters makes them ineffective. This is clearly illustrated by the fact that the dopamine D_2 receptor antagonist, sulpiride, has no effect on K^+-enhanced synthesis in striatal synaptosomes, but causes a substantial increase in striatal slices (El Mestikawy *et al.*, 1986).

10.5.2.3 In vivo *functional tests*

10.5.2.3.1 Behavioural tests. A number of behavioural tests have been developed as a measure of functional receptor changes. One of the best established of these behavioural tests in the rat is clonidine-induced hypoactivity and mydriasis. There is conclusive evidence that these effects are mediated by central α_2-adrenoceptors; however, the brain area responsible for the effect is much less clearly established. There has also been controversy concerning the pre- or postsynaptic location of the receptors. Although degeneration experiments show that α_2-autoreceptors constitute a small proportion of central α_2-adrenoceptors, the remaining receptors are called 'postsynaptic' without distinguishing between presynaptic heteroreceptors and postsynaptic somato-dendritic receptors. There is evidence for presynaptic α_2-heteroreceptors on 5-hydroxytryptaminergic and cholinergic nerve terminals, but there is no electrophysiological evidence that activation of α_2-adrenoceptors can affect neuronal impulse discharge (Szabadi, 1979) other than the autoreceptor-mediated effect on neurones of the locus coeruleus. Although some studies have suggested that the clonidine-induced hypoactivity is mediated by post-synaptic receptors, recent evidence suggests that it is mediated by presynaptic autoreceptors (Heal *et al.*, 1988). Mydriasis is attributed to postsynaptic receptors (Heal, 1990).

Another well characterized behavioural paradigm is the hypothermia and the serotonin behavioural syndrome (motility, forepaw treading, flat body posture, hind-limb abduction, reactivity) induced by 8-hydroxy-2(dipropylamino)tetralin (8-OH-DPAT). There is extensive evidence for the mediation of these behavioural effects by 5-HT_{1A} receptors, the hypothermic response being attributed to presynaptic receptors in the mouse, but this is less certain in the rat. In contrast, in the rat the 5-HT syndrome is attributed to postsynaptic receptors (Goodwin *et al.*, 1987) but in the mouse the behavioural syndrome is induced by 5-HT, but not 8-OH-DPAT, and is thought to involve 5-HT_2 receptors; possible species differences have to be kept in mind therefore.

Attempts to develop behavioural tests which are selective for other receptor subtypes have been less successful (Heal, 1990).

10.5.2.3.2 Neuroendocrine tests. Another set of functional tests for receptors depends on their role in regulating the release of hormones. These include the release of melatonin from the pineal gland and release of hormones from the anterior pituitary; the former is mediated by pineal β-adrenoceptors and the latter by monoamine receptors in the hypothalamus, which control the release of hypothalamic releasing factors. The advantage of these tests is that their underlying mechanisms can be analysed in animal experiments, and they can also be applied in human studies by measuring changes in the concentration of hormones in plasma (see Section 10.2.1).

10.6 CONCLUDING REMARKS

A wide variety of neurochemical tests are now available which have yielded a wealth of information. Much additional information is still needed. To understand the functional significance of changes in a neuronal system, as many different parameters as possible have to be measured.

Many different neuronal systems are activated by stress, and it is the interaction between them, as well as the general pattern of activation, that may be significant rather than the change in any one system. Although the stress response is described as non-specific, the neurochemical changes vary with the nature and intensity of the stress. This may be due both to the specific element of the response and variations in the so-called non-specific response. Extrapolation from stress in animals to stress in humans, from drug effects in normal animals to depressed humans, and from one form of stress to another, have to be treated with caution.

Finally, whereas some of the neurochemical changes in blood and in brain have a physiological function, others are merely by-products of activity. Thus plasma noradrenaline, DBH and chromogranin are indices of sympathetic activity, with no further function after entry into the blood stream; opioid peptides and the various hormones, on the other hand, are messengers with an important physiological role. A similar argument may apply to neurochemical changes in the brain. There has been much discussion about the functional significance of receptor changes which result from chronic stress. It is possible that these receptor changes are merely a by-product of sustained receptor stimulation, the physiologically relevant change being the initiation of third messenger systems, such as early gene transcription. The new techniques of molecular biology are used increasingly to study these problems.

10.7 REFERENCES

Anderson, S.M.P. (1986) The effects of a single and repeated mild stress on central monoaminergic neurones. *DPhil Thesis*, Oxford University

Anderson, S.M.P. and Fillenz, M. (1984) The delayed effect of stress on dopamine synthesis in the rat frontal cortex. *J. Physiol.*, **358**: 47P

Arato, J.B., Banki, C.M., Bissette, G. and Nemeroff, C.B. (1989) Elevated cerebrospinal fluid concentrations of corticotropin-releasing factor in suicide victims. *Biol. Psychiatr.*, **25**: 355–9

Banki, C.M., Bissette, G., Arato, M. *et al.* (1987) Cerebrospinal fluid corticotropin-like immunoreactivity in depression and schizophrenia. *Amer. J. Psychiatr.*, **144:** 873–7

Becker, R.E., Giambalvo, C., Fox, A. and Macho, M. (1983) Endogenous inhibitors of monoamine oxidase present in human cerebrospinal fluid. *Science*, **221:** 476–8

Benedict, C.R., Fillenz, M. and Stanford, S.C. (1979) Noradrenaline in rats during prolonged cold-stress and repeated swim-stress. *Br. J. Pharmacol.* **66:** 421–4

Benveniste, H. (1989) Brain microdialysis. *J. Neurochem.*, **52:** 1667–79

Bertolucci-D'Angio, M., Serrano, A. and Scatton, B. (1990) Mesocorticolimbic dopaminergic systems and emotional states. *J. Neurosci. Meth.*, **34:** 135–42

Biggio, G., Concas, A., Mele, S. and Corda, M.G. (1987) Changes in GABAergic transmission induced by stress, anxiogenic and anxiolytic beta-carbolines. *Brain Res. Bull.*, **19:** 301–8

Birch, P.J., Anderson, S.M.P. and Fillenz, M. (1986) Mild chronic stress leads to desensitisation of presynaptic autoreceptors and a long-lasting increase in noradrenaline synthesis in rat cortical synaptosomes. *Neurochem. Int.*, **9:** 329–36

Blombery, P.A., Kopin, I.J., Gordon, E.K. *et al.* (1980) Conversion of MHPG to vanillyl-mandelic acid: Implications for the importance of urinary MHPG. *Arch. Gen. Psychiatr.*, **154:** 1095–8

Boadle-Biber, M.C., Johannessen, J.N., Narasimhachari, N. and Phan, T.H. (1986) Tryptophan hydroxylase: increase in activity by electrical stimulation of serotonergic neurones. *Neurochem. Int.*, **8:** 83–92

Boadle-Biber, M.C., Corley, K.C., Groves, L. *et al.* (1989) Increase in the activity of tryptophan hydroxylase from cortex and midbrain of Fischer 344 rats in response to acute and chronic sound stress. *Brain Res.*, **482:** 306–16

Boutelle, M.G., Fellows, L.K. and Cook, C. (1992) Enzyme packed bed system for the on-line measurement of glucose, glutamate and lactate in brain microdialysate. *Anal. Chem.* (in press)

Brodie, B.B., Costa, E., Dlabac, A. *et al.* (1966) Application of steady-state kinetics to the estimation of synthesis rate and turnover time of tissue catecholamines. *J. Pharmacol. Exp. Ther.*, **154:** 493–8

Brown, M.J., Jenner, D.A., Allison, D.J. and Dollery, C.J. (1981) Variations in individual organ release of noradrenaline by an improved enzymatic technique: limitations of peripheral venous measurements in the assessment of sympathetic nervous activity. *Clinical Science*, **61:** 585–90

Cannon, W.B. (1915) *Bodily Changes in Pain, Hunger, Fear and Rage*, Appleton & Co, New York

Chappell, P.B., Smith, M.A., Kilts, C.D. *et al.* (1986) Alterations in corticotropin-releasing factor-like immunoreactivity in discrete rat brain regions after acute and chronic stress. *J. Neurosci.*, **6:** 2908–14

Cheetham, S.C., Crompton, M.R., Katona, C.L.E. and Horton, R.W. (1988) Brain 5HT2 receptor binding sites in depressed suicide victims. *Brain Res.*, **443:** 272–80

Curet, O., Dennis, T. and Scatton, B. (1985) The formation of deaminated metabolites of dopamine in the locus coeruleus depends upon noradrenergic neural activity. *Brain Res.*, **335:** 297–301

Davis, P. (1991) Methods of measuring neuropeptides and their metabolism, in J.A. McCubbin, P.G. Kaufmann and C.B. Nemeroff (eds.), *Stress, Neuropeptides and Systemic Disease*, Academic Press, London, pp 303–13

De Bruin, L.A., Schasfoort, E.M.C., Steffens, A.B. and Korf, J. (1990) Effects of stress and exercise on rat hippocampus and striatum extracellular lactate. *Am. J. Physiol.*, **259:** R773–9

Duggan, A.W. (1990) Detection of neuropeptide release in the central nervous system with antibody microprobes. *J. Neurosci. Meth.*, **34:** 47–52

Duggan, A.W., Jarrott, B., Hope, P. *et al.* (1990) Release, spread and persistence of immunoreactive neurokinin A in the dorsal horn of the cat following noxious cutaneous stimulation studies with antibody microprobes. *Neuroscience*, **35:** 195–202

Dykstra, K.H., Hsiao, J.K., Morison, P.F. *et al.* (1992) Quantitative examination of tissue concentration profiles associated with microdialysis. *J. Neurochem.*, **58:** 931–40

Ekins, R. (1990) Measurement of free hormones in blood. *Endocrin. Revs.*, **11**: 5–46

El Mestikawy, S., Glowinski, J. and Hamon, M. (1986) Presynaptic dopamine autoreceptors control tyrosine hydroxylase activation in depolarized striatal dopaminergic terminals. *J. Neurochem.*, **46**: 12–22

Ell, P.J. (1990) Single photon emission computed tomography (SPET) of the brain. *J. Neurosci. Meth.*, **34**: 207–17

Elsworth, J., Dewar, D., Glover, V. *et al.* (1986) Purification and characterisation of tribulin, an endogenous inhibitor of monoamine oxidase and benzodiazepine receptor binding. *J. Neural Transmission*, **67**: 45–56

Fillenz, M., Stanford, S.C. and Benedict, C.R. (1979) Changes in noradrenaline release rate and noradrenaline storage vesicles during prolonged activity of sympathetic neurones, in E. Usdin, I.J. Kopin & J. Banhas (eds.), *Catecholamines: Basic and Clinical Frontiers*, Pergamon, New York, pp 936–8

Fluharty, S.J., Snyder, G.L., Zigmond, M.J. and Stricker, E.M. (1985) Tyrosine hydroxylase activity and catecholamine biosynthesis in the adrenal medulla of rats during stress. *J. Pharmacol. Exp. Ther.*, **233**: 32–8

Frazer, A., Offord, S.J. and Lucki, I. (1988) Regulation of serotonin receptors and responsiveness in the brain, in E. Sanders-Bush (ed.), *The Serotonin Receptors*, The Humana Press, Clifton, NJ, pp 319–62

Giambalvo, G.T. (1984) Purification of endogenous modulators of monoamine oxidase from plasma. *Biochem. Pharmacol.*, **33**: 3929–32

Glover, V., Halket, J.M., Watkins, P.J. *et al.* (1988) Isatin: identity with the purified endogenous monoamine oxidase inhibitor tribulin. *J. Neurochem.*, **51**: 656–9

Glover, V., Clow, A., Bhattacharya, S.K. *et al.* (1989) Monoamine oxidase in stress, in G.R. Van Loon, R. Kvetnansky, R. McCarty and J. Axelrod (eds.), *Stress: Neurochemical and Humoral Mechanisms*, Gordon and Beach, New York, pp 133–41

Gonon, F.G. (1988) Nonlinear relationship between impulse flow and dopamine released by rat midbrain dopaminergic neurons as studied by in vivo electrochemistry. *Neuroscience*, **24**: 19–28

Gonon, F., Buda, M., De Simoni, G. and Pujol, J. (1985) Catecholamine metabolism in the rat locus coeruleus as studied by in vivo differential pulse voltammetry II. *Pharmacol. Behav. Study*, **273**: 207–16

Gonon, F.G., Suaud-Chagny, M.F., Mermet, C.C. and Buda, M. (1991) Relation between impulse flow and extracellular catecholamine levels as studied by in vivo Electrochemistry in CNS, in K. Fuxe and L.F. Agnati (eds.), *Volume Transmission in the Brain: Advances in Neuroscience*, Raven Press, New York, pp 337–50

Goodwin, G.M., De Souza, R.J., Green, A.R. and Heal, D.J. (1987) The pharmacology of the behavioural and hypothermic responses of rats to 8-hydroxy-2-(di-*n*-propylamino)tetralin (8-OH-DPAT). *Psychopharm.*, **91**: 506–11

Gordon, R., Spector, S., Sjoerdsma, A. and Udenfried, S. (1966) Increased synthesis of norepinephrine and epinephrine in the intact rat during exercise and exposure to cold. *J. Pharmacol. Exp. Ther.*, **153**: 440–7

Graham-Jones, S., Fillenz, M. and Gray, J.A. (1983) The effects of footshock and handling on tyrosine hydroxylase activity in synaptosomes and solubilised preparations from rat brain. *Neuroscience*, **9**: 679–86

Haas, D.A. and George, S.R. (1988) Single or repeated mild stress increases synthesis and release of corticotropin-releasing factor. *Brain Res.*, **461**: 230–7

Haycock, J.W. and Haycock, D.A.O. (1991) Tyrosine hydroxylase in rat brain dopaminergic nerve terminals. Multiple site phosphorylation in vivo and in synaptosomes. *J. Biol. Chem.*, **266**: 5650–7

Heal, D.J. (1990) The effects of drugs on behavioural models of central noradrenergic function, in D.J. Heal and C.A. Marsden (eds.), *Pharmacology of Noradrenaline in the CNS*, Oxford University Press, Oxford, pp 266–315

Heal, D.J., Prow, M.R. and Buckett, W.R. (1988) Measurement of brain MHPG levels suggests that clonidine induces hypoactivity and mydriasis via pre-synaptic α_2-adrenoceptors. *Br. J. Pharmacol.*, **94**: 317

Hjemdahl, P. (1986) Measurement of plasma catecholamines by HPLC and the relation of their concentrations to sympathoadrenal activity, in M.H. Joseph, M. Fillenz, I.A. MacDonald and C.A. Marsden (eds.), *Monitoring Neurotransmitter Release during Behaviour*, Ellis Horwood, Chichester, pp 17–32

Ikarashi, Y., Sasahara, T. and Maruyama, Y. (1985) Postmortem changes in catecholamines, indoleamines, and their metabolites in rat brain regions: prevention with 10-kW microwave irradiation. *J. Neurochem.*, **45**: 935–9

Isaac, L., Schoenbeck, R., Bacher, J. *et al.* (1986) Electroconvulsive shock increases endogenous monoamine oxidase inhibitor activity in brain and cerebrospinal fluid. *Neurosci. Lett.*, **66**: 257–62

Iversen, L.L. (1967) *The Uptake and Storage of Noradrenaline in Sympathetic Nerves*, Cambridge University Press, Cambridge

Kalen, P., Nilsson, O.G., Cenci, M.A. *et al.* (1990) Intracerebral microdialysis as a tool to monitor transmitter release from grafted cholinergic and monoaminergic neurones. *J. Neurosci. Meth.*, **34**: 107–15

Katz, F.H. and Shannon, I.L. (1964) Identification and significance of parotid fluid corticosteroids. *Acta Endocrinologica*, **46**: 393–404

Kehr, J. and Ungerstedt, U. (1988) Fast HPLC estimation of aminobutyric acid in microdialysis perfusates: effect of nipecotic and 3-mercaptoproprionic acids. *J. Neurochem.*, **51**: 1308–10

Kendrick, K.M. (1990) Microdialysis measurement of in vivo neuropeptide release. *J. Neurosci. Meth.*, **34**: 35–46

Kirschbaum, C. and Hellhammer, D.H. (1989) Salivary cortisol in psychobiological research: an overview. *Neuropsychobiol.*, **22**: 150–69

Klein, R.L. (1982) Chemical composition of the large noradrenergic vesicles, in R.L. Klein, H. Lagercrantz and H. Zimmerman (eds.), *Neurotransmitter Vesicles*, Academic Press, New York, pp 133–72

Kuchel, O. and Buu, N.T. (1984) Conjugation of norepinephrine and other catecholamines, in M.G. Ziegler and C.R. Lake (eds.), *Norepinephrine, Frontiers of Clinical Neuroscience*, Williams and Wilkins, Baltimore, pp 250–70

Lagercrantz, H. and Fried, G. (1982) Chemical composition of the small noradrenergic vesicles, in R.L. Klein, H. Lagercrantz and H. Zimmerman (eds.), *Neurotransmitter Vesicles*, Academic Press, New York, pp 175–88

Lake, C.R., Chernow, B., Feuerstein, G. *et al.* (1984) The sympathetic nervous system in man: its evaluation and the measurement of plasma NE, in M.G. Ziegler and C.R. Lake (eds.), *Norepinephrine, Frontiers of Clinical Neuroscience*, Williams and Wilkins, Baltimore, pp 1–26

Landsberg, L. and Young, J.B. (1986) Assessment of sympathetic nervous activity from measurements of noradrenaline turnover in rats, in M.H. Joseph, M. Fillenz, I.A. MacDonald and C.A. Marsden (eds.), *Monitoring Neurotransmitter Release During Behaviour*, Ellis Horwood, Chichester, pp 33–47

Lipscombe, D., Kingsamut, S. and Tsien, R.W. (1989) Alpha-adrenergic inhibition of sympathetic neurotransmitter release mediated by modulation of N-type calcium channel gating. *Nature*, **340**: 630–42

Lundberg, J.M. and Hokfelt, T. (1986) Multiple co-existence of peptides and classical transmitters in peripheral autonomic and sensory neurons: functional and pharmacological implications. *Prog. Brain Res.*, **68**: 241–62

Lundberg, J.M., Martinsson, A., Hemsen, A. *et al.* (1985) Co-release of neuropeptide Y and catecholamines during physical exercise in man. *Biochem. Biophys. Res. Commun.*, **133**: 30–6

Lundberg, J.M., Rudehill, A., Sollevi, A. *et al.* (1986) Frequency- and reserpine-dependent chemical coding of sympathetic transmission: differential release of noradrenaline and neuropeptide Y from pig spleen. *Neurosci. Lett.*, **63**: 96–100

Major, L.F., Lerner, P. and Ziegler, M.G. (1984) Norepinephrine and dopamine-β-hydroxylase in cerebrospinal fluid: Indicators of central noradrenergic activity, in M.G. Ziegler and C.R. Lake (eds.), *Norepinephrine, Frontiers of Clinical Neuroscience*, Williams and Wilkins, Baltimore, pp 117–41

Mann, J.J., Arango, V., Marzuk, P.M. *et al.* (1989) Evidence for the 5HT hypothesis of suicide. *Br. J. Psychiatr.*, **155**(Suppl.8): 7–14

Marsden, C.A., Joseph, M.H., Kruk, Z.L. *et al.* (1988) In vivo voltammetry – present electrodes and methods. *Neuroscience*, **25**: 389–400

Maser-Gluth, C., Lorenz, U. and Vecsei, P. (1989) Corticotropin-releasing-factor-like immunoreactivity in human 24 h urine. *Clinical Endocrinology (Oxford)*, **30**: 405–12

Masserano, J.H., Vulliet, P.R., Tank, A.W. and Weiner, N. (1989) The role of tyrosine hydroxylase in the regulation of catecholamine synthesis. *Handb. Exp. Pharmacol.*, **90/II**: 427–69

Maynert, E.W. and Levi, R. (1964) Stress-induced release of brain norepinephrine and its inhibition by drugs. *J. Pharmacol. Exp. Ther.*, **143**: 90–5

McMillen, B.A., Warnak, W., German, D.C. and Shore, P.A. (1980) Effects of chronic desipramine treatment on rat brain noradrenergic responses to α-adrenergic drugs. *Eur. J. Pharmacol.*, **61**: 239–46

Meyerson, L.R., Wennogle, L.P., Abel, M.S. *et al.* (1982) Human brain receptor alterations in suicide victims. *Pharmacol. Biochem. Behav.*, **17**: 159–63

Morgan, J.I. and Curran, T. (1991) Stimulus-transcription coupling in the nervous system: involvement of the inducible proto-oncogenes *fos* and *jun*. *Ann. Revs. Neurosci.*, **14**: 421–51

Mormede, P., Castagne, V., Rivet, J.-M. *et al.* (1990) Involvement of neuropeptide Y in neuroendocrine stress responses. Central and peripheral studies. *J. Neural Transm. Suppl.*, **29**: 65–75

Neff, N.H., Tozer, T.H., Hammer, W. *et al.* (1968) Application of steady-state kinetics to the uptake and decline of 3H-NE in the rat heart. *J. Pharmacol. Exp. Ther.*, **160**: 48–52

Nemeroff, C.B., Widerlov, E., Bissette, G. *et al.* (1984) Elevated concentrations of CSF corticotropin-releasing factor-like immunoreactivity in depressed patients. *Science*, **226**: 1342–4

Nestler, E.J., McMahon, A., Sabban, E.L. *et al.* (1990) Chronic antidepressant administration decreases the expression of tyrosine hydroxylase in the rat locus coeruleus. *Proc. Natl. Acad. Sci. USA.*, **87**: 7522–6

Nisenbaum, L.K., Zigmond, M.J., Sved, A.F. and Abercrombie, E.D. (1991) Prior exposure to chronic stress results in enhanced synthesis and release of hippocampal norepinephrine in response to a novel stressor. *J. Neurosci.*, **11**: 1478–84

O'Connor, D.T., Takiyyuddin, R.J., Parmer, J.H. *et al.* (1991) Sympathoadrenal catecholamine storage and release in humans. Insights from the study of chromogranin A, in M.R. Brown, G.F. Koob and C. Rivier (eds.), *Stress, Neurobiology and Neuroendocrinology*, Marcel Dekker, New York, pp 217–30

Odink, J., Korthals, H. and Knijff, J.H. (1988) Simultaneous determination of the major acidic metabolites of catecholamines and serotonin in urine by liquid chromatography with electrochemical detection after a one-step sample clean-up on Sephadex G-10; influence of vanilla and banana ingestion. *J. Chromatog.*, **424**: 273–83

Ono, T., Ogawa, N. and Mori, A. (1989) The effects of haemorrhagic shock on thyrotropin releasing hormone and its receptors in discrete regions of rat brain. *Regul. Pept.*, **25**: 215–22

Orth, D.N. and Mount, C.D. (1987) Specific high-affinity binding protein for human corticotropin-releasing hormone in normal human plasma. *J. Chromatog.*, **143**: 411–17

Paul, S.M. and Skolnick, P (1984) High affinity binding of antidepressants to biogenic amine transport sites in human brain and platelet: studies in depression, in R.M. Post & J.C. Ballenger (ed.), *Neurobiology of Mood Disorders*, Williams and Wilkins, Baltimore

Peaston, R.T. (1988) Routine determination of urinary free catecholamines by HPLC with electrochemical detection. *J. Chromatog.*, **424**: 263–72

Plotsky, P.M. and Vale, W. (1984) Haemorrhage-induced secretion of corticotropin-releasing factor-like immunoreactivity into the rat hypophyseal portal circulation and its inhibition by glucocorticoids. *Endocrinology (Baltimore)*, **114**: 164–9

Richard, F., Faucon-Biguet, N., Labatut, R. *et al.* (1988) Modulation of tyrosine hydroxylase gene expression in rat brain and adrenals by exposure to cold. *J. Neurosci. Res.* **20**: 32–7

Roth, R.H., Morgenroth III, V.H. and Salzman, P.M. (1975) Tyrosine hydroxylase: allosteric activation induced by stimulation of central noradrenergic neurons. *Arch. Pharmacol.*, **289**: 327–43

Schasfoort, E.M.C., DeBruin, L. and Korf, J. (1988) Mild stress stimulates rat hippocampal glucose utilization transiently via NMDA receptors, as assessed by lactography. *Brain Res.*, **475**: 58–63

Schoffelmeer, A.N.M. and Mulder, A.H. (1982) [3H]-Noradrenaline and [3H]-5-hydroxytryptamine release from rat brain slices and its presynaptic alpha-adrenergic modulation after long-term desipramine pretreatment. *Arch. Pharmacol.*, **318**: 173–80

Schon, F., Allen, J.M., Yeats, J.C. *et al.* (1986) The effect of 6-hydroxydopamine, reserpine and cold stress on neuropeptide Y content of the rat central nervous system. *Neuroscience*, **19**: 1247–50

Selye, H. (1936) A syndrome produced by diverse nocuous agents. *Nature*, **148**: 84–5

Shiomi, H., Watson, S.J., Kelsey, J.E. and Akil, H. (1986) Pretranslational and posttranslational mechanisms for regulating β-endorphin-adrenocorticotropin of the anterior pituitary lobe. *Endocrinology*, **119**: 1793–9

Sibley, D.R., Benovic, J.L., Caron, M.G. and Lefkowitz, J.R. (1987) Regulation of transmembrane signalling by receptor phosphorylation. *Cell*, **48**: 913–22

Skerritt, J.H., Trisdikoon, P. and Johnston, G.A.R. (1987) Increased GABA binding in mouse brain following acute swim stress. *Brain Res.*, **215**: 398–403

Stanford, S.C., Fillenz, M. and Ryan, E. (1984) The effect of repeated mild stress on cerebral cotical adrenoceptors and noradrenaline synthesis in the rat. *Neurosci. Lett.*, **45**: 163–7

Stone, E.A. (1979a) Reduction by stress of norepinephrine-stimulated accumulation of cyclic AMP in rat cerebral cortex. *J. Neurochem.*, **32**: 1335–7

Stone, E.A. (1979b) Subsensitivity to norepinephrine as a link between adaptation to stress and antidepressant therapy: An hypothesis. *Comm. Psychol. Psychiatr. Behav.* **4**: 241–55

Stone, E.A. (1981) Mechanism of stress-induced subsensitivity to norepinephrine. *Pharmacol. Biochem. Behav.*, **14**: 719–23

Stone, E.A. and Platt, J.E. (1982) Brain adrenergic receptors and resistance to stress. *Brain Res.*, **237**: 405–14

Stone, E.A., Slucky, A.V., Platt, J.E. and Trullas, R. (1985) Reduction of the cyclic 3'5'-monophosphate response to catecholamines in rat brain slices after repeated restraint stress. *J. Pharmacol. Exp. Ther.*, **233**: 382–8

Swanson, L.W., Sawchenko, P.E., Rivier, J. and Vale, W.W. (1983) Organisation of ovine corticotropin-releasing factor immunoreactive cells and fibres in the rat brain: an immunohistochemical study. *Neuroendocrinology*, 165–86

Szabadi, E. (1979) Adrenoceptors on central neurones: microelectrophoretic studies. *Neuropharm.*, **18**: 831–43

Takayama, H., Ota, Z. and Ogawa, N. (1986) Effect of immobilization stress on neuropeptides and their receptors in rat central nervous system. *Regul. Pept.*, **25**: 239–48

Thureson-Klein, A. (1983) Exocytosis from large and small dense cored vesicles in noradrenergic nerve terminals. *Neuroscience*, **10**: 245–52

Torda, T., Culman, J., Petrikova, M. and Murgas, K. (1989) Adrenergic and serotonergic receptors in the frontal cortex: effect of stress in rats, in G.R. Van Loon, R. Kvetnansky, R. McCarty and J. Axelrod (eds.), *Stress, Neurochemical and Humoral Mechanisms*, Gordon and Breach, New York, pp 175–88

Tucek, S., Fatranska, M., Kvetnansky, R. and Ricny, J. (1989) Cholinergic aspects of the stress response, in G.R. Van Loon, R. Kvetnansky, R. McCarty and J. Axelrod (eds.), *Stress, Neurochemical and Humoral Mechanisms*, Gordon and Breach, New York, pp 143–72

U'Prichard, D.C. and Kvetnansky, R. (1980) Central and peripheral adrenergic receptors in acute and repeated immobilization stress, in E. Usdin, R. Kvetnansky and I.J. Kopin (eds.), *Catecholamines and Stress: Recent Advances*, Elsevier, North Holland, pp 288–308

van der Kuil, J. and Korf, J. (1991) On-line monitoring of extracellular brain glucose using microdialysis and a NADPH-linked enzymatic assay. *J. Neurochem.*, **57**: 648–54

Weinshilboum, R.M. and Axelrod, J. (1971) Serum dopamine-β-hydroxylase activity. *Circulation Res.*, **28**: 307–15

Westerink, B.H.C. and DeVries, J.B. (1988) Characterization of in vivo dopamine release as determined by brain microdialysis after acute and subchronic implantations: methodological aspects. *J. Neurochem.*, **51**: 683–7

Zigmond, R.E. (1980) The long-term regulation of ganglionic tyrosine hydroxylase by preganglionic nerve activity. *Fed. Proc.*, **39**: 3003–8

Zigmond, R.E., Schwarzschild, M.A. and Rittenhouse, A.R. (1989) Acute regulation of tyrosine hydroxylase by nerve activity and neurotransmitters via phosphorylation. *Ann. Revs. Neurosci.*, **12**: 415–61

Zukowska-Grojek, Z., Konarska, M. and McCarty, R. (1988) Differential plasma catecholamine and neuropeptide Y responses to acute stress in rats. *Life Sci.*, **42**: 1615–24

11

Monoamines in response and adaptation to stress

S. Clare Stanford

Department of Pharmacology
University College London
London, UK

ISBN 0–12–663370–3

11.1. INTRODUCTION

Understanding the neurochemical basis of response and adaptation to stress is clearly regarded as the key to explaining the causes of a wide range of stress-related disorders, as well as their reversal or prevention by drug treatments. To this effect, experiments have commonly studied the influence of drugs on behavioural and neurochemical changes induced by stress. Studies of the effects of anxiolytic and antidepressant drug treatments on central monoamines predominate in this respect. The topics reviewed in this chapter reflect this emphasis.

Experiments generally look at either one or other of the two aspects of stress research: the effects of stress on monoamine neurochemistry or the effects of monoamine status on behavioural responses to stress. This chapter discusses work which has been carried out on peripheral and central monoamines which is relevant to both these questions. It is recognized that, because these questions are interdependent, attempts to establish causation are fraught with difficulties. Nevertheless, research described in this chapter is an obvious first step in this process.

Another problem when discussing research of stress is the lack of conformity of stress paradigms used in different laboratories. This is particularly the case in preclinical studies. Whereas this is an advantage if we are to assess the generality of any findings, it is problematic when trying to decide which models are the most likely to mimic the effects of stress in humans. A particular difficulty concerns use of the terms 'acute' and 'chronic' since criteria are never defined. Neverthless, differences in the impact of acute and chronic stress in humans are widely appreciated, so preclinical analogues are of obvious importance. For the purposes of this chapter, the term 'acute' is taken to represent a single bout of stress lasting minutes to hours, while 'chronic' is restricted to paradigms where stress is applied intermittently, or continuously, over a period of hours to weeks.

11.2 PERIPHERAL CATECHOLAMINES

11.2.1 EFFECTS OF AN ACUTE STRESS CHALLENGE

It was recognized over 80 years ago that secretion of adrenaline was increased by stress (Cannon and de la Paz, 1911). Cannon made it clear that psychological, as well as physical stress had this effect: he found that the increase in the levels of hormone found in 'excited blood', taken from cats within earshot of 'energetic little dogs', was abolished by adrenalectomy. Cannon's concept of the role of adrenaline in the autonomic 'flight or fight' response is widely quoted, but he also proposed a link between adrenaline secretion and the emotional impact of stress. He noted individual differences in the hormonal response to restraint and attributed this to variation in animals' emotional state (Cannon et al., 1911). Within the framework of the fear/flight and anger/fight emergency reaction, Cannon regarded increased secretion of adrenaline as a vital, 'purposive' component of coping (Cannon, 1914).

Subsequent discovery of the sympathetic neurotransmitter, noradrenaline, led to several generations of theories defining the role of peripheral monoamines in the stress response. Unlike adrenaline, which is entirely hormonal in origin, noradrenaline is secreted by both the adrenal medulla and postganglionic sympathetic neurones. This means that it is difficult to distinguish the source of noradrenaline circulating in the plasma. Nevertheless, there was a long-standing belief that secretion of these two monoamines reflected different aspects of sympatho-adrenal function and subserved different functions in stress. A common feature of these ideas was that the relative amounts of adrenaline and noradrenaline secreted during stress depended on either the type of stress, or the type of emotion it provoked. One theory, which is still widely cited, was that fear was linked with adrenaline secretion, while anger was paralleled by an increase in release of both adrenaline and noradrenaline (Ax, 1953). This theory even began to address 'trait' *versus* 'state' aspects of neuroendocrine bases of emotion: lions were thought to secrete more noradrenaline than timid animals such as rabbits and guinea-pigs! A related theory linked adrenaline secretion with 'inwardly directed anger', regarded as akin to anxiety; noradrenaline secretion was thought to predominate during 'outwardly directed anger' or aggression (Funkenstein *et al.*, 1954). Despite evidence to the contrary (von Euler and Hellner, 1952), this led to the premise that adrenaline reflected the emotional impact of stress, while noradrenaline merely effected haemodynamic adjustments to the physical demands of stress.

In line with these proposals, Frankenhaeuser showed that the increase in urinary adrenaline during psychological stress (a selective attention task) was greater than that of noradrenaline (Frankenhaeuser *et al.*, 1968). It is telling that the adrenaline response did not reflect the quality of the emotion: pleasantly arousing stimuli had similar effects (Frankenhaeuser, 1971). However, for subjects with low baseline levels of catecholamine excretion, a stress-induced increase in either adrenaline or noradrenaline was paralleled by improved performance in the task, a greater sense of task control and subjective estimates of reduced stress intensity (Figure 11.1a). Conversely, increasing subjects' actual control over the stress diminished the neurohumoral response (Figure 11.1b). Preclinical experiments have similarly shown that rats with low plasma levels of catecholamines display greater behavioural reactivity (freezing) to the stress of a novel environment than rats with higher levels of secretion (see Dantzer and Mormede, 1985). All these data are consistent with Cannon's belief that monoamine secretion was an essential element of coping, rather than an adverse consequence of stress.

Interestingly, in Frankenhaeuser's experiments, subjects with high baseline rates of catecholamine excretion performed best when they were not stressed. For low basal secretors, the opposite was the case: these subjects performed best when stressed (Figure 11.1a). To explain this, Frankenhaeuser proposed that increased adrenaline secretion attenuates the emotional impact of stress only when the stress is controllable. Mason (1968) also considered that predictability and controllability of stress were important determinants of sympatho-adrenal activation. The idea that low levels of stress enhance performance, and that severe stress impedes it, was not new (see Yerkes and Dodson, 1908). Selye also proposed a similar relationship between glucocorticoid secretion and stress resistance. However, an important aspect of Frankenhaeuser's work was the suggestion that catecholamine responses might follow the same pattern. The possibility that the relationship between stress

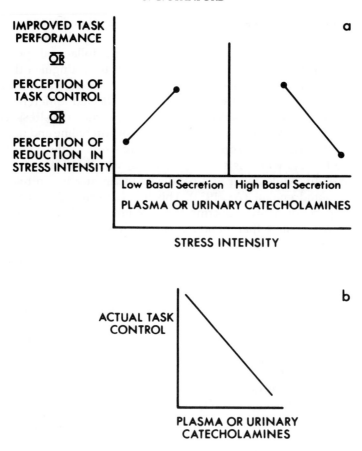

Figure 11.1 Schematic representation of the relation between secretion of peripheral catecholamines and performance in a selective attention task (based on work by M. Frankenhaeuser). (a) For low basal secretors of peripheral catecholamines, the perceived reduction in stress intensity/perceived increase in task control/improved performance in the task are all potentiated as secretion of catecholamines increases. For high basal secretors, the reverse is the case. Also, low basal secretors perform best when stressed whereas the performance of high basal secretors is reduced by stress; (b) A schematic representation of the reduction in the secretion of peripheral catecholamines as controllability of stress is increased

intensity and monoamine function in the brain is similarly described by an 'inverted U-shaped' curve has been considered only relatively recently.

One limitation of all this work was the disregard of the kinetics of peripheral catecholamine secretion and elimination. These factors, which are discussed in more detail in Chapter 10, could explain why an adrenaline response apparently predominates during psychological stress. It is now known that changes in plasma noradrenaline caused by cognitive stress can be similar to those seen during physical stress (Ward *et al.*, 1983) but the increase in noradrenaline (and adrenaline) is more reliable in arterial plasma than in venous samples (Freychuss *et al.*, 1990).

Overall, it seems that there is no clear justification for linking adrenaline or noradrenaline secretion with either specific emotions, or with specific types of stress. In this respect, secretion of peripheral monoamines complies with Selye's concept of a non-specific neuroendocrine response to stress.

Because the broad question underlying all this work was how stress affected plasma or urinary catecholamines, most experiments took only single measurements after a relatively short-lasting stress challenge. If, as is presumed, the sympatho-adrenal system has a vital role in the stress response, then changes in sympatho-adrenal function must be anticipated during adaptation to prolonged or repeated stress.

11.2.2 EFFECTS OF PROLONGED OR REPEATED STRESS

Comparatively few studies have looked at this problem, but some consistent patterns are apparent. Sustained physical exertion in humans (Winder *et al.*, 1978) or sustained restraint (Kornarska *et al.*, 1989) or cold stress in rats (Fillenz *et al.*, 1979) all cause a phasic increase in plasma catecholamines. In detail, there is a *progressive* rise in catecholamine levels which reach a peak and then decline to basal levels before the cessation of stress. Even with repeated brief swim, the peak levels of plasma noradrenaline attained on successive bouts of swimming show a progressive increase, followed by a decline. Drugs which block active reuptake of noradrenaline from the synapse increase by more than 10-fold the peak levels of plasma noradrenaline caused by swim stress, yet noradrenaline is still cleared from the plasma within 2–3 min post-swim (Fillenz *et al.*, 1979). In view of this finding, it is unlikely that the progressive rise in peak levels of noradrenaline during stress is explained by accumulation in the plasma (Benedict *et al.*, 1978; Benedict *et al.*, 1979). The transient increase in plasma noradrenaline more probably reflects a potentiation of release, therefore.

It has been suggested that the subsequent fall in noradrenaline release reflects adaptation to stress (Abercrombie and Jacobs, 1987b). However, during cold stress, the decline coincided with depletion of noradrenaline stores in sympathetic nerve terminals (Fillenz *et al.*, 1979). From this, it was inferred that release is attenuated because the supply of releasable transmitter is exhausted. This suggests that a reduction in plasma catecholamines may not necessarily reflect habituation to the stress: exhaustion of releasable noradrenaline could also produce these changes.

When the stress is repeated over a period of days to weeks, further changes are evident. A progressive reduction in levels of urinary adrenaline, but not noradrenaline, was found during successive bouts of centrifugation stress (Frankenhaeuser, 1971); the effects of more conventional forms of stress have been studied little in humans. In rats experiencing daily restraint (De Turk and Vogel, 1980), cold stress (Ostman-Smith, 1979) or footshock (Kornarska *et al.*, 1989), phasic changes in plasma catecholamines are still seen, but most studies find that peak levels in the urine and plasma are diminished considerably; the extent and rate of attenuation seems to be related to the inter-stress interval, as well as the number and intensity of stress exposures (De Boer *et al.*, 1990). Although we cannot be certain that this reflects stress adaptation, this decline is seen over a period of 3–4 weeks, which is well in excess of that needed to effect induction and transportation to the terminals of factors required to accelerate noradrenaline synthesis (Thoenen *et al.*,

1970). Also, experience of repeated cold stress does not blunt the peripheral catecholamine response to a subsequent acute swim stress (Ostman-Smith, 1979). All this evidence suggests that there is long-latency attenuation of noradrenaline release when stress is repeated.

11.2.3 PERIPHERAL CATECHOLAMINES AND ANXIETY

Early work on catecholamine excretion during stress looked at a wide range of stressful stimuli: aircraft flight, examinations, parachute jumping, athletic competition and even violent films (reviewed by Mason, 1968). Contemporary experiments use exactly these forms of stress to study catecholamine release during 'situational anxiety' (e.g. Neftel et al., 1982). Such interchangeable terminology highlights the striking similarity between the physiological response to stress and the symptoms and signs of anxiety. This similarity has led to the belief that increased release of catecholamines may be one of the causes of anxiety. This proposal deviates markedly from the concept that secretion of peripheral catecholamines is vital for coping with stress.

In the search for abnormal sympatho-adrenal function in anxious patients, there is the lingering problem that different components of this group of disorders may have different neuroendocrine correlates. Nevertheless, studies continue to monitor changes in levels of circulating catecholamines and their metabolites in anxiety. Plasma levels of the noradrenaline metabolite, 3-methoxy-4-hydroxyphenyl glycol (MHPG), have been used as an index of noradrenaline turnover, but reports of any changes, in panic at least, are mixed (cf. Lapierre, 1987, and Uhde et al., 1988). However, MHPG crosses the blood-brain barrier and is not a reliable marker for changes in peripheral sympatho-adrenal function. Increased plasma levels of noradrenaline in anxious patients (Sevy et al., 1989) and during panic (Nesse et al., 1984) have been reported, but there are many negative findings (e.g. Pohl et al., 1987). Also, in panic patients, there is no clear increase in plasma catecholamine levels in patients experiencing a panic attack when compared with pre-attack baseline secretion (reviewed by Woods and Charney, 1990). Some of these negative findings may be because there were inappropriate controls for the study or because patients were taking benzodiazepines (e.g. Schneider et al., 1987) which suppress release of catecholamines (Vogel et al., 1984). A recent study has found a correlation between reduced binding affinity of α_2-adrenoceptors during stress (examinations) and subjects' anxiety, supporting evidence that these receptors may be hyporesponsive in anxiety (Freedman et al., 1990). However, there is also evidence that α_2-adrenoceptor function is increased in panic (reviewed by Charney et al., 1990; Kahn and van Praag, 1992).

Preliminary studies of combat veterans suffering from posttraumatic stress disorder (PTSD) have found higher levels of catecholamine excretion than in patients with other psychiatric diagnoses. Also, compared with PTSD-free veterans, they showed a greater and more prolonged adrenaline response when watching films with combat cues (reviewed by McFall et al., 1989). Notwithstanding this interesting finding, the inconsistent results from studies of anxious patients contrast strikingly with the consistent increase in catecholamine secretion found in stress.

Whereas an increased sympatho-adrenal response may well be linked with situational anxiety, this might merely reflect an acute stress response. PTSD patients may also show potentiated sympatho-adrenal activity during experiences which they find peculiarly stressful. With these exceptions, the inevitable conclusion is that evidence linking anxiety in the clinical context with hyperactivity of the sympatho-adrenal system is unconvincing. The poor efficacy of β-adrenoceptor blockers in treatment of anxiety is consistent with this conclusion although, again, PTSD (Kolb, 1984) and situational anxiety (Hayes and Schultz, 1987; Tyrer, 1988) are possible exceptions.

An alternative approach has been to test whether increased levels of circulating catecholamines induce anxiety. Since there is no evidence that catecholamines cross the blood brain barrier in effective doses (but see Pohl et al., 1990), such work carries the assumption that any somatic or psychological effects of catecholamine infusion can be attributed entirely to actions in the periphery. This is relevant to the possibility, again addressed by Cannon (1927), that somatic changes resembling the stress response act as interoceptive cues and trigger an anxiety attack. It is also consistent with current 'cognitive' explanations for panic: that certain cues cause conditioned anxiety which can progress to a full-blown panic attack.

Early work suggested that adrenaline infusion made normal subjects feel apprehensive, restless and jittery; the effects of noradrenaline infusion were generally similar, but less pronounced (Frankenhaeuser and Jarpe, 1963). However, this work was designed to look at effects of monoamine infusion on performance, so there was no attempt to show how closely these feelings resembled those of an anxiety attack. There is some recent work to suggest that noradrenaline infusion precipitates panic (Pyke and Greenberg, 1986), but the selective β-agonist, isoprenaline, induces panic attacks with a much greater incidence in panic patients than in controls (Pohl et al., 1990). This suggests that dysfunction of β-adrenoceptors (or beyond) in anxious patients, rather than increased catecholamine secretion per se, may be responsible for anxiety. If panic patients are abnormally sensitive to circulating catecholamines this could exaggerate the sympatho-adrenal response to stress which could act, in turn, as an interoceptive cue for anxiety. However, there is some evidence that β-adrenoceptor function is reduced in panic patients which is inconsistent with this theory (Pohl et al., 1990).

11.2.4 NEUROCHEMICAL CHANGES IN EXPERIMENTAL MODELS OF ANXIETY

Many treatments induce anxiety in man and are used experimentally to study this condition. Administration of the α_2-antagonist, yohimbine, is a case in point. This compound increases levels of MHPG and noradrenaline in the periphery. However, it is impossible to distinguish whether this is related to anxiety, or blockade of feedback inhibition of noradrenaline release effected by α_2-adrenoceptors on sympathetic nerve terminals. This does not rule out the possibility that the anxiogenic effects of yohimbine are explained by increased release of noradrenaline (Coupland et al., 1992; see Chapter 3). However, other procedures which induce anxiety in patients, such as lactate or caffeine infusion either do not increase catecholamine secretion or produce increases which do not consistently correspond to patients' levels of anxiety.

Evaluation of psychological status in infrahuman species is obviously problematic (see Chapter 6). Validation of putative anxiogenic effects of drug treatments depends on extrapolation from their known effects in man and on reversal of their behavioural effects by anxiolytic drugs. These limitations have constrained preclinical studies of neurochemical changes concomitant with anxiety. Nevertheless, one series of experiments has investigated the effects of the benzodiazepine inverse agonist, β-carboline-3-carboxylic acid ethyl ester (β-CCE). Such compounds have effects which are opposite to benzodiazepine agonists and produce behavioural changes in monkeys which resemble the stress or fear response; from this, an anxiogenic effect is inferred. These behavioural changes are paralleled by an increase in plasma catecholamines (Ninan et al., 1982; Crawley et al., 1985). Whether such an increase is also found after treatment with other anxiogenic drugs, or under environmental conditions thought to induce anxiety (e.g. conflict or novelty) has not been examined systematically.

11.2.5 ANTIANXIETY DRUGS AND THE STRESS RESPONSE

There are many reports that the anxiolytic benzodiazepines diminish the catecholamine response to stress but have negligible effects on basal secretion (e.g. Vogel et al., 1984). In rats, both acute and repeated administration of benzodiazepines block the adrenaline increase caused by exposure to a novel cage but noradrenaline secretion is less sensitive. However, it is not at all clear that this reflects a diminished stress response since the non-benzodiazepine anxiolytic, buspirone, does not reduce catecholamine secretion caused by stress and can even potentiate it (De Boer et al., 1991).

11.2.6 SUMMARY

An increase in plasma catecholamines has long been linked with the response to stress. However, evidence linking specific neurohumoral responses with either particular forms of stress or stress-related disorders is weak. The inconclusive evidence that there is increased secretion of catecholamines in anxiety illustrates this point. There could be a dysfunction of adrenoceptors in anxious patients, but studies of receptor function have produced ambiguous results: evidence for both increased and reduced functional responses have been reported. However, these changes are not mutually inconsistent if the relation between sympatho-adrenal activation and coping with stress, or suppression of anxiety, is described by an inverse U-shaped curve (Figure 11.2). Consider first, anxiety-prone individuals showing high basal levels of sympatho-adrenal functional activity, which could be due to excessive catecholamine secretion or hyperresponsive adrenoceptor function. In this case, a further increase in sympatho-adrenal activation caused by stress could be sufficiently aversive to precipitate anxiety. The reverse could be the case for individuals with low basal activity caused by inadequate release of catecholamines or hyporesponsive adrenoceptors. At this point of the curve, a blunted

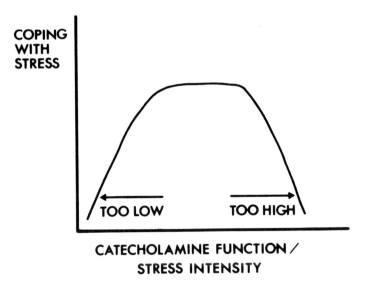

Figure 11.2 Hypothetical relation between stress intensity and anxiety in low and high basal secretors of peripheral catecholamines

catecholamine response to stress would deprive the individual of a mechanism long thought to be crucial in coping with stress.

This proposal is entirely consistent with findings that panic patients have high basal levels of adrenaline (Liebowitz *et al.*, 1985) and low exercise tolerance. It is also consistent with evidence that catecholamine infusion precipitates panic in panic patients and yet, at low doses, improves performance during psychological stress in normal subjects (Frankenhaeuser and Jarpe, 1963). In neither case need the site of dysfunction be peripheral: a disorder in the central nervous system (CNS), responding to somatic effects of catecholamine infusion is equally likely. A consequence of this proposal is that there may be no single marker for anxiety. Future work should consider whether the relationship between catecholamine responses and anxiety status is contingent on basal catecholamine secretion.

11.3 CENTRAL MONOAMINES

11.3.1 NORADRENALINE

Noradrenergic neurones within the brain are derived from two groups of nuclei in the pontine tegmentum (A4 and A6) and the lateral tegmentum (A1, A2, A5 and A7; reviewed by Holets, 1990). The locus coeruleus comprises the A6 cluster of cell bodies (the nucleus locus coeruleus, proper), the subcoeruleus and the A4 nucleus which derives from a dorsolateral extension of the locus coeruleus. In both the rat and the cat, there is evidence for some functional and regional organization of neurones within the locus coeruleus, but extensive fibre collateralization means that individual neurones can project to more than one brain region. The hippocampus, cerebral cortex, olfactory bulb and certain amygdaloid nuclei are all innervated exclusively by neurones derived from the locus coeruleus whereas afferent noradrenergic neurones of the hypothalamus derive mainly, but not exclusively, from the lateral tegmental group. In other brain regions, the locus coeruleus is the major source of neurones, but there is an additional small input from the lateral tegmentum.

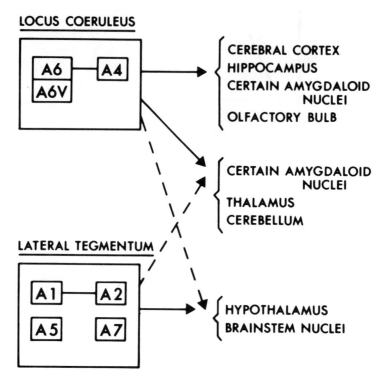

Figure 11.3 Schematic diagram showing major sources of innervation for noradrenegic neurones in different regions of the brain. The major, or exclusive, source of noradrenergic innervation is represented by a solid line; the minor source by a dotted line. A6v: nucleus subcoeruleus

11.3.1.1 Effects of an acute stress challenge

It has been known for some time that phasic, non-noxious stimuli increase the firing rate of noradrenergic neurones in the locus coeruleus (Aston-Jones and Bloom, 1981), but a specific link between firing rate and the stress response was established only recently. Using increased heart rate and plasma noradrenaline as objective criteria for distinguishing a stress response, it was found that an increased firing rate of neurones in the locus neurones in cats was induced by stress only; nonstressful, albeit arousing stimuli had no effect (Abercrombie and Jacobs, 1987a). Monkeys appear to differ in this respect (Grant *et al.*, 1988) and recent findings in studies of the cat challenge this distinction (Sara and Segal, 1991). Despite this controversy, it is generally agreed that aversive stimuli are particularly effective in activating neurones in the locus coeruleus and that such activation is a central component of the 'alarm' reaction.

Activation of neurones in the locus coeruleus results in increased release of noradrenaline in their terminal fields. This was inferred from measurements of noradrenaline turnover and accumulation of its metabolite, MHGP, both of which increase on exposure to stress (e.g. Korf et al., 1973). Although the rate of release is increased, transmitter stores are generally maintained as a result of a concomitant increase in the rate of noradrenaline synthesis. The rates of release and synthesis are coupled because depolarization of the nerve terminals also activates the enzyme tyrosine hydroxylase (reviewed by Fillenz, 1990; see Chapter 10).

In some cases the rate of noradrenaline release exceeds that of synthesis; this culminates in depletion of the stores. Although individual results depend on particular features of the stress, stress controllability is thought to be one factor determining whether or not depletion occurs. Several studies have shown reduced levels of noradrenaline in brains of rats exposed to unavoidable footshock, but not in animals given matched avoidable shock (Tsuda and Tanaka, 1985; Anisman, 1985). Predictability of stress also influences the rate of noradrenaline release. Unsignalled footshock increases MHPG accumulation more than signalled shock (Tsuda et al., 1989). The interpretation of these results was that noradrenaline release is greater during unpredictable than predictable stress, but whether unpredictable shock is more stressful than predictable shock is controversial (Abbott et al., 1984). Finally, the increase in brain MHPG in mice caused by unavoidable footshock is attenuated if animals are able to express aggression towards conspecifics (Tsuda et al., 1988). This suggests that fighting diminishes release of noradrenaline in the brain and alleviates the impact of stress. This supports arguments that contextually irrelevant behaviours such as fighting ('displacement activities') serve as essential coping mechanisms (Anisman, 1985; Dantzer, 1989; see Chapter 7).

When changes in noradrenaline levels are used as an index of the stress response, there is considerable variation from one brain region to another. The amygdala, hypothalamus and locus coeruleus show the greatest and most rapid reduction in noradrenaline levels (Weiss et al., 1982; Glavin et al., 1983) and this change is paralleled by an increase in MHPG accumulation. The reduction in the hypothalamus is consistently reported, but a recent study in mice has shown that changes in other brain regions are strongly dependent on strain (Shanks et al., 1991). Development of in vivo microdialysis has enabled more direct estimates of changes in noradrenaline release within a timeframe which closely matches that of the stress response. Contemporary studies using this technique have confirmed that stress increases noradrenaline outflow in the dialysate of the hypothalamus and amygdala (Yokoo et al. 1990; Tanaka et al., 1991); this is presumed to reflect an increase in noradrenaline release. So far, this technique has not been used to study changes in the locus coeruleus.

Although a reduction of noradrenaline levels has been found in the hippocampus and cortex after inescapable shock (Anisman et al., 1980; Hellhammer et al., 1984) this is not a common finding: the hippocampus and cerebral cortex are comparatively resistant to stress-induced depletion of transmitter stores (Irwin et al., 1986; Tsuda et al., 1989). Yet, there is no doubt that these regions are involved in the stress response. Again, there may be strain-dependent variation in any changes (Shanks *et al.*, 1991), but most reports show increased MHPG accumulation in both the hippocampus and cortex after stress exposure (Glavin *et al.*, 1983; Tanaka *et al.*, 1983). Furthermore, several recent experiments using *in vivo* microdialysis have confirmed that there is an increase in hippocampal and cortical noradrenaline outflow during stress (Rossetti *et al.*, 1990; Thatcher Britton *et al.*, 1992).

In general, the forms of stress used in such work involve a predominantly physical component, such as restraint or footshock. Although experiments studying the influence of controllability suggest that stress with a psychological component has a distinct influence on noradrenaline turnover, there have been few studies investigating the immediate effects of a purely psychological stress. The limited data available suggest that changes in noradrenaline levels are found only in the amygdala (e.g. on hearing other rats being shocked; Iimori *et al.*, 1982). Although *in vivo* microdialysis has not yet been used to investigate the effects of purely psychological stress on noradrenaline release, a clear increase in noradrenaline outflow has been found in the hippocampus after simply handling animals for a limited period (Kalen *et al.*, 1989; Kokaia *et al.*, 1989), or after noise stress (Thatcher Britton *et al.*, 1992).

There has been more interest in the effects of conditioned psychological stress on noradrenaline turnover. In these paradigms, animals first experience an aversive stimulus, usually footshock, in a distinctive environment. Neurochemical changes are then investigated on a subsequent exposure of animals to the cues which they associate with stress (the 'conditioned emotional response'). Early work suggested that whole brain MHPG accumulation was increased under these conditions (Cassens *et al.*, 1980), indicating activation of central noradrenergic neurones. At first, single unit recordings from the locus coeruleus in the cat found that activation of these neurones was peculiar to the response to conditioned aversive stimuli; responses to conditioned food reward were without effect (Rasmussen and Jacobs, 1986). However, a more recent study failed to find such a distinction: instead, locus firing was increased by *changes* in the relation between stimulus and reinforcement (Sara and Segal, 1991). The major inference from this work was that activation of noradrenergic neurones is more a function of the salience or cognitive significance of the stimulus, rather than its aversive nature, and is involved in selective attention rather than stress responses *per se*.

In addition to the problem of defining the role of changes in the activity of noradrenergic neurones in the brain, there is the complication that neurochemical responses to such conditioned stimuli are more apparent in the hypothalamus, amygdala and locus coeruleus than in the hippocampus or cortex (Tsuda *et al.*, 1986; Yokoo *et al.*, 1990). This is interesting because there is no obvious anatomical explanation for the distinctive effect of acute stress in the hypothalamus, amygdala and locus coeruleus by comparison with that in the hippocampus and cortex. Although the innervation of the hypothalamus is exceptional, arising predominantly from the lateral tegmentum, all other brain regions are innervated mainly by neurones from the locus coeruleus. In both the rat (Loughlin *et al.*, 1986) and the cat (Nakazato, 1987), there is a crude topographical organization of neurones in the locus coeruleus, but individual neurones show extensive collaterization and diffuse projections. However, the nature and time course of the response to stress of brainstem noradrenergic nuclei is not the same for all nuclei (Lachuer *et al.*, 1991). Also, the rate of expression of mRNA for the enzyme tyrosine hydroxylase varies from one nucleus to another (Hartman *et al.*, 1992). How such regional variation in functional changes affecting noradrenergic neurones influences noradrenergic transmission in the terminal fields remains to be seen. However, it seems possible that regional differences in functional attributions of noradrenergic neurones in the brain are more sophisticated than believed hitherto.

Another important variable determining whether or not transmitter stores are maintained is the extent to which released transmitter is recycled after neuronal reuptake from the synapse: recent studies have shown that the amount of neuronal uptake of noradrenaline differs markedly from one brain region to another (Thomas *et al.*, 1992). Regional differences in the influences of other neurotransmitters, which are released from adjacent nerve terminals and activate heteroceptors on noradrenergic nerve terminals, are also likely: such axo-axonal interactions modulate impulse-evoked release of noradrenaline. Another possibility is that mechanisms regulating noradrenaline synthesis in the hippocampus and cortex respond more effectively to an increase in demand for releasable store. This could explain why noradrenaline levels are unchanged despite an increase in MHPG accumulation during stress. Finally, there is considerable variation in the detection of peptidergic cotransmitters within noradrenergic neurones of the locus coeruleus (Holets, 1990); differential distribution of cotransmitters could have a profound effect on neuronal function. Whatever the explanation, it follows that stress-induced neurochemical changes in one brain region cannot be used to predict changes in others.

11.3.1.2 Effects of prolonged or repeated stress

Traditionally, the main criterion for a stress response is an increase in secretion of stress hormones, especially catecholamines and glucocorticoids. On this basis, a decline in hormonal secretion when stress is repeated or prolonged is commonly interpreted as indicating stress adaptation. This has led to the general view that attenuation of *any* physiological response reflects adaptation to stress; deciding whether or not this is the case is a major difficulty.

The interpretation of changes in the firing rate of neurones in the locus coeruleus during prolonged or repeated stress is a case in point. In one study in cats, there was a progressive attenuation in the activation of locus neurones during successive bursts of white noise or continuous restraint when these stressors were administered over a period of hours (Abercrombie and Jacobs, 1987b). It was inferred that this change was linked with adaptation to stress because it was paralleled by a decline in markers for peripheral autonomic activation. On the other hand, results from another study in anaesthetized rats suggest the opposite: there was a prolonged increase in basal firing rate in animals which had experienced stress repeated at daily intervals (Pavcovich *et al.*, 1990). There are many methodological differences between these two studies, but an important distinction is the much longer inter-stress interval in the latter. So far, changes in tonic firing frequency during a stress challenge have not been investigated when the stress is repeatedly administered at long intervals. Consequently, we cannot rule out the possibility that exhaustion of factors activating neurones in the locus coeruleus explains the changes seen by Abercrombie and Jacobs (1987b) and we cannot be certain that stress adaptation is linked with a diminished neuronal activation, as they have proposed. This guarded qualification is supported by evidence that release of noradrenaline in the hippocampus is not maintained during noise stress despite sustained release of adrenocorticotropic hormone (ACTH) and corticosterone (Thatcher Britton *et al.*, 1992). Certainly, a decline in the response of central noradrenergic neurones is not consistent with extensive evidence that noradrenergic transmission is increased when stress is repeated.

Many studies suggest that re-exposure to a stress potentiates noradrenaline release. For instance, a footshock which does not normally deplete noradrenaline stores, did reduce noradrenaline levels in the hypothalamus of mice which had experienced a single session of inescapable footshock shock two weeks earlier (Irwin *et al.*, 1986). However, the effects of repeated stress on noradrenaline levels depend on the nature of the stress as well as the brain region studied: if animals first receive escapable shock, then amine depletion in the hippocampus and cortex is prevented on a subsequent exposure. Such 'immunization' by previous experience of escapable shock does not prevent amine depletion in the hypothalamus, however (Anisman *et al.*, 1980). Further evidence that noradrenaline release can be influenced by experience of stress comes from work showing that an acute stress challenge reduced noradrenaline levels in the hippocampus and cortex of chronically stressed rats, but not in naive animals. In the same study, acute stress reduced noradrenaline levels in the hypothalamus of both naive and chronically stressed animals, but the decrease was much greater in the latter group (Adell *et al.*, 1988).

Despite these findings, it is widely recognized that a reduction in tissue concentration of noradrenaline is an unreliable indicator of an increase in noradrenaline release. More direct supporting evidence for potentiation of noradrenaline release when stress is repeated comes from studies of MHPG accumulation *ex vivo*: MHPG accumulation is greater in animals that have had previous experience of repeated stress (Anisman *et al.*, 1987). Finally, measurement of noradrenaline outflow by microdialysis *in vivo* has shown that a brief period of tailshock produces a greater increase in hippocampal noradrenaline outflow in stress-experienced animals when compared with naive controls (Nisenbaum *et al.*, 1991).

Although release may be increased when a stress challenge is repeated, basal levels of noradrenaline in chronically stressed animals are generally not depleted and can even increase. The extent of the increase is time dependent and transmitter levels recover progressively after cessation of the stress (Anisman *et al.*, 1987). One study found such an increase only in the cortex and hippocampus (Adell *et al.*, 1988), but others include the hypothalamus (Anisman *et al.*, 1987). It is most probable that this increase reflects an increase in the rate of synthesis of noradrenaline.

There is extensive evidence that synthesis of noradrenaline is increased when stress is repeated. The possibility that this may bear on animals' emotionality comes from studies of levels of tyrosine hydroxylase in the locus coeruleus in two inbred strains of rats which differ in their behavioural responses to stress: after repeated stress, tyrosine hydroxylase levels were higher in the Maudsley non-reactive strain than in the reactive strain (Blizard, 1988). No such difference in hypothalamic enzyme activity was found in the same rats. This is interesting because, unlike other brain regions, hypothalamic afferents arise mainly in the lateral tegmental area.

Recent studies have used tissue outflow of both 3,4-dihydroxyphenylacetic acid (DOPAC) and accumulation of dihydroxyphenylalanine (DOPA), after inhibition of the enzyme DOPA decarboxylase, as indices of the rate of noradrenaline synthesis (see Chapter 10). During an acute stress challenge both measures were greater in the hippocampus of animals that had previous experience of prolonged stress. Although this suggests that there is activation of tyrosine hydroxylase when animals are stressed, basal synthesis rate was the same in chronically stressed and naive animals. It seems that the neurones are primed to respond to stressful stimuli, but noradrenaline synthesis is unchanged under normal conditions (Nisenbaum et al., 1991; Nisenbaum and Abercrombie, 1992).

It is notable that the forms of stress used to study changes in noradrenaline synthesis and release are relatively severe: prolonged cold stress, repeated restraint and inescapable footshock. Results from such studies all suggest that noradrenaline release and synthesis are increased by stress, even when this is repeated intermittently over periods of days. Generally, the rate of synthesis is closely coupled with release but, under certain circumstances, release can outpace synthesis and, in stress-experienced animals, synthesis can outpace release. Overall, such changes are difficult to reconcile with the concept that activitation of noradrenergic neurones and noradrenaline release are reduced during adaptation to stress. In contrast, the effects of non-noxious stress have been little studied despite the finding that even repeated handling caused a prolonged increase in the activity of tyrosine hydroxylase (Graham-Jones et al., 1983; Stanford et al., 1984). Future studies should address the neglected question of whether enhanced noradrenaline synthesis and release are also apparent after repeated psychological stress.

11.3.1.3 Receptors

Explanation of the effects of stress on central noradrenergic transmission should integrate findings from studies of both transmitter turnover and receptor function. This is difficult because, hitherto, these have usually been investigated in quite separate lines of research. Also, the striking feature of studies of adrenoceptors, unlike those of noradrenaline turnover, is the lack of consistent changes during acute stress. An increase in α_2-adrenoceptors in both cortical and subcortical regions is commonly, but not invariably reported, for instance (Stanford, 1990; U'Prichard and Kvetnansky, 1980).

One interesting caveat is the possibility of rapid changes in β-adrenoceptors during an acute stress (Stanford, 1990). Using radioligand binding to evaluate changes in receptor number and binding affinity, most studies have found no change in β-adrenoceptor binding after an acute stress (reviewed by Stanford, 1990). However, an early report did suggest a reduction in radioligand binding in rat cerebellum in which β-adrenoceptors comprise only the β_2-subtype (U'Prichard and Kvetnansky, 1980). More recent studies suggest that, in the rat, there may also be a rapid reduction in the number of β_2-adrenoceptors on lymphocyte membranes during an acute stress: both physical restraint and psychological stress (novel environment) were effective stimuli for this change (De Blasi *et al.*, 1987). The reduction was explained by sequestration of receptors from cell membranes to an intracellular site inaccessible to lipophilic radioligands (i.e. unable to cross the cell membrane). However, the opposite result has been found in human lymphocytes in which brief stress transiently *increased* the density of β_2-adrenoceptors (Brodde *et al.*, 1984).

One reason why it is difficult to reconcile these conflicting results from studies of rat and human lymphocytes is that it is now thought that binding to lymphocyte receptors is influenced by stress-induced changes in the profile of cell types in the circulation. This change is one of many links between stress and an altered immune response (Khansari *et al.*, 1990). Nevertheless, these data highlight the neglected possibility that stress affects β_1- and β_2-adrenoceptors in different ways. This is both feasible and potentially important because β_1- and β_2-adrenoceptors have different agonist profiles, different distributions within the CNS and there is accumulating evidence that their regulation involves different mechanisms. Since selective ligands for these two receptor subtypes are now available, the possibility of distinct stress-induced changes in these receptors should be addressed, particularly in the brain where dynamic changes in cell composition will not be a confounding problem.

There is no consensus on the effects of repeated stress on any of the adrenoceptor subtypes in subcortical brain regions. Reported changes in α-adrenoceptors in the cortex are also inconsistent. This is certainly a reflection of the variation in stress protocols adopted by different laboratories. Nevertheless, this lack of agreement contrasts strongly with the reports on effects of repeated or prolonged stress on cortical β-adrenoceptors. Many different forms of repeated stress reduce binding of β-adrenoceptor ligands (reviewed by Stanford, 1990); this change has been attributed to a reduction in the number of receptors. Even when binding is not reduced, as after footshock (Stone, 1981), there is evidence that the receptors are functionally desensitized. This is deduced from a reduction in catecholamine-induced production of the β-adrenoceptor second messenger, adenosine $3',5'$cyclic monophosphate (cAMP) *ex vivo* (Stone *et al.*, 1979; see Chapter 10). In this respect, the effect of repeated stress resembles that of antidepressant treatments.

The conclusion that repeated stress causes downregulation of cortical β-adrenoceptors is derived from experiments where the same stress is administered repeatedly. Experiments where animals experience a succession of different forms of stress over a period of days have produced conflicting results. One study found an increase in the number of cortical β-adrenoceptors (Molina *et al.*, 1990), whereas a second found no change in the cortex, but a reduction in the amygdala (Areso and Frazer, 1991). A major methodological difference between these two studies was the combination of ligands used to characterize β-adrenoceptor binding. Recent work suggests that certain combinations of ligands, such as those used by Molina *et al.* (1990), are not selective for β-adrenoceptors (Riva and Creese, 1989). Since the majority of reports suggesting that β-adrenoceptors are reduced by repeated stress have used such ligands, we cannot rule out the possibility that some of the changes hitherto interpreted as β-adrenoceptor downregulation may involve non-β-receptors. This is supported by findings that repeated saline injection, which reduced β-adrenoceptor binding in an early study using [^{3}H]dihydroalprenolol

(Stanford *et al.*, 1984), did not change β-binding when this was measured using the more selective β-adrenoceptor ligand, [^{3}H]CGP 12177 (Davis and Stanford, unpublished). However, this note of caution must be tempered by the reports that repeated stress reduces isoprenaline-induced production of the β-adrenoceptor second messenger, cAMP.

11.3.1.4 *Noradrenaline and behavioural responses to stress*

Opinions are deeply divided about the role of noradrenaline in stress. Some see an increase in noradrenaline release as an important component of coping, while others believe this aggravates adverse effects of stress. These same difficulties inevitably confound our understanding of the role of noradrenaline in anxiety and depression. This is at least partly because, in preclinical experiments, models for these disorders are all based on changes in animals' response to stressful stimuli.

Increased activity of central noradrenergic neurones has long been thought to be a causal factor in anxiety (Redmond and Huang, 1979). This view stems from the behavioural effects of stimulation of the locus coeruleus in monkeys; the behaviours induced by this procedure have some features in common with those induced by threatening stimuli. The possibility that anxiety resembles the stress response, both causally as well as symptomatically, has been reinforced by evidence that presumed anxiogenic drugs increase measures of noradrenaline release in the brain, and that this increase is prevented by anxiolytic benzodiazepines (Tanaka *et al.*, 1990). However, the pharmacological selectivity of compounds used in such work, as well as the interpretation and reliability of the behavioural changes they induce, have all been strongly criticized (Soderpalm and Engel, 1990; Johnston, 1991).

Evidence that anxiolytic benzodiazepines attenuate sensory-evoked firing of neurones in the locus coeruleus (Simson and Weiss, 1989) as well as the release of noradrenaline (Ida *et al.*, 1985; Fung and Fillenz, 1985) supports a role for noradrenaline in anxiety (Figure 11.4). Collectively, this work would suggest that excessive noradrenaline release may be responsible for this disorder. However, the novel non-benzodiazepine anxiolytic, buspirone and generically related compounds, have quite different effects: these drugs increase the firing rate of neurones in the locus coeruleus. The effects of buspirone on measures of terminal release of noradrenaline are unclear: there are reports of both an increase (Sanghera *et al.*, 1990) and a reduction (Broderick and Piercey, 1991).

Changes in central noradrenergic transmission caused by stress have also been linked with depression. Two forms of stress, inescapable shock and forced swimming, have been used extensively in this work. Unpredictable (Murua *et al.*, 1991) and/or inescapable (Maier and Seligman, 1976) stress produces a characteristic behavioural change: 'learned helplessness'. This culminates in the failure of animals to adopt behavioural strategies which enable escape on subsequent exposure to escapable stress. Functional interpretation of learned helplessness is contentious (see Chapter 9) and it is clear that many neurotransmitters are involved in this syndrome (Anisman and Zacharko, 1991). Notwithstanding, Anisman suggests that learned helplessness is a consequence of reduced noradrenergic transmission in the brain: he proposes that severe stress, such as inescapable shock, depletes noradrenaline stores and increases animals' vulnerability to stress by depriving them of an essential coping mechanism (Anisman, 1985). His proposal is supported by evidence that procedures which prevent depletion of noradrenaline also prevent development of learned helplessness, and *vice versa* (Anisman *et al.*, 1980).

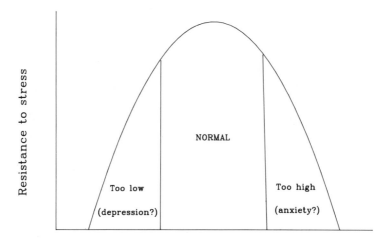

Noradrenergic transmission during stress

Figure 11.4 Schematic representation of the hypothetical relationship between central noradrenergic neurotransmission during stress and individuals' ability to cope with stress

Findings that learned helplessness is prevented by a wide range of antidepressant treatments (Martin *et al.*, 1987) have reinforced the view that this represents an infrahuman analogue of depression. In this respect, Anisman's scheme is consistent with Schildkraut's theory that inadequate noradrenergic transmission in the brain is a causal factor in this disorder (Schildkraut, 1973; see Figure 11.4). However, more recent work has shown that both the behavioural effects of inescapable stress and the efficacy of antidepressants in preventing learned helplessness varies with strain of animal (Anisman and Zacharko, 1991). As a result of this and other factors, Anisman is sceptical about the validity of learned helplessness as a model of depression, but he concedes that cognitive changes arising from the neurochemical effects of inescapable stress are likely to be aggravating factors in depression.

A related line of research has involved measurement of immobility during a forced swim which is widely used to screen putative antidepressant drugs. Although these drugs reliably reduce immobility in the forced swim test (Porsolt *et al.*, 1979), it must be borne in mind that so do many other compounds which have no known antidepressant actions in humans. These experiments have focused on the depletion of noradrenaline in the locus coeruleus, which is a prominent effect of inescapable stress, and related this depletion to immobility in the swim test. The noradrenaline depletion after forced swim has been interpreted as a loss of noradrenaline from axon collaterals which effect feedback inhibition of neuronal firing *via* α_2-adrenoceptors in the locus coeruleus (Cedarbaum and Aghajanian, 1978; Weiss *et al.*, 1986). Weiss has proposed that depletion of noradrenaline by stress releases the locus coeruleus from this feedback inhibition. He supports his theory by showing that local infusion of the α_2-adrenoceptor agonist, clonidine, suppressed immobility in the swim test whereas the antagonist, yohimbine, had the opposite effect (Weiss *et al.*, 1986). Later experiments showed that clonidine depressed the firing rate of locus neurones induced by noxious stimuli while yohimbine increased it; neither drug affected basal firing rate (Simson and Weiss, 1987). Moreover, this increase in firing rate caused by yohimbine was greater in animals that had previous experience of inescapable shock and was paralleled by increased immobility in the swim test (Simson and Weiss, 1988).

Whether or not immobility is a preclinical analogue of depression, Weiss's scheme suggests that *increased* activation of locus neurones is causally related to behavioural changes induced by inescapable stress. Ostensibly, this is the opposite position from that adopted by Anisman. However, the crucial and unresolved question arising from Weiss's work is how noradrenaline depletion in the locus coeruleus affects transmitter release in the terminal fields? The importance of this question is underlined by the common experience that systemic administration of yohimbine in humans is noted for its anxiogenic, rather than depressant, effects (Charney *et al.*, 1990). However, the theory resembles Anisman's concept in that it suggests that inadequate noradrenaline release during stress, albeit in the locus coeruleus only, is linked with behavioural and emotional components of depression.

Another approach has been to consider the behavioural effects of adrenoceptor modulation in response and adaptation to stress. Noting the consistent desensitization of β-adrenoceptors by repeated stress, Stone has suggested that this is a consequence of a stress-induced increase in release of noradrenaline and is an essential component of stress adaptation (Stone, 1979; Platt and Stone, 1982). He has also commented on the similarity in the effects of repeated stress and antidepressant treatments in that they both commonly cause downregulation of cortical β-adrenoceptors (Stanford and Nutt, 1982; Stanford *et al.*, 1983). Stone has proposed that this effect of antidepressants on β-adrenoceptors mimics adaptation to stress and accounts for their therapeutic effects. Supporting evidence comes from experiments showing that only drug regimes which reduce the density of β-adrenoceptors in the cortex reduce immobility in the swim test (Kitada *et al.*, 1986).

One prediction arising from Stone's hypothesis is that acute administration of a β-adrenoceptor antagonist should also reduce immobility in the swim test. However, while such compounds do potentiate the reduction of immobility by antidepressants, they seem to have no effect when given alone (Kitada *et al.*, 1983). Another problem is that several antidepressants act selectively on 5-hydroxytryptaminergic neurones and have no apparent effects on β-adrenoceptor binding or function.

A limitation of all this work is that conclusions are based on parallel changes in neurochemistry and behaviour; the extent to which causal links can be inferred from such work, and how such inferences can be validated further has been discussed elsewhere (Salmon and Stanford, 1992; Stanford and Salmon, 1992) (see Section 11.4). Another relevant factor is the extent to which the procedures used to assess behaviour may themselves affect neurochemical measurements. For instance, stress or antidepressant treatments which do not affect β-adrenoceptors when given alone, cause marked and rapid receptor downregulation when given in combination (Duncan *et al.*, 1985). Even standard procedures such as handling of animals can have a marked, non-additive influence on neurochemical and behavioural changes induced by a subsequent stress (Davis *et al.*, 1992b). Yet, the possibility of such non-additive interactions are rarely taken into account and are only exposed when experiments routinely include 'naive' animals which have not experienced any of the pretreatments being investigated. It is a notable omission that, while experiments usually include either procedural (e.g. handled or vehicle-injected) or naive controls, they rarely include both. Until these factors are examined systematically, it is unlikely that we will reconcile the disparate views on the role of noradrenaline in response and adaptation to stress or stress-related psychopathology.

11.3.1.5 Summary

From our extensive knowledge of the effects of stress on neurochemical control of noradrenergic transmission, a picture of the role of noradrenaline in response and adaptation to stress is beginning to emerge. Noradrenaline release in the brain is increased during stress and, overall, evidence suggests that adaptation to stress depends on neurochemical adjustments which enable sustained or enhanced release of noradrenaline when stress is repeated or prolonged. Changes affecting noradrenergic receptors are also well documented, although a downregulation of β-adrenoceptors in the cerebral cortex is the only consistent long-term change. Despite this wealth of evidence, theories which attempt to integrate these findings place different interpretations on the evidence and it is still not certain that increased noradrenaline release diminishes the adverse impact of stress.

11.3.2 DOPAMINE

Research on the role of dopamine in response and adaptation to stress is confounded by the complex topographical organization of neurones releasing this transmitter. Dopaminergic tracts in the brain which are thought to be directly relevant to behaviour and emotion comprise three major divisions, all derived from the ventral tegmental area (A10 nucleus) and the substantia nigra (A8 and A9 nuclei; Figure 11.5). The nigrostriatal pathway projects to the neostriatum; the mesocortical system projects to the limbic cortex (prefrontal, cingulate and entorhinal cortical zones), while the mesolimbic system projects to subcortical limbic areas such as the amygdala, olfactory tubercle and nucleus accumbens. However, the sources of neurones innervating different forebrain areas show considerable overlap. Also, their fibres show collateralization and can innervate different brain areas (Fallon and Moore, 1978). Only the frontal and entorhinal cortices receive afferents from a single zone of the ventral tegmental area. Nevertheless, the prefrontal cortex, the nucleus accumbens or the olfactory tubercle, and the striatum are the brain regions studied most frequently as representatives of the three major limbs of the CNS dopaminergic network (Figure 11.5).

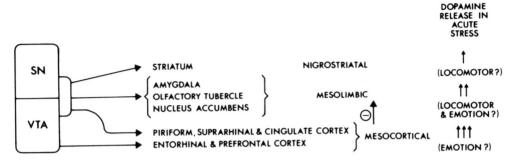

Figure 11.5 The distribution of dopaminergic neurones and long-fibre dopaminergic tracts in the central nervous system. The plan illustrates the indirect inhibition of mesolimbic neurones by mesocortical dopaminergic neurones in the cortex. Also shown is the extent to which dopamine release in different limbs of the forebrain dopaminergic system is increased by acute stress. SN: substantia nigra; VTA: ventral tegmental area

11.3.2.1 *Effects of an acute stress challenge*

Early studies of changes in dopamine turnover suggested that only mesocortical dopaminergic neurones, innervating the prefrontal cortex, were activated by stress (Thierry *et al.*, 1976). This view was subsequently revised to the extent that dopaminergic neurones innervating the prefrontal cortex were thought to be most sensitive to stress, responding to 'moderate' stimuli such as footshock. Neurones supplying other brain regions were generally thought to respond to more 'severe' (e.g. restraint) or prolonged stress only (Bannon and Roth, 1983; Kramarcy *et al.*, 1984).

However, recent methodological refinements have increased the sensitivity of measurements of dopamine release in the brain (see Chapter 10). The use of such techniques has exposed release of dopamine in the mesolimbic system, even in response to moderate stress (Deutch *et al.*, 1985; Claustre *et al.*, 1986). Experiments using *in vivo* microdialysis have provided the most direct and convincing evidence: in both the prefrontal cortex and the nucleus accumbens, increased release of dopamine paralleled the onset and cessation of intermittent tailshock over a period of 30 min (Abercrombie *et al.*, 1989). However, outflow of the dopamine metabolites, dihydroxyphenylacetic acid (DOPAC) and homovanillic acid (HVA), is more variable, a finding which lends support to scepticism about the validity of using metabolite concentrations as indices of dopamine release (Da Silva, 1987).

How long increased release of dopamine is maintained when stress is sustained has been investigated only recently. *In vivo* microdialysis of rat prefrontal cortex or nucleus accumbens during prolonged restraint has revealed that, on reaching a peak after 40–60 min, the concentration of dopamine in the dialysate progressively returns to, or even falls below, baseline levels (Imperato *et al.*, 1991; Puglisi-Allegra *et al.*, 1991a). Reasons for this decline, which awaits confirmation, are currently unclear. It was not paralleled by a reduction in tissue levels of dopamine and so was probably not attributable to exhaustion of releasable transmitter. Habituation to the stress was also considered to be an unlikely explanation because corticosterone release was maintained throughout (Imperato *et al.*, 1991; Puglisi-Allegra *et al.*, 1991a). Possible explanations include: a failure in any of the processes essential for dopamine release; an increase in neuronal or non-neuronal uptake of dopamine from the extracellular space; or a change in the influence on dopamine release of other transmitters or co-transmitters acting at dopaminergic nuclei or terminal heteroreceptors.

Whether or not striatal neurones are activated during stress is another question which has not been fully resolved. Early measures of dopamine turnover generally failed to find any stress-induced changes in the striatum (Kramarcy *et al.*, 1984), but there are exceptions (Dunn and File, 1983). It is possible that the poor sensitivity of early techniques has contributed to these discrepant results. Recent investigations using *in vivo* microdialysis have shown increased outflow of dopamine in striatal dialysate after tailpinch (Boutelle *et al.*, 1990), footshock (Abercrombie *et al.*, 1989; Keefe *et al.*, 1990) and conditioned suppression of motility (Katoh *et al.*, 1991), but not restraint (Imperato *et al.*, 1991). The time course of dopamine release in the striatum is also unclear: one study found a rapid increase after the *onset* of stress (Keefe *et al.*, 1990), while another found a significant increase only after the *cessation* of stress (Abercrombie *et al.*, 1989).

There are several theories to explain why the response of mesocortical dopaminergic neurones to stress is more sensitive than that of mesolimbic or nigrostriatal neurones. One theory rests on findings that mesocortical neurones inhibit activation of subcortical dopaminergic systems. Consequently, a decline in dopaminergic transmission in the cortex, as stress increases in duration or severity, would release mesolimbic neurones from this inhibition (Figure 11.5). There is some supporting evidence for this proposal: a selective lesion of the dopaminergic input to the prefrontal cortex increased footshock-induced DOPAC accumulation in the nucleus accumbens (Deutch *et al.*, 1990). So far, there is no evidence that such disinhibition recruits nigrostriatal neurones during stress (Deutch *et al.*, 1990).

An alternative scheme is that the response of different limbs of the dopaminergic system is related to animals' response to particular forms of stress, rather than to stress severity. It has been proposed that activation of mesocortical neurones determines (or is determined by) animals' ('negative') emotional status, while nigrostriatal neurones modulate the motor response during stress. Mesolimbic neurones are thought to be involved in both these aspects of the stress response (Bertolucci-D'Angio *et al.*, 1990). Such a functional division of different limbs could explain why forced locomotion increases dopamine metabolism in the striatum, a brain region predominantly linked with movement, but not the prefrontal cortex (Bertolucci-D'Angio *et al.*, 1990). It is also consistent with findings that conditioned fear (Herman *et al.*, 1982; Deutch *et al.*, 1985; Ida *et al.*, 1989) or watching conspecifics experiencing stress (Kaneyuki *et al.*, 1991), which evoke emotional reactions to stress, increase dopamine metabolism in the prefrontal cortex but not in other brain regions. Topographically selective influences of different groups of afferent neurones innervating dopaminergic neurones in the ventral tegmentum could effect such regional variation in the dopaminergic response to these different forms of stress (Deutch and Roth, 1990). Yet another proposal is that activation of nigrostriatal neurones is involved in governing selective attention to phasic sensory stimuli, rather than responding to stressful stimuli *per se* (Keller *et al.*, 1983; Strecker and Jacobs, 1985).

Despite the increased release of dopamine during stress, tissue levels of this transmitter are generally unchanged (Tissari *et al.*, 1979), although there are occasional reports of a reduction (Lavielle *et al.*, 1978) or even an increase (Reinhard *et al.*, 1982). As for noradrenaline, tissue stores are most probably maintained by an increase in the rate of synthesis of dopamine. Indirect measures of dopamine synthesis during stress suggest that this is increased by activation of tyrosine hydroxylase, the rate-limiting enzyme in dopamine synthesis (Iuvone and Dunn, 1986). However, a clear understanding of factors regulating dopaminergic tyrosine hydroxylase during stress is elusive, not least because of difficulties in distinguishing enzyme in dopaminergic neurones from that in noradrenergic neurones (Tissari *et al.*, 1979). For this reason, most experiments study tyrosine hydroxylase derived from the striatum, where there are no noradrenergic terminals. Yet, such work has still not produced a clear picture (reviewed by Fillenz, 1990; see Chapter 10). A complicating factor is that it is likely that factors regulating enzyme activity differ in different limbs of the dopaminergic system. In this respect, evidence that most mesocortical neurones, unlike mesolimbic or nigrostriatal neurones, lack both terminal and somatic autoreceptors must be taken into consideration. The impact of other neurotransmitters, operating at the level of the ventral tegmentum (Deutch and Roth, 1990) or through heteroceptors on dopaminergic terminals, will be another important variable.

Finally, changes in dopamine turnover may be influenced by genotype: recent studies in mice have exposed strain differences in the effects of stress on dopamine turnover, but these differences seem to be restricted to the frontal cortex and ventral tegmental area (Cabib *et al.*, 1988; Shanks *et al.*, 1991). Also, there is a possibility that stress-induced changes in dopamine turnover show lateral bias (Carlson *et al.*, 1991). These intriguing results suggest interesting prospects for future research.

11.3.2.2 *Effects of prolonged or repeated stress*

Remarkably little is known about neurochemical changes in dopaminergic neurones on exposure to repeated or prolonged stress despite obvious and long-lasting changes in dopaminergic function. Many studies have shown that behavioural responses to psychostimulants are potentiated after repeated stress, and *vice versa* (reviewed by Kalivas and Stewart, 1991), but the mechanisms underlying these changes are not understood.

There is conflicting evidence on changes in dopamine release when the stress is repeated intermittently over a period of days. Studies using *in vivo* microdialysis have shown that dopamine release in the nucleus accumbens declines progressively with successive bouts of restraint (Kramarcy *et al.*, 1984; Imperato *et al.*, 1992). This was not due to a failure in dopaminergic transmission because, after the last period of restraint, there was an immediate rebound increase in dopamine release when animals were released (Imperato *et al.*, 1992). This finding provoked the suggestion that phasic changes in dopamine release reflect different components of coping with stress and might be linked with 'emotional arousal' rather than responses to aversive stimuli *per se*. In contrast, a recent *in vivo* voltammetric study found that release in the nucleus accumbens increased with each successive restraint. This was not found with tailpinch (Doherty and Gratton, 1992). So far, there is no obvious explanation for these disparate results.

11.3.2.3 Receptors

Surprisingly few studies have looked at stress-induced modulation of dopamine receptor binding. This is certainly related to the poor selectivity of dopamine receptor ligands which commonly bind to receptors for 5-hydroxytryptamine (5-HT) as well. Difficulties in detecting dopamine D_2-receptor binding in the prefrontal cortex is another factor. Yet modulation of dopamine receptors could be an important determinant of the response to stress.

In one experiment, where radioligand binding to 5-HT_2 receptors was ruled out, even a single session of gridshock or tailshock induced a delayed increase in dopaminergic D_2-receptor density in rat prefrontal cortex. However, the functional consequence of this change is open to question because there was also a reduction in binding affinity. Also, these changes were dependent on the assay conditions and were not found in other brain regions (MacLennan *et al.*, 1989).

After repeated restraint in rats, dopamine D_1-receptor binding was unchanged throughout the brain but dopamine D_2-receptor density was increased in the caudate nucleus and nucleus accumbens as well as in the prefrontal cortex (Friedhoff *et al.*, 1986). Again increased dopamine D_2-receptor density in the prefrontal cortex was paralleled by a reduction in binding affinity. A recent study of binding *in vivo* in mice after repeated restraint has produced quite different findings. Here, dopamine D_1-receptor binding was unchanged in the nucleus accumbens but increased in the striatum. Also dopamine D_2-receptor binding was reduced in the nucleus accumbens and striatum (Puglisi-Allegra *et al.*, 1991b). Since any of the major methodological differences between the *in vitro* study in rats and *in vivo* study in mice could be responsible for these conflicting results, it is clear that this important aspect of neurochemical adaptation to stress needs further investigation.

11.3.2.4 Dopamine and behavioural responses to stress

One approach to unravelling the role of dopamine in stress has been to compare changes in dopamine turnover in strains of animals distinguished by their different behavioural reactivity to stress. In one such study, footshock-induced DOPAC accumulation in the frontal cortex of the stress-reactive strain of mice, BALB/c, did not differ from that in the less reactive C57 BL/6 strain (Herve *et al.*, 1979). However, a later study using the same strains of mice found a greater increase in dopamine metabolism in the prefrontal cortex of 'emotional' BALB/c mice when they were exposed to a novel environment (Tassin *et al.*, 1980). In this latter study, locomotor activity also differed between the two strains, but it remains to be seen whether dopamine turnover is related to either the emotional response to stress or the locomotor activity in either strain.

Changes in dopamine metabolism have also been compared in freely-moving Wistar Roman High-Avoidance (RHA) and Low-Avoidance (LHA) rat strains. Differences in the behaviour of these rats consistently suggest that the RHA strain has greater resistance to stress. Using *in vivo* voltammetry to measure DOPAC accumulation in the prefrontal cortex, both psychological stress (exposure to a Y-maze) and physical stress (immobility) increased dopamine metabolism in the RHA strain, but not the LHA strain (Bertolucci-D'Angio *et al.*, 1990). This suggests that rats with the greatest resistance to stress also show the greater increase in mesocortical dopamine turnover, and that dopamine release in this brain region could be another mechanism for coping with stress. So far, the disparity between conclusions from experiments in the mouse and rat cannot be reconciled, but this is mainly because of the dearth of information on changes in dopaminergic transmission which parallel behavioural responses to stress.

Unlike noradrenaline or 5-HT, evidence linking a dysfunction of dopaminergic transmission with either anxiety or depression is fragmented and largely inconclusive (e.g. Gurguis and Uhde, 1990). The strongest evidence rests on findings that antianxiety or antidepressant treatments can cause marked neurochemical changes in dopaminergic neurones. However, drug-induced neurochemical changes might be unrelated to their psychotropic effects, or they could reflect recruitment of compensatory mechanisms rather than reversal of factors causing the disorder.

When changes in dopamine release are deduced from accumulation of dopamine metabolites, anxiolytic benzodiazepines consistently prevent the increase in dopamine turnover in rat frontal cortex and nucleus accumbens caused by stress (Fadda *et al.*, 1978; Giorgi *et al.*, 1987). This action is apparent at drug doses which have no effect on basal release of dopamine and which influence the response to psychological (conditioned fear: Ida *et al.*, 1989); watching conspecifics experiencing footshock (Kaneyuki *et al.*, 1991) as well as physical stress. This suppression of dopamine turnover is itself blocked by the benzodiazepine receptor antagonist, flumazenil, and therefore seems to involve activation of the benzodiazepine binding site on the GABA$_A$ receptor.

In addition to this action of benzodiazepines, it is now apparent that a wide range of established and putative non-benzodiazepine anxiolytic agents from different generic groups all prevent this stress-induced increase in dopamine turnover. Such compounds range from ligands for the benzodiazepine receptor, but which are not themselves benzodiazepines, to drugs acting at noradrenaline and 5-HT receptors. On the basis of such results, several authors have speculated that there is a causal link between the adverse emotional aspects of stress, anxiety and increased dopamine release in the prefrontal cortex (Fadda *et al.*, 1978; Kaneyuki *et al.*, 1991). However, it is unclear how these conclusions can be reconciled with those from studies of strains of rats which differ in their behavioural reactivity to stress (discussed above).

Contrasting with this evidence, the opposite conclusions were reached in a recent study using *in vivo* dialysis to measure dopamine outflow. Here, diazepam reduced basal release without affecting the increased release caused by restraint (Imperato *et al.*, 1990). The inference that diazepam does not influence the role of dopamine in response to restraint stress challenges the view that anxiolysis involves attenuation of dopaminergic responses.

So far, the possibility cannot be discounted that these discrepant results are explained by different stress paradigms or procedures for drug administration. One particular concern is the relative merit of measures of dopamine turnover obtained *ex vivo* and *in vivo*, particularly when dopamine metabolites are used as indices of transmitter release. Until these questions are resolved, claims of links between changes in dopamine release and stress responses, or between changes in dopamine release and the behavioural effects of antianxiety drugs must be regarded with caution.

An alternative approach, to examine the effects of drug treatments which cause behavioural changes consistent with anxiety, has looked at a range of benzodiazepine inverse agonists. These are compounds which bind to the benzodiazepine receptor, but have the opposite pharmacological and behavioural effects. Again, there is a worrying disparity between the effects of these agents in subcortical regions when changes in dopamine turnover estimated *ex vivo* (Tam and Roth, 1990) are compared with those measured *in vivo* (Bertolucci-D'Angio *et al.*, 1990). However, these agents consistently increase DOPAC accumulation (Claustre *et al.*, 1986; Ida and Roth, 1987) and dopamine outflow measured by *in vivo* microdialysis (Bradberry *et al.*, 1991a) in the prefrontal cortex. There is also evidence that these compounds increase the activity of tyrosine hydroxylase in the prefrontal cortex (Knorr *et al.*, 1989). All these changes suggest an inverse agonist-induced increase in dopamine turnover. However, in the absence of behavioural measurements, it is not possible to determine whether these neurochemical changes parallel an anxiogenic drug effect.

Only recently have behavioural and neurochemical changes been investigated in the same animals and much of this work has been concerned with drug-induced changes in locomotor activity. In one study, investigating behavioural responses to novelty, behavioural testing preceded neurochemical measurements by up to two weeks and dopamine release was measured by *in vivo* microdialysis in anaesthetized animals (Bradberry *et al.*, 1991b). It remains to be seen whether dopamine release and behavioural responses to stress are related when these are measured concurrently in freely-moving animals.

11.3.2.5 *Summary*

Our understanding of the role of dopamine in response and adaptation to stress is somewhat fragmented. However, recent techniques have provided evidence that release of this transmitter in stress may have a more important role than suspected hitherto. As yet, we know little about the effects of stress on dopamine receptors, or about how neurochemical modulation of dopaminergic transmission is affected by repeated stress. Links between changes in dopaminergic transmission and behavioural responses to stress have also received little attention. However, developments in techniques for monitoring changes in dopamine release *in vivo* are beginning to address this problem and may indicate that the role of dopamine in stress is as important as that of the other monoamines.

11.3.3 5-HYDROXYTRYPTAMINE

Neurones releasing 5-hydroxytryptamine (5-HT) arise from groups of nuclei found in the midline brainstem and they project throughout the CNS. Ascending fibres are derived from a rostral cluster of nuclei which includes the dorsal and median raphe. Fibres arising from these two nuclei have distinct morphologies: those from the dorsal raphe have fine, varicose axons which innervate terminal fields in a diffuse manner and do not appear to make specialized synaptic contacts. Those from the median raphe are beaded in appearance and make distinct, asymmetrical synaptic contacts in the terminal fields. This and other evidence supports the concept of dual, albeit parallel, ascending serotonergic systems in the brain (Tork, 1990).

The innervation of terminal fields is diffuse and overlapping, with individual fibres showing extensive collateralization. 5-HT-containing terminals are particularly dense in frontal and sensory areas of the cerebral cortex, the basal ganglia and limbic regions such as the hippocampus, hypothalamus and amygdala (reviewed by Jacobs and Azmitia, 1992; Figure 11.6). Many brain regions receive neurones from both the dorsal and median raphe but there is some topographical specialization of raphe efferents. For instance, the basal ganglia, amygdala and nucleus accumbens are innervated mainly by neurones from the dorsal raphe, while the median raphe projects predominantly to limbic areas.

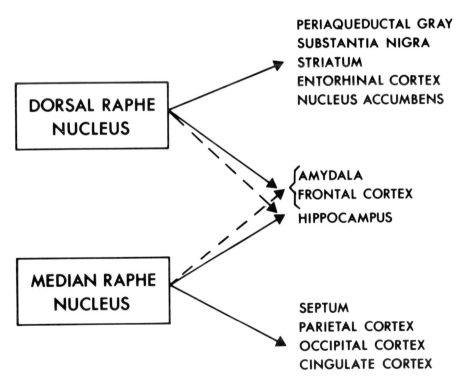

Figure 11.6 The distribution of 5-HT containing neurones in the brain. Solid lines indicate exclusive, or most functionally important, links. Dotted lines represent less significant links

11.3.3.1 Effects of an acute stress challenge

Electrophysiological studies on the effects of stress on neurones which release 5-HT have concentrated mainly on the dorsal and, although less so, the median raphe nuclei (reviewed by Jacobs and Azmitia, 1992). In unanaesthetized cats the firing frequency of most, but not all, cells is strongly dependent on animals' state of motor arousal. This is deduced from the parallel changes in muscle tone and firing frequency caused by centrally acting drugs or phasic sensory stimuli. Such stimuli cause an excitation followed by inhibition of firing of raphe neurones. Unlike noradrenergic neurones, there is no apparent decline in firing frequency on repetition of the stimulus. Stress, such as continuous white noise, restraint, or exposure to dogs increases single unit activity in the dorsal raphe but this increase is no larger than that associated with behavioural arousal. This is despite obvious behavioural signs of stress, as well as raised plasma catecholamines and tachycardia. Such findings have led some to conclude that serotonergic neurones merely relay information about animals' state of arousal and facilitate orientation of attention towards pertinent sensory stimuli.

Despite this somewhat ambiguous relationship between stress and neuronal activity, it is clear that stressful stimuli increase release of 5-HT in the terminal fields. This was first deduced from the accelerated release of [³H]5-HT from prelabelled stores induced by footshock (Thierry *et al.*, 1968) but subsequently confirmed using a wide range of methods to estimate transmitter turnover. Recent experiments using *in vivo* microdialysis have shown increased outflow of 5-HT after both non-noxious or noxious stress ('plus-maze', Wright *et al.*, 1991; tail-pinch, Boutelle *et al.*, 1990). Even handling rats, or stroking their fur, causes a measurable increase in transmitter outflow in the hippocampus (Kalen *et al.*, 1989; Kokaia *et al.*, 1989). Consistent with electrophysiological evidence, when stress is continuous, as in immobilization, outflow of 5-HT is maintained above basal levels albeit at lower levels than those seen immediately after the onset of stress (Shimizu *et al.*, 1992).

Increased release of 5-HT has also been inferred from measurement of its metabolite, 5-hydroxyindoleacetic acid (5-HIAA) in brain tissue after a period of stress (Joseph and Kennett, 1980; Hellhammer *et al.*, 1984). A recent *in vivo* voltammetric study has also reported accumulation of 5-HIAA during immobilization stress (Houdouin *et al.*, 1991). However, the reliability of 5-HIAA levels as an index of 5-HT release has been questioned (Kalen *et al.*, 1989; Shimizu *et al.*, 1992). A further consideration is that effects of stress on 5-HIAA accumulation may vary appreciably between different brain regions and animal strains (Shanks *et al.*, 1991).

Many studies continue to investigate the effects of stress on tissue levels of 5-HT but, predictably, results are inconsistent. Most reports indicate that there is no change (Joseph and Kennet, 1980). However, footshock transiently reduces 5-HT levels in the mouse prefrontal cortex and hypothalamus (Dunn, 1988) and inescapable tailshock depletes 5-HT levels in rat cortex (Hellhammer *et al.*, 1984); others have found an increase after restraint (Torda *et al.*, 1990).

Any change in tissue concentrations of 5-HT must be presumed to depend on the balance between release and synthesis. A common finding has been a stress-induced increase in brain levels of tryptophan, the substrate for the enzyme tryptophan hydroxylase (Curzon *et al.*, 1974), but there are exceptions (Adell *et al.*, 1988). Since availability of tryptophan limits the activity of tryptophan hydroxylase and, consequently, the rate of synthesis of 5-HT, potentiation of release of 5-HT used to be thought to depend on this increase in the levels of tryptophan in the brain (Joseph and Kennett, 1983). How increased brain levels of tryptophan are brought about is unknown but activation of the sympathetic limb of the autonomic nervous system may be involved (Dunn and Welch, 1991).

A second mechanism regulating 5-HT synthesis which is dependent on neuronal activation was identified only recently (Boadle-Biber *et al.*, 1989). This culminates in an increase in the maximal activity of tryptophan hydroxylase, with no change in affinity for enzyme or cofactor. The magnitude of the increase in enzyme activity seems to relate to stress intensity and is blocked by alkaline phosphatase; this indicates that enzyme activation involves phosphorylation. Contrasting with early theory, this mechanism is presumed to be independent of any changes in brain tryptophan. Interestingly enzyme activation is induced by intracerebral infusion of corticotropin releasing factor and requires glucocorticoid secretion from the adrenals (Singh *et al.*, 1992).

11.3.3.2 Effects of prolonged or repeated stress

Surprisingly few studies have looked at the effects of repeated stress on central 5-HT turnover. Increased levels of 5-HT and 5-HIAA in cortical and subcortical brain regions 20 h after the last of a course of daily periods of immobilization have been reported (Adell *et al.*, 1988), but this is not a consistent finding (Kennett *et al.*, 1985). When rats which had experienced repeated immobilization were given a subsequent restraint challenge, levels of 5-HT were reduced in most brain regions, and there was a further increase in levels of 5-HIAA (Adell *et al.*, 1988). However, others have found that the increase in 5-HIAA is the same, regardless of whether or not animals have previous experience of the stress (Pol *et al.*, 1992). So far, *in vivo* microdialysis has not been used to investigate this problem.

There is evidence that repeated stress causes long-latency neurochemical changes in 5-HT containing neurones. Repeated sound stress caused a long-lasting increase in tryptophan hydroxylase activity which, unlike that seen after a single stress, was not prevented by alkaline phosphatase and may involve an increased synthesis of enzyme protein (Boadle-Biber *et al.*, 1989). Mechanisms which trigger this change have not yet been investigated.

11.3.3.3 Receptors

Again, surprisingly little is known about the effects of stress on receptors for 5-HT. This is certainly due to the large number of different receptor subtypes for this neurotransmitter and difficulties in distinguishing between them pharmacologically. A comprehensive review of the pharmacological characterization of these receptor subtypes, their localization, molecular structure and coupling to second messenger systems is to be found in Frazer et al. (1990).

There were no changes in forebrain binding to the 5-HT$_{1A}$ receptor subtype, which is postsynaptic in the frontal cortex and hippocampus, after repeated footshock (Ohi et al., 1989). However, this finding conflicts with measures of changes in 5-HT$_{1A}$ receptor function. In rats, the 5-HT$_{1A}$ agonist, 5-methoxy-N,N-dimethyltryptamine (5-MeODMT), induces a specific cluster of behaviours: 'the 5-HT syndrome'. Prominent features of this behavioural syndrome include reciprocal forepaw treading, 'wet dog' shakes and 'Straub tail'. Although 5-MeOMDT binds to other 5-HT$_1$ receptor subtypes, its affinity for these sites is much lower than that for the 5-HT$_{1A}$ subtype and in rats the behavioural syndrome is thought to be evoked by activation of 5-HT$_{1A}$ receptors. An increased incidence of certain aspects of this behavioural syndrome has been found after repetition of either restraint or footshock (Kennett et al., 1985; Cancela et al., 1990), suggesting that 5-HT$_{1A}$ receptor function is increased by repeated stress. On the other hand, prolonged cold stress had the opposite effect: 5-HT$_{1A}$-mediated behaviours were attenuated, but agonist induced changes in plasma glucose and corticosterone were unaffected (Zamfir et al., 1992). This could suggest that there is regional variation in the effects of stress on 5-HT$_{1A}$ receptors and that any changes depend on the form of stress. Although evidence cited above suggests that 5-HT$_{1A}$ receptor density is unaffected by repeated stress, this does not rule out the possibility that changes in agonist-induced behaviours involve a change in coupling to its cAMP-generating second messenger system. Alternatively, there could be functional changes downstream of the receptor, possibly in other neurotransmitter systems. This disparity between indices of receptor density and function has also been noted after chronic administration of antidepressants (Lund et al., 1992).

Whether or not these functional changes are relevant to stress adaptation is even less clear: using signs such as locomotor activity, defecation and weight loss to indicate adaptation to stress, one study has shown this to be apparent after three days of intermittent footshock, while the changes in 5-HT$_{1A}$ receptor function were not seen until after five days of treatment (Ohi et al., 1989). If animals are exposed to a range of different stressors, to prevent habituation, then there are no changes in 5-MeODMT-induced behaviour unless they were also given an antidepressant (Molina et al., 1990). This finding was interpeted as evidence that 5-HT$_{1A}$ receptors are involved in adaptation to stress, and that repeated unpredictable stress precludes recruitment of adaptative changes in the function of this receptor subtype (see also Deakin et al., 1992).

In contrast to the general lack of effect of stress on 5-HT$_{1A}$ receptor binding, inescapable shock induces marked changes in 5-HT$_{1B}$ receptor binding. The density of these receptors, which act as presynaptic autoreceptors on 5-hydroxytryptaminergic nerve terminals in the rat, is increased in the cortex, hippocampus and septum, but reduced in the hypothalamus. These changes are not found if animals receive an equivalent amount of avoidable shock (Edwards et al., 1991). The behavioural implications of these findings are discussed in Section 11.3.3.4.

An increase in the binding of [$_3$H]ketanserin, used to define 5-HT$_2$ receptors, was found in rat frontal cortex after a 2h period of restraint (Torda et al., 1990). Interestingly, this change was abolished by either lesion of noradrenergic neurones with 6-OHDA, or pretreatment with the β-receptor antagonist, propranolol, but not by chemical lesion of 5-HT containing neurones. Stress-induced modulation of

5-HT$_2$ receptors seems independent of changes in 5-HT release, but is influenced by activation of β-adrenoceptors, therefore. These findings highlight some unusual features of 5-HT$_2$ receptors and their regulation which have even provoked suggestions that 5-HT is not the endogenous ligand for this binding site (Apud, 1991). In contrast to this increase in the density of 5-HT$_2$ receptors seen after restraint, there were no changes in binding in mouse cortex after a brief swim (up to 20 min, Kawanami et al., 1992; Davis et al., 1992a; Figure 11.7).

Measurements of changes in 5-HT$_2$ receptor function have further confused, rather than resolved these conflicting results. One measure rests on findings that metabolism of phosphoinositides is stimulated by activation of 5-HT$_2$ receptors; it is thought that this process underlies the receptor second messenger transduction system. Although there were no changes in receptor binding, when accumulation of inositol phosphates was used as an index of 5-HT$_2$ receptor function, this was reduced after periods of swim of up to 20 min (Kawanami et al., 1992).

Although a brief swim had no immediate effect on 5-HT receptor binding or 5-MeODMT-induced head-twitches, there was a significant increase in cortical 5-HT$_2$ receptor binding seven days later (Figure 11.7; Davis et al., 1992a). Moreover, this increase was paralleled by changes in behavioural measures of 5-HT$_2$ receptor function (Figure 11.7). In the mouse, unlike the rat, 5-MeODMT acts at 5-HT$_2$ receptors to evoke a characteristic head-twitch; the frequency of head twitches is a commonly used index of 5-HT$_2$ receptor function. In line with the binding data, the frequency of 5-MeODMT-induced head-twitches was increased in animals which had received a swim test seven days earlier (Figure 11.7). Although it is unlikely that receptor changes in the cortex are solely responsible for the potentiated 5-MeODMT-mediated behaviour, these findings support the possibility that increased 5-HT$_2$ receptor density explains the long-latency changes in indices of receptor function.

As for acute stress, few studies have looked at the effects of prolonged or repeated stress on 5-HT receptors. There were no changes in cortical 5-HT$_2$ receptor binding after exposure to repeated footshock (Ohi et al., 1989) or repeated once-daily saline injection (Davis et al., 1992a). Notwithstanding evidence that the stress of repeated saline injection did not affect 5-HT$_2$ receptors, this procedure did abolish the long-latency increase in 5-HT$_2$ receptor binding and function caused by a brief swim (Davis et al., 1992a; Figure 11.7). This raises important questions about the effects of animals' previous experience of stress on long-latency changes in 5-HT receptors. These questions have important implications for both clinical and research aspects of neurochemical changes underlying adaptation to stress. So far, the possibility that stress has long-latency effects on brain neurochemistry has largely been ignored.

11.3.3.4 5-Hydroxytryptamine and behavioural responses to stress

As with other monoamines, a link between changes in animals' emotional status and their behaviour in various experimental paradigms has been proposed. Regardless of whether or not these claims are justified, all behavioural paradigms have the common feature of exposing animals to stressful stimuli. Again, two clusters of experiments are prominent. First are those involving non-noxious stress such as novelty or conflict. These produce behavioural changes commonly prevented by

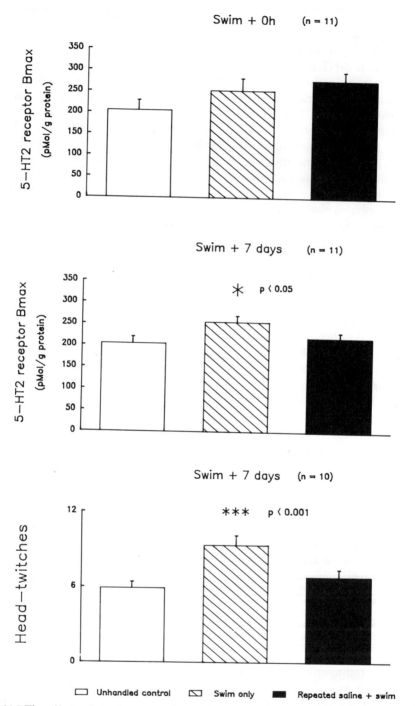

Figure 11.7 The effects of previous experience of repeated stress on changes in 5-HT$_2$ receptor binding and function induced by a brief swim (6 min). Top: 5-HT$_2$ binding density (Bmax) immediately after the swim. Centre: 5-HT$_2$ binding density seven days after the swim. Bottom: 5-HT$_2$ receptor function (5-MeODMT-induced head-twitches) 7 days after the swim

antianxiety drugs and are used to screen such compounds (Kahn *et al.*, 1988). Second are those involving uncontrollable and unpredictable noxious stress, such as footshock or swim, which culminate in escape deficits (learned helplessness and behavioural despair) commonly used to screen putative antidepressant compounds.

There is no clearcut evidence that 5-hydroxytryptaminergic transmission is increased in anxiety (Kahn *et al.*, 1988), but it has been suggested that 5-HT$_1$ receptors may be hyperresponsive in this disorder (Kahn and van Praag, 1988, 1992). In preclinical experiments, it has long been held that attenuation of 5-hydroxytryptaminergic transmission modifies behavioural responses to stress in ways which are consistent with an anxiolytic effect in humans (Chopin and Briley, 1987). This is supported by evidence that depletion of 5-HT with *para*chlorophenylalanine (pCPA; Kahn *et al.*, 1988) or 5,7-DHT (Briley *et al.*, 1990) generally mimics the behavioural effects of antianxiety drugs. However, this effect is not invariable and experiments investigating the effects of anxiolytic or anxiogenic drugs on turnover of 5-HT and behavioural responses to stress have produced confusing results.

In terms of their anxiolytic actions the azaspirodecanone derivatives, which are 5-HT$_{1A}$ agonists, are the most important group of compounds because at least one of these compounds, buspirone, has established clinical efficacy (reviewed by Blackburn, 1992). Yet, in preclinical experiments, reversal of stress-induced behavioural changes by these drugs is far less clear than with anxiolytic benzodiazepines. In part, this may be because 5-HT$_{1A}$ agonists have an inverted U-shaped dose-effect relationship with higher levels of drug having little effect in animal models (Thiebot and Martin, 1991). Another feature of 5-HT$_{1A}$ agonists is that they rarely reduce glucocorticoid or catecholamine responses to stress (Rittenhouse *et al.*, 1992; see Chapter 8). In some instances, these responses are even potentiated (Taylor *et al.*, 1989; Korte *et al.*, 1990). It is thought that changes in glucocorticoid secretion, at least, are mediated by postsynaptic receptors and that stress controllability is an important determinant of whether this occurs or not (Korte *et al.*, 1992). Clearly, this finding calls into question the general validity of using these hormonal measures as indicators of the overall impact of stress.

Somatodendritic 5-HT$_{1A}$ receptors are found in the raphe nuclei and, in the dorsal raphe at least, their activation depresses neuronal firing (Sprouse and Aghjanian, 1988). Recent experiments have shown that 5-HT$_{1A}$ agonists both reduce firing of dorsal raphe neurones and release of 5-HT from terminals in the hippocampus (Broderick and Piercey, 1991). 5-HT$_{1A}$ receptors are also found postsynaptically in terminal fields, with a particularly high density in the hippocampus, cortex and hypothalamus. Since agonist actions at somatodendritic and postsynaptic receptors would have opposite effects on 5-hydroxytryptaminergic transmission, it has been extremely difficult to explain the anxiolytic actions of 5-HT$_{1A}$ agonists.

The traditional explanation for the anxiolytic effects of these compounds rests on presumed activation of the 5-HT$_{1A}$ autoreceptor which would attenuate release of 5-HT from the terminals. Interestingly, in preclinical experiments, 'anxiolytic' effects are found only if buspirone is infused into the median raphe; infusion into the dorsal raphe is without effect (Carli *et al.*, 1989). If the effects of these drugs depends on diminished neuronal activity, it is hard to explain why lesions of serotonergic neurones attenuate the anticonflict effects of these drugs (Eison *et al.*, 1986) and why many 5-HT antagonists are inactive in such paradigms. However, behavioural changes resulting from infusion of the 5-HT$_{1A}$ agonist, tandospirone, into the hippocampus suggest that the anticonflict effects of these drugs could involve postsynaptic 5-HT$_{1A}$ receptors (Kataoka et al., 1991).

There are many further reasons why it is hard to unravel the mechanisms underlying the behavioural effects of 5-HT$_{1A}$ receptor ligands. First, clinical anxiolysis is only apparent after several weeks of administration. This calls into question the involvement of any neurochemical or behavioural effects seen after acute administration of these drugs. Secondly, they generally show poor selectivity, binding not only to 5-HT receptors subtypes but also to receptors for other neurotransmitters, notably α-adrenoceptors and dopamine D$_2$-receptors. There is also evidence that they are full agonists at presynaptic 5-HT$_{1A}$ receptors, but partial agonists at postsynaptic sites (reviewed by Tunnicliffe, 1991). Finally, it is thought that pre- and postsynaptic actions may require different levels of receptor occupancy: the number of spare receptors at these two sites could have profound effects on behavioural outcome.

There is also evidence that 5-HT$_2$ and 5-HT$_3$ receptor antagonists suppress behavioural changes induced by stress (Gleeson et al., 1989; Costall et al., 1990). However, as yet, claims for anxiolytic effects of 5-HT$_2$ and 5-HT$_3$ receptor ligands have not been confirmed in human studies. It is hard to distinguish between 5-HT$_2$ and 5-HT$_{1C}$ receptors pharmacologically, but these receptors are also arousing interest because of evidence that a drug binding to these sites, 1-(m-chlorophenyl)piperazine (mCPP), causes behavioural changes in preclinical models which predict an anxiogenic effect in man (Curzon et al., 1991) as well as inducing panic attacks in humans (Charney et al., 1987). However, this compound also binds, albeit at a tenfold lower affinity, to other 5-HT$_1$ receptor subtypes and 5HT$_3$ receptors, so actions via these sites cannot be excluded.

Like anxiety, evidence that 5-HT function is disrupted in depression is equivocal. In preclinical studies, whether or not 5-HT prevents or exacerbates behavioural changes induced by inescapable stress is also open to question. However, the prevention of learned helplessness by antidepressants is commonly taken to indicate that increased extraneuronal levels of 5-HT prevent both depression and behavioural effects of inescapable stress. An early report that learned helplessness was reversed by local infusion of 5-HT into the septum and frontal cortex was consistent with this view (Sherman and Petty, 1980). Also, lesions of 5-HT-releasing neurones prevent the reversal of learned helplessness by antidepressants which selectively block 5-HT uptake (Martin et al., 1987). The effects of other groups of antidepressants are unaffected, however (Soubrie et al., 1986). This suggests that behavioural effects of these compounds can be produced by, but are not dependent on, their action at 5-hydroxytryptaminergic nerve terminals. There is some confusion, however: para-chlorophenylalanine (pCPA), which blocks synthesis of 5-HT, prevents development of escape deficits induced by inescapable footshock (Edwards et al., 1986), but lesions with 5,7-DHT do not (Soubrie et al., 1986).

Recent experiments have investigated links between 5-HT receptors and behavioural changes induced by learned helplessness. This is not paralleled by changes in 5-HT$_{1A}$ receptors but is abolished by the 5-HT$_{1A}$ partial agonist, buspirone (Thiebot and Martin, 1991). There are conflicting data on whether or not other, more selective 5-HT$_{1A}$ receptor ligands abolish learned helplessness but, again, this could be because of their complex dose-effect profile. 5-HT$_{1A}$ agonists can also reduce immobility in swim stress (reviewed by Lucki, 1991). All these actions are consistent with clinical evidence that 5-HT$_{1A}$ agonists may have antidepressant effects, as well as established anxiolytic effects, in man (Schweizer and Rickels, 1991). Whether the effects of 5-HT$_{1A}$ ligands on behavioural responses to inescapable stress are mediated by presynaptic receptors in the raphe nuclei or by postsynaptic receptors in the terminal fields, or even whether they involve actions at receptors for other neurotransmitters is, as yet, unresolved (Thiebot and Martin, 1991). These are important questions because, so far, it is not possible to distinguish whether the effects are explained by reduced or increased 5-hydroxytryptaminergic activation of target cells.

In contrast to 5-HT$_{1A}$ receptor binding, learned helplessness is paralleled by marked changes in 5-HT$_{1B}$ receptor binding in the rat: the density of this receptor subtype is increased in the cortex, hippocampus and septum, but reduced in the hypothalamus. Animals experiencing an equivalent amount of shock without developing learned helplessness did not show these changes (Edwards et al., 1991). It has been suggested that increased density of 5-HT$_{1B}$ receptors underlies learned helplessness, therefore. Interpretation of these findings is as yet unclear. Some 5-HT$_{1B}$ receptors are terminal autoreceptors and depress release of 5-HT when activated by this transmitter. However, 5-HT$_{1B}$ receptors are also found on post-synaptic elements so a direct effect, independent of changes in 5-HT release, is possible (Heal et al., 1992). A direct involvement of 5-HT$_{1B}$ receptors in helplessness seems unlikely because the 5-HT$_{1B}$ agonist, 4-(4-methylpiperazin-1-yl)-7-(trifluoro-methyl)pyrrolo[1,2-a]quinoxaline dimaleate (CGS 12066B), does not influence helpless behaviour. However, it does prevent the reversal of learned helplessness by selective 5-HT uptake blockers (Martin and Puech, 1991). This suggests that, although 5-HT$_{1B}$ receptors may not be directly involved in development or reversal of helpless behaviour, they can influence the effects of antidepressant drugs, probably through their effects on release of 5-HT.

5-HT$_3$ antagonists also reduce escape deficits (Martin et al., 1992). However, these drugs act indirectly by reducing the release of several other neurotransmitters, so development or reversal of learned helplessness cannot be specifically attributed to a direct effect on 5-hydroxytryptaminergic transmission.

11.3.3.5 Summary

Whereas it seems clear that 5-HT release is increased by stress, there is as yet no clear explanation for the role of 5-HT in behavioural changes induced by stress. This is mainly attributable to the dearth of information on neurochemical modulation of these neurones. Differences between the effects of acute and repeated stress have hardly been considered, for instance, and there is no clear understanding of how stress affects 5-HT receptors. Moreover, in the few studies that have been carried out, measurements of receptor binding and receptor function often lead to conflicting conclusions.

Only one theory integrates evidence from anatomical and behavioural experiments with that from contemporary studies of different 5-HT receptor subtypes: the '5-HT receptor imbalance theory' (Deakin *et al.*, 1992). This proposes that responses to acute stress, or to cues for aversive stimuli, are effected by activation of dorsal raphe neurones. Release of 5-HT in forebrain structures (basal ganglia, amygdala and cortex) effects anticipatory avoidance behaviour while release in the periaqueductal grey suppresses the flight/fight reaction. These processes are seen to involve predominantly $5-HT_2$ or $5-HT_{1C}$ and, possibly, $5-HT_3$ receptors. It is argued further that excessive 5-HT activity in the forebrain could explain anxiety, while a deficit in the periaqueductal grey could explain panic. In contrast, experience of repeated stressful stimuli is proposed to activate median raphe neurones. $5-HT_{1A}$ receptors, tentatively in the hippocampus, are proposed to 'disengage' the processes causing the cognitive effects of repeated or prolonged stress. This could underlie adaptation to stress and a failure of this mechanism could lead to depression.

This interesting theory has much experimental support, but confirmation awaits improved understanding of the effects of stress on 5-HT turnover and its receptors in different brain regions, as well as evidence that anxiolytic and antidepressant compounds have appropriate neurochemical selectivity.

11.4 LINKING NEUROCHEMICAL AND BEHAVIOURAL RESPONSES TO STRESS

Recording parallel changes in neurochemistry and behaviour usually leads to the presumption that they are causally linked, and that the former explains the latter. However, until we can confirm that this is the case, we are unlikely to identify factors which increase individuals' vulnerability to stress, or to develop ways of protecting susceptible individuals from long-lasting adverse effects of stress.

A first approach to validating these claims of causation would be to show that inferences concerning the relationship between neurochemistry and behaviour in different groups of animals are replicated in studies of individual differences (Salmon and Stanford, 1992; Stanford and Salmon, 1992). A further aspect of studies of the neurochemical basis of stress resistance is that the forms of stress used in such work most commonly involve physically noxious stimuli. Foregoing sections of this chapter highlight how little is known about the effects of psychological stressors. The limited information available leads to the conclusion that the neurochemical effects of one form of stress cannot be used to predict the effects of another. Consequently, it is doubtful whether forms of noxious physical stress are appropriate paradigms for testing the neurochemical basis of resistance to psychological stress. Preceding chapters, at least, make it clear that stress with features of 'loss', disappointment or novelty are those most commonly encountered by humans. Whereas these forms of stress are used routinely to study behaviour, their neurochemical effects are rarely considered. Despite these limitations, theories linking neuronal function and stress-related behaviour are not in short supply.

An elaborate theory supported exhaustively by experimental data has addressed this problem. This theory, which draws on evidence including that discussed in this chapter, attends particularly to the effects of stressful stimuli which induce anxiety in humans (Gray, 1982, 1987). As discussed above, such forms of stress are known to activate central noradrenergic and 5-hydroxytryptaminergic neurones. The septohippocampal system, which is thought to govern 'behavioural inhibition', is proposed to be a crucial target for this increase in monoaminergic transmission. What remains to be seen is how neurochemical changes which determine the functional activity of monoaminergic neurones account for the changes in behaviour seen in response to different forms of stress, i.e. what presynaptic (modulation of transmitter release) and postsynaptic (modulation of different synaptic receptors) processes could underlie normal and abnormal function of the behavioural inhibition system?

One theory where this has been attempted makes a specific claim about the role of β-adrenoceptors in response and adaptation to stress. Since every type of neurotransmitter seems to be involved in the stress response, attribution of stress-induced behaviour to a single neurochemical response might be regarded as unlikely. Yet, based on experimental findings, Stone (1979) has suggested that downregulation and/or desensitization of β-adrenoceptors in the cerebral cortex underlies stress adaptation (Section 11.3.1.4). A corollary of this inference is that differences in β-adrenoceptor function also underlie stress resistance.

If this is the case, then a clear prediction can be made: that within a group of animals, those with the fewest number of, or least responsive, β-adrenoceptors will show the greatest resistance to stress. Experimentally, this means that there should be a negative correlation between β-adrenoceptors and resistance to stress. A further test of the stringency of this prediction is that this relationship must be shown to hold for all forms of stress, including those most commonly encountered by humans.

In preliminary experiments aimed at testing this prediction, the reverse was found, however. When animals were stressed by novelty (exposure to the open field; Salmon and Stanford, 1989) or frustration ('nonreward'; Stanford and Salmon, 1989), individuals with the greatest number of cortical β-adrenoceptors showed the greatest behavioural resistance to stress. This was deduced from the positive correlation between β-adrenoceptor density and behavioural measures of resistance to stress.

The possibility that the severity of stress is an important determinant of the relationship between β-adrenoceptors and behaviour was suggested by the results of a subsequent experiment. This again used nonreward but the frequency of exposure of animals to non-rewarded trials was increased ('massed extinction'). Under these experimental conditions, a negative correlation between receptor binding and behaviour was obtained: animals with the fewest number of β-adrenoceptors now showed the greatest behavioural resistance to stress (Marsland et al., 1990). It is possible that increasing the frequency of nonrewarded trials intensified the severity of the stress, thereby satisfying conditions relevant to Stone's theory.

An alternative explanation for the change in the relationship between neurochemistry and behaviour in these different paradigms is that the relationship between β-adrenoceptors and behaviour is influenced by animals' history of stress or drug treatment. This is supported by experiments in which the effects of drugs on behavioural responses to non-reward were not paralleled by changes in binding, although there was still a significant correlation between behavioural and neurochemical measures (Marsland *et al.*, 1990). This result points to a dissociation between neurochemical coding of individual differences ('trait') and effects of experience ('state'). So far, validation of these proposals is at a preliminary stage and it is important to establish that changes in binding reflect changes in β-adrenoceptor function. If this proves to be the case then the following questions should be addressed:

(1) *Do the gradient and intercept of the regression of behaviour on neurochemistry seen in individual differences predict neurochemical and behavioural changes caused by experience of drugs or stress?* This question can be answered by evaluating individual differences in neurochemistry and behaviour *in the same animals* and testing whether a change in either one of these parameters predicts changes in the other (Figure 11.8).

(2) *Is the gradient of the regression of individuals' behaviour on neurochemistry fixed, or is it modified by experience of stress or drugs?* To find that the regression of behaviour

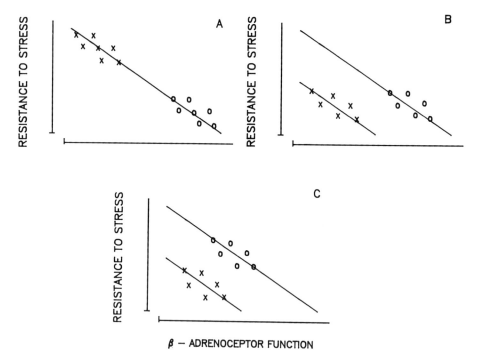

Figure 11.8 Schematic scattergrams showing individual differences and, in addition, possible effects of procedures which increase resistance to stress and/or modify β-adrenoceptor function. The gradient has been drawn to conform with that predicted by Stone's hypothesis and is assumed to be unaffected by treatment. 0: untreated subjects. X: drug pretreatment designed to reduce β-adrenoceptor function. (a) A single regression line explains both individual differences and differences between group means; (b) β-adrenoceptor function is modified but the same resistance to stress is maintained; (c) mean β-adrenoceptor function and resistance to stress are each modified in a way that cannot be predicted by the regression. (From Salmon and Stanford, 1992.)

on neurochemical measures is constant regardless of animals' experience of drugs or stress would suggest that these experiences have no influence on the neurochemical coding of behaviour. Conversely, a change in the gradient would suggest that the relationship between neurochemistry and behaviour is determined by animals' history of drugs or stress (Figure 11.9). If, for instance, the coding is different for non-noxious and noxious stress, as suggested by experiments described above, this could explain why increased noradrenergic function is coupled with stress resistance (or amelioration of the adverse effects of stress) in some experiments and reduced stress resistance (or exacerbation of the adverse effects of stress) in others (see also Figure 11.4).

(3) *Do the gradient and intercept of the regression of neurochemistry on behaviour seen in individual differences predict genetically-based neurochemical and behavioural differences? Alternatively, is the slope of the regression influenced by genotype?* A familial link in trait aspects of resistance to stress (e.g. susceptibility to panic) is widely recognized and there are several inbred strains of rodents characterized by differences in their resistance to stress. The basis of these differences can be investigated by testing whether the neurochemical and behavioural differences between strains correspond with predictions from the regression of neurochemistry on behaviour seen within strains. Alternatively, a change in gradient would suggest a genetic influence on the neurochemical coding of behaviour.

The strategy outlined above has focused on β-adrenoceptors but recognizes that any change in the relationship between neurochemistry and behaviour will depend on the influence of other transmitters. Further, the role of any neurotransmitter in stress resistance could be the primary focus of study and could be investigated by substitution of measurements of β-adrenoceptor function by the appropriate neurochemical parameter.

Finally, an ultimate objective must be to measure presynaptic function (transmitter release) as well as postsynaptic function (receptor activation) simultaneously; technology which might enable this to be done is now available. All these methodological refinements will allow progressively more rigorous analysis of the relationship between central neurotransmission and behaviour. This is especially relevant to explanations of drugs used to treat stress-related disorders. Currently, such explanations constantly refer to their effects on brain neurochemistry, yet there is rarely any attempt to validate links between these effects and changes in behaviour or emotion. Although the strategy outlined here cannot confirm such causal links, it provides an alternative way of testing them.

11.5 CONCLUDING REMARKS

While none would dispute that all the monoamines are involved in the response to stress and the actions of drugs which diminish its impact, the organization of this chapter reflects a regrettable limitation of research in this field: that most investigations tend to concentrate on individual monoamines. Consequently, few have attempted to explain the integrated impact of central monoamine systems in response and adaptation to stress.

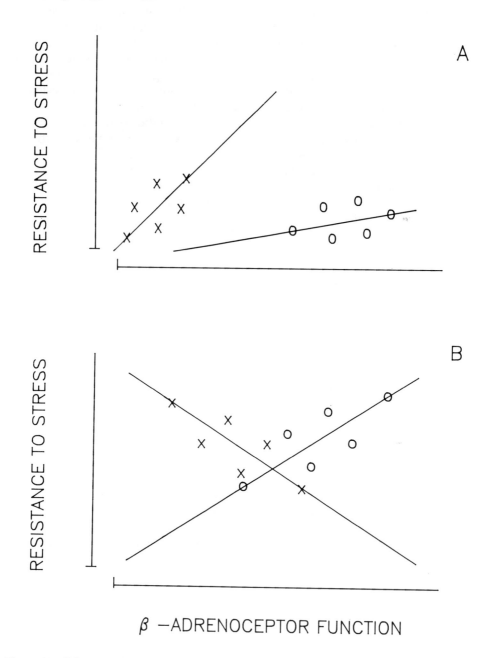

Figure 11.9 Schematic diagram showing individual differences and mean treatment effects. In contrast to Figure 11.8, the gradient of the regression for individual differences is now affected also. 0: untreated animals; X: previous experience of drugs which affect β-adrenoceptor number. (a) The gradient is positive in both cases, but the magnitude is changed; (b) the gradient has changed from positive to negative; greater β-adrenoceptor number predicts greater resistance to stress in the untreated group, but lower resistance to stress after treatment with drugs. (From Salmon and Stanford, 1992.)

What remains to be seen is how individual elements of monoamine transmission contribute to this scheme: how modulation of pre- and postsynaptic receptors, coupled with rapid or delayed changes in transmitter synthesis and release, accounts for the behavioural and emotional impact of stress. Although discoveries such as the existence of cotransmitters and families of neurotransmitter receptors might seem to complicate this task, it offers the opportunity of explaining the neurochemical basis of resistance to stress in more sophisticated ways than have been possible hitherto.

ACKNOWLEDGEMENT

I am indebted to Doreen Gettins for her invaluable help with the preparation of this chapter.

11.6 REFERENCES

Abbott, B.B., Schoen, L.S. and Badia, P. (1984) Predictable and unpredictable shock: behavioral measures of aversion and physiological measures of stress. *Psychol. Bull.*, **96**: 45–71

Abercrombie, E.D. and Jacobs, B.L. (1987a) Single-unit response of noradrenergic neurons in the locus coeruleus of freely moving cats. I. Acutely presented stressful and non stressful stimuli. *J. Neurosci.*, **7**: 2837–43

Abercrombie E.D. and Jacobs, B.L. (1987b) Single-unit response of noradrenergic neurons in the locus coeruleus of freely moving cats. II Adaptation to chronically presented stressful stimuli. *J. Neurosci.*, **7**: 2844–8

Abercrombie, E.D., Keefe, K.A., DiFrischia, D.S. and Zigmond, M.J. (1989) Differential effect of stress on in vivo dopamine release in striatum, nucleus accumbens, and medial frontal cortex. *J. Neurochem.*, **52**: 1655–8

Adell, A., Garcia-Marquez, C., Amario, A. and Gelpi. E. (1988) Chronic stress increases serotonin and noradrenaline in rat brain and sensitizes their responses to a further acute stress. *J. Neurochem.*, **50**: 1678–81

Anisman, H. (1985) Vulnerability to depression: contribution of stress, in M.G. Ziegler and C.R. Lake (eds.), *Frontiers of Clinical Neurosciences: Vol 2. Norepinephrine*, Williams and Williams, Baltimore and London, pp 407–31

Anisman, H. and Zacharko, R.M. (1991) Multiple neurochemical and behavioral consequences of stressors: implications for depression, in S.E. File (ed.), *Psychopharmacology of Anxiolytics and Antidepressants*, Pergamon, New York, pp 57–82

Anisman, H., Pizzino, A. and Sklar, L.S. (1980) Coping with stress, norepinephrine depletion and escape performance. *Brain Res.*, **191**: 583–8

Anisman, H., Irwin, J., Bowers, W. *et al.* (1987) Variations of norepinephrine concentrations following chronic stressor application. *Pharmacol. Biochem. Behav.*, **26**: 653–9

Apud, J.A. (1991) The 5-HT$_2$ receptor in brain: recent biochemical and molecular biological developments and new perspectives in its regulation. *Pharmacol. Res.*, **23**: 217–32

Areso, M.P. and Frazer, A. (1991) Effect of repeated administration of novel stressors on central beta adrenoceptors. *J. Neural Transm.*, **86**: 229–35

Aston-Jones, G. and Bloom, F.E. (1981) Norepinephrine-containing locus coeruleus neurons in behaving rats exhibit pronounced responses to non-noxious environmental stimuli. *J. Neurosci.*, **8**: 887–900

Ax, A.F. (1953) The physiological differentiation between fear and anger in humans. *Psychosom. Med.*, **15**: 433–42

Bannon, M.J. and Roth, R.H. (1983) Pharmacology of mesocortical dopamine neurons. *Pharmacol. Revs.*, **35**: 53–68

Benedict, C.R., Fillenz, M. and Stanford, C. (1978) Changes in plasma noradrenaline concentration as a measure of release rate. *Br. J. Pharmacol.*, **64**: 305–9

Benedict, C.R., Fillenz, M. and Stanford, C. (1979) Noradrenaline release in rats during prolonged cold-stress and repeated swim-stress. *Br. J. Pharmacol.*, **66**: 521–4

Bertolucci-D'Angio, M., Serrano, A. and Scatton, B. (1990) Mesocorticolimbic dopaminergic systems and emotional states. *J. Neurosci. Meth.*, **34**: 135–42

Blackburn, T.P. (1992) 5-HT receptors and anxiolytic drugs, in C.A. Marsden and D.J. Heal (eds.), *Central Serotonin Receptors and Psychotropic Drugs*, Blackwell Scientific, Oxford, pp 175–97

Blizard, D.A. (1988) The locus ceruleus: a possible neural focus for genetic differences in emotionality. *Experientia*, **44**: 491–5

Boadle-Biber, M.C., Corley, K.C., Graves, L. *et al.* (1989) Increase in the activity of tryptophan hydroxylase from cortex and midbrain of male Fischer 344 rats in response to acute or repeated sound stress. *Brain Res.*, **482**: 306–16

Boutelle, M.G., Zetterstrom, T., Pei, Q. *et al.* (1990) *In vivo* neurochemical effects of tail pinch. *J. Neurosci. Methods*, **34**: 151–7

Bradberry, C.W., Lory, J.D. and Roth, R.H. (1991a) The anxiogenic β-carboline FG7142 selectively increases dopamine release in rat prefrontal cortex as measured by microdialysis. *J. Neurochem.*, **56**: 748–52

Bradberry, C.W., Gruen, R.J., Berridge, C.W. and Roth, R.H. (1991b) Individual differences in behavioral measures: correlations with nucleus accumbens dopamine measured by microdialysis. *Pharmacol. Biochem. Behav.*, **39**: 877–82

Briley, M., Chopin, P. and Moret, C. (1990) Effect of serotonergic lesion on 'anxious' behaviour measured in the elevated plus-maze test in the rat. *Psychopharmacol.*, **101**: 187–9

Brodde, O-E., Daul, A. and O'Hara, N. (1984) β-Adrenoceptor changes in human lymphocytes, induced by dynamic exercise. *Naunyn-Schmiedeberg's Arch. Pharmacol.*, **325**: 190–2

Broderick P.A. and Piercey, M.F. (1991) 5-HT$_{1A}$ agonists uncouple somatodendritic impulse flow and terminal release. *Brain Res. Bull.*, **27**: 693–6

Cabib, S., Kempf, E., Schleef, C. *et al.* (1988) Effects of immobilization stress on dopamine and its metabolites in different brain areas of the mouse: role of genotype and stress duration. *Brain Res.*, **441**: 153–60

Cancela, L., Volosin, M. and Molina, V.A. (1990) Opioid involvement in the adaptive change of 5-HT$_1$ receptors induced by restraint. *Eur. J. Pharmacol.*, **176**: 313–19

Cannon, W.B. (1914) The emergency function of the adrenal medulla in pain and the major emotions. *Am. J. Physiol.*, **33**: 356–72

Cannon, W.B. (1927) The James-Lange theory of emotions: a critical examination and an alternative theory. *Am. J. Psychol.*, **39**: 106–24

Cannon, W.B. and de la Paz, D. (1911) Emotional stimulation of adrenal secretion. *Am. J. Physiol.*, **28**: 64–70

Cannon, W.B., Shohl, A.T. and Wright, W.S. (1911) Emotional glycosuria. *Am. J. Physiol.*, **29**: 280–7

Carli, M., Prontera, C. and Samanin, R. (1989) Evidence that central 5-hydroxytryptaminergic neurones are involved in the anxiolytic activity of buspirone. *Br. J. Pharmacol.*, **96**: 829–36

Carlson, J.N., Fitzgerald, L.W., Keller, R.W. and Glick, S.D. (1991) Site and region dependent changes in dopamine activation with various durations of restraint stress. *Brain Res.*, **550**: 313–18

Cassens, G., Roffman, M., Kuruc, A. *et al.* (1980) Alterations in brain norepinephrine metabolism induced by environmental stimuli previously paired with inescapable shock. *Science*, **209**: 1138–40

Cedarbaum, J.M. and Aghajanian, G.K. (1978) Activation of locus coeruleus neurons by peripheral stimuli: modulation by a collateral inhibitory mechanism. *Life Sci.*, **23**: 1383–92

Charney, D.S., Woods, S.W., Price, L.H. *et al.* (1990), Noradrenergic dysregulation in panic disorder in J.C. Ballenger (ed.), *Neurobiology of Panic Disorder*, Wiley-Liss, New York, pp 91–105

Charney, D.S., Woods, S.W., Goodman, W.K. and Heninger, G.R. (1987) Serotonin function in anxiety. *Psychopharmacol.*, **92**: 14–24

Chopin, P. and Briley, M. (1987) Animal models of anxiety: the effect of compounds that modify 5-HT neurotransmission. *TIPS*, **8:** 383–8

Claustre, Y., Rivy, J.P., Dennis, T. and Scatton, B. (1986) Pharmacological studies on stress-induced increase in frontal cortical dopamine metabolism in the rat. *J. Pharmacol. Exp. Ther.*, **238:** 693–700

Costall, B., Naylor, R.J. and Tyers, M.B. (1990) The psychopharmacology of 5-HT$_3$ receptors. *Pharm. Ther.*, **47:** 181–202

Coupland, N., Glue, P. and Nutt, D.J. (1992) Challenge tests: an assessment of noradrenergic and GABA systems in depression and anxiety disorders. *Molec. Aspects Med.*, **13:** 221–47

Crawley, J.N., Ninan, P.T., Packar, D. *et al.*, (1985) Neuropharmacological antagonism of the β-carboline induced 'anxiety' response in rhesus monkeys. *J. Neuropharmacol.*, **5:** 477–85

Curzon, G., Joseph, M.H. and Knott, P.J. (1974) Effects of immobilisation and food deprivation on rat brain tryptophan metabolism. *J. Neurochem.*, **19:** 1967–74

Curzon, G., Gibson, E.L., Kennedy, A.J. *et al.*, (1991) Anxiogenic and other effects of mCPP, a 5-HT$_{1C}$ agonist, in M. Briley and S.E. File (eds.), *New Concepts in Anxiety*, MacMillan Press, Basingstoke, pp 154–67

Dantzer, R. and Mormede, P. (1985) Stress in domestic animals: a psychoneuroendocrine approach, in G.P. Moberg (ed.), *Animal Stress*, American Physiology Society, Bethesda, pp 81–95

Dantzer, R. (1989) Neuroendocrine correlates of control and coping, in A. Steptoe and A. Appels (eds.), *Stress, Personal Control and Health*, Wiley, Chichester, pp 277–94

Da Silva, S. (1987) Does brain 3,4-dihydroxyphenylacetic acid reflect dopamine release? *J. Pharm. Pharmacol.*, **39:** 127–9

Davis S., Heal, D.J., Luscombe, G.P. and Stanford, S.C. (1992a) Short and long-latency effects of a six-min swim test on central 5-HT$_2$ receptors. *Br. J. Pharmacol.*, **105**

Davis, S., Heal, D.J., Salmon, P. and Stanford S.C. (1992b) Is vehicle-injection an adequate control for evaluating drug effects? *Clin. Neuropharm.*, **15:** 400B

Deakin, J.F.W., Graeff, F.G. and Guimaraes, F.S. (1992) 5-HT receptor subtypes and the modulation of aversion, in C.A. Marsden and D.J. Heal (eds.), *Central Serotonin Receptors and Psychotropic Drugs*, Blackwell Scientific, Oxford, pp 147–74

De Blasi, A., Fratelli, M., Wielosz, M. and Lipartiti, M. (1987) Regulation of beta adrenergic receptors on rat mononuclear leukocytes by stress: receptor redistribution and down-regulation are altered with aging. *J. Pharmacol. Exp. Ther.*, **240:** 228–33

De Boer, S.F., Koopmans, S.J., Slangen, J.L. and van der Gugten, J. (1990) Plasma catecholamines, corticosterone and glucose responses to repeated stress in rats: effect of interstressor interval length. *Physiol. Behav.*, **47:** 1117–24

De Boer, S.F., Slangen, J.L. and van der Gugten, J. (1991) Effects of buspirone and chlordiazepoxide on plasma catecholamine and corticosterone levels in stressed and nonstressed rats. *Pharmacol. Biochem. Behav.*, **38:** 299–308

Deutch, A.Y. and Roth, R.H. (1990) The determinants of stress-induced activation of the prefrontal cortical dopamine system. *Prog. Brain Res.*, **85:** 367–403

Deutch, A.Y., Tam, S-Y. and Roth, R.H. (1985) Footshock and conditioned stress increase 3,4-dihydroxyphenylacetic acid (DOPAC) in the ventral tegmental area but not the substantia nigra. *Brain Res.*, **333:** 143–6

Deutch, A.Y., Clark, W.A. and Roth, R.H. (1990) Prefrontal cortical dopamine depletion enhances the responsiveness of mesolimbic dopamine neurons to stress. *Brain Res.*, **521:** 311–15

DeTurk, K.H. and Vogel, W.H. (1980) Factors influencing plasma catecholamine levels in rats during immobilization. *Pharmacol. Biochem. Behav.*, **13:** 129–31

Doherty, M.D. and Gratton, A. (1992) High-speed chronoamperometric measurements of mesolimbic and nigrostriatal dopamine release associated with repeated daily stress. *Brain Res.*, **586:** 295–302

Duncan, G.E., Paul, I.A., Harden, T.K. *et al.*, (1985) Rapid down regulation of beta adrenergic receptors by combining antidepressant drugs with forced swim: a model of antidepressant-induced neural adaptation. *J. Pharmacol. Exp. Ther.*, **234:** 402–8

Dunn, A.J. (1988) Changes in plasma and brain tryptophan and brain serotonin and 5-hydroxyindoleacetic acid after footshock stress. *Life Sci.*, **42:** 1847–53

Dunn, A.J. and File, S.E. (1983) Cold restraint and dopamine metabolism in frontal cortex, nucleus accumbens and striatum. *Physiol. Behav.*, **31**: 511–13

Dunn, A.J. and Welch, J. (1991) Stress- and endotoxin-induced increases in brain tryptophan and serotonin metabolism depend on sympathetic nervous system activity. *J. Neurochem.*, **57**: 1615–22

Edwards, E., Johnson, J., Anderson, D. *et al.* (1986) Neurochemical and behavioral consequences of mild, uncontrollable shock: effects of PCPA. *Pharmacol. Biochem. Behav.*, **25**: 415–21

Edwards E., Harkins, K., Wright, G. and Henn, F.A. (1991) 5-HT$_{1B}$ receptors in an animal model of depression. *Neuropharmacol.*, **30**: 101–5

Eison, A.S., Eison, M.S., Stanley, M. and Riblet, L.A. (1986) Serotonergic mechanisms in the behavioral effects of buspirone and gepirone. *Pharmacol. Biochem. Behav.*, **24**: 701–7

Fadda, F., Argiolas, A., Melis, M.R. *et al.* (1978) Stress-induced increase in 3,4-dihydroxyphenylacetic acid (DOPAC) levels in the cerebral cortex and in n. accumbens: reversal by diazepam. *Life Sci.*, **23**: 2219–24

Fallon, J.H. and Moore, R.Y. (1978) Catecholamine innervation of the basal forebrain. *J. Comp. Neurol.*, **180**: 545–80

Fillenz, M. (1990) Regulation of catecholamine synthesis: multiple mechanisms and their significance. *Neurochem. Int.*, **17**: 303–20

Fillenz, M., Stanford, S.C. and Benedict, C.R. (1979) Changes in noradrenaline release rate and noradrenaline storage vesicles during prolonged activity of sympathetic neurones, in E. Usdin, I.J. Kopin and J. Barchas (eds.), *Catecholamines, Basic and Clinical Frontiers*, Pergamon Press, Oxford, pp 936–9

Frankenhaeuser, M. (1971) Behavior and circulating catecholamines. *Brain Res.*, **31**: 241–62

Frankenhaeuser, M. and Jarpe, G. (1963) Psychophysiological changes during infusion of adrenaline in various doses. *Psychopharmacol.*, **4**: 424–32

Frankenhaeuser, M., Mellis, I., Rissler, A. *et al.* (1968) Catecholamine secretion as related to cognitive and emotional reaction patterns. *Psychosom. Med.*, **30**: 109–20

Frazer, A., Maayani, S. and Wolfe, B.B. (1990) Subtypes of receptors for serotonin. *Ann. Rev. Pharmacol. Toxicol.*, **30**: 307–48

Freedman, R.R., Embury, J., Migaly, P. *et al.* (1990) Stress-induced desensitization of α_2-adrenergic receptors in human platelets. *Psychosomat. Med.*, **52**: 624–30

Freyschuss, U., Fagius, J., Wallin, B.G. *et al.* (1990) Cardiovascular and sympathoadrenal responses to mental stress: a study of sensory intake and rejection reactions. *Acta. Physiol. Scand.*, **139**: 173–83

Friedhoff, A.J., Rosengarten, H. and Stone, E.A. (1986) Increase in D2 receptor density in rat brain after repeated stress. *Soc. Neurosci. Abs.*, **12**: 192

Fung, S.-C. and Fillenz, M. (1985) Studies on the mechanism of modulation of [^{3}H]noradrenaline release from rat hippocampus synaptosomes by GABA and benzodiazepine receptors. *Neurochem. Int.*, **7**: 95–101

Funkstein, D.H., King, S.H. and Drolette, M. (1954) The direction of anger in a laboratory stress-inducing situation. *Psychosom. Med.*, **16**: 404–13

Giorgi, O., Corda, M.G. and Biggio, G. (1987) The anxiolytic β-carboline ZK 93423 prevents the stress-induced increase in dopamine turnover in the prefrontal cortex. *Eur. J. Pharmacol.*, **134**: 327–31

Glavin, G.B., Tanaka, M., Tsuda, A. *et al.* (1983) Regional rat brain noradrenaline turnover in response to restraint stress. *Pharmacol. Biochem. Behav.*, **19**: 287–90

Gleeson, S., Ahlers, S.T., Mansbach, R.S. *et al.* (1989) Behavioral studies with anxiolytic drugs. VI. Effects on punished responding of drugs interacting with serotonin receptor subtypes. *J. Pharamacol. Exp. Ther.*, **250**: 809–17

Grahame-Jones, S., Fillenz, M. and Gray, J.A. (1983) The effects of footshock and handling on tyrosine hydroxylase activity in synaptosomes and solubilised preparations from rat brain. *Neurosci.*, **9**: 679–86

Grant, S.J., Aston-Jones, G. and Redmond, D.E. (1988) Responses of primate locus coeruleus neurons to simple and complex sensory stimuli. *Brain Res. Bull.*, **21**: 401–10

Gray, J.A (1982) *The Neuropsychology of Anxiety*, Oxford University Press, Oxford

Gray, J.A. (1987) *The Psychology of Fear and Stress*, Cambridge University Press, Cambridge

Gurguis, G.N.M. and Uhde, T.W. (1990) Effect of yohimbine on plasma homovanillic acid in panic disorder patients and normal controls. *Biol. Psychiatr.*, **28**: 292–6

Hartman, R.D., Liaw, J-J., He, J-R. and Barraclough, C.A. (1992) Effects of reserpine on tyrosine hydroxylase mRNA levels in locus coeruleus and medullary A1 and A2 neurons analyzed by in situ hybridization histochemistry and quantitative image analysis methods. *Molec. Brain Res.*, **13**: 223–9

Hayes, P.E. and Schultz, S.C. (1987) Beta-blockers in anxiety disorders. *J. Affective Disord.*, **13**: 119–30

Heal, D.J., Luscombe, G.P. and Martin, K.F. (1992) Pharmacological identification of 5-HT receptor subtypes using behavioural models, in C.A. Marsden and D.J. Heal (eds.), *Central Serotonin Receptors and Psychotropic Drugs*, Blackwells, London, pp 56–99

Hellhammer, D.H., Rea, M.A., Bell., M. *et al.* (1984) Learned helplessness: effects on brain monoamines and the pituitary-gonadal axis. *Pharmacol. Biochem. Behav.*, **21**: 481–5

Herman, J.P., Guillonneau, D., Dantzer, R. *et al.* (1982) Differential effects of inescapable footshocks and of stimuli previously paired with inescapable footshocks on dopamine turnover in cortical and limbic areas of the rat. *Life Sci.*, **30**: 2207–14

Herve, D., Tassin, J.P., Barthelemy, C. *et al.* (1979) Difference in the reactivity of the mesocortical dopaminergic neurones to stress in the BALB/c and C57 BL/6 mice. *Life Sci.*, **25**: 1659–64

Holets, V.R. (1990) The anatomy and function of noradrenaline in the mammalian brain, in D.J. Heal and C.A. Marsden (eds.) *The Pharmacology of Noradrenaline in the Central Nervous System*, Oxford University Press, Oxford, pp 1–40

Houdouin, F., Cespuglio, R., Gharib, A. *et al.* (1991) Detection of the release of 5-hydroxyindole compounds in the hypothalamus and the n. raphe dorsalis throughout the sleep-waking cycle and during stressful situations in the rat: a polygraphic and voltammetric approach. *Brain Res.*, **85**: 153–62

Ida, Y. and Roth, R.H. (1987) The activation of mesoprefrontal dopamine neurones by FG7142 is absent in rats treated chronically with diazepam. *Eur. J. Pharmacol.*, **137**: 185–90

Ida, Y., Tanaka, M., Tsuda, A. *et al.* (1985) Attenuating effect of diazepam on stress-induced increases in noradrenaline turnover in specific brain regions of rats: antagonism by Ro 15–1788. *Life Sci.*, **37**: 2491–8

Ida, Y., Tsuda, A., Sueyoshi, K. *et al.* (1989) Blockade by diazepam of conditioned fear-induced activation of rat meso-prefrontal dopamine neurones. *Pharmacol. Biochem. Behav.*, **33**: 477–9

Iimori, K., Tanaka, M., Kohno, Y. *et al.* (1982) Psychological stress enhances noradrenaline turnover in specific brain regions in rats. *Pharmacol. Biochem. Behav.*, **16**: 637–46

Imperato, A., Puglisi-Allegra, S., Zocchi, A. *et al.* (1990) Stress activation of limbic and cortical dopamine release is prevented by ICS 205–930 but not by diazepam. *Eur. J. Pharmacol.*, **175**: 211–14

Imperato, A., Puglisi-Allegra, S., Casolini, P. and Angelucci, L. (1991) Changes in brain dopamine and acetylcholine release during and following stress are independent of the pituitary-adrenocortical axis. *Brain Res.*, **538**: 111–17

Imperato, A., Angelucci, L., Casolini, P. *et al.* (1992) Repeated stressful experiences differently affect limbic dopamine release during and following stress. *Brain Res.*, **577**: 194–9

Irwin, J., Ahluwalia, P. and Anisman, H. (1986) Sensitization of norepinephrine activity following acute and chronic footshock. *Brain Res.*, **379**: 98–103

Iuvone, P.M. and Dunn, A.J. (1986) Tyrosine hydroxylase activation in mesocortical 3,4-dihydroxyphenylethylamine neurones following footshock. *J. Neurochem.*, **47**: 837–44

Jacobs, B.L. and Azmitia, E.C. (1992) Structure and function of the brain serotonin system. *Pharmacol. Revs.*, **72**: 165–229

Johnston, A.L. (1991) The implication of noradrenaline in anxiety, in M. Briley and S.E. File (eds.) *New Concepts in Anxiety*, MacMillan Press, London, pp 347–65

Joseph, M.H. and Kennett, G.A. (1980) Brain tryptophan and 5-HT function in stress. *Br. J. Pharmacol.*, **73**: 267

Jospeh, M.H. and Kennett, G.A. (1983) Stress-induced release of 5-HT in the hippocampus and its dependence on increased tryptophan availability: an in vivo electrochemical study. *Brain Res.*, **270**: 251–7

Kahn, R.S. and van Praag, H.M. (1988) A serotonin hypothesis of panic disorder. *Human Psychopharmacol.*, **3**: 285–8

Kahn, R.S. and van Praag, H.M. (1992) Panic disorder: a biological perspective. *Eur. J. Neuropsychopharmacol.*, **2**: 1–20

Kahn, R.S., van Praag, H.M., Wetzler, S. *et al.* (1988) Serotonin and anxiety revisited. *Biol. Psychiatr.*, **23**: 189–208

Kalen, P., Rosegren, E., Lindvall, O. and Bjorklund, A. (1989) Hippocampal noradrenaline and serotonin release over 24 hours as measured by the dialysis technique in freely moving rats: correlation to behavioural activity state, effect of handling and tail-pinch. *Eur. J. Neurosci.*, **1**: 181–8

Kalivas, P.W. and Stewart, J. (1991) Dopamine transmission in the initiation and expression of drug- and stress-induced sensitization of motor activity. *Brain Res. Revs.*, **16**: 223–44

Kawanami, T., Morinobu, S., Totsuka, S. and Endoh, M. (1992) Influence of stress and antidepressant treatment on 5-HT-stimulated phosphoinositide hydrolysis in rat brain. *Eur. J. Pharmacol.*, **216**: 385–92

Kaneyuki, H., Yokoo, H., Tsuda, A. *et al.* (1991) Psychological stress increases dopamine turnover selectively in mesoprefrontal dopamine neurons of rats: reversal by diazepam. *Brain Res.*, **557**: 154–61

Kataoka, Y., Shibata, K., Miyazaki, A. *et al.* (1991) Involvement of the dorsal hippocampus in mediation of the antianxiety action of tandispirone, a 5-hydroxytryptamine$_{1A}$ agonist anxiolytic. *Neuropharmacol.*, **30**: 475–80

Katoh, A., Nabeshima, T. and Kameyama, T. (1991) Interaction between enkephalinergic and dopaminergic systems in stressful situations. *Eur. J. Pharmacol.*, **193**: 95–9

Keefe, K.A., Stricker, E.M., Zigmond, M.J. and Abercrombie, E.D. (1990) Environmental stress increases extracellular dopamine in striatum of 6-hydroxydopamine treated rats: in vivo microdialysis studies. *Brain Res.*, **527**: 350–3

Keller, R.W., Stricker, E.M. and Zigmond, M.J. (1983) Environmental stimuli but not homeostatic challenges produce apparent increases in dopaminergic activity in the striatum: an analysis by *in vivo* voltammetry. *Brain Res.*, **279**: 159–70

Kennett, G.A., Dickinson, S.L. and Curzon, G. (1985) Enhancement of some 5-HT-dependent behavioural responses following repeated immobilization in rats. *Brain Res.*, **330**: 253–63

Khansari, D.N., Murgo, A.J., and Faith, R.E. (1990) Effects of stress on the immune system. *Immunol. Today*, **11**: 170–5

Kitada, Y., Miyauchi, T., Kanazawa, Y. *et al.* (1983) Involvement of α- and β1-adrenergic mechanisms in the immobility-reducing action of desipramine in the forced swimming test. *Neuropharmacol.*, **22**: 1055–60

Kitada, Y., Miyauchi, T., Kosasa, T. and Satoh, S. (1986) The significance of β-adrenoceptor down regulation in the desipramine action in the forced swimming test. *Naunyn-Schmiedeberg's Arch. Pharmacol.*, **333**: 31–5

Knorr, A.M., Deutch, A.Y. and Roth, R.H. (1989) The anxiogenic β-carboline FG7142 increases in vivo and in vitro tyrosine hydroxylation in the prefrontal cortex. *Brain Res.*, **495**: 355–61

Kokaia, M., Kalen, P., Bengzon, J. and Lindvall, O. (1989) Noradrenaline and 5-hydroxytryptamine release in the hippocampus during seizures induced by hippocampal kindling stimulation: an *in vivo* microdialysis study. *Neurosci.*, **32**: 647–56

Kolb, L.C. (1984) The post-traumatic stress disorders of combat: a subgroup with a conditioned emotional response. *Military Med.*, **149**: 237–43

Konarska, M., Stewart, R.E. and McCarty, R. (1989) Habituation of sympathetic-adrenal medullary responses following exposure to chronic intermittent stress. *Physiol. Behav.*, **45**: 255–61

Korf, J., Aghajanian, G.K. and Roth, R.H. (1973) Increased turnover of norepinephrine in the rat cerebral cortex during stress: role of the locus coeruleus. *Neuropharmacol.*, **12**: 933–8

Korte, S.M., Smit, J., Bouws, G.A.H. *et al.* (1990) Behavioral and neuroendocrine response to psychosocial stress in male rats: the effects of the 5-HT$_{1A}$ agonist ipsapirone. *Hormones & Behav.*, **24**: 554–67

Korte, S.M., Bouws, G.A.H. and Bohus, B. (1992) Adrenal hormones in rats before and after stress-experience: effects of ipsapirone. *Physiol. Behav.*, **51**: 1129–33

Kramarcy, N.R., Delanoy, R.L. and Dunn, A.J. (1984) Footshock treatment activates catecholamine synthesis in slices of mouse brain regions. *Brain Res.*, **290**: 311–19

Lachuer, J., Gaillet, S., Barbagli, B. *et al.* (1991) Differential early time course activation of the brainstem catecholaminergic groups in response to various stresses. *Neuroendocrinology*, **53**: 589–96

Lapierre, Y.D. (1987) Clinical and biological correlates of panic states. *Prog. Neuro-Psychopharmacol. Biol. Psychiatr.*, **11**: 91–6

Lavielle, S., Tassin, J-P., Thierry, A-M. *et al.* (1978) Blockade by benzodiazepines of the selective high increase in dopamine turnover induced by stress in mesocortical dopaminergic neurones of the rat. *Brain Res.*, **168**: 585–94

Liebowitz, M.R., Fyer, A.J., Gorman, J.M. *et al.* (1985) Specificity of lactate infusions in social phobia versus panic disorders. *Am. J. Psychiatry.*, **142**: 947–50

Loughlin, S.E., Foote, S.L. and Bloom F.E. (1986) Efferent projections of nucleus locus coeruleus: topographic organization of cells of origin demonstrated by three-dimensional reconstruction. *Neurosci.*, **18**: 291–306

Lucki, I. (1991) Behavioral studies of serotonin receptor agonists as antidepressant drugs. *J. Clin. Psychiatry*, **52**: 24–31

Lund, A., Mjellem-Joly, N. and Hole, K. (1992) Desipramine administered chronically influences 5-hydroxytryptamine$_{1A}$-receptors, as measured by behavioural tests and receptor binding in rats. *Neuropharmacol.*, **31**: 25–32

Maclennan, A.J., Pelleymounter, M.A., Atmadja, S. *et al.* (1989) D$_2$ dopamine receptors in the rat prefrontal cortex: a characterization and alteration by stress. *Brain Res.*, **477**: 300–7

Maier, S.F. and Seligman, M.E.M. (1976) Learned helplessness. Theory and evidence. *J. Exp. Psychol. Gen.*, **105**: 3–46

Marsland, A.L., Salmon, P., Terry, P. and Stanford, S.C. (1990) Effects of propranolol on, and noradrenergic correlates of, the response to nonreward. *Pharmacol. Biochem. Behav.*, **35**: 41–6

Martin, P. and Puech, A.J. (1991) Is there a relationship between 5-HT$_{1B}$ receptors and the mechanisms of action of antidepressant drugs in the learned helplessness paradigm in rats? *Eur. J. Pharmacol.*, **192**: 193–6

Martin, P., Soubrie, P. and Simon, P. (1987) The effect of monoamine oxidase inhibitors compared with classical tricyclic antidepressants on learned helplessness paradigm. *Prog. Neuro-Psychopharmacol. Biol. Psychiat.*, **11**: 1–7

Martin, P., Gozlan, H. and Puech, A.J. (1992) 5-HT$_3$ receptor antagonists reverse helpless behaviour in rats. *Eur. J. Pharmacol.*, **212**: 73–8

Mason, J.W. (1968) A review of psychoendocrine research on the sympathetic-adrenal medullary systems. *Psychosom Med.*, **30**: 631–53

McFall, M.E., Murburg, M.M., Roszell, D.K. and Veith, R.C. (1989) Psychophysiologic and neuroendocrine findings in posttraumatic stress disorder: a review of theory and research. *J. Anxiety Disord.*, **3**: 243–57

Molina, V.A., Volosin, M., Cancela, L. *et al.* (1990) Effect of chronic variable stress on monoamine receptors: influence of imipramine administration. *Pharmacol. Biochem. Behav.*, **35**: 335–40

Murua, V.S., Gomez, R.A., Andrea, M.E. and Molina, V.A. (1991) Shuttle-box deficits induced by chronic variable stress: reversal by imipramine administration. *Pharmacol. Biochem. Behav.*, **38**: 125–30

Nakazaro, T. (1987) Locus coeruleus neurons projecting to the forebrain and the spinal cord in the cat. *Neurosci.*, **23**: 529–38

Neftel, K.A., Adler, R.H., Kapell, L. *et al.* (1982) Stage fright in musicians: a model illustrating the effect of beta-blockers. *Psychosom. Med.*, **44**: 461–9

Nesse, R.M., Cameron, O.G., Curtis, G.C. *et al.* (1984) Adrenergic function in patients with panic anxiety. *Arch. Gen. Psychiatr.*, **41**: 771–6

Ninan, P.T., Insel, T.M., Cohen, R.M. *et al.* (1982) Benzodiazepine receptor-mediated experimental 'anxiety' in primates. *Science*, **218**: 1332–4

Nisenbaum, L.K. and Abercrombie, E.D. (1992) Enhanced tyrosine hydroxylation in hippocampus of chronically stressed rats upon exposure to a novel stressor. *J. Neurochem.*, **58**: 276–81

Nisenbaum, L.K., Zigmond, M.J., Sved, A.F. and Abercrombie, E.D. (1991) Prior exposure to chronic stress results in enhanced synthesis and release of hippocampal norepinephrine in response to a novel stressor. *J. Neurosci.*, **11**: 1478–84

Ohi, K., Mikuni, M. and Takahashi, K. (1989) Stress adaptation and hypersensitivity in 5-HT neuronal systems after repeated foot shock. *Pharmacol. Biochem. Behav.*, **34**: 603–8

Ostman-Smith, I. (1979) Adaptive changes in the sympathetic nervous system and some effector organs of the rat following long term exercise or cold acclimation and the role of cardiac sympathetic nerves in the genesis of compensatory cardiac hypertrophy. *Acta. Physiol. Scand.*, **S477**: 1–118

Pavcovich, L.A., Cancela, L.M., Volosin, M. *et al.* (1990) Chronic stress-induced changes in locus coeruleus neuronal activity. *Brain Res. Bull.*, **24**: 293–6

Platt, J.E. and Stone, E.A. (1982) Chronic restraint stress elicits a positive antidepressant response in the forced swim test. *Eur. J. Pharmacol.*, **82**: 179–81

Pohl, R., Yeragani, V., Balon, R. *et al.* (1990) Isoproterenol-induced panic: a beta-adrenergic model of panic anxiety, in J.C. Ballenger (ed.), *Neurobiology of Panic Disorder*, Wiley-Liss, New York, pp 107–20

Pol, O., Campmany, L., Gil, M. and Amario, A. (1992) Behavioral and neurochemical changes in response to acute stressors: influence of previous chronic exposure to immobilization. *Pharmacol. Biochem. Behav.*, **42**: 407–12

Porsolt, R.D., Bertin, A., Blavet, N. *et al.* (1979) Immobility induced by forced swimming in rats: effects of agents which modify central catecholamine and serotonin activity. *Eur. J. Pharmacol.*, **57**: 201–10

Puglisi-Allegra, S., Imperato, A., Angelucci, L. and Cabib, S. (1991a) Acute stress induces time-dependent responses in dopamine mesolimbic system. *Brain Res.*, **554**: 217–22

Puglisi-Allegra, S., Kempf, E., Schleef, C. and Cabib, S. (1991b) Repeated stressful experiences differently affect brain dopamine receptor subtypes. *Life Sci.*, **48**: 1263–8

Pyke, R.E. and Greenberg, H.S. (1986) Norepinephrine challenges in panic patients. *J. Clin. Psychopharmacol.*, **6**: 279–85

Rasmussen, K. and Jacobs, B.L. (1986) Single unit activity of locus coeruleus neurons in the freely moving rat. II Conditioning and pharmacologic studies. *Brain Res.*, **371**: 335–44

Redmond, D.E. and Huang, Y.H. (1979) II. New evidence for a locus coeruleus – norepinephrine connection with anxiety. *Life Sci.*, **25**: 2149–62

Reinhard, J.F., Bannon, M.J. and Roth, R.H. (1982) Acceleration by stress of dopamine synthesis and metabolism in prefrontal cortex: antagonism by dopamine. *Naunyn-Schmiedeberg's Arch. Pharmacol.*, **318**: 374–7

Rittenhouse, P.A., Bakkum, E.A., O'Connor, P.A. *et al.* (1992) Comparison of neuroendocrine and behavioral effects of ipsapirone, a 5-HT$_{1A}$ agonist, in three stress paradigms: immobilization, forced swim and conditioned fear. *Brain Res.*, **580**: 205–14

Riva, M.A. and Creese, I. (1989) Comparison of two putatively selective radioligands for labeling central nervous system β-adrenergic receptors: inadequacy of [^{3}H]dihydroalprenolol. *Molec. Pharmacol.*, **36**: 201–10

Rossetti, Z.L., Portas, C., Pani, L. *et al.* (1990) Stress increases noradrenaline release in the rat frontal cortex: prevention by diazepam. *Eur. J. Pharmacol.*, **176**: 229–31

Salmon, P. and Stanford, S.C. (1989) β-Adrenoceptor binding correlates with behaviour of rats in the open field. *Psychopharmacol.*, **98**: 412–16

Salmon, P. and Stanford, S.C. (1992) Research strategies for decoding the neurochemical basis of resistance to stress. *J. Psychopharmacol.*, **6**: 1–7

Sanghera, M.K., Coke, J.A., Williams, H.L. and McMillen, B.A. (1990) Ipsapirone and 1-(2-pyrimidinyl)-piperazine increase rat locus coeruleus noradrenergic activity. *Brain Res. Bull.*, **24**: 17–22

Sara, S.J. and Segal, M. (1991) Plasticity of sensory responses of locus coeruleus neurons in the behaving rat: implications for cognition. *Prog. Brain Res.*, **88**: 571–85

Schildkraut, J.J. (1973) Neuropharmacology of the affective disorders. *Ann. Rev. Pharmacol.*, **13**: 427–54

Schneider, P., Evans, L., Ross-Lee, L. *et al.* (1987) Plasma biogenic amine levels in agoraphobia with panic attacks. *Pharmacopsychiatr.*, **20**: 102–4

Schweizer, E., and Rickels, K. (1991) Serotonergic anxiolytics: a review of their clinical efficacy, in R.J. Rodgers and S.J. Cooper (eds.), $5\text{-}HT_{1a}$ *Agonists, $5\text{-}HT_3$ Antagonists and Benzodiazepines*, Wiley, Chichester, pp 365–76

Sevy S., Papadimitriou G.N., Surmont D.W., Goldman S. and Mendlewicx J. (1989) Noradrenergic function in generalized anxiety disorder, major depressive disorder and healthy subjects. *Biol. Psychiatr.*, **25**: 141–52

Shanks, N., Zalcman, S., Zacharko, R.M. and Anisman, H. (1991) Alterations of central norepinephrine, dopamine and serotonin in several strains of mice following acute stressor exposure. *Pharmacol. Biochem. Behav.*, **38**: 69–75

Sherman, A.D. and Petty, F. (1980) Neurochemical basis of the action of antidepressants on learned helplessness. *Behavioral & Neural Biol.*, **30**: 119–34

Shimizu, N., Take, S., Hori, T. and Oomura, Y. (1992) In vivo measurement of hypothalamic serotonin release by intracerebral microdialysis: significant enhancement by immobilization stress in rats. *Brain Res. Bull.*, **28**: 727–34

Simson, P.E., and Weiss, J.M. (1987) Alpha-2 receptor blockade increases responsiveness of locus coeruleus neurons to excitatory stimulation. *J. Neurosci.*, **7**: 1732–40

Simson, P.E. and Weiss, J.M. (1988) Altered activity of the locus coeruleus in an animal model of depression. *Neuropsychopharmacol.*, **1**: 287–95

Simson, P.E. and Weiss, J.M. (1989) Peripheral, but not local or intracerebroventricular, administration of benzodiazepines attenuates evoked activity of locus coeruleus neurons. *Brain Res.*, **490**: 236–42

Singh, V.B., Hoa-Phan, T., Corley, K.C. and Boadle-Biber, M.C. (1992) Increase in cortical and midbrain tryptophan hydroxylase activity by intracerebroventricular administration of corticotropin releasing factor: block by adrenalectomy, by RU 38486 and by bilateral lesions to the central nucleus of the amygdala. *Neurochem. Int.*, **20**: 81–92

Soderpalm, B. and Engel, J.A. (1990) $\alpha 1$- and β-adrenoceptor stimulation potentiate the anticonflict effect of a benzodiazepine. *J. Neural Transm.*, **79**: 155–67

Soubrie, P., Martin. P., El Mestikawy, S. *et al.* (1986) The lesion of serotonergic neurons does not prevent antidepressant-induced reversal of escape failures produced by inescapable shocks in rats. *Pharmac. Biochem Behav.*, **25**: 1–6

Sprouse, J.S. and Aghajanian, G.K. (1988) Responses of hippocampal pyramidal cells to putative serotonin $5\text{-}HT_{1A}$ and $5\text{-}HT_{1B}$ agonists: a comparative study with dorsal raphe neurones. *Neuropharmacol.*, **27**: 707–15

Stanford, S.C. (1990) Central adrenoceptors in response and adaptation to stress, in D.J. Heal and C.A. Marsden (eds.) *The Pharmacology of Noradrenaline in the Central Nervous System*, Oxford University Press, Oxford, pp 379–422

Stanford, S.C. and Nutt, D.J. (1982) Comparison of the effects of repeated electroconvulsive shock on α_2- and β-adrenoceptors in different regions of rat brain. *Neurosci.*, **7**: 1753–7

Stanford, S.C. and Salmon, P. (1989) Neurochemical correlates of behavioural responses to frustrative nonreward in the rat: implications for the role of central noradrenergic neurones in behavioural adaptation to stress. *Exp. Brain Res.*, **75**: 133–8

Stanford, S.C. and Salmon, P. (1992) β-Adrenoceptors and resistance to stress: old problems and new possibilities. *J. Psychopharmacol.*, **6**: 15–19

Stanford, S.C., Nutt, D.J. and Cowen, P.J. (1983) Comparison of the effects of chronic desmethylimipramine administration on α_2- and β-adrenoceptors in different regions of rat brain. *Neurosci.*, **8**: 161–4

Stanford, S.C. Fillenz, M. and Ryan, E. (1984) The effect of repeated mild stress on cerebral cortical adrenoceptors and noradrenaline synthesis in the rat. *Neurosci. Lett.*, **45**: 163–7

Stone, E.A. (1979) Subsensitivity to norepinephrine as a link between adaptation to stress and antidepressant therapy: an hypothesis. *Res. Comm. Psychol. Psychiatr. & Behav.*, **4**: 241–55

Stone, E.A. (1981) Mechanism of stress-induced subsensitivity to norepinephrine. *Pharmacol. Biochem. Behav.*, **14**: 719–23

Strecker, R.E. and Jacobs, B.L. (1985) Substantia nigra dopaminergic unit activity in behaving cats: effect of arousal on spontaneous discharge and sensory evoked activity. *Brain Res.*, **361**: 339–50

Tam, S.Y. and Roth, R.H. (1990) Modulation of mesoprefrontal dopaminergic neurones by central benzodiazepine receptors: Pharmacological characterization. *J. Pharmacol. Exp. Ther.*, **252**: 989–96

Tanaka, M., Kohno, Y., Nakagawa, R. *et al.* (1983) Regional characteristics of stress-induced increases in brain noradrenaline release in rats. *Pharmacol. Biochem. Behav.*, **19**: 543–7

Tanaka, M., Tsuda, A., Yokoo, H. *et al.* (1990) Involvement of the brain noradrenaline system in emotional changes caused by stress in rats. *Ann. N.Y. Acad. Sci.*, **597**: 159–74

Tanaka, T., Yokoo, H., Mizoguchi, K. *et al.* (1991) Noradrenaline release in the rat amygdala is increased by stress: studies with intracerebral microdialysis. *Brain Res.*, **544**: 174–6

Tassin, J.P., Herve, D., Blanc, G. and Glowinski, J. (1980) Differential effects of a two minute open field session on dopamine utilization in the frontal cortices of BALB/c and C57 BL/6 mice. *Neurosci. Lett.*, **17**: 67–71

Thatcher Britton, K., Segal, D.S., Kuczenski, R. and Hauger, R. (1992) Dissociation between in vivo hippocampal norepinephrine response and behavioral/neuroendocrine responses to noise stress in rats. *Brain Res.*, **574**: 125–30

Taylor, J., Harris, N., Krieman, M. and Vogel, W.H. (1989) Effects of buspirone on plasma catecholamines, heart rate, and blood pressure in stressed and non stressed rats. *Pharmacol. Biochem. Behav.*, **34**: 349–53

Thiebot, M.-H. and Martin, P. (1991) Effects of benzodiazepines, 5-HT$_{1A}$ agonists and 5-HT$_3$ antagonists in animal models sensitive to antidepressant drugs, in R.J. Rodgers and S.J. Cooper (eds.), *5-HT$_{1A}$ Agonists, 5-HT$_3$ Antagonists and Benzodiazepines*, Wiley, Chichester, pp 159–94

Thierry, A-M., Fekete, M. and Glowinski, J. (1968) Effects of stress on the metabolism of noradrenaline dopamine and serotonin (5-HT) in the central nervous system of the rat. 2. Modification of serotonin metabolism. *Eur. J. Pharmacol.*, **4**: 384–9

Thierry, A.M., Tassin J.P., Blanc G. and Glowinski J. (1976) Selective activation of the mesocortical DA system by stress. *Nature*, **263**: 242–4

Thoenen, H., Mueller, R.A. and Axelrod, J. (1970) Phase difference in the induction of tyrosine hydroxylase in cell body and nerve terminals of sympathetic neurones. *Proc. Natn. Acad. Sci. USA*, **65**: 58–62

Thomas, D.N., Nutt, D.J. and Holman, R.B. (1992) Effects of acute and chronic electroconvulsive shock on noradrenaline release in the rat hippocampus and frontal cortex. *Br. J. Pharmacol.*, **106**: 430–44

Tissari, A.H., Argiolas, A., Fadda, F. *et al.* (1979) Footshock stress accelerates non-striatal dopamine synthesis without activating tyrosine hydroxylase. *Naunyn-Schmiedeberg's Arch. Pharmacol.*, **308**: 155–7

Torda, T., Murgas, K., Cechova, E. *et al.* (1990) Adrenergic regulation of [^{3}H]ketanserin binding sites during immobilization stress in the rat frontal cortex. *Brain Res.*, **527**: 198–203

Tork, I. (1990) Anatomy of the serotonergic system. *Ann. N. Y. Acad. Sci.*, **100**: 9–35

Tsuda, A. and Tanaka, M. (1985) Differential changes in noradrenaline turnover in specific regions of rat brain produced by controllable and uncontrollable shocks. *Behav. Neurosci.*, **99**: 802–17

Tsuda, A., Tanaka, M., Ida, Y. *et al.* (1986) Effects of preshock experience on enhancement of rat brain noradrenaline turnover induced by psychological stress. *Pharmacol. Biochem. Behav.*, **24**: 115–19

Tsuda, A., Tanaka, M., Ida, Y. *et al.* (1988) Expression of aggression attenuates stress-induced increases in rat brain noradrenaline turnover. *Brain Res.*, **474**: 174–80

Tsuda, A., Ida, Y., Satoh, H. *et al.* (1989) Stressor predictability and rat brain noradrenaline metabolism. *Pharmacol. Biochem. Behav.*, **32**: 569–72

Tunnicliffe, G. (1991) Molecular basis of buspirone's anxiolytic action. *Pharmacol. Toxicol.*, **69**: 149–56

Tyrer, P. (1988) Current status of β-blocking drugs in the treatment of anxiety disorders. *Drugs*, **36**: 773–83

Uhde, T.W., Joffe, R.T., Jimerson, D.C. and Post, R.M. (1988) Normal urinary free cortisol and plasma MHPG in panic disorder: clinical and theoretical implications. *Biol. Psychiatr.*, **23**: 575–85

U'Prichard, D.C. and Kvetnansky, R. (1980) Central and peripheral adrenergic receptors in acute and repeated immobilization stress, in E. Usdin, R. Kvetnansky and I.J. Kopin (eds.), *Catecholamines and Stress: recent advances*, Elsevier, Holland, pp 299–308

Vogel, W.H., Miller, J., DeTurk, K.H. and Routzahn, B.K. (1984) Effects of psychoactive drugs on plasma catecholamines during stress in rats. *Neuropharmacol.*, **23**: 1105–8

Von Euler, U.S. and Hellner, S. (1952) Excretion of noradrenaline and adrenaline in muscular work. *Acta. Physiol. Scand.*, **26**: 183–91

Ward, M.M., Mefford, I.N., Parker, S.D. *et al.* (1983) Epinephrine and norepinephrine responses in continuously collected human plasma to a series of stressors. *Psychosom. Med.*, **45**: 471–86

Weiss, J.M., Bailey, W.H., Goodman, P.A. *et al.* (1982) A model for neurochemical study of depression, in A. Levy and M.Y. Spiegelstein (eds.), *Behavioral Models and the Analysis of Drug Action*, Elsevier, Amsterdam, pp 195–223

Weiss, J.M., Simson, P.G., Hoffman, L.J. *et al.* (1986) Infusion of adrenergic receptor agonists and antagonists into the locus coeruleus and ventricular system of the brain. *Neuropharmacol.*, **25**: 367–84

Winder, W.W., Hagberg, J.M., Hickson, R.C. *et al.* (1978) Time course of sympathoadrenal adaptation to endurance exercise training in man. *J. Applied Physiol.*, **45**: 370–6

Woods, S.W. and Charney, D.S. (1990) Biologic responses to panic anxiety elicited by nonpharmacologic means, in J.C. Ballenger (ed.), *Neurobiology of Panic Disorder*, Wiley-Liss, New York, pp 205–17

Wright, I.K., Upton, N. and Marsden, C.A. (1991) Effect of diazepam on extracellular 5-HT in the ventral hippocampus observed in rats on the X-maze model of anxiety using in vivo microdialysis. *Br. J. Pharmacol.*, **104**: 69

Yerkes, R.M. and Dodson, J.D. (1908) The relation to the strength of the stimulus of the rapidity of habit formation. *J. Comp. Neurol. Psychol.*, **18**: 459–82

Yokoo, H., Tanaka, M., Yoshida, M. *et al.* (1990) Direct evidence of conditioned fear-elicited enhancement of noradrenaline release in the rat hypothalamus assessed by intracranial microdialysis. *Brain Res.*, **536**: 305–8

Zamfir, O., Broqua, P., Baudrie, V. and Chaouloff, F. (1992) Effects of cold-stress on some 5-HT$_{1A}$, 5-HT$_{1C}$ and 5-HT$_2$ receptor-mediated responses. *Eur. J. Pharmacol.*, **219**: 261–9

12

The role of GABA in the regulation of the stress response

WIN SUTANTO AND E. RON DE KLOET

DIVISION OF MEDICAL PHARMACOLOGY,
LEIDEN-AMSTERDAM CENTER FOR DRUG RESEARCH
UNIVERSITY OF LEIDEN
2300 RA LEIDEN
THE NETHERLANDS

12.1 INTRODUCTION

In his description of the 'general adaptation syndrome', Hans Selye regarded stress as a disturbance of homeostasis produced by diverse noxious agents. This concept interprets the cascade of neurochemical events triggered by stress as restoring homeostasis. Among other changes, stressful stimuli activate the hypothalamic-pituitary-adrenal (HPA) axis. This results in increased secretion of corticotropin releasing factor (CRF) from the hypothalamus, adrenocorticotropin (ACTH) from the anterior pituitary and corticosteroids from the adrenal cortex. Studies of the stress response, including CRF and ACTH secretion, have revealed complex interactions between steroid hormones, neurotransmitters and neuromodulators. These interactions have been the subject of numerous recent reports (for reviews see: McEwen et al., 1986; Dallman et al., 1987; Jones and Gillham, 1988; de Kloet, 1991; see also Chapter 13).

The neurotransmitter γ-aminobutyric acid (GABA) is a major inhibitory neurotransmitter found in approximately 65% of neurones in the central nervous system. An inhibitory effect in the HPA-axis is a notable action of GABA, but precise mechanisms underlying this inhibition and its importance in the stress response are not clear. GABA also influences the functional activity of the anterior pituitary and the adrenal cortex. Numerous reports on this subject have accumulated in the last two decades and are reviewed elsewhere (Yoneda et al., 1983; McCann and Rettori, 1986; Otero-Losada, 1988a,b; Hillhouse and Milton, 1989; Miguez and Aldegunde, 1990b;). The present chapter concentrates on studies of the role of GABA in the stress response that have been carried out within the last ten years, and discusses neurochemical, behavioural and clinical aspects of the field.

12.2 GABA IN RESPONSE AND ADAPTATION TO STRESS

Neurochemical changes in several neurotransmitter pathways (in both the peripheral and central nervous system) are provoked by stress. Studies of the involvement of GABA in response and adaptation to stress have considered both changes in the metabolism of this transmitter and changes in GABA receptors.

12.2.1 CHANGES IN THE GABAERGIC SYSTEM FOLLOWING STRESS

Neurochemical changes affecting the GABAergic system, as judged by measurements of GABA concentrations or the activity of the GABA synthetic enzyme, glutamate decarboxylase (GAD), have been assessed during the application of a variety of different forms of stress. These include repeated electrical shock (Green et al., 1978) and acute immobilization stress (Yoneda et al., 1983). In the olfactory bulb, a brief period of immobilization reduced tissue concentrations of GABA and accelerated the turnover of this neurotransmitter in vivo (Otero-Losada, 1989). Although [^{3}H]GABA uptake was also reduced in this brain region, this was not a consistent finding in others: [^{3}H]GABA uptake in the corpus striatum was increased by stress, for instance. In the frontal cortex, hippocampus and medio-basal hypothalamus there was no change in either [^{3}H]GABA uptake or other measures of GABA turnover, such as the activity of the synthetic enzyme GAD, or of the degradative enzyme GABA transaminase. While the GABAergic system in the corpus striatum seems to be that most affected by acute stress, changes in GABA turnover are found in the frontal cerebral cortex after chronic stress (Otero-Losada,

1988a). In consequence, it has been suggested that the cortical GABAergic system may have an important role in adaptation to stress.

12.2.2 CHANGES IN THE GABA$_A$ RECEPTOR FOLLOWING STRESS

The cloning, expression, physico-chemical and biochemical properties of the GABA$_A$ receptor complex are described in several recent reviews (DeFeudis, 1990; Olsen and Tobin, 1990). Briefly, the GABA$_A$ receptor comprises an integral membrane chloride channel lined by peptide subunits. The specific combination of subunits determines the pharmacology of the receptor which not only binds GABA to open the chloride channel, but can also have recognition sites for a number of CNS depressant agents: barbiturates, benzodiazepines (BDZs), alcohol and steroids. Binding studies using specific ligands for different components of the GABA$_A$ receptor have revealed stress-induced alterations in binding parameters at both the GABA and benzodiazepine recognition sites. These studies are summarized in Table 12.1.

Exposure of neurones to GABA increases neuronal influx of chloride ions. Consequently, chloride influx is widely used as an index of GABA$_A$ receptor function *ex vivo* and is reduced in tissues from animals given an acute stress challenge. GABA$_A$ receptor function is also reduced in several brain areas (cerebral cortex, hippocampus and striatum) after inescapable footshock, a procedure which is well known to induce deficits in subsequent acquisition of escape behaviour ('learned helplessness'; Drugan *et al.*, 1989; see Chapter 9) and is commonly interpreted as an animal model of depression. This reduction in GABA$_A$ receptor function is mimicked by acute administration of inhibitors of GABA$_A$ receptors *in vivo* and is prevented by agonist benzodiazepines. It is also prevented by repeated stress or repeated administration of drugs which attenuate activation of the GABA$_A$ receptor and cause a long-lasting downregulation of these receptors (Biggio *et al.*, 1990). These findings suggest that adaptive changes in GABA$_A$ receptor function are apparent after repeated exposure to stressful stimuli.

Alterations in BDZ receptor binding and function as a result of stress have been the subject of numerous reports. In general, stress increases BDZ receptor binding (for a review see Trullas *et al.*, 1987), but this can be prevented by pretreatment with agonist benzodiazepines. The stress-induced increase in benzodiazepine receptors is inhibited by adrenalectomy but restored by corticosteroid replacement. This suggests an important, albeit probably indirect, role for corticosteroids in modulation of the GABAergic stress response.

The above summary conceals many inconsistencies in the results, however. In mice, defeat stress increased binding to benzodiazepine receptors in various brain regions when this was measured using the antagonist ligand [^{3}H]flumazenil *in vivo*. However, a decrease in BDZ receptor binding *in vivo* has been reported in several brain regions after both a single or repeated swim stress (hippocampus, cerebral cortex, hypothalamus, midbrain, striatum; Weizman *et al.*, 1989, 1990b). This change, like that reported above, was prevented by adrenalectomy (Weizman et al, 1990b). Further disparities are apparent when binding was estimated using the radioligand, [^{3}H]flunitrazepam, *in vitro*. With this protocol, defeat stress did not

Table 12.1 Changes in $GABA_A$–BDZ receptors following various stress stimuli: a summary. + = increase; − = decrease; MBH = mediobasal hypothalamus

Stress	Brain area (or Peripheral)	Changes	References
Social stress (defeat stress)	Cortex Cerebellum Hypothalamus	+ $GABA_A$ mRNAs No change	Kang et al., 1991
	Cortex Cerebellum Hypothalamus	+ [³H]Ro15-1788 binding in vivo; no change in [³H]flunitrazepam binding in vitro	Miller et al., 1987
Swim stress	Cortex, Midbrain Hippocampus Hypothalamus, Striatum	− [³H]Ro15-1788 binding in vivo	Weizman et al., 1989
Forced swim stress	Cortex	+ [³H]Flunitrazepam binding in vitro	Rago et al., 1989
	Kidney	+ [³H]Ro5-4864 binding in vitro	
Swim stress	Cerebral cortex	No change in muscimol induced ³⁶Cl⁻ flux	Tuominen and Korpi, 1991
Immobilization stress	Olfactory bulb	+ GABA turnover in vivo; − [³H]GABA uptake	Otero-Losada, 1989
	Corpus striatum	+ [³H]GABA uptake	
	Frontal cortex Hippocampus, MBH	No change in [³H]GABA uptake, GAD/GAT activities	
Ether stress	Cortex Hypothalamus	+ [³H]GABA uptake − [³H]GABA uptake; + GAD activity	Acosta et al., 1990
Inescapable tail shock (learned helplessness)	Cortex, Striatum, Hippocampus	− ³⁶Cl⁻ flux	Drugan et al., 1988
	Kidney, Heart	− BDZ receptors	
Repeated stress/ anxiety disorders	Platelet	− BDZ receptors	Dar et al., 1991
Food deprivation	Kidney, Heart Adrenals, cerebellum	− BDZ receptors	Weizman et al., 1990a
Neonatal handling	Whole brain	+ [³H]GABA binding	Bolden et al., 1990
Gastric hyperacidity	Whole brain	+ GABA content	Hara et al., 1991

increase benzodiazepine binding, as described above, and an increase in binding was found after swim stress (Rago et al., 1989).

The duration of these changes has been given little consideration. However, one report suggests that a brief stressful event (a single needle injection) blocks multiple actions of diazepam for up to a month, at least (Antelman et al., 1988). Since diazepam is known to influence receptor-mediated effects of GABA (see Section

12.4.4), this finding suggests the likelihood of long-term effects arising from acutely stressful events.

Not only do changes in GABA$_A$–BDZ receptor binding depend on the nature of the stress and the protocol for the binding assay, but there is evidence that the age of the animal is another important variable. Stress-induced increases in binding of the benzodiazepine receptor ligand, [^{3}H]flumazenil, are attenuated in aged animals, for example (Barnhill *et al.*, 1991). Since it is likely that alterations in BDZ receptor binding are related to individuals' physiological and behavioural responses, it is possible that these changes could be relevant to the decreased ability to adapt to stress with advancing age.

In addition to the above changes at the GABA$_A$-BDZ receptor complex, changes have also been found in a BDZ binding site which is found intracellularly in mitochondrial membranes and which has different pharmacological properties from BDZ binding sites associated with the GABA$_A$ receptor. Inescapable footshock decreases binding to this receptor site (Drugan *et al.*, 1988). Since this change is unaffected by adrenalectomy, hypophysectomy or chemical sympathectomy, the HPA axis does not seem to influence this change (Drugan *et al.*, 1988). Interestingly, such changes may be subject to sexual dimorphism: the reduction in female rats (23%) was considerably less than that in males (55%; Drugan *et al.*, 1991). It has also been shown that repeated stress reduced binding to this BDZ binding site on platelet membranes. In this context, it is notable that abnormally low levels of platelet benzodiazepine receptor binding have been reported in patients suffering from generalized anxiety disorders (Dar *et al.*, 1991).

12.3 THE CENTRAL GABAERGIC SYSTEM AND THE HPA-AXIS

GABA influences the activity of the HPA-axis, and therefore the stress response, at several levels. It is a neurotransmitter in the hypothalamus (Decavel and van den Pol, 1990) and a variety of techniques have allowed the tracing of GABAergic pathways projecting to targets in the hypothalamic-pituitary complex (Vincent *et al.*, 1982; Apud *et al.*, 1989). These pathways presumably mediate the GABAergic control of anterior pituitary hormone secretion (Wass, 1983; McCann and Retori, 1986). In addition, such studies have helped explain the neurochemical and behavioural effects of drugs such as benzodiazepines, barbiturates and GABA antagonists, all of which modify the effects of GABA at the GABA$_A$ receptor complex (see Section 12.4).

12.3.1 CRF AND ACTH

GABAergic innervation is found in all the hypothalamic nuclei. It also appears that a subpopulation of CRF-containing neurones in the parvocellular division of the hypothalamic periventricular nucleus (pvPVN) also contain GABA; this indicates possible co-release of these compounds, presumably in the median eminence (Meister *et al.*, 1988). Moreover, CRF neurones in anterior portions of the periventricular nucleus (PVN) and pvPVN are found close to dopaminergic neurones. Symmetrical synapses between some of these same neurones suggest possible interactions between CRF, dopamine and GABA in mediation of responses to stress (Thind and Goldsmith, 1989).

Although GABA does not modify basal release of CRF from the hypothalamus, it does inhibit the stimulation of CRF release by acetylcholine and 5-HT *in vitro* (Jones, 1979; Hillhouse and Milton, 1989). This suggests a role for GABA in modulation of CRF release. An involvement of GABA in hormonal responses to stress is further indicated by various studies of the control of ACTH secretion by this neurotransmitter. Inhibition of GABA degradation or uptake, or blockade of GABA transmission, can alter ACTH secretion from the anterior pituitary (McCann and Rettori, 1986). For instance, drugs such as sodium valproate, which prevent metabolic degradation of GABA, reduce ACTH secretion (Dornhorst *et al.*, 1983; Jones *et al.*, 1984). In rats, intraventricular administration of GABA inhibits the rise in ACTH release induced by surgical trauma (Makara and Stark, 1974) while the GABA antagonists picrotoxin and bicuculline stimulate release of this hormone. This stimulation is not affected by complete hypothalamic deafferentation, suggesting that the locus of this effect of GABA is confined within the hypothalamus. Although GABA$_A$ receptors are found in the human anterior pituitary (Grandison *et al.*, 1982), GABA does not seem to affect ACTH secretion through direct actions in the pituitary (Jones *et al.*, 1984; Miguez and Aldegunde, 1990a). However, there is evidence that it inhibits basal secretion of ACTH *via* an effect on monoaminergic neurotransmission (Miguez and Aldegunde, 1990b).

The anxiolytic benzodiazepine, diazepam, which enhances the actions of GABA at GABA$_A$ receptors, inhibits stress-induced increases in the secretion of anterior pituitary hormones, including ACTH. This effect is mediated primarily through actions at GABA$_A$ receptors in the hypothalamus and possibly the anterior pituitary (de Souza *et al.*, 1990). However, diazepam also binds to the mitochondrial benzodiazepine receptor. These receptors are characterized by binding of the antagonist isoquinoline derivative, [1-(2-chlorophenyl)-N-methyl-(1-methylpropyl)-3-isoquinoline-carboxamide] (PK 11195) and also influences the function of the HPA-axis in rats (Calogero *et al.*, 1990). Release of hypothalamic CRF, stimulated by the benzodiazepine derivative, 4'-chlorodiazepam (Ro 5-4864) *in vitro*, was antagonized by PK 11195. PK 11195 also stimulated the activity of pituitary cells, releasing adrenocorticotropin, cultured *in vitro*.

12.3.2 CORTICOSTEROIDS

Intracerebroventricular administration of GABA reduces serum corticosterone levels: this is a consequence of the inhibition of basal ACTH secretion, mentioned above (Miguez and Aldegunde, 1990b). However, metabolic degradation of GABA yields several metabolites, including γ-hydroxybutyric acid (GHB). The effects of GHB on HPA-secretion contrast with those of its parent compound. Intraperitoneal administration of GHB to male rats results in a significant increase in plasma corticosterone (Miguez and Aldegunde, 1990a) through an action independent of the GABA$_A$-BDZ receptor complex. Since neither GHB nor GABA have any direct effects on adrenocortical cells influencing corticosterone production, it seems that the effects of these compounds on corticosterone secretion are mediated centrally.

Agonist benzodiazepines acting at the $GABA_A$ receptor complex (Waddington, 1978) are well known for their anxiolytic actions. Apart from these anxiolytic actions, BDZs active at the $GABA_A$ receptor have endocrine effects including effects on plasma corticosterone. Some BDZs stimulate steroidogenesis in corticosteroid producing cells (Muhkin et al., 1989), an effect mediated by the mitochondrial benzodiazepine receptor. A study by Pohorecky et al. (1988) showed that diazepam has a biphasic effect on plasma corticosterone in the rat in vivo: low doses decreased, while high doses increased, plasma corticosterone. The former action of diazepam is believed to be mediated at the hypothalamic and/or pituitary level rather than at the adrenal cortex.

12.4 MODULATION OF $GABA_A$ RECEPTOR FUNCTION BY STRESS AND DRUGS

Actions of GABA in the CNS have been linked with those of many other neuroactive compounds. Endogenous substances known to influence, and to be influenced by, GABA include steroid hormones, neuropeptides, purines, and excitatory amino acids. In addition, many psychoactive drugs, such as anxiolytic/hypnotic and anti-depressant drugs influence GABA function.

12.4.1 STEROIDS

Both mineralocorticoid and glucocorticoid hormones are commonly thought to interact with specific intracellular receptors. Radioligand binding studies, immunocytochemistry and cRNA/mRNA hybridization techniques have made it possible to determine the physico-chemical properties, topography and gene expression of these steroid receptors in the CNS (for a review see van Eekelen et al., 1988).

Glucocorticoid receptors (GR) are widely distributed throughout the brain and are activated by glucocorticoid hormones. These include corticosterone and cortisol; the predominant hormone depends on individual species. Mineralocorticoid receptors (MR) respond to the mineralocorticoid hormones, aldosterone and deoxycorti-costerone, and are localized in the septo-hippocampus, anterior hypothalamus and circumventricular regions. MR have also been characterized in the kidney where aldosterone selectivity is maintained by the enzyme 11β-hydroxysteroid dehydro-genase (Edwards et al., 1988; Sakai et al., 1992). The interaction of mineralocorticoids with aldosterone-selective MR in the anterior hypothalamus and circumventricular organs, as in the kidneys, controls sodium homeostasis. In contrast, hippocampal MR bind aldosterone and glucocorticoid hormones (corticosterone or cortisol) with equal affinity. The two types of receptor, MR and GR, mediate different effects, however. Thus, interaction of corticosterone and/or cortisol with MR in the hippo-campus is thought to be responsible for the tonic influences of steroid hormones on neuronal energy metabolism and excitability. In contrast, the interaction of gluco-corticoids with GR appears to mediate stress-activated regulation of neuronal energy metabolism and circuitry. As such, this receptor has a crucial role in terminating the stress response and in controlling subsequent adaptive behaviour (McEwen et al., 1986; Dallman et al., 1987; de Kloet, 1991).

From the above, it can be seen that some of the effects of glucocorticoids are slow (but long-lasting) and gene-mediated (Joëls and de Kloet, 1992a,b), while others are rapid. The latter appear to be mediated by GABA$_A$ receptors. This is possible because the GABA$_A$ receptor complex possesses recognition sites for certain steroids, particularly for metabolites of corticosterone and progesterone (Lambert et al., 1987; Schumacher and McEwen, 1989; Sutanto et al., 1989; see Section 12.4.1). These steroid metabolites originate both from the periphery and from metabolism in the brain and hence can be termed 'neurosteroids' (Le Goascogne et al., 1987). It is also possible that steroids such as corticosterone may have a direct action at a membrane-bound steroid receptor which is pharmacologically distinct from the GABA$_A$ receptor (Orchinik et al., 1991). So far, this latter membrane receptor has been described only in the brain of the amphibian Taricha granulosa, which is known to have rapid behavioural responses to corticosterone.

12.4.1.1 Corticosterone

There is extensive evidence that stress-related modifications in GABA$_A$ receptor function are induced by corticosteroids (Otero-Losada, 1988a). Adrenalectomy reduces binding of radiolabelled benzodiazepine agonists to GABA$_A$ receptors in certain brain regions (cerebral cortex, cerebellum and hippocampus); corticosterone replacement reverses this effect. Moreover, nanomolar concentrations of corticosterone enhance binding of GABA agonists in these brain regions. This increase is explained by increased binding affinity for the GABA$_A$ receptor (Majeswka et al., 1985).

The hypothalamus may well be different in this respect because, in this brain region, agonist binding was increased by adrenalectomy (Kendall et al., 1982). Also corticosterone lowered GABA levels and [^{3}H]GABA uptake in the mediobasal hypothalamus and corpus striatum, but not in the cerebral frontal cortex. Although these changes suggest increased turnover of GABA, the activities of enzymes involved in GABA biosynthesis and metabolism, glutamic acid decarboxylase and GABA-transaminase, were unaffected. In a later study corticosterone administration did affect the activities of these enzymes, however (Otero-Losada, 1988b). This corticosterone-induced increase in GABA turnover and reduction [^{3}H]GABA uptake could be linked with the anti-convulsant role of corticosteroids (Majewska et al., 1985). It is possible that corticosteroids may act as a 'natural anti-convulsant' which attenuates stress-induced over-stimulation of neurones in the brain.

The corticosterone-induced increase in turnover of GABA in the basomedial hypothalamus is likely to potentiate the effects of GABA. GABA inhibits CRF secretion (see Section 12.2.1) and this will lead to reduced secretion of ACTH and corticosterone. Hence, potentiation by corticosterone of the inhibitory action of GABA in the hypothalamus could participate in a feedback mechanism regulating HPA-function.

BDZ receptor binding, among other sites on the GABA$_A$ receptor complex, is subject to steroid-induced alterations in the CNS. Thus BDZ receptor binding, as labelled *in vivo* with the BDZ receptor antagonist [^{3}H]flumazenil, was increased in the cortex, hypothalamus and hippocampus in mice one week after adrenalectomy. This was explained by an increase in receptor number with no change in the binding affinity. These changes were reversed by physiological doses of corticosterone or chronic treatment with the mineralocorticoid, deoxycorticosterone, but not by chronic dexamethasone, aldosterone or dihydroprogesterone (Miller *et al.*, 1988). This effect of corticosterone or deoxycorticosterone (DOC) is presumably mediated by the hippocampal (and perhaps hypothalamic) mineralocorticoid receptor (MR).

The hippocampus may be a key area in terms of the relationship between corticosterone and BDZ receptors. Elevation of plasma corticosterone can be seen after sudden withdrawal of chronic treatment with the benzodiazepine, diazepam, and can be induced experimentally by bilateral intracerebral placement of diazepam micropellets into the dorsal and ventral hippocampus (Eisenberg, 1987). Secondly, one of the greatest changes in BDZ binding following adrenalectomy occurs in the hippocampus, the brain area with the highest density of MR (Miller *et al.*, 1988).

The implication that corticosterone-GABA–BDZ interactions may be important in the maintenance of basal activities and the regulation of the stress response (de Boer *et al.*, 1990), and that this is mediated *via* MR, is plausible but requires further study. Results from our laboratory have shown that steroids which preferentially bind to MR (corticosterone, DOC, spironolactone) increase ligand binding to the chloride channel of the GABA$_A$ receptor in the cortex and hippocampus, while those which compete for binding to the channel (allopregnanolone and allotetrahydroDOC) enhance binding of ligands to benzodiazepine receptors and do not bind to MR or GR: Sutanto *et al.*, 1989).

12.4.1.2 *Sex steroids: oestrogens and testosterone*

Gonadal steroids have profound influences on the activity of the GABAergic system. These interactions are of interest because, in rats and mice, the oestrous cycle influences the HPA response to stress. It appears that proestrous rats show greater responsiveness to restraint stress and proestrous mice show increased sensitivity to stress, compared with that at other stages of the cycle. In ovariectomized rats, stress responsiveness is significantly greater in oestradiol treated animals and this effect is attenuated by progesterone (Viau and Meaney, 1991).

Interestingly, in mice, responses to the anti-aversive effects of diazepam are also influenced by the oestrous cycle; this reflects modulation of GABA$_A$ receptor function by ovarian steroids (Carey *et al.*, 1992). Notwithstanding these interesting findings, interactions between sex steroid hormones and GABAergic modulation of the stress response have been little studied.

12.4.1.3 *Progesterone, deoxycorticosterone and their metabolites*

It has been over 50 years since Selye described the anaesthetic and sedative properties of several 3α-hydroxy A-ring reduced pregnane steroids, including the main progesterone metabolites, 3α-hydroxy-5α-pregnan-20-one (allopregnanolone) and 3α-hydroxy-5β-pregnan-20-one (pregnanolone), and the major deoxy-corticosterone (DOC) metabolite, 3α-21-dihydroxy-5α-pregnan-20-one (allotetra-hydroDOC) (Purdy *et al.*, 1990a,b). In addition to their sedative-hypnotic properties and anaesthetic effects (Kavaliers and Wiebe, 1987), these steroids also have profound behavioural effects when given to laboratory animals (see Section 12.5.2).

Progesterone, which can arise from both ovarian and adrenal cortical sources, is known to affect GABA receptor function in various regions of the CNS. The steroid is anxiolytic in humans (Nott *et al.*, 1976) and, in the rat, potentiates anti-conflict effects of benzodiazepines (Rodriguez-Sierra *et al.*, 1984). On the basis of electrophysiological studies, it has been proposed that the anxiolytic effects of progesterone are explained by its ability to enhance the actions of GABA and suppress the effects of the amino acid, glutamate (Smith *et al.*, 1987). The interaction of these steroids with the central GABAergic system has also been studied, some examples of which will be described below.

Prompted by the potential physiological importance of these steroid metabolites, various methods have been developed to measure them in plasma and brain. Allopregnanolone is now known to be a major progesterone metabolite in both endocrine tissues and the brain (Holzbauer *et al.*, 1984; Jung-Testas *et al.*, 1989; Purdy *et al.*, 1991). Using contemporary, sensitive assays for this compound, rapid (less than 5 min) and robust (4- to 20-fold) increases of allopregnanolone and allotetrahydroDOC were detected in the brain (hypothalamus and cerebral cortex) and in the plasma of rats after exposure to swim stress (Purdy *et al.*, 1991). Allopregnanolone is present in the cortex even in adrenalectomized rats, suggesting that this steroid is synthesized in the CNS independently of a peripheral source of substrate (Baulieu *et al.*, 1987).

These neuroactive steroids are, in turn, known to modulate $GABA_A$ receptor binding and function. In a series of radioligand binding studies, both allopregnanolone and allotetrahydroDOC, at nanomolar to low micromolar concentrations, stimulated binding of both [^{3}H]flunitrazepam to the benzodiazepine site and [^{3}H]muscimol to the GABA recognition site on the $GABA_A$ receptor. Evidence suggests that they also bind to the chloride channel of the $GABA_A$ receptor complex (Sutanto *et al.*, 1989; Hiemke *et al.*, 1991). These steroids also potentiate chloride ($^{36}Cl^-$) flux in rat cortical synaptosomes *in vitro* when the receptor is activated by the GABA agonist, muscimol (Majewska *et al.*, 1986; Morrow *et al.*, 1987). In many respects, the interaction of these two steroids with the $GABA_A$–BDZ-receptor complex resembles that of barbiturates. Thus, in electrophysiological studies, these steroids produce a prolongation of GABA-mediated inhibitory post-synaptic currents recorded at synapses between rat hippocampal neurones in culture (Harrison *et al.*, 1987a,b) or in a mouse spinal neurone preparation (Lambert *et al.*, 1987). In contrast, in an electrophysiological study using whole-cell voltage-clamp recordings from isolated cerebral cortical neurones of neonatal rats, two related steroids synthesized *de novo* in the CNS, pregnenolone sulphate and dehydroepiandrosterone sulphate, reversibly inhibited GABA-induced current, behaving as receptor antagonists (Majewska *et al.*, 1988, 1990; Majewska and Schwartz, 1987). Reasons for these contrasting findings are unclear although the antagonist action of pregnenalone sulphate and dehydroepiandrosterone sulphate occur at high concentrations only and may not occur under normal physiological conditions.

12.4.2 PEPTIDES

The activity of the GABAergic system can be influenced by many neuropeptides including cholecystokinin and caerulin (Nagahama, 1989; Sano et al., 1989), somatostatin (Dichter et al., 1990) and the endogenous brain peptides, melanocyte-stimulating hormone inhibiting factor (MIF-1) and Tyr-MIF-1 (Miller et al., 1989). Some of these are even found in the same neurones as GABA (e.g. CRF and somatostatin; Meister et al., 1988; Legido et al., 1990). The association between GABA and some hypothalamic/anterior pituitary peptides has been described earlier (see Section 12.3.1)

Corda et al. (1990) have shown that an active subunit of ACTH, $ACTH_{1-24}$ and melanocyte-stimulating hormone produces a dose-dependent decrease of punished licking in the conflict test in rats. This 'proconflict' effect is blocked by diazepam or its antagonist, flumazenil. Another example of an anterior pituitary hormone which is closely linked to the GABAergic system in the regulation of stress is thyrotropin releasing hormone (TRH). Gastric ulceration in the rat, induced by cold restraint stress, is aggravated by bilateral microinjections of TRH into the central nucleus of the amygdala. Pretreatment with the agonist benzodiazepines, chlordiazepoxide or midazolam, prevents this effect of TRH (Ray et al., 1990).

A peptide action on the GABAergic system is also shown by the work of Klusa and co-workers (Klusa et al., 1990). It is now well known that behavioural tasks such as the elevated plus-maze, licking-conflict test and forced-swimming stress increase the density of hippocampal [^{3}H]flunitrazepam binding sites, decrease the density of [^{3}H]muscimol binding sites in the hippocampus and increase plasma corticosterone levels. The pentapeptide, thymopentin, does not on its own affect behaviour in these tests but pretreatment with this peptide prevents the changes in GABA-BDZ receptor density and the elevation in plasma corticosterone caused by the stress. Thymopentin may have a 'stress-protective' action, therefore. This action does not seem to be exerted directly at the $GABA_A$ receptor complex since thymopentin does not affect the binding of [^{3}H]flunitrazepam or muscimol in vitro or in vivo. It is more likely that it affects the release and/or metabolism of GABA.

$GABA_A$ receptor function can also be modulated by an endogenous peptide, diazepam binding inhibitor (DBI). DBI was initially proposed to be an endogenous ligand for neuronal BDZ recognition sites on the $GABA_A$ receptor. However, this peptide now appears to have multiple actions in the brain and peripheral tissues and stimulates steroid biosynthesis through an action at the mitochondrial benzodiazepine receptor (Costa and Guidotti, 1985, 1991; Papadopoulos et al., 1991). Levels of DBI and CRF in the cerebrospinal fluid are both elevated in depressed patients (see Section 12.6). On this basis, it has been proposed that DBI may have a role in coordinating responses to stress and may be a causal factor in the pathophysiology of depression (Roy et al., 1989).

12.4.3 PURINES, EXCITATORY AMINO ACIDS AND OTHER AGENTS

Many purine analogues interact with BDZ receptors in the rat brain (Kelley *et al.*, 1989, 1990a,b) and some of them exhibit anxiolytic activity. Phosphatidylserine, extracted from bovine cerebral cortex, modulates the central BDZ receptor sites: *in vivo* administration of this compound reduces [^{3}H]flunitrazepam binding in certain brain areas, such as that seen when rats were subjected to an acute swim stress (Levi de Stein *et al.*, 1989).

It has been observed that microinjection of the GABA$_A$ antagonist, bicuculline methiodide, into the hypothalamus causes cardiorespiratory and behavioural changes resembling those seen in emotional stress. The role of excitatory amino acids (EEAs) in these changes have been studied recently. Of particular interest are the actions of glutamate at two of its different glutamate receptor subtypes: the N-methyl-D-aspartate (NMDA)-receptor and the kainate receptor (Murase *et al.*, 1989; Soltis and DiMicco, 1991a,b; Zeevalk and Nicklas, 1991). Like bicuculline methiodide, injection of NMDA or kainic acid, which act as agonists at their respective receptors, into the dorsomedial hypothalamus, produced dose-related increases in heart rate and blood pressure (Soltis and DiMicco, 1991a). Moreover, the effects of bicuculline methiodide were attenuated by administration of the non-selective antagonist, kynurenic acid (Soltis and DiMicco, 1991b).

The tachycardia effects of bicuculline were attenuated by both the NMDA-receptor antagonist, 2-amino-5-phosphonopentanoic acid, or the non-NMDA excitatory amino acid receptor antagonist 6-cyano-7-nitroquinoxaline-2,3-dione. This suggests that cardiovascular effects caused by blockade of GABAergic inhibition in the dorsomedial hypothalamus depend upon activation of local NMDA and non-NMDA excitatory amino acid receptors.

12.4.4 BENZODIAZEPINES, BARBITURATES AND ALCOHOL

Various types of compounds and hormonal treatments influence GABA$_A$ receptors and function. As mentioned above, as well as binding to benzodiazepine receptors on the GABA$_A$ receptor, the benzodiazepine derivative 4'-chlorodiazepam (Ro 5-4864) is thought to bind at a specific mitochondrial BDZ receptor which is pharmacologically distinct from the GABA$_A$ receptor. However, 4'-chlorodiazepam binding to cortical synaptosomes is modulated by allotetrahydroDOC and pregnanediol and the latter effect is GABA-dependent. 4'-Chlorodiazepam antagonized the enhancement of GABA-stimulated ^{36}Cl$^-$ uptake by both steroids (Belelli *et al.*, 1990b). This study suggests that a GABA-linked, 4'-chlorodiazepam binding site is coupled to the putative progesterone metabolite recognition site on the GABA$_A$ receptor (as distinct from the Ro 5-4864 site on the mitochondrial BDZ receptor) and confirms the GABA mimetic properties of pregnanediol and allotetrahydroDOC.

The activity of the GABA$_A$ receptor is also subject to modulation by a variety of hypnotic agents including depressant barbiturates (Keane and Biziere, 1987). The precise nature of the interaction between these drugs and the GABA$_A$–BDZ receptor complex is not yet clear. A clue may lie in the barbiturate-like effects of steroids on the GABA$_A$ receptor as mentioned previously (see Section 12.4.1).

The association between alcohol and stress has been known for a long time (see Pohorecky, 1990, for a review). Several possible mechanism(s) for alcohol-stress interaction have been proposed; these include mediation *via* endogenous opioids, noradrenergic neurones and components of the HPA-axis. However, ethanol appears to be a potent modulator of the GABAergic system. The responses vary depending on brain region, the particular paradigm studied, dose and route of administration of alcohol and the animal's sensitivity to alcohol. However, the study of Bower and Wehner (1989) showed that ethanol, given either *in vitro* (to brain preparations) or *in vivo* (injected into the animal) enhances GABA-stimulated [^{3}H]flunitrazepam binding in various brain regions in the mouse (cortex, cerebellum). Interestingly, changes in the cerebellum were diminished by adrenalectomy. Administration of ethanol to animals induces a number of physiological and behavioural changes. For example, ethanol induces motor impairment in the tilting plane test, on activity in the plus-maze test of anxiety, hypoactivity, hypothermia and hyperglycaemia (Wood *et al.*, 1989; Tuominen and Korpi, 1991). The ethanol antagonist, Ro 15-4513, which binds to a subtype of the GABA$_A$ receptor complex localized in the cerebellum, attenuates ethanol-induced hypoactivity but not hypothermia or hyperglycaemia. It appears that hypoactivity (i.e. the behavioural effect) of ethanol is GABA-mediated, while the physiological effects are not, or at least not by GABA$_A$ receptors sensitive to Ro 15-4513 (Wood *et al.*, 1989).

12.5 PHYSIOLOGICAL AND BEHAVIOURAL EFFECTS OF DRUGS WHICH MODIFY GABA RECEPTOR FUNCTION

Drugs which modify GABA receptor function show profound physiological and behavioural effects in the rat and human. Different forms of stress or drugs often produce divergent behavioural, physiological and biochemical effects (reviewed by Fernandez-Teruel *et al.*, 1991). This section summarizes the recent findings on the effects of specific non-steroidal and steroidal agonist and antagonist ligands for central or peripheral GABA and/or BDZ sites. Their effects in both physiological and behavioural paradigms are considered.

12.5.1 NON-STEROIDAL COMPOUNDS

The association between the GABA$_A$/BDZ receptor complex and alterations in neuroendocrine function is most clearly illustrated by the effects of administration of benzodiazepine receptor inverse agonists. These are compounds which bind to the benzodiazepine recognition site, but have effects which are opposite to those of benzodiazepine agonists. In particular, benzodiazepine inverse agonists induce behavioural and neuroendocrine changes characteristic of stress and anxiety. In contrast, the benzodiazepine agonist, diazepam, inhibits stress-induced increases in anterior pituitary hormone secretion (De Souza, 1990). Another anxiolytic benzodiazepine, chlordiazepoxide, reduces ulcer development induced by cold-restraint; this effect is antagonized by infusion of the BDZ receptor antagonist, flumazenil, into the amygdala (Sullivan *et al.*, 1989).

In a neuroendocrine study (de Boer *et al.*, 1990), neither the BDZ receptor agonist, chlordiazepoxide, nor the antagonist flumazenil, when given alone affected basal levels of plasma corticosterone or catecholamines (adrenaline and noradrenaline). However, the antagonist potentiated the corticosterone elevation caused by novel environment and suppressed the concomitant rise in plasma noradrenaline. At moderate doses, chlordiazepoxide attenuated the novelty-elicited rise in plasma corticosterone and adrenaline without affecting the rise in noradrenaline. These effects of chlordiazepoxide were blocked by flumazenil pretreatment, and indicate the involvement of the central BDZ receptor systems in the regulation of adrenomedullary and adrenocortical, but not sympathetic, responses to non-noxious stress.

Drugs which modify GABA function have profound effects on stress-induced analgesia such as that produced by inescapable footshock. The benzodiazepine agonists (e.g. clonazepam and diazepam) decrease the analgesic effect, while the antagonist, flumazenil, increases it. All the BDZ receptor ligands block the naloxone-induced antagonism of footshock analgesia (Rovati *et al.*, 1990).

Cardiovascular (and behavioural) manifestations of emotional stress are produced when drugs affecting GABA-mediated synaptic inhibition are injected into the posterior (dorsomedial) hypothalamic nucleus. Injection of the $GABA_A$ receptor agonist, muscimol, into this site abolishes stress-induced tachycardia. These findings support a role for activation of neurones in this brain area in the generation of stress-induced cardiovascular changes and for the control of this mechanism by local GABA receptors (Lisa *et al.*, 1989; Soltis and DiMicco, 1991a,b). Moreover, microinjection of bicuculline into the hypothalamus produces respiratory, cardiovascular and behavioural changes resembling those caused by emotional stress (Soltis and DiMicco, 1991a; see also Section 12.4.3).

A recent study using *in vivo* microdialysis has investigated the effects of perfusion with nipecotic acid (an inhibitor of GABA uptake) in conscious rats. This procedure, when coupled with microinjection of muscimol into the dorsomedial hypothalamus significantly reduced the tachycardia induced by air puff stress and was accompanied by an elevation in extracellular GABA. It appears, therefore, that extracellular levels of endogenous GABA in the dorsomedial hypothalamus may regulate the cardiovascular response to stress (Anderson and DiMicco, 1990).

4'-Chlorodiazepam (Ro 5-4864), has proconvulsant actions, potentiates shock-induced suppression of drinking and reduces activity in the social interaction test. On the other hand, studies using the central BDZ antagonist, flumazenil, or the inverse agonist n-butyl-β-carboline-3-carboxylate (β-CCB), report facilitation of retention of a step-down inhibitory avoidance task. The agonists, clonazepam and diazepam, had the opposite effect while 4'-chlorodiazepam had no effect (Izquierdo *et al.*, 1990). Flumazenil, given pre-training, antagonized the effect of the inverse agonists and, at higher doses, enhanced the retention of habituation to a buzzer, but not that to an open field. These findings have led to the suggestion that there is an endogenous mechanism mediated by BDZ agonists (and which is sensitive to inverse agonists), which normally down-regulates acquisition of certain behaviours. This mechanism is activated only when the tasks involve a certain degree of stress or anxiety (i.e. inhibitory avoidance) and not in less stressful or anxiogenic tasks (i.e. habituation to an open field).

12.5.2. STEROIDS

Corticosteroids and sex steroids have a profound influence on the GABAergic system (see Section 12.4.1) and behaviour influenced by this receptor. Corticosterone, for example, increases social interaction, a paradigm widely used to screen anxiolytic drugs (File *et al.*, 1978). Of particular interest is one of the progesterone metabolites, allopregnanolone, which is among the most potent of all known ligands for the GABA$_A$ receptor complex. Allopregnanolone, and some other steroid metabolites, may constitute endogenous barbiturate-like modulators of central GABA receptors and play an important role in the response of the CNS to stress. They may also be important in psychiatric disorders of suspected endocrine aetiology, such as premenstrual syndrome (Majewska *et al.*, 1986; Morrow *et al.*, 1987, 1988).

Progesterone has an anticonflict effect in ovariectomized rats (Rodriguez-Sierra *et al.*, 1984). It is thought that this may be due to the conversion of progesterone to allopregnanolone in the brain (Jung-Testas *et al.*, 1989; Krieger and Scott, 1984; Penning *et al.*, 1985); the progesterone metabolites, allopregnanolone, pregnanolone and allotetrahydroDOC all show various behavioural effects when administered to laboratory animals, including anti-conflict actions (Crawley *et al.*, 1986), anti-convulsant (Belelli *et al.*, 1989, 1990a) and analgesic actions (Kavaliers and Wiebe, 1987).

The findings described above clearly indicate that both steroids and non-steroidal compounds affect neuroendocrine, cardiovascular, nociceptive and behavioural paradigms by interacting with the GABA$_A$ receptor. Most of these studies have been carried out in the rat; in the human, compounds which modulate GABA receptor functions have been used therapeutically. Their use in disorders which are commonly linked with stress are described below.

12.6 GABA FUNCTION IN PSYCHIATRIC ILLNESS

Alterations in GABAergic neurotransmission and GABA$_A$ receptor function have been closely associated with physiological and pathophysiological conditions and have been hypothesized as the molecular basis of anxiety. GABAergic drugs have been used for therapeutic purposes; these include sodium valproate and γ-vinyl GABA (reviewed by Hammond and Wilder, 1985). BDZ-like agents and barbituric acid derivatives, acting on the GABAergic system or receptors, are used as antiepileptic drugs (Porter, 1989) and are potential treatment strategies in post-traumatic epilepsy (Wilmore, 1990).

In depression, the cerebrospinal fluid levels of CRF are elevated, together with those of diazepam binding inhibitor, DBI, a neuromodulatory peptide for GABA neurotransmission (see Section 12.4.2). DBI may have a role in co-ordinating responses to stress and in the pathophysiology of depression (Roy *et al.*, 1989; Barbaccia *et al.*, 1990; Vezzani *et al.*, 1991). Certain other substances, in conjunction with the GABAergic system, are associated with neurological and psychiatric disorders. Neuropeptide Y, for example, is a 36 amino acid tyrosine-rich peptide which co-exists with GABA, among other neurotransmitters, in discrete brain areas. The peptide has been linked with Alzheimer's disease, eating disorders and major depressive illness (Wahlestedt *et al.*, 1989).

GABAergic neurotransmission is involved in the modulation of mood; GABA concentrations in the CSF and plasma of depressed patients are lower than those in normal subjects (Gold *et al.*, 1980; Petty and Schlesser, 1981; Berrettini *et al.*, 1982). There is preliminary evidence that treatment of depressed patients with GABA mimetic agents, such as progabide, relieves symptoms of this disorder (Lloyd and Morselli, 1987). These authors have also shown that, in rats, antidepressants increased binding to a second GABA binding site, the GABA$_B$-receptor. This is a presynaptic receptor which effects feedback inhibition of release of GABA and other neurotransmitters. This could explain why the increase in dopamine-modulated secretion of growth hormone induced by the GABA$_B$ receptor agonist, baclofen, is blunted in depressed patients (Marchesi *et al.*, 1991). Not all studies find an increase in the density of GABA$_B$ receptors after chronic administration of antidepressants, however. Finally, in a pilot study, the agonist for the mitochondrial benzodiazepine receptor, PK 11195, was beneficial in patients suffering from anxiety and depression without causing sedation (Ansseau *et al.*, 1991).

12.7 CONCLUDING REMARKS

This chapter has reviewed the role of gamma aminobutyric acid (GABA) in the regulation of the stress response. In response to stress, the GABAergic system undergoes alterations which depend upon the nature and intensity of the stress, the condition and age of the animal, and the brain (or peripheral) areas investigated.

Increased secretion of HPA-hormones by stress is well documented and commonly used as an index of the stress response. GABA influences the activity of the HPA-axis at several levels: it inhibits the release of CRF from the hypothalamus and possibly affects the secretion of ACTH from the anterior pituitary gland and corticosteroidogenesis in the adrenal cortex. In turn, the GABAergic system is influenced by (1) steroid hormones (and their metabolites), (2) peptides, and (3) purines and excitatory amino acids.

Modifications of GABAergic function by benzodiazepines and GABA$_A$ receptor antagonists change physiological and behavioural responses to stress. Catecholamines and various neurotransmitters also affect the GABAergic activity in response to stress. Finally, alterations in GABA turnover and GABA receptor function are closely associated with pathophysiological disorders; many of which are effectively treated by agents which modify GABAergic neurotransmission.

ACKNOWLEDGEMENTS

We are grateful to Drs Jonathan Fry and Margaret Carey for critically reading the first version of this manuscript.

12.8 REFERENCES

Acosta, G.B., Otero-Losada, M.E. and Rubio, M.C. (1990) Chemical stress and GABAergic central system. *Gen. Pharmacol.*, **21**: 517–20

Anderson, J.J. and DiMicco, J.A. (1990) Effect of local inhibition of γ-aminobutyric acid uptake in the dorsomedial hypothalamus on extracellular levels of γ-aminobutyric acid and on stress-induced tachycardia: a study using microdialysis. *J. Pharmacol. Exp. Ther.*, **255**: 1399–1407

Ansseau, M., von Frenckell, R., Cerfontaine, J.L. and Paprt, P. (1991) Pilot study of PK 11195, a selective ligand for the peripheral-type benzodiazepine binding sites, in patients with anxious or depressive symptomatology. *Pharmacopsychiatry*, **24**: 8–12

Antelman, S.M., Knopf, S., Kocan, D. *et al.* (1988) One stressful event blocks multiple actions of diazepam for up to at least a month. *Brain Res.*, **445**: 380–5

Apud, J.A., Cocchi, D., Locatelli, V. *et al.* (1989) Biochemical and functional aspects on the control of prolactin release by the hypothalamo-pituitary GABAergic system. *Psychoneuroendo.*, **14**: 3–17

Barbaccia, M.L., Berkovich, A., Guarneri, P. and Slobodyansky, E. (1990) DBI (diazepam binding inhibitor): the precursor of a family of endogenous modulators of GABA$_A$ receptor function. History, perspectives and clinical implications. *Neurochem. Res.*, **15**: 161–8

Barnhill, J.G., Miller, L.G., Greenblatt, D.J. *et al.* (1991) Benzodiazepine receptor binding response to acute and chronic stress is increased in aging animals. *Pharmacology*, **42**: 181–7

Baulieu, E.E., Robel, P., Vatier, O. *et al.* (1987) Neurosteroids: pregnenolone and dehydroepiandrosterone in the brain, in K. Fuxe, L. Agnati (eds.), *Receptor-Receptor Interactions: A New Intramembrane Integrative Mechanism*, MacMillan, London, pp 89–104

Belelli, D., Bolger, M.B. and Gee, K.W. (1989) Anticonvulsant profile of the progesterone metabolite 5α-pregnan-3α-ol-20-one. *Eur. J. Pharmacol.*, **166**: 325–9

Belelli, D., McCauley, L. and Gee, K.W. (1990a) Anticonvulsant steroids and the GABA/benzodiazepine receptor-chloride ionophore complex. *Neurosci. Biobehav. Rev.*, **14**: 315–22

Belelli, D., McCauley, L. and Gee, K.W. (1990b) Heterotropic cooperativity between putative recognition sites for progesterone metabolites and the atypical benzodiazepine Ro 5-4864. *J. Neurochem.*, **55**: 83–7

Berrettini, W.H., Nurnberger, J.I., Hare, T. *et al.* (1982) Plasma and CSF GABA in effective illness. *Br. J. Psychiatr.*, **141**: 483–7

Biggio, G., Concas, A., Corda, M.G. *et al.* (1990) GABAergic and dopaminergic transmission in the rat cerebral cortex: effect of stress, anxiolytic and anxiogenic drugs. *Pharmacol. Ther.*, **48**: 121–42

Bolden, S.W., Hambley, J.W., Johnston, G.A. and Rogers, L.J. (1990) Neonatal stress and long-term modulation of GABA receptors in the brain. *Neurosci. Lett.*, **111**: 258–62

Bower, B.J. and Wehner, J.M. (1989) Interaction of ethanol and stress with the GABA/BDZ receptor in LS and SS mice. *Brain Res. Bull.*, **23**: 53–9

Calogero, A.E., Kamilaris, T.C., Bernardini, R. *et al.* (1990) Effects of peripheral benzodiazepine receptor ligands on hypothalamic-pituitary-adrenal axis function in the rat. *J. Pharmacol. Exp. Ther.*, **253**: 729–37

Carey, M.P., Billing, A.E. and Fry, J.P. (1992) Fluctuations in responses to diazepam during the oestrous cycle in the mouse. *Pharmacol. Biochem. Behav.*, **41**: 719–25

Corda, M.G., Orlandi, M. and Fratta, W. (1990) Proconflict effect of ACTH1-24: interaction with benzodiazepines. *Pharmacol. Biochem. Behav.*, **36**: 631–4

Costa, E. and Guidotti, A. (1985) Endogenous ligands for benzodiazepine recognition sites. *Biochem. Pharmacol.*, **34**: 3399–403

Costa, E. and Guidotti, A. (1991) Diazepam binding inhibitor (DBI): a peptide with multiple biological actions. *Life Sci.*, **49**: 325–44

Crawley, J.N., Glowa, J.R., Majewska, M.D. and Paul, S.M. (1986) Anxiolytic activity of an endogenous adrenal steroid. *Brain Res.*, **398**: 382–5

Dallman, M.F., Akana, S.F., Cascio, C.S. *et al.* (1987) Characterization of corticosterone feedback regulation of ACTH secretion. *Ann. N.Y. Acad. Sci.*, **512**: 402–15

Dar, D.E., Weizman, A., Karp, L. *et al.* (1991) Platelet peripheral benzodiazepine receptors in repeated stress. *Life. Sci.*, **48:** 341–6

De Boer, S.F., van der Gugten, J. and Slangen, J.L. (1990) Brain benzodiazepine receptor-mediated effects on plasma catecholamines and corticosterone concentrations in rats. *Brain Res. Bull.*, **24:** 843–7

Decavel, C. and van den Pol, A.N. (1990) GABA: a dominant neurotransmitter in the hypothalamus. *J. Comp. Neurol.*, **302:** 1019–37

DeFeudis, F.V. (1990) Overview-GABA$_A$ receptors. *Ann. N.Y. Acad. Sci.*, **585:** 231–40

De Kloet, E.R. (1991) Brain corticosteroid receptor balance and homeostatic control. *Frontiers in Neuroendocrinol.*, **12:** 95–164

De Souza, E.B. (1990) Neuroendocrine effects of benzodiazepines. *J. Psychiat. Res.*, 24 (Suppl. 2): 111–19

Dichter, M.A., Wang, H.L. and Reisine, T. (1990) Electrophysiological effects of somatostatin-14 and somatostatin-28 on mammalian central nervous system neurons. *Metabolism*, **39** (Suppl. 2): 86–90

Dornhorst, A., Jenkins, J.S., Lamberts, S.W.J. *et al.* (1983) The evaluation of sodium valproate in the treatment of Nelson's syndrome. *J. Clin. Endocrinol. Metab.*, **56:** 985–91

Drugan, R.C. and Holmes, P.V. (1991) Central and peripheral benzodiazepine receptors: involvement in an organism's response to physical and psychological stress. *Neurosci. Biobehav. Rev.*, **15:** 277–98

Drugan, R.C., Basile, A.S., Crawley, J.N. *et al.* (1988) Characterization of stress-induced alterations in [^{3}H]Ro5-4864 binding to peripheral benzodiazepine receptors in rat heart and kidney. *Pharmacol. Biochem. Behav.*, **30:** 1015–20

Drugan, R.C., Morrow, A.L., Weizman, R. *et al.* (1989) Stress-induced behavioural depression in the rat is associated with a decrease in GABA receptor-mediated chloride ion flux and brain benzodiazepine receptor occupancy. *Brain Res.*, **487:** 45–51

Drugan, R.C., Holmes, P.V. and Stringer, A.P. (1991) Sexual dimorphism of stress-induced changes in renal and peripheral benzodiazepine receptors in rat. *Neuropharmacology*, **30:** 413–16

Edwards, C.R.W., Stewart, P.M., Burt, D. *et al.* (1988) Tissue localization of 11β-hydroxysteroid dehydrogenase. Paracrine protector of the mineralocorticoid receptor. *The Lancet*, October 29, 986–9

Eisenberg, R.M. (1987) Diazepam withdrawal as demonstrated by changes in plasma corticosterone: a role for the hippocampus. *Life Sci.*, **40:** 817–25

Fernandez-Teruel, A., Escorihuela, R.M., Tobena, A. and Driscoll, P. (1991) Stress and putative endogenous ligands for benzodiazepine receptors: the importance of characteristics of the aversive situation and of differential emotionality in experimental animals. *Experentia*, **47:** 1051–6

File, S.E., Vellucci, S.V. and Wendland, S. (1978) Corticosterone – an anxiogenic or an anxiolytic agent? *J. Pharm. Pharmacol.*, **31:** 300–5

Gold, B.I., Bowers, M.B., Roth, R.H. and Sweeney, D.W. (1980) GABA levels in cerebrospinal fluid of patients with psychiatric disorders. *Am. J. Psychiatr.*, **137:** 362–4

Grandison, L., Cavagrini, F., Schmid, R. *et al.* (1982) GABA and benzodiazepine-binding sites in human anterior pituitary. *J. Clin. Endocrinol. Metab.*, **54:** 597–603

Green, A.R., Peralta, E., Hong, J.S. *et al.* (1978) Alterations in GABA metabolism and met-enkephalin content in rat brain following repeated electrical shocks. *J. Neurochem.*, **31:** 607–18

Hammond, E.J. and Wilder, B.J. (1985) Minireview – Gamma vinyl GABA. *Gen. Pharmacol.*, **16:** 441–7

Harrison, N.L. and Simmonds, M.A. (1984) Modulation of the GABA receptor complex by a steroid anaesthetic. *Brain Res.*, **323:** 287–92

Harrison, N.L., Majewska, M.D., Harrington, J.W. and Barker, J.L. (1987a) Structure-activity relationships for steroid interaction with the GABA$_A$ receptor complex. *J. Pharmacol. Exp. Ther.*, **241:** 346–53

Harrison, N.L., Vicini, S. and Barker, J.L. (1987) A steroid anaesthetic prolongs inhibitory postsynaptic currents in cultured rat hippocampal neurons. *J. Neurosci.*, **7:** 604–9

Hiemke, C., Jussofie, A. and Jüptner, M. (1991) Evidence that 3α-hydroxy-5α-pregnan-20-one is a physiologically relevant modulator of GABA-ergic neurotransmission. *Psychoneuroendo.*, **16**: 517–23

Hillhouse, E.W. and Milton, N.G.N. (1989) Effect of noradrenaline and GABA on the secretion of CRF-41 and AVP from rat hypothalamus *in vitro*. *J. Physiol.*, **122**: 719–23

Holzbauer, M., Birmingham, M.K., De Nicola, A.F. and Oliver, J.T. (1984) *In vivo* secretion of 3α-hydroxy-5α-pregnan-20-one, a potent anaesthetic steroid, by the adrenal gland of the rat. *J. Ster. Biochem.*, **22**: 97–102

Izquierdo, I., Pereira, M.E. and Medina, J.H. (1990) Benzodiazepine receptor ligand influences on acquisition: suggestion of an endogenous modulatory mechanism mediated by benzodiazepine receptors. *Behav. Neural. Biol.*, **54**: 27–41

Joëls, M. and de Kloet, E.R. (1992a) Control of neuronal excitability by corticosteroid hormones. *Trends in Neurosci.*, **15**: 25–30

Joëls, M. and de Kloet, E.R. (1992b) Coordinative mineralocorticoid and glucocorticoid receptor-mediated control of responses to serotonin in rat hippocampus. *Neuroendocrinology*, **55**: 344–50

Jones, M.T. (1979) Control of adrenocortical secretion, in V.H.T. James (ed.), *The Adrenal Gland*, Raven Press, New York, pp 93–130

Jones, M.T. and Gillham, B. (1988) Factors involved in the regulation of adrenocorticotropic/β-lipotrophic hormone. *Physiol. Rev.*, **68**: 743–818

Jones, M.T., Gillham, B., Altaher, A.R.H. *et al.* (1984) Clinical and experimental studies on the role of GABA in the regulation of ACTH secretion: A review. *Psychoneuroendocrinol.*, **9**: 107–23

Jung-Testas, I., Hu, Z.Y., Baulieu, E.E. and Robel, P. (1989) Neurosteroids: biosynthesis of pregnenolone and progesterone in primary cultures of rat glial cells. *Endocrinology*, **125**: 2083–91

Kang, I., Thompson, M.L., Heller, J. and Miller, L.G. (1991) Persistent elevation of GABA receptor subunit mRNAs following social stress. *Brain Res. Bull.*, **26**: 809–12

Kavaliers, M. and Wiebe, J.P. (1987) Analgesic effects of the progesterone metabolite, 3α-hydroxy-5α-pregnan-20-one, and possible mode of action in mice. *Brain Res.*, **415**: 393–8

Keane, P.E. and Biziere, K. (1987) The effects of general anaesthetics on GABAergic synaptic transmission. *Life Sci.*, **41**: 1437–41

Kelley, J.L., McLean, E.W., Ferris, R.M. and Howard, J.L. (1989) Benzodiazepine receptor binding activity of 6,9-disubstituted purines. *J. Med. Chem.*, **32**: 1020–4

Kelley, J.L., McLean, E.W., Ferris, R.M. and Howard, J.L. (1990a) Benzodiazepine receptor binding activity of 8-substituted-9-(3-substituted-benzyl)-6-(dimethylamino)-9H-purines. *J. Med. Chem.*, **33**: 196–202

Kelley, J.L., McLean, E.W., Ferris, R.M. and Howard, J.L. (1990b) Benzodiazepine receptor binding activity of 9-(1-phenylethyl)purines. *J. Med. Chem.*, **33**: 1910–14

Kendall, D.A., McEwen, B.S. and Enna, S.J. (1982) The influence of ACTH and corticosterone on ³H-GABA receptor binding in rat brain. *Brain Res.*, **23**: 365–74

Klusa, V., Kiivet, R.A., Muceniece, R. *et al.* (1990) Thymopentin antagonizes stress-induced changes of GABA/benzo-diazepine receptor complex. *Regul. Pept.*, **27**: 355–65

Krieger, N.R. and Scott, R.G. (1984) 3α-Hydroxysteroid oxidoreductase in rat brain. *J. Neurochem.*, **42**: 887–90

Lambert, J.J., Peters J.A. and Cottrell, G.A. (1987) Actions of synthetic and endogenous steroids on the GABA$_A$ receptor. *Trends in Pharmacological Sci.*, **8**: 224–7

Legido, A., Reichlin, S., Dichter, M.A. and Buchhalter, J. (1990) Expression of somatostatin and GABA immunoreactivity in cultures of rat hippocampus. *Peptides*, **11**: 103–9

Le Goascogne, G., Robel, P., Gouézou, M. *et al.* (1987) Neurosteroids: Cytochrome P-450$_{SCC}$ in rat brain. *Science*, **237**: 1212–15

Levi de Stein, M., Medina, J.H. and De Robertis, E. (1989) *In vivo* and *in vitro* modulation of central type benzodiazepine receptors by phosphatidylserine. *Brain Res. (Mol. Brain Res.)*, **5**: 9–15

Lisa, M., Marmo, E., Wible, J.H. Jr. and DiMicco, J.A. (1989) Injection of muscimol into posterior hypothalamus blocks stress-induced tachycardia. *Am. J. Physiol.*, **257**: R246–51

Lloyd, K.G. and Morselli, P.L. (1987) Psychopharmacology of GABAergic drugs, in Y. Meltzer (ed.), *Psychopharmacology: The Third Generation of Progress*, Raven Press, New York, pp 183–95

Majewska, M.D. and Schwartz, R.D. (1987) Pregnenolone-sulphate: an endogenous antagonist of the γ-aminobutyric acid receptor complex in brain?. *Brain Res.*, **404:** 355–60

Majewska, M.D., Bisserbe, J-C. and Eskay, R.L. (1985) Glucocorticoids are modulators of GABA$_A$ receptors in brain. *Brain Res.*, **339:** 178–82

Majewska, M.D., Harrison, N.L., Schwartz, R.D. *et al.* (1986) Steroid hormone metabolites are barbiturate-like modulators of the GABA receptors. *Science*, **232:** 1004–7

Majewska, M.D., Mienville, J-M. and Vicini, S. (1988) Neurosteroid pregnenolone sulfate antagonizes electrophysiological responses to GABA in neurons. *Neurosci. Lett.*, **90:** 279–84

Majewska, M.D., Demirgoren, S., Spivak, C.E. and London, E.D. (1990) The neurosteroid dehydroepiandrosterone sulfate is an allosteric antagonist of the GABA$_A$ receptor. *Brain Res.*, **526:** 143–6

Makara, G.B. and Stark, E. (1974) Effect of gamma-aminobutyric acid (GABA) and GABA antagonistic drugs on ACTH release. *Neuroendocrinology*, **16:** 178–90

Marchesi, C., Chiodera, P., De Ferri, A. *et al.* (1991) Reduction of GH response to the GABA-B agonist baclofen in patients with major depression. *Psychoneuroendo.*, **16:** 475–9

McCann, S.M. and Rettori, V. (1986) Gammaamino butyric acid (GABA) controls anterior pituitary hormone secretion, in G. Racagni and A.O. Donoso (eds.), *GABA and Endocrine Function*, Raven Press, New York, pp 173–89

McEwen, B.S., de Kloet, E.R. and Rostene, W. (1986) Adrenal steroid receptors and actions in the nervous system. *Physiol. Rev.*, **66:** 1121–88

Meister, B., Hökfelt, T., Geffard, M. and Oertel, W. (1988) Glutamic acid decarboxylase- and GABA-like immunoreactivities in corticotropin-releasing factor-containing parvocellular neurons of the hypothalamic paraventricular nucleus. *Neuroendocrinology*, **48:** 516–26

Miguez, M.I. and Aldegunde, M. (1990a) Effect of naloxone on the secretion of corticosterone induced by gamma-hydrobutyric acid in male rats. *J. Neuroendocrinol.*, **2:** 501–3

Miguez, M.I. and Aldegunde, M. (1990b) Effect of GABA on corticosterone secretion: involvement of the noradrenergic system. *Life Sci.*, **46:** 875–80

Miller, L.G., Thompson, M.L., Greenblatt, D.J. *et al.* (1987) Rapid increase in brain benzodiazepine receptor binding following defeat stress in mice. *Brain Res.*, **414:** 395–400

Miller, L.G., Greenblatt, D.J., Barnhill, J.G. *et al.* (1988) Modulation of benzodiazepine receptor binding in mouse brain by adrenalectomy and steroid replacement. *Brain Res.*, **446:** 314–20

Miller, L.G., Kastin, A.J. and Roy, R.B. (1989) MIF-1 and Tyr-MIF-1 augment muscimol-stimulated chloride uptake in cerebral cortex. *Brain Res. Bull.*, **23:** 413–15

Morrow, A.L., Suzdak, P.D. and Paul, S.M. (1987) Steroid hormone metabolites potentiate GABA receptor-mediated chloride ion flux with nanomolar potency. *Eur. J. Pharmacol.*, **142:** 483–5

Morrow, A.L., Suzdak, P.D. and Paul, S.M. (1988) Benzodiazepine, barbiturate, ethanol and hypnotic steroid hormone modulator of GABA-mediated chloride ion transport in rat brain synaptoneurosomes. *Adv. Biochem. Psychopharmacol.*, **45:** 247–61

Muhkin, A.G., Papadopoulos, V., Costa, E. and Krueger, K.E. (1989) Mitochondrial benzodiazepine receptors regulate steroid biosynthesis. *Proc. Natl. Acad. Sci. USA*, **86:** 9813–16

Murase, K., Ryu, P.D. and Randic, M. (1989) Excitatory and inhibitory amino acid and peptide-induced responses in acutely isolated rat spinal dorsal horn neurons. *Neurosci. Lett.*, **103:** 56–63

Nagahama, H. (1989) Acute and long-lasting effects of peripheral injection of caerulein and CCK-8 on the central GABAergic system in mice. *Peptides*, **10:** 1247–51

Nott, P.N., Franklin, M., Armitage, C. and Gelder, M.G. (1976) Hormonal changes and mood in puerperium. *Br. J. Psychiatry*, **128:** 379–83

Olsen, R.W. and Tobin, A.J. (1990) Molecular biology of GABA$_A$ receptors. *FASEB J.*, **4:** 1469–80

Orchinik, M., Murray, T.F. and Moore, F.L. (1991) A corticosteroid receptor in neuronal membranes. *Science*, **252:** 1848–51

Otero-Losada, M.E. (1988a) Acute stress and GABAergic function in the rat brain. *Br. J. Pharmacol.*, **93**: 483–90

Otero-Losada, M.E. (1988b) Changes in the central GABAergic system after acute treatment with corticosterone. *Naunyn-Schmiedeberg's Arch. Pharmacol.*, **337**: 669–74

Otero-Losada, M.E. (1989) Acute stress and GABAergic function in the rat brain. *Br. J. Pharmacol.*, **96**: 507–12

Papadopoulos, V., Berkovich, A., Grueger, K.E. *et al.* (1991) Diazepam binding inhibitor and its processing products stimulate mitochondrial steroid biosynthesis via an interaction with mitochondrial benzodiazepine receptors. *Endocrinology*, **129**: 1481–8

Penning, T.M., Sharp, R.B. and Krieger, N.R. (1985) Purification and properties of 3α-hydroxysteroid dehydrogenase from rat brain cytosol. Inhibition by nonsteroidal anti-inflammatory drugs and progestins. *J. Biol. Chem.*, **260**: 15266–72

Petty, F. and Schlesser, M.A. (1981) Plasma GABA in affective illness. *J. Affective Disord.*, **3**: 339–43

Pohorecky, L.A., Cotler, S., Carbone, J.J. and Roberts, P. (1988) Factors modifying the effect of diazepam on plasma corticosterone levels in rats. *Life Sci.*, **43**: 2159–67

Pohorecky, L.A. (1990) Interaction of ethanol and stress: research with experimental animals – an update. *Alcohol-Alcohol*, **25**: 263–76

Porter, R.J. (1989) Mechanisms of action of new antiepileptic drugs. *Epilepsia*, **30** (Suppl. 1): S29–34

Purdy, R.H., Moore Jr., P.H., Narasimha Rao, P. *et al.* (1990a) Radioimmunoassay of 3α-hydroxy-5α-pregnan-20-one in rat and human plasma. *Steroids*, **55**: 290–6

Purdy, R.H., Morrow, A.L., Blinn, J.R. and Paul, S.M. (1990b) Synthesis, metabolism, and pharmacological activity of 3α-hydroxy steroids which potentiate GABA-receptor-mediated chloride ion uptake in rat cerebral cortical synaptoneurosomes. *J. Med. Chem.*, **33**: 1572–81

Purdy, R.H., Morrow, A.L, Moore, P.H. and Steven, P.M. (1991) Stress-induced elevations of γ-aminobutyric acid type A receptor-active steroids in the rat brain. *Proc. Natl. Acad. Sci. USA*, **88**: 4553–7

Rago, L., Kiivet, R.A., Harra, J. and Pold, M. (1989) Central and peripheral-type benzodiazepine receptors: similar regulation by stress and GABA receptor agonists. *Pharmacol. Biochem. Behav.*, **32**: 879–83

Ray, A., Henke, P.G. and Sullivan, R.M. (1990) Effects of intra-amygdalar thyrotropin releasing hormone (TRH) and its antagonism by atropine and benzodiazepine during stress ulcer formation in rats. *Pharmacol. Biochem. Behav.*, **36**: 597–601

Rodriguez-Sierra, J.F., Howard, J.L., Pollard, G.T. and Hendrickes, S.E. (1984) Effect of ovarian hormones on conflict behavior. *Psychoneuroendocrinology*, **9**: 293–300

Roy, A., Pickar, D., Gold, P. *et al.* (1989) Diazepam-binding inhibitor and corticotropin-releasing hormone in cerebrospinal fluid. *Acta Psychiatr. Scand.*, **80**: 287–91

Rovati, L.C., Sacerdote, P., Fumagalli, P. *et al.* (1990) Benzodiazepines and their antagonists interfere with opioid-dependent stress-induced analgesia. *Pharmacol. Biochem. Behav.*, **36**: 123–6

Sakai, R.R., Lakshmi, V., Monder, C. and McEwen, B.S. (1992) Immunocytochemical localization of 11β-hydroxysteroid dehydrogenase in hippocampus and other brain regions of the rat. *J. Neuroendocrinology*, **4**: 101–6

Sano, I., Taniyama, K. and Tanaka, C. (1989) Cholecystokinin, but not gastrin, induces gamma-aminobutyric acid release from myenteric neurons of the guinea pig ileum. *J. Pharmacol. Exp. Ther.*, **248**: 378–83

Schumacher, M. and McEwen, B.S. (1989) Steroid and barbiturate modulation of the GABA$_A$ receptor. Possible mechanisms. *Mol. Neurobiol.*, **3**: 275–304

Smith, S.S., Waterhouse, B.D., Chapin, J.K. and Woodward, D.J. (1987) Progesterone alters GABA and glutamate responsiveness: a possible mechanism for its anxiolytic action. *Brain Res.*, **400**: 353–9

Soltis, R.P. and DiMicco, J.A. (1991a) GABA$_A$ and excitatory amino acid receptors in dorsomedial hypothalamus and heart rate in rats. *Am. J. Physiol.*, **260**: R13–20

Soltis, R.P. and DiMicco, J.A. (1991b) Interaction of hypothalamic GABA$_A$ and excitatory amino acid receptors controlling heart rate in rats. *Am. J. Physiol.*, **261**: R427–33

Sullivan, R.M., Henke, P.G., Ray, A. *et al.* (1989) The GABA/benzodiazepine receptor complex in the central amygdalar nucleus and stress ulcers in rats. *Behav. Neural. Biol.*, **51**: 262–9

Sutanto, W., Handelmann, G., de Bree, F. and de Kloet, E.R. (1989) Multifaceted interaction of corticosteroids with the intracellular receptors and with membrane GABA$_A$ receptor complex in the rat brain. *J. Neuroendocrinol.*, **1**: 243–7

Thind, K.K. and Goldsmith, P.C. (1989) Corticotropin-releasing factor neurons in the periventricular hypothalamus of juvenile macaques. Synaptic evidence for a possible companion neurotransmitter. *Neuroendocrinology*, **50**: 351–8

Trullas, R., Havoundjian, H., Zamir, N. *et al.* (1987) Environmentally-induced modification of the benzodiazepine/GABA receptor coupled chloride ionophore. *Psychopharmacology*, **91**: 384–90

Tuominen, K. and Korpi, E.R. (1991) Lack of effects of handling habituation and swimming stress on ethanol-induced motor impairment and GABA$_A$ receptor function. *Acta Physiol. Scand.*, **141**: 409–13

Van Eekelen, J.A.M., Jiang, W., de Kloet, E.R. and Bohn, M.C. (1988) Distribution of the mineralocorticoid and the glucocorticoid receptor mRNAs in the rat hippocampus. *J. Neurosci. Res.*, **21**: 88–94

Vezzani, A., Serafini, R., Stasi, M.A. *et al.* (1991) Epileptogenic activity of two peptides derived from diazepam binding inhibitor after intrahippocampal injection in rats. *Epilepsia*, **32**: 597–603

Viau, V. and Meaney, M.J. (1991) Variations in the hypothalamic-pituitary-adrenal response to stress during the estrous cycle in the rat. *Endocrinology*, **129**: 2503–11

Vincent, S.R., Hökfelt, T. and Wu, J-Y. (1982) GABA neuron systems in hypothalamus and the pituitary gland. *Neuroendocrinology*, **34**: 117–23

Waddington, J.L. (1978) Behavioural evidence for GABAergic activity of the benzodiazepine flurazepam. *Eur. J. Pharmacol.*, **51**: 417–22

Wahlstedt, C., Ekman, R. and Widerlov, E. (1989) Neuropeptide Y (NPY) and the central nervous system: distribution effects and possible relationship to neurological and psychiatric disorders. *Prog. Neuropsychopharmaol. Biol. Psychiatr.*, **13**: 31–54

Wass, J.A.H. (1983) Growth hormone neuroregulation and the clinical relevance of somatostatin, in M.F. Scanlon (ed.), *Clinics in Endocrinology and Metabolism 12*, Saunders, London, pp 695–724

Weizman, R., Weizman, A., Kook, K.A. *et al.* (1989) Repeated swim stress alters brain benzodiazepine receptors measured *in vivo*. *J. Pharmacol. Exp. Ther.*, **249**: 701–7

Weizman, A., Bidder, M., Fares, F. and Gavish, M. (1990a) Food deprivation modulates γ-aminobutyric acid receptors and peripheral benzodiazepine binding sites in rats. *Brain Res.*, **535**: 96–100

Weizman, A., Weizman, R., Kook, K.A. *et al.* (1990b) Adrenalectomy prevents the stress-induced decrease in *in vivo* [^{3}H]Ro15-1788 binding to GABA$_A$ benzodiazepine receptors in the mouse. *Brain Res.*, **519**: 347–50

Wilmore, L.J. (1990) Post-traumatic epilepsy: cellular mechanism and implications for treatment. *Epilepsia*, **31** (Suppl. 3): S67–73

Wood, A.L., Healey, P.A., Menendez, J.A. *et al.* (1989) The intrinsic and interactive effects of Ro 15-4513 and ethanol on locomotor activity, body temperature, and blood glucose concentration. *Life Sci.*, **45**: 1467–73

Yoneda, Y., Kanmori, K., Ida, S. and Kuriyama, K. (1983) Stress-induced alterations in metabolism of GABA in rat brain. *J. Neurochem.*, **40**: 350–6

Zeevalk, G.D. and Nicklas, W.J. (1991) Mechanism underlying initiation of excitotoxicity associated with metabolic inhibition. *J. Pharmacol. Exp. Ther.*, **257**: 870–8

13

Role(s) of neuropeptides in responding and adaptation to stress: A focus on corticotropin-releasing factor and opioid peptides

DIANE M. HAYDEN-HIXSON[1] AND CHARLES B. NEMEROFF[2]

[1]INTEGRATED TOXICOLOGY PROGRAM AND DEPARTMENT OF PHARMACOLOGY
DUKE UNIVERSITY MEDICAL CENTER
DURHAM, NORTH CAROLINA, USA
[2]DEPARTMENT OF PSYCHIATRY
EMORY UNIVERSITY SCHOOL OF MEDICINE
ATLANTA, GEORGIA, USA

ISBN 0–12–663370–3

13.1 INTRODUCTION

Darwinian theory assumes that evolution selects for organisms that successfully adapt to environmental challenge. The process of adapting to environmental challenge is, in the broadest sense, the process of adapting to stress. Successful adaptation involves a series of physiological adjustments that support behavioural responses appropriate to the eliciting stressful stimulus. Acutely, these adjustments constitute a 'generic' response to stress that is integrated, rapid and nonspecific. Generic stress responding is adaptive because it confers on an organism the ability to respond to extreme or life-threatening challenges. While most real or perceived challenges seldom constitute a serious threat to the health of an organism, the failure to maintain systemic function within narrowly-defined limits does. It is not surprising, therefore, that organisms have also evolved complex regulatory mechanisms that allow them to generate stress responses appropriate to a broad range of eliciting stimuli.

Accumulated data indicate that neuropeptide circuits in the brain play important roles in these complex regulatory mechanisms. The list of neuropeptides hypothesized to be involved in modulating stress responding is extensive. It includes corticotropin-releasing factor/hormone (CRF/CRH), opioid peptides (dynorphins, β-endorphin, enkephalins), thyrotropin-releasing factor/hormone (TRF/TRH), neurotensin, cholecystokinin, substance P, neuropeptide Y, somatostatin- and bombesin-related peptides, vasopressin, and oxytocin.

This chapter is an attempt at a meaningful compromise between an oversimplified and comprehensive review of the role of neuropeptides in stress responding and adaptation. Two widely acccepted hypotheses are examined: first, that CRF plays important roles in initiating and integrating the whole-organism response to stress and secondly, that functional interactions between CRF and opioid peptides underlie successful adaptation to stress. Readers interested in broader treatments of the role of neuropeptides in stress responding and adaptation should consult recent reviews by Bissette (1991), Brown (1991a) and Gardiner and Bennett (1989).

This chapter attempts to synthesize an enormous body of literature. It integrates major reviews, and past and current original research papers on the roles of CRF and opioid peptides in stress responding and adaptation. The materials reviewed in this chapter have an inherent bias since most stress research uses rodents. However, data from other mammalian species are presented for comparison where available.

The neuroanatomical data are simplified as follows. Collections of brain stem nuclei with distinct functions (e.g. the inferior and superior collicular nuclei) and those with defined anatomical subdivisions (e.g. trigeminal, reticular, tegmental or raphe nuclei) are designated by collective terms (e.g. collicular nuclei, olivary nuclei). References to all other broad nuclear groupings (e.g. lateral thalamic or medial amygdaloid nuclei) follow accepted conventions. 'Neocortex' includes areas designated only as frontal, parietal, occipital or temporal cortex. 'Allocortex' includes areas specifically identified as entorhinal, piriform, or cingulate cortex. While facilitating tabulation of the data, these conventions obscure important distinctions in the neuroanatomy of the CRF and opioid peptide systems. Appendices 13.1–3 refer the interested reader to sources providing more comprehensive treatment.

13.2 STRESS AND CRF-MEDIATED MECHANISMS: AN OVERVIEW

In the decade following the isolation and characterization of the neuropeptide corticotropin-releasing factor (CRF; Spiess et al., 1981), it has become widely accepted that CRF plays a pivotal role in regulating adaptive and, possibly, pathological responses to stress. Considerable evidence suggests that CRF plays an important role in initiating and integrating centrally- and peripherally-mediated endocrine, autonomic and behavioural responses to stressful experiences. In fact, CRF-modulation of global stress responding is generally considered both essential and adaptive in the acute (fight or flight) time domain (Dunn and Berridge, 1990; Owens and Nemeroff, 1991).

This interpretation is underscored by what appear to be evolutionarily-conserved patterns of CRF-mediated mechanisms related to stress. CRF or CRF-like peptides exhibit 50–90% gene structure homology, similar pharmacological profiles and nearly identical biological actions in all vertebrate species examined thus far. However, recent clinical and preclinical evidence suggests that sustained activation of CRF-mediated mechanisms can have profound pathophysiological consequences in susceptible individuals. Specifically, dysfunctional CRF-mediated mechanisms are implicated in the pathogenesis of hypertension (Hattori et al., 1986) and a number of psychiatric disorders, including anxiety, posttraumatic stress disorder and eating and affective disorders (Gold and Chrousos, 1985; Banki et al., 1987; Nemeroff et al., 1984; Smith et al., 1989).

13.3 NEUROANATOMY OF THE CRF NEURONAL SYSTEMS

The neurobiological substrates of CRF-mediated modulation of adaptive and pathological responses to stress appear to involve at least two anatomically and functionally distinct neuronal systems. The first system, classically referred to as the 'stress axis', consists of highly circumscribed neuroendocrine circuits. Neuroendocrine CRF neurones originate in the hypothalamus, primarily from the medial parvocellular division of the paraventricular nucleus (pvPVN), and terminate in the external layer of the median eminence (Kawata et al., 1983; Swanson and Sawchenko, 1983). CRF released from the nerve terminals of these cells gains access to the anterior pituitary after passing through fenestrated capillaries into the hypophyseal (pituitary) portal system. In the anterior pituitary, CRF functions as a classical hypothalamic releasing factor, stimulating the synthesis and release of adrenocorticotropin (ACTH) and other pro-opiomelanocortin (POMC)-derived peptides from pituitary corticotrophs. In fact, CRF is considered to be, perhaps, the single most important physiological activator of the hypothalamic-pituitary-adrenal (HPA) axis in response to most physical and psychological stressors (Jones and Gillham, 1988; Rivier and Plotsky, 1986).

The second CRF system consists of diffuse neuronal circuits that lie outside of the neuroendocrine zone of the hypothalamus and pituitary. Nonendocrine CRF-responsive cells are differentially localized in brain areas participating in affective and behavioural arousal, autonomic regulation, sensory information processing, memory, and learning (Swanson et al., 1983; Sakanata et al., 1987a). CRF neuropeptide, its messenger ribonucleic acid (mRNA) and binding sites or receptors exhibit largely complementary, widespread and heterogeneous distributions in the central nervous system (CNS) of all mammalian species examined to date (see Appendix 13.1). In general, nonendocrine CRF-responsive neuronal circuits are preferentially localized in hypothalamic, limbic, neocortical and brainstem areas activated by stress.

Brain CRF receptors are predominantly localized in the neocortex and limbic system in both primates and rodents. In general, both rodents and primates show a poor correspondence between CRF neuropeptide-containing cells and CRF receptors in the neocortex, but a very good correspondence in the limbic system. Species differences in the anatomical distributions of nonendocrine CRF circuits do exist, however. Overall, CRF receptor densities are considerably higher in the limbic system of primates than of rodents (Millan et al., 1986; De Souza, 1987). This species difference coincides with a dramatic increase in the total number of limbic sites that are interconnected in higher, as compared to lower, mammals (Hauger et al., 1989).

As previously stated, high CRF receptor densities occur throughout the neocortex in both rodents and primates. In general, relatively uniform distributions of CRF receptors occur throughout the neocortex of rats and marmosets, but not cynomolgus monkeys. In the rat, CRF receptor densities are somewhat higher in the anterior cingulate cortex, the somatosensory area of the frontoparietal cortex and the auditory area of the temporal lobe than in other cortical areas. In the cynomolgus monkey, CRF receptors are more dense in prefrontal, orbital and insular cortices than in other neocortical brain areas (Hauger et al., 1989).

Neurones containing CRF can be identifed by immunohistochemical techniques and are heterogeneously distributed throughout the cerebral cortex in both rodents and higher mammals (Lewis et al., 1989). Cells containing CRF are diffusely distributed in the rat neocortex, except in the prefrontal and insular cortices where they are relatively enriched (Swanson et al., 1983). Most neurones showing CRF immunoreactivity (CRF-ir) in the rat cortex are bipolar in shape and appear to be interneurones. Compared with the rat, CRF-immunoreactive neurones in squirrel monkey cortical tissue are more widely distributed, morphologically more diverse and more enriched overall (Lewis et al., 1989). In primates, CRF-ir neurones are especially enriched in the cingulate cortex and, to a lesser degree, the association areas of the occipital, prefrontal, parietal and temporal cortices (Lewis et al.,

1989). Collectively these findings suggest that CRF-mediated stress-response mechanisms are as important, if not more so, in primates as in lower mammals.

13.4 FUNCTIONAL DISTINCTIONS BETWEEN THE CRF NEURAL SYSTEMS

It has become increasingly evident that neuroendocrine and nonendocrine CRF-responsive circuits have complementary but distinct functions in stress-related responding. They are complementary in that both play dynamic, and possibly critical, roles in simultaneously activating and coordinating compensatory homeostatic mechanisms following real or perceived threat to the organism. They are distinct in that the endocrine CRF system functions primarily to mediate the organism's peripheral responses to stress, while the nonendocrine CRF system mediates the organism's central responses (Lenz et al., 1987; Valentino and Wehby, 1988).

Functional distinctions between these two systems are clearly evident in the case of the regulatory changes that normally follow perturbations of the HPA axis. Both CRF systems exhibit extensive overlap with corticosteroid-responsive neural circuits in the CNS (Liposits et al., 1987). However, only the neuroendocrine CRF system appears to be sensitive to changes in the circulating levels of corticosteroids. Adrenalectomy (Aguilera et al., 1986), chronic stress (Hauger et al., 1988), chronic corticosteroid administration (Hauger et al., 1987) and chronic CRF infusion (Wynn et al., 1988) all selectively downregulate CRF receptors in the anterior pituitary, but not the CNS. Adrenalectomy upregulates CRF transcription (Imaki et al., 1991b), synthesis (Alonso et al., 1988) and release (Berkenbosch and Tilders, 1988) in the hypothalamic-pituitary system, but not in nonendocrine neural circuits (Imaki et al., 1991b).

Exogenous corticosteroid administration decreases CRF content (Jessop et al., 1990), downregulates CRF transcription (Imaki et al., 1991b) and dose-dependently inhibits CRF release from hypothalamic CRF neurones which project to the pituitary (Calogero et al., 1988). In contrast, exogenous corticosteroid administration either increases (Swanson and Simmons, 1989), or has no effect on, CRF content (Bagdy et al., 1990) and transcription (Imaki et al., 1991b) in nonendocrine neurones.

Functional distinctions between the neuroendocrine and nonendocrine CRF systems are also indicated by the different effects of peripheral and central CRF administration on endocrine, cardiovascular and visceral function and on locomotor, grooming, defensive withdrawal and conditioned avoidance behaviours. Furthermore, centrally-administered CRF has similar effects on cardiovascular and gastrointestinal function, locomotor activity, exploratory behaviour, feeding, grooming and active avoidance in adrenalectomized, hypophysectomized and intact animals (Eaves et al., 1985) (see Appendix 13.2). Finally, central, but not peripheral, administration of CRF antagonists blocks the effects of centrally-administered CRF on sympathetic and adrenomedullary tone (Brown et al., 1983), cardiovascular function (Fisher, 1989), and gastric acid secretion (Lenz et al., 1988) (see Appendix 13.3).

13.5 CRF AS A NEUROTRANSMITTER IN THE CNS

CRF released from nonendocrine cells exerts neurotransmitter-like actions and modulates CNS activity during stressful experiences. A neurotransmitter role for endogenous CRF in the CNS is consistent with the following observations. With few exceptions, similar distributions of CRF-immunoreactive cells or fibres, CRF mRNA, and CRF binding sites/receptors occur in the CNS of all species examined thus far (see Appendix 13.1). Findings that CRF is enriched in synaptosomes, compared with other samples derived by cell fractionation, suggest that this peptide is localized mostly in nerve terminals (Cain et al., 1991; see Chapter 10). Potassium-stimulated, calcium-dependent release of CRF from brain tissue is demonstrated in vitro (Smith et al., 1986). Direct CRF administration produces regionally-specific alterations in the firing rate of CNS neurones in vivo (Valentino and Foote, 1988) and in vitro (Aldenhoff et al., 1983).

Further, indirect evidence for a neurotransmitter role for endogenous CRF is found in accumulated data indicating that many stress-induced neurobiological events are functionally related to the integrity of nonendocrine CRF circuits. The central administration of CRF elicits a number of physiological and behavioural changes that mimic those elicited by stressful experiences (see Appendix 13.3). Parallels exist between the different activational effects of acute versus chronic exposure to CRF and to stressors (Krahn et al., 1990). Similarities also exist in the direction, magnitude and duration of behavioural and physiological responses as a function of CRF dose or stressor intensity (for a review see Dunn and Berridge, 1990). CRF antisera or antagonists, on the other hand, attenuate or reverse the activational effects of both centrally-administered CRF and stressful experiences (Rivier and Vale, 1983; Kalin et al., 1988) (see Appendix 13.3). Finally, acute as well as chronic exposure to stressors increase cerebrospinal fluid (CSF) concentrations of CRF (Britton et al., 1984) and result in anatomically heterogeneous alterations in CRF content, mRNA and receptor levels within the CNS. Collectively, these findings suggest that endogenous CRF acts as a neurotransmitter, regulating neural activity in central afferent and efferent pathways activated by stress.

13.6 EXOGENOUS CRF MIMICS STRESS-INDUCED NEUROBIOLOGICAL EVENTS

In laboratory animals, intracerebroventricular (ICV) administration of CRF results in a complete spectrum of physiological and behavioural changes symptomatic of stress (for reviews see Cole and Koob, 1991; Owens and Nemeroff, 1991). Most notably, these changes include increases in emotionality (Diamant and De Wied, 1991), cardiovascular and adrenomedullary function (Diamant and De Wied, 1991), oxygen consumption, glucose mobilization and utilization (e.g. Rothwell, 1990), as well as electroencephalographic activation (Ehlers, 1990). Like stress, CRF induces context-dependent changes in motor activity (Sherman and Kalin, 1986), decreases in sexual activity (Sirinathsinghji, 1986), and decreases in food consumption (Morley, 1987). As a general rule, the effects of ICV-administered CRF are biphasic, with low doses having effects that are consistent with an increase in attention or arousal, and high doses increasing anxiety or fearfulness (see Appendix 13.3).

13.6.1 PHYSIOLOGICAL ADJUSTMENTS

The common physiological adjustments induced by stress and centrally-administered CRF can be subdivided into five distinct categories: endocrine, autonomic, visceral, biochemical and electrophysiological effects. While similarities between the effects of stress and ICV-administered CRF have been extensively studied only in rodents, the limited data available from other species suggest that the findings in rodents generalize to other mammals.

13.6.1.1 Changes in endocrine function

Both stress and ICV administration of CRF activate the HPA axis in a range of species (rats: Rivier and Plotsky, 1986; mice: Dunn and Berridge, 1987; monkeys: Insel et al., 1984; sheep: Donald et al., 1983). In rodents, both stress and ICV CRF inhibit the hypothalamic-pituitary gonadal (HPG) axis (Gambacciani et al., 1986; Rivier et al., 1986), the hypothalamic-pituitary thyroid axis (Mitsuma et al., 1987) and growth hormone secretion (Ono et al., 1984). Stress and ICV-administered CRF also inhibit HPG axis function in primates. Unlike rodents, inhibitory actions of CRF on HPG axis function in primates appear to be peripherally-mediated (secondary to HPA axis activation) rather than centrally-mediated (Gindoff and Ferin, 1987).

13.6.1.2 Changes in autonomic function

Parallel autonomic changes induced by stress and CRF include increases in adrenomedullary tone in rats (Brown, 1986), monkeys (Kalin et al., 1983b) and dogs (Brown and Fisher, 1983). They also include increases in mean arterial blood pressure and heart rate in several species (Diamant and De Wied, 1991; Brown and Fisher, 1983; Kalin et al., 1983a; Scoggins et al., 1984).

13.6.1.3 Changes in visceral function

Acute stress and ICV-administered CRF have similar effects on visceral function in rats (Garrick et al., 1988), mice (Buéno and Gué, 1988) and dogs (Buéno and Fioramonti, 1986). These include inhibition of gastric emptying and acid secretion and the stimulation of large bowel transit and faecal excretion. However, with few exceptions, these effects appear to be secondary to HPA axis or adrenomedullary activation, rather than centrally-mediated (see Dunn and Berridge, 1990). Notable exceptions to this generalization include HPA axis-independent, ulceroprotective effects of CRF administered peripherally (Murison et al., 1989) and ICV (Bakke et al., 1990) in rats and gastric emptying by CRF administered ICV in mice (Sheldon et al., 1990) and dogs (Buéno and Fioramonti, 1986).

13.6.1.4 Changes in electrophysiological activity

Stress and ICV-administered CRF result in similar electrophysiological changes including encephalographic activation (Weiss et al., 1986), increased tonic discharges in the locus coeruleus (Valentino and Foote, 1988) and excitation of hippocampal (Aldenhoff et al., 1983), cortical, hypothalamic, thalamic and septal (Eberley et al., 1983) neurones. Based on glucose utilization (as measured by [^{14}C]-2-deoxyglucose), CRF increases neuronal activity in the median eminence, lateral hypothalamus, locus coeruleus, raphe and red nuclei, and inferior olive, but decreases it in the prefrontal cortex, nucleus accumbens and dorsal tegmentum (Sharkey et al., 1989).

13.6.1.5 Biochemical changes

Stress and ICV-administered CRF induce parallel biochemical changes in catecholaminergic, GABAergic and opioid peptidergic function in the CNS. Like stress, CRF induces regionally-specific increases in CNS concentrations of both dopamine and noradrenaline metabolites in pigeons (Barrett et al., 1989), mice (Dunn and Berridge, 1987) and rats (Butler et al., 1990) in vivo. Both stress and CRF stimulate γ-aminobutyric acid (GABA; Sirinathsinghji and Heavens, 1989), methionine enkephalin and dynorphin release (Sirinathsinghji et al., 1989) from the neostriatum and globus pallidus of rats.

13.6.2 BEHAVIOURAL CHANGES

Stress and ICV-administered CRF produce a number of qualitatively similar behavioural effects. These include: potentiated acoustic startle (Swerdlow et al., 1989), increased behavioural inhibition (Kalin et al., 1989) and grooming (Morley and Levine, 1982). They also include alterations in locomotor activity/exploratory behaviour (Diamant and De Wied, 1991) and decreases in social interaction (Dunn and File, 1987), punished responding (Thatcher-Britton and Koob, 1986) and food intake (Tarjan et al., 1991). Finally, sexual behaviour (Sirinathsinghji, 1987) and sleep topography (Holsboer et al., 1988) are similarly disrupted by stress and exogenous CRF.

13.7 SIGNIFICANCE OF CRF-INDUCED EFFECTS

A large number of the effects of ICV-administered CRF do not appear to be mediated by CRF-induced activation of the HPA axis. In most cases, central and peripheral CRF administration have qualitatively different effects (Dunn and Berridge, 1990). Furthermore, neither administration of the synthetic corticosteroid, dexamethasone (Britton et al., 1986b), nor hypophysectomy (removal of the pituitary gland; Berridge and Dunn, 1989) alter the profile of stress-like changes produced by ICV-administered CRF in untreated, intact animals (see Appendix 13.2).

In many cases where stress and CRF induce similar physiological or behavioural changes, CRF antagonists or CRF immunoneutralization attenuate or reverse the effect of both stress and CRF (Dunn and Berridge, 1990; Owens and Nemeroff, 1991) (see Appendix 13.3). These findings are important because they rule out the possibility that ICV-administered CRF produces stress-like effects simply by acting as a nonspecific stressor.

Interestingly, many of the effects of stress and direct CNS administration of CRF in laboratory animals resemble the signs and symptoms of anorexia nervosa, anxiety, and major depression in humans. These similarities include: disturbed eating and sleep patterns, decreased libido, hypogonadism and psychomotor retardation (DSM III-R; American Psychiatric Association, 1987).

In summary, stress responding and adaptation appear to be modulated by widespread and evolutionarily-conserved, neurobiological actions of endogenous CRF in both neuroendocrine and nonendocrine circuits. The neuroendocrine system exerts effects that are relatively slow in onset and long in duration. The nonendocrine system exerts effects that are relatively fast in onset and of variable duration. The evolutionarily-conserved neurobiological actions of CRF range from adaptive activational effects to potentially deleterious or maladaptive anxiogenic effects in susceptible individuals.

13.8 FUNCTIONAL CRF-OPIOID PEPTIDE INTERACTIONS IN STRESS ADAPTATION

The CNS opioid peptide neuronal systems, like CRF neuronal systems, are widely implicated in the neurobiological correlates of stress responding and adaptation (De Souza and Appel, 1991). A number of physiological and psychological stressors activate both CRF and opioid peptide circuits in the CNS (Lightman and Young, 1987; Brady et al., 1990). In fact, it has been hypothesized that direct and indirect functional interactions between CRF- and opioid peptide-mediated neuronal mechanisms underlie stress-adaptation (Grossman, 1991; Rasmussen, 1991). A number of different lines of experimental evidence appear to support this hypothesis.

13.8.1 NEUROANATOMICAL EVIDENCE

Neuroanatomical studies demonstrate overlapping distributions of CRF and opioid peptide circuits in brain areas activated by stress (Pretel and Piekut, 1990) (see Appendix 13.4). Colocalization of CRF with enkephalin and CRF with dynorphin occur in the paraventricular nucleus (Roth et al., 1983; Pretel and Piekut, 1990). Enkephalin and CRF colocalization also occurs in the bed nucleus of the stria terminalis, in the medial preoptic, lateral and dorsal hypothalamic areas, as well as in the periventricular, supraoptic hypothalamic and subincertal nuclei (Sakanaka et al., 1989). Finally, enkephalin and CRF have a similar distribution throughout the afferent cerebellar system. Colocalization occurs in climbing and mossy fibre projections to the cerebellum, as well as neurones in the medial accessory olive, prepositus nucleus and scattered cells throughout the reticular formation (Cummings and King, 1990). Overlaps occur in the gigantocellular, subtrigeminal and lateral reticular nuclei, the pallidus, the raphe nuclei obscurus and magnus, the medial vestibular nuclei and the locus coeruleus (Cummings and King, 1990).

Notable overlapping distributions of opioid and CRF responsive neurones also occur in brain regions subserving autonomic function. These include the parabrachial nucleus (Lind and Swanson, 1984), the autonomic fields in the medial preoptic, anterior and paraventricular hypothalamic nuclei, the locus coeruleus and raphe nuclei, amygdala, hippocampus, bed nucleus of the stria terminalis, septum, cingulate and insular cortices (De Souza and Appel, 1991).

13.8.2 BIOCHEMICAL EVIDENCE

Biochemical studies show complex mutual regulatory actions of CRF and opioids on release *in vivo* (Nikolarakis *et al.*, 1989) and *in vitro* (Bronstein and Akil, 1990). In general, CRF stimulates opioid release and opioids inhibit CRF release. Centrally-administered CRF stimulates dynorphin, met-enkephalin, and β-endorphin release (Sirinathsinghji *et al.*, 1989). Centrally-administered morphine and dynorphin inhibit CRF release (Tsagarakis *et al.*, 1989), while β-endorphin has biphasic effects on CRF secretion, stimulating it at low doses, and inhibiting it at high doses (Buckingham, 1986).

13.8.3 PHYSIOLOGICAL EVIDENCE

Physiological studies show that opioid peptides, like CRF, influence endocrine and autonomic function. Common actions of opioids and CRF include effects on HPA, HPG and thyroid function (see Appendix 13.3). Opioids alter HPA function in rodents, humans and lower primates (for a review see Chatterton, 1990). However, important species differences exist in the endocrine actions of opioids. Central (Gunion *et al.*, 1991) and peripheral (Ignar and Kuhn, 1990) opioid administration result in dose-dependent, naloxone-reversible activation of the HPA axis in rats. In humans (Garland and Zis, 1990), sheep (Parrott and Thornton, 1989) and some lower primates, opioid agonists and antagonists have effects opposite to those seen in rats: inhibition and stimulation, respectively (Pfeiffer *et al.*, 1985). Evidence suggests that species differences in opioid modulation of HPA function relate to differential distributions of μ- and κ-opioid receptor subtypes in the CNS (for a review see De Souza and Appel, 1991).

Opioids also affect HPG function in mammals. In both primates (Williams *et al.*, 1990) and rats (Almeida *et al.*, 1988), opioid antagonists reverse or attenuate CRF-induced inhibition of HPG axis function. In rats, stress-induced suppression of HPG function can be reversed by either ICV administration of CRF antagonists (Rivier *et al.*, 1986) or β-endorphin antagonists (Parrott and Thornton, 1989). This suggests that CRF and β-endorphin may both be obligatory intermediates in stress-induced suppression of the HPG axis in rats. Finally, opioids, like stress and CRF, have complex, species-dependent effects on thyroid function (for a review see Chatterton, 1990).

Like stress and CRF, opioid peptides modulate cardiovascular (Sarne *et al.*, 1991), sympathetic nervous system and adrenomedullary (Van Loon *et al.*, 1981) as well as visceral (Lenz, 1989) function. In the cases of cardiovascular, sympathetic and adrenomedullary (Fisher, 1991b) but not visceral (Taché *et al.*, 1990), function, centrally-administered opioids alleviate or reverse many of the effects of stress and CRF. Dynorphin and dynorphin-related peptides alleviate increases in cardiovascular and adrenomedullary function that follow ICV CRF administration

(Overton and Fisher, 1989), while intrahypothalamic administration of μ- and κ-opioid agonists blunts stress-induced tachycardia (Kiritsy-Roy *et al.*, 1986).

Several apparent inconsistencies in opioid-mediated effects do exist however. In the absence of CRF or stress, central enkephalin and β-endorphin administration increases adrenomedullary function (Van Loon *et al.*, 1981). In contrast, enkephalins increase (Pfeiffer *et al.*, 1983), and β-endorphin decreases (Sitsen *et al.*, 1982) cardiovascular function. Also paradoxically, peripheral naloxone administration prevents increases in blood pressure and heart rate that normally occur following restraint and heat stress (Rhee and Hendrix, 1989) or ICV CRF administration (Saunders and Thornhill, 1986).

13.8.4 BEHAVIOURAL EVIDENCE

Finally, behavioural studies provide evidence of CRF-opioid interactions in stress-related responding. Like CRF, centrally-administered opioid peptides influence locomotor activity, exploratory, sexual and ingestive behaviour, as well as grooming, conditioned responding, acoustic startle and antinociception/stress-induced analgesia. Indirect blockade of opioid peptide neural circuits, by passive immunoneutralization with opioid antisera, or direct blockade by opioid antagonist administration, attenuates or reverses the effects of stress, CRF and/or opioids on a number of different behaviours (see Appendix 13.3).

Like CRF, evidence suggests that opioid peptides have an important role in regulating sexual behaviour (Pfaus and Gorzalka, 1987). In female rats, direct administration of opiate antagonists or β-endorphin antisera into the mesencephalic grey or arcuate-ventromedial hypothalamic area significantly or totally reverses the inhibitory effects of CRF administration on lordosis behaviour (Sirinathsinghji, 1986). Likewise, in male rats, the opiate antagonist, naloxone, blocks CRF-induced disruption of sexual behaviour (Sirinathsinghji, 1987).

Naloxone also antagonizes a number of other behaviours induced by CRF or stress. It reverses decrements in exploratory behaviour induced by CRF or stress in rats (Koob *et al.*, 1984) and mice (Berridge and Dunn, 1986). In rats, naloxone blocks CRF-induced increments in grooming behaviour (Dunn *et al.*, 1987). In mice, intrathecally-administered CRF blocks morphine-induced antinociception, and the antagonistic effects of CRF are blocked by β-opioid receptor antagonists (Song and Takemori, 1991). Collectively, these findings suggest that many of the stress-related effects of endogenous CRF involve a simultaneous, or possibly CRF-mediated, activation of opioidergic mechanisms. Furthermore, they suggest that the activation of opioidergic mechanisms may in turn feedback to amplify or attenuate CRF-induced mechanisms. If these hypothesized functional interactions are in fact critical for stress-adaptation, then dysfunctions in either of these systems could be associated with psychopathological clinical features. Unfortunately, CNS interactions are difficult to prove in humans because of the 'imitations imposed by the noninvasive techniques that are currently available. However, CNS interactions can be inferred from abnormal neurochemical profiles.

Perhaps the strongest support for the existence of important functional interactions between CRF and opioids comes from anorexia nervosa, an eating disorder in which stressful life situations are frequently considered to be predisposing or triggering factors. Its essential features include: reduced food intake, increased physical activity, disturbed sleep patterns and amenorrhea (DSM-III-R). Clinical and basic research findings support the hypothesis that dysfunctional CRF-opioid interactions underlie the pathogenesis of this disorder.

Food intake is altered in anorexia nervosa and anorexics appear to have CRF or opioid peptide system dysfunction. Women with anorexia nervosa have basal plasma β-endorphin levels that are significantly lower than in normal people. Also, when compared with normal people, β-endorphin levels in anorexia patients show an exaggerated response to clonidine and a depressed response to naloxone. At the same time, anorexia nervosa patients have elevated levels of CRF in the cerebrospinal fluid and elevated plasma cortisol levels; they exhibit dexamethasone nonsuppression and blunted ACTH responses to both CRF and insulin-induced hypoglycaemia (Baranowska, 1990). Collectively, these findings suggest that both the opioid and CRF systems are dysregulated in patients with anorexia nervosa. In short, hypoactive opioid systems and hyperactive CRF systems are associated with this disorder.

In male and female rats, food deprivation produces significant decreases in mRNA for both CRF in the paraventricular nucleus and for pro-opiomelanocortin (POMC: the precursor for β-endorphin) in the arcuate nucleus of the hypothalamus (Brady et al., 1990). Food deprivation also significantly reduces the β-endorphin content of the anterior pituitary (Hotta et al., 1991). These effects are clearly consistent with the decrease in CSF β-endorphin levels in anorexia nervosa (Kaye et al., 1987a).

In contrast to the effects of food deprivation, ICV infusion of CRF for seven days in rats results in reduced food intake and reduced body weight gain, decreased hypothalamic CRF content and increased pituitary β-endorphin levels (Hotta et al., 1991). These changes might seem inconsistent with the increased levels of CRF and depressed levels of β-endorphin seen in anorexia nervosa. However, as previously described, the interactions between the opioids and the neuroendocrine CRF system in rats and primates are quite different: β-endorphin inhibits CRF release (Garland and Zis, 1990) and stimulates feeding (Nader and Barrett, 1989) in primates. Hence, unlike in rodents, hypoactivity of the arcuate β-endorphin system in primates could result in increased CRF synthesis and release and decreased feeding: the CRF-opioid profile observed in anorexia nervosa. Future research should determine whether or not this hypothesized rodent-primate distinction is valid.

13.9 COMPLEXITIES, CONTRADICTIONS AND INCONSISTENCIES

Unfortunately, any simplistic synthesis of the evidence supporting the hypothesis that functional CRF-opioid interactions underlie stress adaptation is misleading. No review of the role of opioid peptides in stress responding is complete until the words 'complex', 'contradictory' and 'inconsistent' have been introduced. The CRF and opioid peptide systems implicated in stress responding each consist of anatomically and functionally distinct neuroendocrine and nonendocrine components. In terms of complexity, that is about all they have in common.

13.9.1 COMPLEXITY AND THE HYPOTHALAMO-ADENOHYPOPHYSEAL SYSTEMS

As previously stated, the endocrine CRF system is highly restricted and localized. It consists of one family of closely related peptides. Approximately 90% of the CRF-containing projections to the external lamina of the median eminence originate in the medial parvocellular subdivision of the paraventricular nucleus of the hypothalamus (for a review see Jones and Gillham, 1988).

In contrast, the endocrine opioid system is neither restricted nor localized and consists of three families of related peptides: pro-opiomelanocortin (POMC), pro-dynorphin (pro-DYN) and pro-enkephalin (pro-ENK). POMC-derived opioid (β-, α-, and γ-endorphin) projections to the external lamina of the median eminence originate in the arcuate nucleus of the hypothalamus (Finley et al., 1981). Putative pro-DYN opioid which yields dynorphin A and B (DYN A and B) and α- and β-neoendorphin is found in neurones which project to the median eminence and originate in the paraventricular nucleus (Roth et al., 1983), and the arcuate, retrochiasmatic and supraoptic hypothalamic nuclei (Neal and Newman, 1989). Pro-ENK opioid (met- and leu-enkephalin) projections to the median eminence originate from the paraventricular nucleus, the periventricular region and the arcuate nucleus of the hypothalamus (Fallon and Leslie, 1986), but note that leu-enkephalin can also be derived from cleavage of the proDYN peptide, leumorphin. CRF released in the median eminence exerts most, if not all, its action on CRF receptors in the anterior lobe of the pituitary. In contrast, opioids released in the median eminence exert most, if not all, their effects locally (Illes, 1989). CRF and opioid projections to the posterior pituitary are not included in this discussion.

Unlike CRF, which is not synthesized in the pituitary, opioid peptides are synthesized throughout the pituitary: POMC opioids are synthesized in the anterior and intermediate lobes. β-endorphin appears to be the only physiologically relevant species in pituitary tissue. Pro-DYN opioids are synthesized in the anterior and posterior lobes, with the largest number of pro-DYN cells in the posterior lobe. Pro-ENK opioids are synthesized in all three lobes of the pituitary. A more complete description of the endocrine opioid system can be found in a review by Millan and Herz (1985).

CRF receptors are found in the intermediate and anterior lobes of the pituitary, with the highest concentration in the anterior lobe (Hauger et al., 1989). In contrast, opioid receptors are found in all three lobes of the pituitary, with the highest concentration in the posterior pituitary (Simantov and Snyder, 1977). In short, it appears that in the anterior pituitary, the stress response involves CRF receptor activation and opioid peptide release, but not opioid receptor activation or CRF secretion.

Of the opioid peptides released by the pituitary, β-endorphin appears to be physiologically the most important in stress. Hypophysectomy (removal of the pituitary gland) significantly decreases β-endorphin levels in the peripheral circulation, but levels of pro-DYN and pro-ENK opioid peptide in the circulation are unaffected (Millan and Herz, 1985).

While the endocrine component of the CRF system is clearly crucial for stress responding and adaptation, its relevance to the rapid modulation of physiology and behaviour is questionable. The HPA axis stress response (involving corticosteroid release from the adrenal gland) is relatively slow in onset and long in duration and, in most species, plasma corticosteroid levels are not significantly elevated until 10–15 min after stress onset. In contrast, circulating (plasma) β-endorphin levels become significantly elevated within seconds to minutes, depending on the stressor (Antoni, 1986). Furthermore, evidence suggests pituitary POMC-derived peptides gain access to the brain via vascular and CSF backflow, in addition to access through the general circulation (De Kloet *et al.*, 1981). Taken together, these findings suggest that the opioid endocrine system plays a critical role in acute responses to stressors.

13.9.2 COMPLEXITY AND THE NONENDOCRINE SYSTEMS

Like the endocrine peptide systems, the CNS nonendocrine opioid system is more complex than the CRF nonendocrine system. All three families of opioid peptides participate in the nonendocrine system(s) in the brain. Where CRF binds to one type of high affinity receptor in the CNS, opioid peptides bind with differing affinities to three different classes/types of opioid receptors: mu (μ), delta (δ) and kappa (κ). Some researchers also recognize a fourth class of sigma (σ), but recent evidence suggests that it is non-opioid in function. In addition, multiple, functionally distinct subclasses of μ- (μ_1 and μ_2) and κ- (κ_1, κ_2/ϵ, and κ_3), and possibly δ- (Negri *et al.*, 1991) opioid receptors appear to exist. Functional coupling between the different opioid receptor types may also occur. For example, κ-agonists (e.g. bremazocine, dynorphin$_{1-13}$) can function as μ-antagonists, while κ-agonists or μ-antagonists also function as δ-inverse agonists (antagonizing the effects of δ antagonists) under some experimental conditions (Holaday *et al.*, 1991).

The endogenous ligand for μ-receptors is not known. It binds β-endorphin and met-enkephalin with comparable high affinity. Mu receptors are implicated in pain regulation and sensorimotor integration. In the CNS, μ-receptor-activated effects include supraspinal analgesia, catalepsy, hypothermia, respiratory depression, as well as changes in dopamine and acetylcholine turnover, prolactin and growth hormone release and gastrointestinal transit (e.g. Mansour *et al.*, 1988).

Enkephalins are the endogenous ligands for δ-receptors which participate in olfactory information processing, motor integration, and cognitive functioning (Mansour *et al.*, 1988). Dynorphin and β-endorphin are endogenous ligands for the non-ϵ and ϵ- κ-opioid receptors, respectively. Opioid κ-receptors are implicated in the central regulation of ingestive behaviours, pain perception, and neuroendocrine functioning (Mansour *et al.*, 1988).

Both the CRF and opioid neuroendocrine systems have anatomical distributions that are consistent with a role in initiating and coordinating the whole organism's response to stress. Likewise, findings that μ- and κ-opioid agonists have opposing effects on electrophysiological activity, fluid balance, motor behaviour and thermoregulation are consistent with the hypothesis that opioid peptides play critical roles in stress responding and adaptation (for a review see Mansour *et al.*, 1988).

One possibility is that the opioid, but not the CRF, neuroendocrine system is involved in modulating the set point for stress responding; by analogy, the CRF systems function as an on/off light switch while opioid systems act as a dimmer attachment.

13.9.3 CONTRADICTIONS AND INCONSISTENCIES

Until recently, highly specific agonists and antagonists for μ-, δ-, and κ-opioid receptors did not exist. It is not surprising, therefore, that reports of contradictory findings and inconsistencies occur in opioid stress research. Drug type, site of administration, dose, chronicity, the species, time of day, age, experiential history and sex of the animals used all influence the effects of opioid peptides on stress responding (Kepler et al., 1991). While these factors also contribute to the inconsistencies observed in studies with CRF (Dunn and Berridge, 1990), they appear to be even more critical for opioid peptides.

Although somewhat overwhelming, the existence of a large number of critical variables argues for, rather than against, an important role of opioid peptides in stress responding and adaptation. For example, μ- and κ-receptor distributions are conserved across all species of mammals in the brainstem and spinal cord, but exhibit large interspecific differences in the midbrain and telencephalon. This finding suggests that phylogenic divergence has occurred in brain areas most critical for species-typical behaviours that are shaped by the experiential history of the animal.

13.10 CONCLUDING REMARKS

This review has focused on the roles of CRF and opioid peptides in stress responding and adaptation. It has described a high degree of correspondence between the neural circuits for these peptides, especially within the neuroanatomical substrates of stress responding. It has implicated endogenous CRF and opioids in stress responding by providing evidence for biochemical changes in brain CRF and opioid CNS circuits following stressful experiences. It summarized the convincing pharmacological evidence of CRF- and opioid-induced regulation of a number of physiological adjustments and behavioural responses considered to be reliable markers of an organism's response to stress (i.e. plasma glucose, luteinizing hormone, ACTH and catecholamine levels, exploratory behaviour, feeding etc.). It has also reviewed evidence showing that stress-like responses to exogenous CRF and opioid peptides occur through mechanisms involving the activation of peptide-specific receptors in the central nervous system. Finally, we presented convincing evidence of reciprocal functional interactions between CRF and opioid neural circuits in stress responding and adaptation.

ACKNOWLEDGEMENTS

This work was supported by NIEHS-ES-07031 (DHH), NIMH-MH-42088 (CBM), and the John D. and Catherine T. MacArthur Foundation for Mental Health Research Network 1: The Psychobiology of Depression and Affective Disorders (CBN).

13.11 REFERENCES

Aguilera, G., Wynn, P.C., Harwood, J.P. et al. (1986) Receptor-mediated actions of corticotropin-releasing factor in pituitary gland and nervous system. Neuroendocrinology, **43**: 79–88

Aguilera, G., Millan, M.A., Hauger, R.L. and Catt, K.J. (1987) Corticotropin-releasing factor receptors: Distribution and regulation in brain, pituitary, and peripheral tissues. Ann. N.Y. Acad. Sci., **512**: 48–66

Aldenhoff, J.B., Gruol, D.L., Rivier, J. et al. (1983) Corticotropin releasing factor decreases postburst hyperpolarizations and excites hippocampal neurons. Science, **221**: 875–7

Almeida, O.F.X., Nikolarakis, K.E. and Herz, A. (1988) Evidence for the involvement of endogenous opioids in the inhibition of luteinizing hormone by corticotropin-releasing factor. Endocrinology, **122**: 1034–41

Alonso, G., Siaud, P. and Assenmacher, I. (1988) Immunocytochemical ultrastructural study of hypothalamic neurons containing corticotropin-releasing factor in normal and adrenalectomized rats. Neuroscience, **34**: 553–65

American Psychiatric Association (1987) Diagnostic and Statistical Manual of Mental Disorders (3rd Edn), American Psychiatric Association Press, Washington, DC

Amir, S., Brown, Z.W. and Amit, Z. (1980) The role of endorphins in stress: evidence and speculations. Neurosci. Biobehav. Rev., **4**: 77–86

Antoni, F.A. (1986) Hypothalamic control of adrenocorticotropin secretion: advances since the discovery of 41-residue corticotropin-releasing factor. Endocrine Rev., **7**: 351–78

Ayesta, F.J. and Nicolarakis, K.E. (1989) Peripheral but not intracerebroventricular corticotropin-releasing hormone (CRH) produces antinociception which is not opioid mediated. Brain Res., **503**: 219–24

Bagdy, G., Calogero, A.E., Szemeredi, K. et al. (1990) Effects of cortisol treatment on brain and adrenal corticotropin-releasing hormone (CRH) content and other parameters regulated by CRH. Reg. Peptides, **31**: 83–92

Bakke, H.K., Bogsnes, A. and Murison, R. (1990) Studies on the interaction between ICV effects of CRF and CNS noradrenaline depletion. Physiol. Behav., **47**: 1253–60

Banki, C.M., Bissette, G., Arato, M. et al. (1987) CSF corticotropin-releasing factor-like immunoreactivity in depression and schizophrenia. Am. J. Psychiatry, **144**: 873–7

Baranowska, B. (1990) Are disturbances in opioid and adrenergic systems involved in the hormonal dysfunction of anorexia nervosa? Psychoneuroendocrinology, **15**: 371–9

Barmack, N.H. and Young III, W.S. (1990) Optokinetic stimulation increases corticotropin-releasing factor mRNA in inferior olivary neurons of rabbits. J. Neurosci., **10**: 631–40

Barrett, J.E., Zhang, L., Ahlers, S.T. and Wojnicki, F.H. (1989) Acute and chronic effects of corticotropin-releasing factor on schedule-controlled responding and neurochemistry of pigeons. J. Pharmacol. Exp. Ther., **250**: 788–94

Battaglia, G., Webster, E.I. and De Souza, E.B. (1987) Characterization of corticotropin-releasing factor receptor-mediated adenylate cyclase activity in the rat central nervous system. Synapse, **1**: 572–81

Berkenbosch, F. and Tilders, F.J.H. (1988) Effect of axonal transport blockade on corticotropin releasing factor immunoreactivity in the median eminence of intact and adrenalectomized rats: Relationship between depletion rate and secretory activity. Brain Res., **442**: 312–20

Berridge, C.W. and Dunn, A.J. (1986) Corticotropin-releasing factor elicits naloxone sensitive stress-like alterations in exploratory behavior in mice. Reg. Peptides, **16**: 83–93

Berridge, C.W. and Dunn, A.J. (1989) CRF and restraint-stress decrease exploratory behavior in hypophysectomized mice. Pharmacol. Biochem. Behav., **34**: 517–19

Beyer, H.S., Matta, S.G. and Sharp, B.M. (1988) Regulation of the messenger ribonucleic acid for corticotropin-releasing factor in the paraventricular nucleus and other brain sites of the rat. Endocrinology, **123**: 2117–23

Bissette, G. (1991) Neuropeptides involved in stress and their distribution in the mammalian central nervous system, in J.A. McCubbin, P.G. Kaufmann and C.B. Nemeroff (eds.), Stress, Neuropeptides, and Systemic Disease, Academic Press, New York, pp 55–72

Brady, L.S., Smith, M.A., Gold, P.W. and Herkenham, M. (1990) Altered expression of hypothalamic neuropeptide mRNAs in food-restricted and food-deprived rats. *Neuroendocrinology*, **52**: 441–7

Britton, K.T., Lyon, M., Vale, W. and Koob, G.F. (1984) Stress-induced secretion of corticotropin-releasing factor immunoreactivity in rat cerebrospinal fluid. *Soc. Neurosci. Abstr.*, **10**: 94

Britton, D.R., Varela, M., Garcia, A. and Rosenthal, M. (1986a) Dexamethasone suppresses pituitary-adrenal but not behavioral effects of centrally administered CRF. *Life Sci.*, **38**: 211–16

Britton, K.T., Lee, G., Dana R., Risch, S.C. and Koob, G.F. (1986b) Activating and 'anxiogenic' effects of corticotropin-releasing factor are not inhibited by blockade of the pituitary-adrenal system with dexamethasone. *Life Sci.*, **39**: 1281–6

Bronstein, D.M. and Akil, H. (1990) *In vitro* release of hypothalamic β-endorphin (βE) by arginine vasopressin, corticotropin-releasing hormone and 5-hydroxytryptamine: Evidence for release of opioid active and inactive βE forms. *Neuropeptides*, **16**: 33–40

Brown, M. (1986) Corticotropin releasing factor: Central nervous system sites of action. *Brain Res.*, **399**: 10–14

Brown, M.R. (1991a) Neuropeptide-mediated regulation of the neuroendocrine and autonomic response to stress, in J.A. McCubbin, P.G. Kaufmann and C.B. Nemeroff (eds.), *Stress: Neuropeptides, and Systemic Disease*, Academic Press, New York, pp 73–93

Brown, M.R. (1991b) Brain peptide regulation of autonomic nervous and neuroendocrine functions, in M.R. Brown, G.F. Koob and C. Rivier (eds.), *Stress: Neurobiology and Neuroendocrinology*, Marcel Dekker, New York, pp 193–215

Brown, M.R. and Fisher, L.A. (1983) Central nervous system effects of corticotropin releasing factor in the dog. *Brain Res.*, **280**: 75–9

Brown, M.R. and Gray, T.S. (1988) Peptide injections into the amygdala of conscious rats: effects on blood pressure, heart rate and plasma catecholamines. *Reg. Peptides*, **21**: 95–106

Brown, M.R., Fisher, L.A., Spiess, J. *et al.* (1982a) Corticotropin-releasing factor: actions on the sympathetic nervous system and metabolism. *Endocrinology*, **111**: 928–31

Brown, M.R., Fisher, L.A., Rivier, J. *et al.* (1982b) Corticotropin-releasing factor: effects on the sympathetic nervous system and oxygen consumption. *Life Sci.*, **30**: 207–10

Brown, M.R., Fisher, L.A. and Vale, W.W. (1983) Sympathetic nervous system and adrenocortical interactions. *Soc. Neurosci. Abstr.*, **9**: 391

Buckingham, J.C. (1986) Stimulation and inhibition of corticotropin releasing factor secretion by beta endorphin. *Neuroendocrinology*, **42**: 148–52

Buéno, L. and Fioramonti, J. (1986) Effects of corticotropin-releasing factor, corticotropin and cortisol on gastrointestinal motility in dogs. *Peptides*, **7**: 73–7

Buéno, L. and Gué, M. (1988) Evidence for the involvement of corticotropin-releasing factor in the gastrointestinal disturbances induced by acoustic and cold stress in mice. *Brain Res.*, **441**: 1–4

Butler, P.D., Weiss, J.M., Stout, J.C. and Nemeroff, C.B. (1990) Corticotropin-releasing factor produces fear-enhancing and behavioral activating effects following infusion into the locus coeruleus. *J. Neurosci.*, **10**: 176–83

Cain, S.T., Owens, M.J. and Nemeroff, C.B. (1991) Subcellular distribution of corticotropin-releasing-factor-like immunoreactivity in rat central nervous system. *Neuroendocrinology*, **54**: 36–41

Calogero, A.E., Gallucci, W.T., Gold, P.W. and Chrousos, G.P. (1988) Multiple feedback regulatory loops upon rat hypothalamic corticotropin-releasing hormone secretion: potential clinical implications. *J. Clin. Invest.*, **82**: 767–74

Cha, C.I. and Foote, S.L. (1988) Corticotropin-releasing factor in olivocerebellar climbing-fiber system of monkey (*Saimiri sciureus* and *Macaca fascicularis*): parasagittal and regional organization visualized by immunohistochemistry. *J. Neurosci.*, **8**: 4121–37

Chai, S.Y., Tarjan, E., McKinley, M.J., Paxinos, G. and Mendelsohn, F.A.O. (1990) Corticotropin-releasing factor receptors in the rabbit brain visualized by *in vitro* autoradiography. *Brain Res.*, **512**: 60–9

Chatterton, R.T. (1990) The role of stress in female reproduction: Animal and human considerations. *Int. J. Fertil.*, **35**: 8–13

Cole, B.J. and Koob, G.F. (1991) Corticotropin-releasing factor, stress, and animal behavior, in J.A. McCubbin, P.G. Kaufmann and C.B. Nemeroff (eds.), *Stress, Neuropeptides, and Systemic Disease*, Academic Press, New York, pp 119–148.

Cummings, S. (1989) Distribution of corticotropin-releasing factor in the cerebellum and precerebellar nuclei of the cat. *J. Comp. Neurol.*, **289**: 657–75

Cummings, S. and King, J.S. (1990) Coexistence of corticotropin releasing factor and enkephalin in cerebellar afferent systems. *Synapse*, **5**: 167–74

Cummings, S., Elde, R., Ells, J. and Lindall, A. (1983) Corticotropin-releasing factor immunoreactivity is widely distributed within the central nervous system of the rat: an immunohistochemical study. *J. Neurosci.*, **3**: 1355–68

Cummings, S., Sharp, B. and Elde, R. (1988) Corticotropin-releasing factor in cerebellar afferent systems: a combined immunohistochemistry and retrograde transport study. *J. Neurosci.*, **8**: 543–54

Cummings, S.L., Young III, W.S., Bishop, G.A. *et al.* (1989) Distribution of corticotropin-releasing factor in the cerebellum and precerebellar nuclei of the opossum: A study utilizing immunohistochemistry, *in situ* hybridization histochemistry, and receptor auto-radiography. *J. Comp. Neurol.*, **280**: 501–21

De Kloet, E.R., Palkovits, M. and Mezey, E. (1981) Opiocortin peptides: localization, source and avenues of transport. *Pharmac. Ther.*, **12**: 321–52

De Souza, E.B. (1987) Corticotropin-releasing factor receptors in the rat central nervous system: characterization and regional distribution. *J. Neurosci.*, **7**: 88–100

De Souza, E.B. and Appel, N.M. (1991) Distribution of brain and pituitary receptors involved in mediating stress responses, in M.R. Brown, G.F. Koob and C. Rivier (eds.), *Stress: Neurobiology and Neuroendocrinology*, Marcel Dekker, New York, pp 91–117

De Souza, E.B. and Kuhar, M.J. (1986) Corticotropin-releasing factor receptors in the pituitary gland and central nervous system: Methods and overview. *Meth. Enzymology*, **124**: 560–90

Diamant, M. and De Wied, D. (1991) Autonomic and behavioral effects of centrally administered corticotropin-releasing factor in rats. *Endocrinology*, **129**: 446–54

Donald, R.A., Redekopp, C., Cameron, V. *et al.* (1983) The hormonal actions of corticotropin-releasing factor in sheep: effect of intravenous and intracerebroventricular injection. *Endocrinology*, **113**: 866–70

Druge, G., Raedler, A., Greten, H. and Lenz, H.J. (1989) Pathways mediating CRF-induced inhibition of gastric acid secretion in rats. *Am. J. Physiol.*, **256**: G214–19

Dunn, A.J. and Berridge, C.W. (1987) Corticotropin-releasing factor administration elicits a stress-like activation of cerebral catecholaminergic systems. *Pharmacol. Biochem. Behav.*, **27**: 685–91

Dunn, A.J. and Berridge, C.W. (1990) Physiological and behavioral responses to corticotropin-releasing factor administration: is CRF a mediator of anxiety or stress responses? *Brain Res. Rev.*, **15**: 71–100

Dunn, A.J. and File, S.E. (1987) Corticotropin-releasing factor has an anxiogenic action in the social interaction test. *Horm. Behav.*, **21**: 193–202

Dunn, A.J., Berridge, C.W., Lai Y.I. and Yachabach, T.L. (1987) CRF-induced excessive grooming behavior in rats and mice. *Peptides*, **8**: 841–4

Eaves, M., Thatcher-Britton, K., Rivier, J. *et al.* (1985) Effects of corticotropin releasing factor on locomotor activity in hypophysectomized rats. *Peptides*, **6**: 923–6

Eberly, L.B., Dudley, C.A. and Moss, R.L. (1983) Iontophoretic mapping of corticotropin-releasing factor (CRF) sensitive neurons in the rat forebrain. *Peptides*, **4**: 837–41

Ehlers, C.L. (1990) CRF effects on EEG activity: Implications for the modulation of normal and abnormal brain states, in E.B. De Souza and C.B. Nemeroff (eds.), *Corticotropin-Releasing Factor: Basic and Clinical Studies of a Neuropeptide*, CRC Press, Boca Raton, FL, pp 233–52.

Fallon, J.H. and Leslie, F.M. (1986) Distribution of dynorphin and enkephalin peptides in the rat brain. *J. Comp. Neurol.*, **249**: 293–336

Fellmann, D., Bugnon, C., Bresson, J.L. *et al.* (1984) The CRF neuron: immunocytochemical study. *Peptides*, **5**(Suppl. 1): 19–33

Finley, J.C.W., Linström, P. and Petrusz, P. (1981) Immunocytochemical localization of β-endorphin-containing neurons in the rat brain. *Neuroendocrinology*, **33**: 28–42

Fischman, A.J. and Moldow, R.L. (1982) Distribution of CRF-like immunoreactivity in the rabbit. *Peptides*, **3**: 841–3

Fisher, L.A. (1989) Corticotropin-releasing factor: endocrine and autonomic integration of responses to stress. *Trends in Pharmacol. Sci.*, **10**: 189–93

Fisher, L.A. (1991a) Stress and cardiovascular physiology in animals, in M.R. Brown, G.F. Koob and C. Rivier (eds.), *Stress: Neurobiology and Neuroendocrinology*, Marcel Dekker, New York, pp 463–74

Fisher, L.A. (1991b) Corticotropin-releasing factor and autonomic and cardiovascular responses to stress, in J. A. McCubbinn, P.G. Kaufmann and C.B. Nemeroff (eds.), *Stress: Neuropeptides and Systemic Disease*, Academic Press, New York, pp 95–118

Fisher, L.A., Jessen, G. and Brown, M.R. (1983) Corticotropin-releasing factor (CRF): mechanism to elevate mean arterial pressure and heart rate. *Reg. Peptides*, **5**: 153–61

Foote, S.L. and Cha, C.I. (1988) Distribution of corticotropin-releasing factor-like immunoreactivity in brainstem of two monkey species (*Saimiri sciureus* and *Macaca fascicularis*): an immunohistochemical study. *J. Comp. Neurol.*, **276**: 239–64

Gambacciana, M., Yen, S.S.C. and Rasmussen, D.D. (1986) GnRH release from the mediobasal hypothalamus: *in vitro* regulation by oxytocin. *Neuroendocrinol.*, **43**: 533–6

Gardiner, S.M. and Bennett, T. (1989) Brain neuropeptides: actions on central cardiovascular control mechanisms. *Brain Res. Rev.*, **14**: 79–116

Garland, E.J. and Zis, A.P. (1990) Effect of vasopressin and naloxone alone and in combination on cortisol secretion after dexamethasone pretreatment. *Horm. Res.*, **34**: 249–53

Garrick, T., Veiseh, A., Sierra, A., Weiner, H. and Tach, Y. (1988) Corticotropin-releasing factor acts centrally to suppress stimulated gastric contractility in the rat. *Reg. Peptides*, **21**: 173–81

Gindoff, P.R. and Ferin, M. (1987) Endogenous opioid peptides modulate the effect of corticotropin-releasing factor on gonadotropin release in the primate. *Endocrinology*, **121**: 837–42

Gindoff, P.R., Xiao, E., Luckhaus, J. and Ferin, M. (1989) Dexamethasone treatment prevents the inhibitory effect of corticotropin-releasing hormone on gonadotropin release in the primate. *Neuroendocrinology*, **49**: 202–6

Gold, P.W. and Chrousos, P. (1985) Clinical studies with corticotropin-releasing factor: implications for the diagnosis and pathophysiology of depression, Cushing's disease, and adrenal insufficiency. *Psychoneuroendocrinology*, **10**: 401–19

Grossman, A.B. (1991) Regulation of human pituitary responses to stress, in M.R. Brown, G.F. Koob and C. Rivier (eds.), *Stress: Neurobiology and Neuroendocrinology*, Marcel Dekker, New York, pp 151–71

Gunion, M.W., Rosenthal, M.J., Morley, J.E. *et al.* (1991) μ-Receptor mediates elevated glucose and corticosterone after third ventricle injection of opioid peptides. *Am. J. Physiol.*, **261**: R70–81

Harbuz, M., Russell, J.A., Sumner, B.E.H. *et al.* (1991) Rapid changes in the content of proenkephalin A and corticotrophin releasing hormone mRNAs in the paraventricular nucleus during morphine withdrawal in urethane-anaesthetized rats. *Molec. Brain Res.*, **9**: 285–91

Hargreaves, K.M., Dubner, R. and Costello, A.H. (1989) Corticotropin-releasing factor (CRF) has a peripheral site of action for antinociception. *Eur. J. Pharmacol.*, **170**: 275–9

Hargreaves, K.M., Flores, C.M., Dionne, R.A. and Mueller, G.P. (1990) The role of pituitary β-endorphin in mediating corticotropin-releasing factor-induced antinociception. *Am. J. Physiol.*, **258**: E235–42

Hattori, T., Hashimoto, K. and Ota, Z. (1986) Brain corticotropin releasing factor in the spontaneously hypertensive rat. *Hypertension*, **8**: 1027–31

Hauger, R.L., Millan, M.A., Catt, K.J. and Aguilera, G. (1987) Differential regulation of brain and pituitary corticotropin-releasing factor receptors by corticosterone. *Endocrinology*, **120**: 1527–33

Hauger, R.L., Millan, M.A., Lorang, M. *et al.* (1988) Corticotropin-releasing factor receptors and pituitary adrenal responses during immobilization stress. *Endocrinology*, **123**: 396–405

Hauger, R.L., Millan, M., Harwood, J.P. *et al.* (1989) Receptors for corticotropin releasing factor in the pituitary and brain: Regulatory effects of glucocorticoids, CRF, and stress, in S. Breznitz and O. Zinder (eds.), *Molecular Biology of Stress*, Alan R. Liss, New York, pp 3–17

Herman, J.P., Schäfer, K.H., Sladek, C.D. *et al.* (1989) Chronic electroconvulsive shock treatment elicits up-regulation of CRF and AVP mRNA in select populations of neuroendocrine neurons. *Brain Res.*, **501**: 235–46

Holaday, J.W., Porreca, F. and Rothman, R.B. (1991) Functional coupling among opioid receptor types, in F.G. Estafanous (ed.), *Opioids in Anesthesia II*, Boston, Butterworth-Heinemann, pp 50–60

Holsboer, F., Von Bardeleben, U. and Steiger, A. (1988) Effects of intravenous corticotropin-releasing hormone upon sleep-related growth hormone surge and sleep EEG in man. *Neuroendocrinology*, **48**: 32–8

Hotta, M., Shibasaki, T., Yamauchi, N. *et al.* (1991) The effects of chronic central administration of corticotropin-releasing factor on food intake, body weight, and hypothalamic-pituitary-adrenocortical hormones. *Life Sci.*, **48**: 1483–91

Ignar, D.M. and Kuhn, C.M. (1990) Effects of specific *mu* and *kappa* opiate tolerance and abstinence on hypothalamo-pituitary-adrenal axis secretion in the rat. *J. Pharmacol. Exp. Ther.*, **255**: 1287–95

Illes, P. (1989) Modulation of transmitter and hormone release by multiple neuronal opioid receptors. *Rev. Physiol. Biochem. Pharmacol.*, **112**: 139–233

Imaki, T., Nahon, J.L., Sawchenko, P.E. and Vale, W. (1989) Widespread expression of corticotropin-releasing factor messenger RNA and immunoreactivity in the rat olfactory bulb. *Brain Res.*, **496**: 35–44

Imaki, J., Imaki, T., Vale, W. and Sawchenko, P.E. (1991a) Distribution of corticotropin-releasing factor mRNA and immunoreactivity in the central auditory system of the rat. *Brain Res.*, **547**: 28–36

Imaki, T., Nahon, J.L., Rivier, C., Sawchenko, P.E. and Vale, W. (1991b) Differential regulation of corticotropin-releasing factor mRNA in rat brain regions by glucocorticoids and stress. *J. Neurosci.*, **11**: 585–99

Insel, T.R., Aloi, J.A., Goldstein, D. *et al.* (1984) Plasma cortisol and catecholamine responses to intracerebroventricular administration of CRF to rhesus monkeys. *Life Sci.*, **34**: 1873–8

Jessop, D.S., Chowdrey, H.S. and Lightman, S.L. (1990) Differential effects of glucocorticoids on corticotropin-releasing factor in the rat pituitary neurointermediate lobe and median eminence. *Eur. J. Neurosci.*, **2**: 109–11

Jingami, H., Matsukura, S., Numa, S. and Imura, H. (1985) Effects of adrenalectomy and dexamethasone administration on the level of prepro-corticotropin-releasing factor messenger ribonucleic acid (mRNA) in the hypothalamus and adrenocorticotropin/β-lipotropin precursor mRNA in the pituitary in rats. *Endocrinology*, **117**: 1314–20

Jones, M.T. and Gillham, B. (1988) Factors involved in the regulation of adrenocorticotropic hormone/beta-lipotropic hormone. *Physiol. Rev.*, **68**: 743–818

Joseph, S.A. and Knigge, K.M. (1983) Corticotropin-releasing factor: immunocytochemical localization in rat brain. *Neurosci. Lett.*, **35**: 135–41

Ju, G. and Han, Z. (1989) Coexistence of corticotropin-releasing factor and neurotensin within oval nucleus neurons in the bed nuclei of the stria terminalis in the rat. *Neurosci. Lett.*, **99**: 246–50

Kalin, N.H. (1990) Behavioral and endocrine studies of corticotropin-releasing hormone in primates, in E.B. De Souza and C.B. Nemeroff (eds.), *Corticotropin-Releasing Factor: Basic and Clinical Studies of a Neuropeptide*, CRC Press, Boca Raton, FL, pp 275–89

Kalin, N.H., Gonder, J.C. and Shelton, S.E. (1983a) Effects of synthetic ovine CRF on ACTH, cortisol and blood pressure in sheep. *Peptides*, **4**: 221–3

Kalin, N.H., Shelton, S., Kraemer, G. and McKinney, W. (1983b) Corticotropin-releasing factor causes hypotension in rhesus monkeys. *Lancet*, **2**: 1042

Kalin, N.H., Sherman, J.E. and Takahashi, L.K. (1988) Antagonism of endogenous CRH systems attenuates stress-induced freezing behavior in rats. *Brain Res.*, **457**: 130–5

Kalin, N.H., Shelton, S.E. and Barksdale, C.M. (1989) Behavioral and physiologic effects of CRH administered to infant primates undergoing maternal separation. *Neuropsychopharmacology*, **2**: 97–104

Katakami, H., Arimura, A. and Frohman, L.A. (1985) Involvement of hypothalamic somatostatin in the suppression of growth hormone secretion by central corticotropin-releasing factor in conscious male rats. *Neuroendocrinology*, **41**: 390–3

Kawata, M., Hashimoto, K., Takahara, J. and Sano, Y. (1983) Immunohistochemical identification of neurons containing corticotropin-releasing factor in the rat hypothalamus. *Cell Tiss. Res.*, **230**: 239–46

Kaye, W.H., Berretini, W., Gwirstman, H. *et al.* (1987a) Reduced cerebrospinal fluid levels of immunoreactive proopiomelancortin related peptides (including beta-endorphin) in anorexia nervosa. *Life Sci.*, **41**: 2147–55

Kepler, K.L., Standifer, K.M., Paul, D. *et al.* (1991) Gender effects and central opioid analgesia. *Pain*, **45**: 87–94

Kiang, J.G. and Wei, E.T. (1985) CRF-evoked bradycardia in urethane-anesthetized rats is blocked by naloxone. *Peptides*, **6**: 409–13

Kiritsy-Roy, J.A., Appel, N.A., Bobbitt, F.G. and Van Loon, G.R. (1986) Effects of mu-opioid receptor stimulation in the hypothalamic paraventricular nucleus on basal and stress-induced catecholamine secretion and cardiovascular responses. *J. Pharmacol. Exp. Ther.*, **239**: 814–22

Kitahama, K., Luppi, P.-H., Tramu, G. *et al.* (1988) Localization of CRF-immunoreactive neurons in the cat medulla oblongata: their presence in the inferior olive. *Cell Tiss. Res.*, **251**: 137–43

Koob, G.F. (1985) Stress, corticotropin-releasing factor, and behavior, in R.B. Williams Jr. (ed.), *Perspectives on Behavioral Medicine, Vol. 2; Neuroendocrine Control and Behavior*, Academic Press, Orlando, FL, pp 39–52

Koob, G.F. (1991) Behavioral responses to stress, in M.R. Brown, G.F. Koob and C. Rivier (eds.), *Stress, Neurobiology, and Neuroendocrinology*, Marcel Dekker, New York, pp 255–71

Koob, G.F., Swerdlow, N., Seeligson, M. *et al.* (1984) Effects of alpha-flupenthixol and naloxone on CRF-induced locomotor activation. *Neuroendocrinology*, **39**: 459–64

Krahn, D.D., Gosnell, B.A. and Majchrzak, M.J. (1990) The anorectic effects of CRH and restraint stress decrease with repeated exposures. *Biol. Psychiatry*, **27**: 1094–102

Lenz, H.J. (1989) Regulation of duodenal bicarbonate secretion during stress by corticotropin-releasing factor and β-endorphin. *Proc. Natl. Acad. Sci. USA*, **86**: 1417–20

Lenz, H.J., Raedler, A., Greten, H. and Brown, M.R. (1987) CRF initiates biological actions within the brain that are observed in response to stress. *Am. J. Physiol.*, **21**: R34–9

Lenz, H.J., Raedler, A., Greten, H. *et al.* (1988) Stress-induced gastrointestinal secretion and motor responses in rats are mediated by endogenous corticotropin-releasing factor. *Gastroenterology*, **95**: 1510–17

Lewis, D.A., Foote, S.L. and Cha, C.I. (1989) Corticotropin-releasing factor immunoreactivity in monkey neocortex: An immunohistochemical analysis. *J. Comp. Neurol.*, **290**: 599–613

Lightman, S.L. and Young, W.S. (1987) Vasopressin, oxytocin, dynorphin, enkephalin and corticotropin-releasing factor mRNA stimulation in the rat. *J. Physiol.*, **394**: 23–29

Lima L. and Sourkes T.L. (1987) Effect of corticotropin-releasing factor on adrenal DBH and PNMT activity. *Peptides*, **8**: 437–41

Lind, R.W. and Swanson, L.W. (1984) Evidence for corticotropin-releasing factor and leu-enkephalin in the neural projection from the lateral parabrachial nucleus to the median preoptic nucleus: a retrograde transport, immunohistochemical double labeling study in the rat. *Brain Res.*, **321**: 217–24

Liposits, Z., Paull, W.K., Wu, P., Jackson, I.M. and Lechan, R.M. (1987) Hypophysiotropic thyrotropin releasing hormone (TRH) synthesizing neurons. Ultrastructure, adrenergic innervation and putative neurotransmitter action. *Histochemistry*, **87**: 407–12

Lupica, C.R. and Dunwiddie, T.V. (1991) Differential effects of mu- and delta-receptor selective opioid agonists on feedforward and feedback GABAergic inhibition in hippocampal brain slices. *Synapse*, **8**: 237–48

Mansour, A., Khachaturian, H., Lewis, M.E. *et al.* (1987) Autoradiographic differentiation of mu, delta, and kappa opioid receptors in the rat forebrain and midbrain. *J. Neurosci.*, **7**: 2445–64

Mansour, A., Khachaturian, H., Lewis, M.E. *et al.* (1988) Anatomy of CNS opioid receptors. *Trends in Neurosci.*, **11**: 308–14

Merchenthaler, I. (1984) Corticotropin-releasing factor (CRF)-like immunoreactivity in the rat central nervous system. Extrahypothalamic distribution. *Peptides*, **5**: 53–69

Merchenthaler, I., Vigh, S., Petrusz, P. and Schally, A.V. (1982) Immunocytochemical localization of corticotropin-releasing factor (CRF) in the rat brain. *Am. J. Anat.*, **165**: 385–96

Merchenthaler, I., Hynes, M.A., Vigh, S. *et al.* (1983) Immunocytochemical localization of corticotropin releasing factor (CRF) in the rat spinal cord. *Brain Res.*, **275**: 373–7

Mezey, E. and Palkovits, M. (1991) Time dependent changes in CRF and its mRNA in the neurons of the inferior olive following surgical transection of the olivocerebellar tract in the rat. *Mol. Brain Res.*, **10**: 55–9

Millan, M.A., Jacobowitz, D.M., Hauger, R.L. *et al.* (1986) Distribution of corticotropin-releasing factor receptors in primate brain. *Proc. Nat. Acad. Sci. USA.*, **83**: 1921–5

Millan, M.J. (1986) Multiple opioid systems and pain. *Pain*, **27**: 303–47

Millan, M.J. and Herz, A. (1985) The endocrinology of opioids. *Int. Rev. Neurobiol.*, **26**: 1–83

Mitsuma, T., Nogimori, T. and Hirooka, Y. (1987) Effects of growth hormone-releasing hormone and corticotropin-releasing hormone on the release of thyrotropin-releasing hormone from the rat hypothalamus *in vitro*. *Exp. Clin. Endocrinol.*, **90**: 365–8

Morley, J.E. (1987) Neuropeptide regulation of appetite and weight. *Endocrine Rev.*, **8**: 256–87

Morley, J.E. and Levine, A.S. (1982) Corticotropin releasing factor, grooming and ingestive behavior. *Life Sci.*, **31**: 1459–64

Morley, J.E., Levine, A.S. and Silvis, S.E. (1982) Minireview: central regulation of gastric acid secretion: the role of neuropeptides. *Life Sci.*, **31**: 399–410

Murison, R., Overmier, J.B., Hellhamer, D.H. and Carmora, M. (1989) Hypothalamo-pituitary-adrenal manipulations and stress ulceration in rats. *Psychoneuroendocrinology*, **14**: 331–8

Nader, M.A. and Barrett, J.E. (1989) Effects of corticotropin-releasing factor (CRF), tuftsin and dermorphin on behavior of squirrel monkeys maintained by different events. *Peptides*, **10**: 1199–204

Neal, C.R. and Newman, S.W. (1989) Prodynorphin peptide distribution in the forebrain of the Syrian hamster and rat: a comparative study with antisera against dynorphin A, dynorphin B., and the C-terminus of the prodynorphin precursor molecule. *J. Comp. Neurology*, **288**: 353–86.

Negri, L., Potenza, R.L., Corsi, R. and Melchiorri, P. (1991) Evidence for two subtypes of δ opioid receptors in rat brain. *Eur. J. Pharmacol.*, **196**: 335–6

Nemeroff, C.B., Widerlov, E., Bissette, G. *et al.* (1984) Elevated concentration of CSF corticotropin-releasing factor-like immunoreactivity in depressed patients. *Science*, **226**: 1342–4

Nikolarakis, K.E., Pfeiffer, A., Stalla, G.K. and Herz, A. (1989) Facilitation of ACTH secretion by morphine is mediated by activation of CRF releasing neurons and sympathetic neuronal pathways. *Brain Res.*, **498**: 385–8

North, R.A. (1979) Opiates, opioid peptides, and single neurones. *Life Sci.*, **24**: 1527–46

North, R.A. (1986) Opioid receptor types and membrane channels. *Trends in Neurosci.*, **9**: 114–17

Olson, G.A., Olson, R.D. and Kastin, A.J. (1987) Endogenous opiates: 1986. *Peptides*, **8**: 1125–64

Ono, N., Lumpkin, M.D., Samson, W.K. *et al.* (1984) Intrahypothalamic action of corticotropin-releasing factor (CRF) to inhibit growth hormone and LH release in the rat. *Life Sci.*, **35**: 1117–23

Overton, J.M. and Fisher, L.A. (1989a) Central nervous system actions of corticotropin-releasing factor on cardiovascular function in the absence of locomotor activity. *Reg. Peptides*, **25**: 315–24

Owens, M.J. and Nemeroff, C.B. (1991) The physiology and pharmacology of corticotropin-releasing factor. *Pharmacol. Rev.*, **43**: 425–73

Palkovits, M., Brownstein, M.J. and Vale, W. (1983) Corticotropin-releasing factor (CRF) immunoreactivity in hypothalamic and extrahypothalamic nuclei of sheep brain. *Neuroendocrinology*, **37**: 302–5

Palkovits, M., Brownstein, M.J. and Vale, W. (1985) Distribution of corticotropin-releasing factor in rat brain. *Fed. Proc.*, **44**: 215–19

Palkovits, M., Leranth, C., Gorcs, T. and Young III, W.S. (1987) Corticotropin-releasing factor in the olivocerebellar tract of rats: demonstration by light- and electron-microscopic immunohistochemistry and in situ hybridization histochemistry. *Proc. Natl. Acad. Sci. USA*, **84**: 3911–15

Parrott, R.F. and Thornton, S.N. (1989) Opioid influences on pituitary function in sheep under basal conditions and during psychological stress. *Psychoneuroendocrinology*, **14**: 451–9

Paull, W.K., Phelix, C.F., Copeland, M. *et al.* (1984) Immunohistochemical localization of corticotropin releasing factor (CRF) in the hypothalamus of the squirrel monkey, *Saimiri sciureus*. *Peptides*, 5(Suppl.1): 45–51

Pfaus, J.G. and Gorzalka, B.B. (1987) Opioids and sexual behavior. *Neurosci. Biobehav. Rev.*, **11**: 1–34

Pfeiffer, A. and Herz, A. (1984) Endocrine actions of opioids. *Horm. Metab. Res.*, **16**: 386–97

Pfeiffer, A., Feurstein, G. Kopin I.J. and Fader, A.I. (1983) Cardiovascular and respiratory effects of mu-, delta- and kappa-opiate antagonists microinjected into the anterior hypothalamic brain area of awake rats. *J. Pharmacol Exp. Ther.*, **225**: 735–41

Pfeiffer, A., Herz, A., Loriaux, D.L. and Pfeiffer, D.G. (1985) Central kappa- and mu-opiate receptors mediate ACTH-release in rats. *Endocrinology*, **116**: 2688–90

Powers, R.E., De Souza, E.B., Walker, L.C. *et al.* (1987) Corticotropin-releasing factor as a transmitter in the human olivocerebellar pathway. *Brain Res.*, **415**: 347–52

Pretel, S. and Piekut, D. (1990) Coexistence of corticotropin-releasing factor and enkephalin in the paraventricular nucleus of the rat. *J. Comp. Neurol.*, **294**: 192–201

Raff, H., Skelton, M., Merrill, D.C. and Cowley Jr., A.W. (1986) Vasopressin responses to corticotropin-releasing factor and hyperosmolarity in conscious dogs. *Am. J. Physiol.*, **251**: R1235–9

Rasmussen, D.D. (1991) The interaction between mediobasohypothalamic dopaminergic and endorphinergic neuronal systems as a key regulator of reproduction: an hypothesis. *J. Endocrinol. Invest.*, **14**: 323–52

Rhee, H.M. and Hendrix, D.W. (1989) Effects of stress intensity and modality on cardiovascular system: An involvement of opioid system, in S. Breznitz and O. Zinder (eds.), *Molecular Biology of Stress*, Alan R. Liss, New York, pp 87–96

Rivier, C. (1989) Involvement of endogenous corticotropin-releasing factor (CRF) in modulating ACTH and LH secretion function during exposure to stress, alcohol or cocaine in the rat, in S. Breznitz and O. Zinder (eds.), *Molecular Biology of Stress*, Alan R. Liss, New York, pp 31–47

Rivier, C. (1991) Neuroendocrine mechanisms of anterior pituitary regulation in the rat exposed to stress, in M.R. Brown, G.F. Koob and C. Rivier (eds.), *Stress: Neurobiology and Neuroendocrinology*, Marcel Dekker, New York, pp 119–36

Rivier, C.L. and Plotsky, P.M. (1986) Mediation by corticotropin-releasing factor (CRF) of adenohypophysial hormone secretion. *Ann. Rev. Physiol.*, **48**: 475–94

Rivier, C. and Vale, W. (1983) Modulation of stress-induced ACTH release by corticotropin-releasing factor, catecholamines and vasopressin. *Nature*, **305**: 325–7

Rivier, C. and Vale, W. (1984) Influence of corticotropin-releasing factor on reproductive functions in the rat. *Endocrinology*, **114**: 914–21

Rivier, C. and Vale, W. (1985) Effect of the long-term administration of corticotropin-releasing factor on the pituitary-adrenal and pituitary-gonadal axis in the male rat. *J. Clin. Invest.*, **75**: 689–94

Rivier, C., Rivier, J. and Vale, W. (1986) Stress-induced inhibition of reproductive functions: role of endogenous corticotropin-releasing factor. *Science*, **231**: 607–9

Roth, K.A., Weber, E., Barchas, J.D. *et al.* (1983) Immunoreactive dynorphin-(1-8) and corticotropin-releasing factor in subpopulation of hypothalamic neurons. *Science*, **219**: 189–91

Rothwell, N.J. (1990) Central effects of CRF on metabolism and energy balance. *Neurosci. Behav. Rev.*, **14**: 263–71

Sakanaka, M.S., Shibasaki, T. and Lederis, K. (1986) Distribution and efferent projections of corticotropin-releasing factor-like immunoreactivity in the rat amygdaloid complex. *Brain Res.*, **382**: 213–38

Sakanaka, M., Shibasaki, T. and Lederis, K. (1987a) Corticotropin-releasing factor-like immunoreactivity in the rat brain as revealed by a modified cobalt-glucose oxidase-diaminobenzidine method. *J. Comp. Neurol.*, **260**: 256–98

Sakanaka, M., Shibasaki, T. and Lederis, K. (1987b) Corticotropin-releasing factor-containing afferents to the inferior colliculus of the rat brain. *Brain Res.*, **414**: 68–76

Sakanaka, M., Magari, S., Shibasaki, T. and Inoue, N. (1989) Co-localization of corticotropin-releasing factor- and enkephalin-like immunoreactivities in nerve cells of the rat hypothalamus and adjacent areas. *Brain Res.*, **487**: 357–62

Sancibrian, M., Serrano, J.S. and Minamo, F.J. (1991) Opioid and prostaglandin mechanisms involved in the effects of GABAergic drugs on body temperature. *Gen. Pharmacol.*, **22**: 259–62

Sarne, Y., Flitstein, A. and Oppenheimer, E. (1991) Anti-arrhythmic activities of opioid agonists and antagonists and their stereoisomers. *Br. J. Pharmacol.*, **102**: 696–8

Saunders, W.S. and Thornhill, J.A. (1986) Pressor, tachycardic and behavioral excitatory responses in conscious rats following ICV administration of ACTH and CRF are blocked by naloxone pretreatment. *Peptides*, **7**: 597–601

Scoggins, B.A., Coghlan, J.P., Denton, D.A. *et al.* (1984) Intracerebroventricular infusions of corticotropin-releasing factor (CRF) and ACTH raise blood pressure in sheep. *Clin. Exp. Pharm. Phys.*, **11**: 365–8

Sharkey, J., Appel, N.M. and De Souza, E.B. (1989) Alterations in local cerebral glucose utilization following central administration of corticotropin-releasing factor in rats. *Synapse*, **4**: 80–7

Sheldon, R.J., Jiang, Q., Porreca, F. and Fisher, L.A. (1990) Gastrointestinal motor effects of corticotropin-releasing factor in mice. *Reg. Peptides*, **28**: 137–51

Sherman, J.E. and Kalin, N.H. (1986) ICV-CRH potently affects behavior without altering antinociceptive responding. *Life Sci.*, **39**: 433–41

Sherman, J.E. and Kalin, N.H. (1988) ICV-CRH alters stress-induced freezing behavior without affecting pain sensitivity. *Pharmacol. Biochem. Behav.*, **30**: 801–7

Shibasaki, T., Yamauchi, N., Kato, Y. *et al.* (1988) Involvement of corticotropin-releasing factor in stress-induced anorexia and reversion of the anorexia by somatostatin in the rat. *Life Sci.*, **43**: 1103–10

Siggins, G.R. (1990) Electrophysiology of corticotropin-releasing factor in nervous tissue, in E.B. De Souza and C.B. Nemeroff (eds.), *Corticotropin-Releasing Factor: Basic and Clinical Studies of a Neuropeptide*, CRC Press, Boca Raton, FL, pp 205–16

Simantov, R. and Snyder, S. (1977) Opiate receptor binding in the pituitary gland. *Brain Res.*, **124**: 178–84

Sirinathsinghji, D.J.S. (1986) Regulation of lordosis behaviour in the female rat by corticotropin-releasing factor, β-endorphin/corticotropin and luteinizing hormone-releasing hormone neuronal systems in the medial preoptic area. *Brain Res.*, **375**: 49–56

Sirinathsinghji, D.J.S. (1987) Inhibitory influence of corticotropin releasing factor on components of sexual behaviour in the male rat. *Brain Res.*, **407**: 185–90

Sirinathsinghji, D.J.S. and Heavens, R.P. (1989) Stimulation of GABA release from the rat neostriatum and globus pallidus in vivo by corticotropin-releasing factor. *Neurosci. Lett.*, **100**: 203–9

Sirinathsinghji, D., Nikolarakis, K.E. and Herz, A. (1989) Corticotropin-releasing factor stimulates the release of methionine-enkephalin and dynorphin from the neostriatum and globus pallidus of the rat: *in vivo* and *in vitro* studies. *Brain Res.*, **490**: 276–91

Sitsen, J.M., Van, Fee, J.M. and De Jong, W. (1982) Cardiovascular and respiratory effects of beta-endorphin in anesthetized and conscious rats. *J. Cardiovasc. Pharmacol.*, **4**: 883–8

Skofitsch G. and Jacobowitz, D.M. (1985) Distribution of corticotropin-releasing factor-like immonoreactivity in the rat brain by immunohistochemistry and radioimmunoassay: comparison and characterization of bovine and rat/human CRF antisera. *Peptides*, **6**: 319–36

Smith, M.A., Bissette, G., Slotkin, T.A. *et al.* (1986) Release of corticotropin-releasing factor from rat brain regions *in vitro*. *Endocrinology*, **118**: 1997–2001

Smith, M.A., Davidson, J., Ritchie, J.C. *et al.* (1989) The corticotropin-releasing hormone test in patients with post-traumatic stress disorder. *Biol. Psychiatry*, **26**: 349–55

Song, Z.H. and Takemori, A.E. (1991) Antagonism of morphine antinociception by intrathecally administered corticotropin-releasing factor in mice. *J. Pharmacol. Exp. Therap.*, **256:** 909–12

Spadaro, F., Berridge, C.W., Baldwin, H.A. and Dunn, A.J. (1990) Corticotropin-releasing factor acts via a third ventricle site to reduce exploratory behavior in rats. *Pharmacol. Biochem. Behav.*, **36:** 305–9

Spiess, J., Rivier, J., Rivier, C. and Vale, W. (1981) Primary structure of corticotropin-releasing factor from ovine hypothalamus. *Proc. Natl. Acad. Sci. USA.*, **78:** 6517–21

Sutton, R.E., Koob, G.F., Le Moal, M. *et al.* (1982) Corticotropin releasing factor produces behavioural activation in rats. *Nature*, **297:** 333–4

Swanson, L.W. and Sawchenko, P.E. (1983) Hypothalamic integration and organization of the paraventricular and supraoptic nuclei. *Ann. Rev. Neurosci.*, **6:** 269–324

Swanson, L.W. and Simmons, D.M. (1989) Differential steroid hormone and neural influences on peptide mRNA levels in CRH cells of the paraventricular nucleus: a hybridization histochemical study in the rat. *J. Comp. Neurol.*, **285:** 413–35

Swanson, L.W., Sawchenko, P.E., Rivier, J. and Vale, W.W. (1983) Organization of ovine corticotropin-releasing factor immunoreactive cells and fibers in the rat brain: an immunohistochemical study. *Neuroendocrinology*, **36:** 165–86

Swerdlow, N.R., Britton, K.T. and Koob, G.F. (1989) Potentiation of acoustic startle by corticotropin-releasing factor (CRF) and by fear are both reversed by α-helical CRF (9-41) *Neuropsychopharmacology*, **2:** 285–92

Taché, Y., Goto, Y., Gunion, M.M. *et al.* (1983) Inhibition of gastric acid secretion in rats by intracerebral injections of corticotropin-releasing factor. *Science*, **222:** 935–37

Taché, Y., Goto, Y., Gunion, M.M. *et al.* (1984) Inhibition of gastric acid secretion in rats and dogs by corticotropin-releasing factor. *Gastroenterology*, **86:** 281–6

Taché, Y., Maeda-Hagiwara, M. and Turkelson, C.M. (1987) Central nervous system actions of corticotropin-releasing factor to inhibit gastric emptying in rats. *Am. J. Physiol.*, **253:** G241–5

Taché, Y., Gunion, M.M. and Stephens, R. (1990) CRF: Central nervous system action to influence gastrointestinal function and role in the gastrointestinal response to stress, in E.B. de Souza and C.B. Nemeroff (eds.), *Corticotropin-Releasing Factor: Basic and Clinical Studies of a Neuropeptide*, CRC Press, Boca Raton, FL, pp 299–307

Tarjan, E., Denton, D.A. and Weisinger, R.S. (1991) Corticotropin-releasing factor enhances sodium and water intake/excretion in rabbits. *Brain Res.*, **542:** 219–24

Thatcher-Britton, K. and Koob, G.F. (1986) Alcohol reverses the proconflict effect of corticotropin-releasing factor. *Reg. Peptides*, **16:** 315–20

Thind, K.K. and Goldsmith, P.C. (1989) Corticotropin-releasing factor neurons innervate dopamine neurons in the periventricular hypothalamus in juvenile macaques. *Neuroendocrinology*, **50:** 351–8

Thompson, R.C., Seasholtz, A.F. and Herbert, E. (1987) Rat corticotropin-releasing hormone gene: sequence and tissue-specific expression. *Mol. Endocrinol.*, **1:** 363–70

Tsagarakis, S., Navara, P., Rees, L.H. *et al.* (1989) Morphine directly modulates the release of stimulated corticotropin-releasing factor-41 from rat hypothalamus *in vitro*. *Endocrinology*, **124:** 2330–5

Uhl, G.R., Goodman, R.R., Kuhar, M.J. *et al.* (1979) Immunocytochemical mapping of enkephalin containing cell bodies, fibers, and nerve terminals in the brain stem of the rat. *Brain Res.*, **166:** 75–94

Valentino, R.J. (1989) Corticotropin-releasing factor: Putative neurotransmitter in the noradrenergic locus ceruleus. *Psychopharmacology Bull.*, **25:** 306–11

Valentino, R.J. and Foote, S.L. (1988) Corticotropin-releasing hormone increases tonic but not sensory-evoked activity of noradrenergic locus coeruleus neurons in unanesthetized rats. *J. Neurosci.*, **8:** 1016–25

Valentino, R.J. and Wehby, R.G. (1988) Corticotropin-releasing factor: evidence for a neurotransmitter role in the locus coeruleus during haemodynamic stress. *Neuroendocrinol.*, **48:** 674–7.

Van Loon, G.R., Appel, N.M. and Ho, D. (1981) Beta-endorphin-induced stimulation of central sympathetic outflow: beta-endorphin increases plasma concentrations of epinephrine, norepinephrine and dopamine in rats. *Endocrinology*, **109**: 46–53

Veldhuis, H.D. and De Wied, D. (1984) Differential behavioral actions of corticotropin-releasing factor (CRF). *Pharmacol. Biochem. Behav.*, **21**: 707–13

Weiss, S.R.B., Post, R.M., Gold, P.W. *et al.* (1986) CRF-induced seizures and behavior: interaction with amygdala kindling. *Brain Res.*, **372**: 345–51

Williams, C.L., Peterson, J.M., Villar, R.G. and Burks, T.F. (1987) Corticotropin-releasing factor directly mediates colonic responses to stress. *Am. J. Physiol.*, **253**: G582–6

Williams, C.L., Nishihara, M., Thalabard, J.-C. *et al.* (1990) Corticotropin-releasing factor and gonadotropin-releasing hormone pulse generator activity in the rhesus monkey. *Neuroendocrinology*, **52**: 133–7

Wynn, P.C., Harwood, J.P., Catt, K.J. and Aguilera, G. (1988) Corticotropin-releasing factor (CRF) induces densitization of the rat pituitary CRF receptor-adenylate cyclase complex. *Endocrinology*, **122**: 351–8

Xiao, E., Luckhaus, J., Niemann, W. and Ferin, M. (1989) Acute inhibition of gonadotropin secretion by corticotropin-releasing hormone in the primate: are the adrenal glands involved? *Endocrinology*, **124**: 1632–7

Young III, W.S., Mezey, E. and Siegel, R. (1986a) Quantitative in situ hybridization histochemistry reveals increased levels of corticotropin-releasing factor mRNA after adrenalectomy in rats. *Neurosci. Lett.*, **70**: 198–203

Young III, W.S., Walker, L.C., Powers, R.E. *et al.* (1986b) Corticotropin-releasing factor mRNA is expressed in the inferior olives of rodents and primates. *Mol. Brain Res.*, **1**: 189–92

Zukin, R.S., Tempel, A. and Eghbali, M. (1986) Selective radioligands for characterization and neuroanatomical distribution studies of brain opioid receptors, in R.M. Brown, D.H. Clouet and D.P. Friedman (eds.), *Opiate Receptor Subtypes and Brain Function*, NIDA Res. Monograph, Washington, DC, pp 28–47

APPENDIX 13.1: CRF NEURAL CIRCUITS IN THE MAMMALIAN BRAIN

Region	mRNA	Peptide	Receptors
Neocortex (all laminae)			
rat	Laminae II & III > V	ci-v;fi-v	rv-d
monkey	> IV & VI	ci-v;fi-v	rd
human		ci;fi	rv-d
sheep		cø-i;fø-i	
rabbit			rv-d
Allocortex (all laminae)			
rat	yes	ci-v;fi	rv-d
monkey		ci-d;fi-d	rd-d+
sheep		ci-v;fi-v	
rabbit			rv-d
Olfactory Tubercle/bulb			
rat	yes	cø-i;fø-i	rv-d
monkey			rv-d
sheep		ci;fi	
rabbit			ri-d
Basal Ganglia			
rat		cø-i;fi-d	ri-v
monkey			ri-d
human			
sheep		cø-i;fø-i	rv
rabbit			rø-d

APPENDIX 13.1 – (*cont.*)

Region	mRNA	Peptide	Receptors
Septal Area			
Nucleus accumbens			
rat		ci-d;fv	rv
monkey			rv
sheep		cø;fø	
rabbit			rv
Nucleus diagonal band (Broca)			
rat		ci-v;fi	ri
sheep		ci;fi	
rabbit			rø
Septal nuclei			
rat		ci-v;fi-d	ri-v
monkey			ri-v
sheep		ci;fi	
rabbit			rø-d
Bed nucleus stria terminalis			
rat	yes	cv-d;fi-d	ri-v
monkey			ri-v
sheep		cø-i;fø-1	
rabbit			rv
Amygdala (Whole)			
rat		ci-v;fi-d	ri-v
monkey			rø-d
sheep		cv;fv	
Basolateral nucleus			
rat		ci;fi	rv
rabbit			rd
Central nucleus			
rat	yes	cd;fø-i	ri-v
Medial nucleus			
rat		ci-v;fd	ri-v
rabbit			rv
Thalamus			
Lateral			
rat		ci-v;fi	rv
monkey		cv;fv	rv
sheep		cø-v;fd	
rabbit			ri-v
Medial			
rat	yes	cø-v;fi	ri-v
monkey		cv-d;fd	rv
sheep		cø;fø	
rabbit			ri-v
Anterior			
rat			rv
monkey		cv	
sheep		cø-v;fø-v	
rabbit			rv

APPENDIX 13.1 – (*cont.*)

Region	mRNA	Peptide	Receptors
Habenula (medial plus lateral)			
rat		ci;fi	rv
sheep		ci-v;fi-v	
Hypothalamus			
Lateral			
rat		ci-v;fv	rv
monkey		cø;fd	
sheep		ci-v;fi-v	
rabbit			ro
Medial			
rat		ci-v;fv-d	ri-v
monkey		cø;fd	
sheep		ci-d;fi-v	
rabbit			ri-v
Anterior			
rat		ci-v	ri-v
monkey		cø	ri-v
sheep		ci-v;fi-v	
rabbit			ri-v
Paraventricular nucleus			
rat	yes	ci-d+;fi-v	rø-v
monkey		cd+;fd+	
sheep		ci-d;fi-v	
rabbit			rø
Arcuate nucleus			
rat		ci-v;fv	rø-v
monkey		cø	rd
rabbit			rø
Mammillary peduncle			
rat		ci-v;fi	ri-v
monkey		cø	ri-v
sheep		cø-i;fø-i	
rabbit			ri-v
Hippocampus (whole)			
sheep		ci-v;fi-v	
Dentate gyrus			
rat		cø-fi-v	ri-v
sheep		cø-i;fø-i	
rabbit			rø-d
Subiculum			
rat		cv;fi-v	ri-v
rabbit			rø
Ammons horn			
rat		cv;fi-v	
rabbit			rø-v

APPENDIX 13.1 – *(cont.)*

Region	mRNA	Peptide	Receptors
Midbrain			
Superior colliculus			
rat		ci;fi	rv
monkey			rv
sheep		cø;fø	
rabbit			rø-d
Inferior colliculus			
rat	yes	ci-v;fi	rv-d
monkey			rv
sheep		cø;fø	
rabbit			rø
Raphe nuclei			
rat	yes	ci-v;fi-v	rø-i
cat		cv	
sheep		ci-v;fi-v	
rabbit			rø
opossum		ci;fi	
Central gray			
rat		ci-v;fv-d	ri-v
sheep		ci-v;fi-v	
rabbit			rø
Substantia nigra			
rat		cø;fi	ri-v
monkey		cv-d;fv-d	
sheep		cø;fø	
rabbit			rø
Medulla-Pons			
Inferior olive			
rat	yes	ci-d;fv	rv
monkey	yes	cv-d;fv-d	
cat		ci-d;fd	
human	yes	cv-d;fi-v	
sheep		ci-d;fi-d	
rabbit	yes		ro
opossum	yes	ci-d;fi-v	rv
Superior olive			
rat	yes	ci;fi	rv-d
sheep		cø;fø	
Lateral reticular nucleus			
rat	yes	ci-v;fi	rv
monkey		cd;fv	
sheep		cø-i;fø-i	
rabbit			rø
opossum		ci-v;fi-v	
Paragigantocellular reticular nucleus			
rat		ci-v;fi	rv
sheep		cø-i;fø-i	
rabbit			rø
opossum		ci-v;fi-v	

APPENDIX 13.1 – (*cont.*)

Region	mRNA	Peptide	Receptors
Spinal trigeminal nucleus			
rat	none	cv-d;fi-d	rv-d
monkey		fv-d	
cat		ci	
sheep		cø-i;fø-i	
rabbit			rø
Cuneate nucleus			
rat		ci-v;fi-v	rv-d
monkey		fv	
cat		ci	
rabbit			rø
Gracile nucleus			
rat		ci;fi	rv-d
rabbit			rø
Nucleus of the solitary tract			
rat	yes	cv(fv)	rv
monkey		ci(fv)	
cat		cø	
sheep		cø-i;fø-i	
rabbit			rø
Lateral vestibular nucleus			
rat		ci;fi	ri
cat		ci;fi	
rabbit			rø
Medial vestibular nucleus			
rat		ci;fi	rv
cat		ci;fi	
rabbit			rø
opossum		ci;fi	
Facial nucleus			
rat		fi	rd
cat		ci	
rabbit			rø
Lateral cervical nucleus			
rat			rv-d
rabbit			rø
Vestibulocochlear nucleus			
rat		fi	rv-d
rabbit			rø
Cochlear nucleus			
rat	yes	fi	rv
rabbit			rø

APPENDIX 13.1 – (*cont.*)

Region	mRNA	Peptide	Receptors
Dorsal tegmental nucleus			
rat		cv;fi	rv
monkey		cd;fv	
cat		cv	
sheep		cø;fø	
rabbit			rø
Ventral tegmental nucleus			
rat		cv;fv	rv
monkey		cd;fi	
cat		cv	
human		ci;fi	
sheep			rø
rabbit			
Parabrachial nucleus			
rat		ci-v;fi-v	rv
monkey		ci;fv	rv
sheep		sheep	
rabbit			rø
Pontine nucleus			
rat		ci-v;fi-v	ri-v
rabbit			rø
Hypoglossal nucleus			
rat		ci;fi	rv-d
monkey		fv	
cat		ci;fi	
rabbit			rø
opossum		ci;fi	
Locus Coeruleus			
rat	none	ci-v;fv	ri-v
monkey		ci;fd	ri-v
cat		ci;fi	
rabbit			rø
opossum		ci;fi	
Cerebellum (all lobules)			
rat		cø;fv	rv-d
monkey		cø;fd	rv-d
cat		cø;fv	
human		cø;fv	ri-d
sheep		cø;fi	
rabbit			rv-d
opossum		cø;fi-d	ri-d
Interpositus nucleus			
rat		cø;fv	rv
monkey		cø;fd	
cat		cø;fi-v	
opossum		cø;fi	

APPENDIX 13.1 – (cont.)

Region	mRNA	Peptide	Receptors
Medial nucleus			
rat		cø;fv	rv-d
monkey		cø-i;fi-d	
cat		cø;fi-v	
opossum		cø;fi	
Paraflocculus			
rat			rd
cat		cø;fi-v	
sheep		cø;fi-v	
opossum		cø;fi-v	

Abbreviations *Structures* – c = cells; f = fibres; r = receptors. *Density* – ø = none; i = isolated; v = variable; d = dense

References

Rat
Aguilera *et al.*, 1986
Aguilera *et al.*, 1987
Battaglia *et al.*, 1987
Beyer *et al.*, 1988
Cummings *et al.*, 1983
De Souza, 1987
De Souza and Appel, 1991
De Souza and Kuhar, 1986
Fellmann *et al.*, 1984
Harbuz *et al.*, 1991
Hauger *et al.*, 1988,1989
Herman *et al.*, 1989
Imaki *et al.*, 1989
Imaki *et al.*, 1991a,b
Jingami *et al.*, 1985
Joseph and Knigge 1983
Ju and Han, 1989
Merchenthaler, 1984
Merchenthaler *et al.*, 1982
Merchenthaler *et al.*, 1983
Mezey and Palkovits, 1991
Palkovits *et al.*, 1985
Palkovits *et al.*, 1987
Sakanaka *et al.*, 1986
Sakanaka *et al.*, 1987a,b
Skofitsch and Jacobowitz, 1985
Swanson and Simmons, 1989
Swanson *et al.*, 1983
Thompson *et al.*, 1987
Young *et al.*, 1986a,b

Monkey
Cha and Foote, 1988
DeSouza and Kuhar, 1986
Foote and Cha, 1988
Lewis *et al.*, 1989
Millan *et al.*, 1986
Paull *et al.*, 1984
Thind and Goldsmith, 1989
Young *et al.*, 1986b

Cat
Cummings; 1989
Cummings *et al.*, 1988
Kitahama *et al.*, 1988

Human
Powers *et al.*, 1987
Young *et al.*, 1986b

Sheep
Cummings *et al.*, 1988
Palkovits *et al.*, 1983

Rabbit
Barmack and Young, 1990
Chai *et al.*, 1990
Fischman and Moldow, 1982

Opossum
Cummings and King, 1990
Cummings *et al.*, 1989

APPENDIX 13.2: HPA AXIS-DEPENDENT AND HPA-AXIS-INDEPENDENT ACTIONS OF CRF IN STRESS RESPONDING

RESPONSE	TREATMENT					REFERENCE
	CRF (ICV)	CRF (PERI)	HYPX	ADRX	DEX	
Endocrine changes						
Plasma ACTH	+	+		ø	ø/B	26, 27, 29
Plasma corticosteroids	+	+		B	ø/B	2, 3, 4, 39, 42
Plasma growth hormone	−	ø				19, 25
Plasma luteinizing hormone	−	−/ø		ø/A	ø/B	16, 25, 28, 41, 42
Plasma vasopressin	+	+			B	6
Adrenomedullary/ Sympathetic neuronal function						
Plasma noradrenaline	+	ø/+			ø	2, 9, 23
Plasma adrenaline	ø/+	ø/+			ø	2, 9, 23
Cardiovascular function						
Blood pressure	ø/+	−/ø	ø	ø	ø	5, 14, 26
Heart rate	ø/+	ø/+	ø	ø	ø	5, 10, 14, 26
Bradycardia	+		B		B	20
Metabolism & energy balance						
Thermogenesis	+/−	ø	ø	ø	ø	8, 30
Plasma glucose	ø/+		ø	ø	ø	7, 33
Gastrointestinal function						
Gastric acid secretion	−	−	ø	ø/B	B	11, 36, 37, 40
Gastric emptying	−	−	ø	ø/A	ø	15, 22, 31, 38
Gastric ulceration	−	+/−				2
Small bowel transit	−	−/ø	ø			20, 22, 31, 40
Large bowel transit	+	+	ø			22, 40
Bicarbonate secretion	+	+	B	ø	ø	21
Behavioural						
Motor activity						
familiar environment	+	ø	ø		ø	3, 4, 13, 32, 35
novel environment	+/−	ø	ø			2, 35, 39
Ingestive behaviour						
Feeding	−	ø	ø	A	ø	3, 10, 25
Drinking	−				ø	10, 25
Sexual behaviour						
Females	−	ø				34
Grooming	ø/+	ø	ø	ø	ø	1, 3, 12, 24, 32
Conditioned responding						
punished	−/ø	+/−			ø	4, 39
nonpunished	−	+		ø	ø	4, 39
Nociceptive	ø	ø	ø/B	ø	B	1, 7, 17, 18, 32

Key
− : decreased; + : increased; o : no effect; A : attenuated; b : blocked.

Abbreviations
ICV : intracerebroventricular;   PERI : peripheral;   HYPX : Hypophysectomy;   ADRX : adrenalectomy; DEX : dexamethasone.

APPENDIX 13.2 – (cont.)

References

[1]Ayesta and Nikoarakis, 1989
[2]Bakke et al., 1990
[3]Britton et al., 1986a
[4]Britton et al., 1986b
[5]Brown and Fisher, 1983
[6]Brown and Gray, 1988
[7]Brown et al., 1982a
[8]Brown et al., 1982b
[9]Brown et al., 1983
[10]Diamant and De Wied, 1991
[11]Druge et al., 1989
[12]Dunn and Berridge, 1987
[13]Eaves et al., 1985
[14]Fisher et al., 1983
[15]Garrick et al., 1988
[16]Gindoff et al., 1989
[17]Hargreaves et al., 1989
[18]Hargreaves et al., 1990
[19]Katakami et al., 1985
[20]Kiang and Wei, 1985
[21]Lenz, 1989

[22]Lenz et al., 1988
[23]Lima and Sourkes, 1987
[24]Morley and Levine, 1982
[25]Ono et al., 1984
[26]Raff et al., 1986
[27]Rivier and Vale, 1983
[28]Rivier and Vale, 1984
[29]Rivier and Vale, 1985
[30]Rothwell, 1990
[31]Sheldon et al., 1990
[32]Sherman and Kalin, 1988
[33]Shibasaki et al., 1988
[34]Sirinathsinghji, 1986
[35]Sutton et al., 1982
[36]Taché et al., 1983
[37]Taché et al., 1984
[38]Taché et al., 1987
[39]Velduis and De Wied, 1984
[40]Williams et al., 1987
[41]Williams et al., 1990
[42]Xiao et al., 1989

APPENDIX 13.3: A COMPARISON OF CHANGES INDUCED BY STRESS, CRF AND OPIOIDS

RESPONSE	STRESS	CRF	OPIOIDS	Ab or αhCRF PLUS		OPIATE Ab or antag PLUS		
				STRESS	CRF	STRESS	CRF	OPIOID
Endocrine								
Plasma ACTH	↑	↑	↑	a-b	b	ø-b		b
Plasma CS	↑	↑	↑	b				b
Plasma GH	↓	↓	↓↑	a-b				a
Plasma LH	↓	↓	↓ø	a-b		b	b-a-ø	b
Plasma AVP	↑	ø↑	↓ø			ø-b		b
Plasma glucagon		↑	↑					
Adrenomedullary function								
Plasma noradrenaline	↑	↑	↓↑	a-ø	b-ø	a		
Plasma adrenaline	↑	↑	↓↑	b	b-ø	a		
Cardiovascular function								
Blood pressure	↑	↑	↓↑	a	b	b	b	b
Heart rate	↑	↑	↓↑	a	b	b	b	b
Baroreceptor reflex	↓	↓	↑	b	b	b	b	
Energy balance								
BAT thermogenesis		↓↑	↑					
O₂ consumption		↑	↑		ø		b	b
Plasma glucose	↑	↑	↑	b	b			
Gastrointestinal function								
Gastric acid secretion	↓	↓↑	↓	a-b		b	ø-a-b	ø-b
Gastric emptying	↓	↓	↓	a-b	b		ø-b	b
Gastric ulceration	↑	↓	↓					
Small bowel transit	↓	↓	↓	b			ø-b	b
Large bowel transit	↑	↑	↓	b			ø	b
Bicarbonate secretion	↑	↑	ø	a-b	b	b	b	

APPENDIX 13.3 – (cont.)

RESPONSE	STRESS	CRF	OPIOIDS	Ab or αhCRF PLUS		OPIATE Ab or antag PLUS		
				STRESS	CRF	STRESS	CRF	OPIOID
Electrophysiological								
EEG		↑	↑					ø-b
Locus coeruleus firing	↑	↑						b
Hippocampal firing		↑	↓ø↑				b	ø-b
Amygdala firing		↑	↓↑					a
Lateral septum firing	↓							
Neurochemical								
Dopamine release	↑	↑	↓ø↑				ø	b
Noradrenaline release	↑	↑	↓					b
5-HT release	↑	ø	↓ø↑				ø	a-b
GABA release	↑	↑	↓ø					ø-a
Behavioural								
Motor activity								
Familiar environment	↑	↑	↓↑	b	b		b-ø	
Novel environment	↓↑	↓ø↑	↓↑	ø-b	b	b	b	b
Ingestive behaviour								
Feeding	↓↑	↓ø↑	↓↑	a-b	a-b	b	b	
Drinking	↓↑	↓ø↑	↓ø↑	a-b		a		
Sexual behaviour								
Males	↓	↓↑	↓				b	b-ø
Females	↓	↓	↓ø↑				a-b	b
Grooming	↑	ø↑	ø↑				ø-b	
Conditioned responding								
punished	↓	↓	↑	ø-a	b			b
nonpunished	↓	↓	↓	ø	b			b
Acoustic startle	↑	ø↑		a	b	a		
Nociceptive	↓ø	↓ø	↓			b	b	øb

Key

↑ : increase; ↓ : decrease; a : attenuated; b : blocked; ø : no change.

Abbreviations

Ab : antibody; αhCRF : α-helical CRF; ant : antagonist.

Adapted from

Amir *et al.*, 1980
Brown, 1991a,b
Chatterton, 1990
Cole and Koob, 1991
Dunn and Berridge, 1990
Ehlers, 1990
Fisher, 1991a,b
Grossman, 1991
Illes, 1989
Kalin, 1990
Koob, 1985, 1991
Levine *et al.*, 1985
Lupica and Dunwiddle, 1991
Millan, 1986
Millan and Herz, 1985
Morley, 1987

Morley *et al.*, 1982
Nader and Barrett, 1989
North, 1979, 1986
Olson *et al.*, 1987
Pfaus and Gorzalka, 1987
Pfeiffer and Herz, 1984
Rhee and Hendrix, 1989
Rivier, 1989, 1991
Sancibrian *et al.*, 1991
Siggins, 1990
Spadaro *et al.*, 1990
Swerdlow *et al.*, 1989
Taché *et al.*, 1990
Tarjan *et al.*, 1991
Valentino, 1989

APPENDIX 13.4: CRF AND OPIOID PEPTIDE CELL, FIBRE AND RECEPTOR DISTRIBUTIONS

Region	Cell or Fibre (*) Density				Receptor density (†) fmol/mg protein			
	CRF	β-END	ENK	DYN	μ	δ	κ	CRF
Neocortex	0-1(1-2)	0(0)	0(0-1)	1-3(1)	49(3)	71(2)	42(1)	80
Piriform Cortex	1-2(1)	0(0)	0(1-2)	0-3(1)	nd(2)	nd(2)	(2)	29
Cingulate Cortex	1-2(1)	0(0)	0(0-1)	2-3(1-2)	127			44
Entorhinal Cortex	1-2(1)	0(0)	0-1(0-2)	0-2(0-2)	(2)	(2)	(2)	76
Olfactory Tubercle	0-1(0-1)	0(0)	0-1(1-3)	1(1-3)	(1)	(3)	(3)	66
Basal Ganglia	0-1(0-1)	0(0)	0-2(2-4)	0-3(1-3)	170(4)	180(4)	58(3)	46
Nucl.Accumbens	1-3(2)	0(1-3)	1(1-3)	1-3(1-4)	183(4)	125(4)	(3)	65
Septum	1-2(1-2)	0(3)	0-3(1-3)	0-2(1-3)	127(3)	nd(1)	(1)	20
BNST	2-3(1)	0(4)	0-4(1-4)	0-3(0-4)	nd(2)	nd(2)	82(3)	22
Amygdala								
Medial	1-2(4)	0(3)	0(0-3)	0-1(1-2)	68(3)	nd(2)	(2)	26
Lateral	1(1)	0(2)	0(3)	0-2(0-2)	101(3)	nd	(3)	51
Central	2-4(1-3)	0(4)	0-3(2-4)	0-4(2-4)	nd	nd	(2)	22
Hippocampus	1(1)						21	20
Dentate gyrus	(1-2)	0(0)	2(2)	0-2(2)	37(3)	nd(1)	nd(1)	
Ammons horn	2(1-2)	0(0)	0(0-2)	0(0-3)	42(3)	nd(2)	42(1)	
Subiculum	(1-2)		0(2)	0(1)				31
Habenula nuclei	0(0-1)	0(0)	0-2(1-3)	0-2(0-3)	122(3)	nd(1)	56(3)	34
Thalamus								
Lateral	1-2(0-1)	0(0-2)	0-2(0-3)	0-2(0-3)	49	nd	52	38
Medial	0-2(1-2)	0(0-4)	0-1(1-2)	0-1(1-2)	71(4)	nd(1)	96(2)	21
Anterior			0(1-2)	0-4(0-4)	44	nd		
Hypothalamus								
POA	1-2	0(3)	0-1(2-3)	0-2(2-3)	(1)	(1)	(4)	28
Lateral	1-2	0(3)	0(2-3)	0-4(1-4)	(1)	nd	(2)	24
MPOA/AHA	1-2	0(2-3)	0(2-3)	0-2(0-4)				25
DMH/VMH	1-2(1-2)	0-1(1-2)	0-3(0-3)	2-3(1-4)	nd	nd	51(3)	21
PVN	1-3(1-2)	0(4)	0(2-4)	0-4(2-4)	nd	nd	(2)	33
Arcuate nucleus	1-3(2)	4(4)	3(3)	2-4(2-4)	nd	nd	(2)	0
ME/RCA	0(3-4)	0(0-2)		4(4)				
Midbrain								
Collicular nuclei	1(1)	0-1(1-2)	0-2(0-3)	0(0-3)	99(4)	nd(1)	78(2)	43
Raphe nuclei	2(1-2)	0(1-3)	0-4(0-4)	0-2(1-3)	(2)	nd	(2)	8
Central gray	1-2(2-3)	0(4)	1-2(3-4)	1-4(3-4)	38(1)	nd	56(2)	22
Substantia nigra	(1)	0(0-1)	0(1-3)	0-2(2-4)	(2-3)	(0-1)	(0-1)	nd
Medulla-Pons								73
Olivary nuclei	0-4(1)		0-1(0-3)	0(0-3)				55
Reticular nuclei	1-2(1-2)	0-1(2)	0-2(0-4)	0-3(0-4)	(1)	nd	(1)	96
Trigeminal nuclei	(1)	0-1(2)	2(4)	2-3(3)	(3)	nd	(2)	88
Cuneate nucleus	2(1)		0(0)	0(0)				86
Gracile nucleus			0(0-1)	0(0)				42
Nucl. solitary tract	2(2)	0(3)	2(3-4)	2-4(3-4)	(4)	(1)	(3)	42
Vestibular nuclei	1(1)		0-2(0-1)	0(0-1)				50
Facial nucleus		0(2)	0(2-3)	0(2)				151
Cochlear nuclei	(1)		1(2)	0(2)				95
Tegmental nuclei	0-2(1-2)		0-4(0-3)	0-2(0-2)				51
Parabrachial nucl.	1-2(1-2)	0(3)	2-3(3-4)	2(4)	(3)	nd	(2)	59
Hypoglossal nucl.	1(1)	0-1(2)	0(1-2)	0(1-2)				65
Locus Coeruleus	1-2(2)	0(2)	0(3)	0(2)	92	nd	78	9
Cerebellum (whole)	0(0-2)		0(0-1)	0(0-1)	nd	nd	33	99
Corpus callosum			0(0-1)	0(0-1)	9	3	7	8

Key
1 = isolated; 2 = low; 3 = moderate; 4 = dense; nd = not detected.
(*) = fibre density; (†) = relative density.

APPENDIX 13.4 – *(cont.)*

Abbreviations

DYN = dynorphin; END = endorphin; ENK = enkephalin
Region abbreviations, see: Abbreviations pp. xvii–xviii

Adapted from

CRF
Aguilera *et al.*, 1987
Cummings *et al.*, 1983
De Souza, 1987
De Souza and Appel, 1991
Palkovits *et al.*, 1983, 1985
Swanson *et al.*, 1983

Opioids
Fallon and Leslie, 1986
Finley *et al.*, 1981
Mansour *et al.*, 1987, 1988
Neal and Newman, 1989
Uhl *et al.*, 1979
Zukin *et al.*, 1986

Part Five
Future prospects

14

Emotional effects of physical exercise

PETER SALMON

DEPARTMENT OF CLINICAL PSYCHOLOGY,
UNIVERSITY OF LIVERPOOL,
WHELAN BUILDING,
PO BOX 147,
LIVERPOOL L69 3BX, UK

14.1 INTRODUCTION

Claims for the psychological benefits of physical exercise regimes include dramatic improvements in depression and other emotional problems and increased resilience to life stress. Much of the scientific literature has been concerned with how these benefits might result from unique properties of exercise training – in particular, its improvement of physical fitness – which set it apart from processes with which stress research is normally concerned. In contrast, this chapter will argue that physical exercise might provide a way of investigating mechanisms which are fundamental in the control of emotional distress and sensitivity to stress.

There is, of course, no abrupt division between physical activity which is part of daily life and activity which can be termed 'physical exercise'. Indeed, in improving

ISBN 0–12–663370–3

cardiovascular health, physically hard manual or domestic work is probably just as beneficial as formal exercise (Paffenbarger and Hyde, 1988). However, this review will reflect the literature on emotional effects of exercise in its overwhelming concern with formal exercise regimes. A further restriction will be a focus on aerobic exercise, which involves the activity of large muscle groups and which can be sustained for long periods such as in running, swimming, cycling or dancing. Regular training at high intensity over about 12 weeks can increase aerobic 'fitness': that is, the body's capacity for such work. The best measure of this is a person's oxygen uptake at maximal intensity exercise (VO_2max), but this value is usually estimated from heart rate responses to submaximal exercise. Much less information is available about pyschological effects of anaerobic exercise, which entails rapid muscular effort such as in weight-lifting and which, because of oxygen deficit in the muscle groups involved, cannot be sustained. Training in such procedures has sometimes been used as a control activity for aerobic training on the grounds that it does not increase aerobic fitness. Since, as will become clear, changes in aerobic fitness do not explain the emotional effects of aerobic exercise, anaerobic exercise may prove to share some of the same effects in relation to stress as does aerobic exercise.

14.2 EFFECTS OF SINGLE SESSIONS OF PHYSICAL EXERCISE

In regular exercisers, at least, strenuous exercise increases positive mood or relieves negative mood (Berger and Owen, 1983; Boutcher and Landers, 1988; Morris *et al.*, 1988). In samples not selected for regular exercise habits, enforced *mild* walking or voluntary participation in whatever form and duration of exercise they choose is correlated, day-to-day, with better positive mood and, although not so clearly, with less negative mood (Thayer, 1987; Watson, 1988; McIntyre *et al.*, 1990). This type of result has led some to conclude that exercise is intrinsically mood-improving (Morgan, 1985), but acute effects of strenous exercise in people who are not regular exercisers have been more variable. Negative mood has even been increased and positive mood decreased, particularly in less fit subjects (Boutcher and Landers, 1988; Steptoe and Bolton, 1988; Steptoe and Cox, 1988). These results, although inconsistent with the positive experience of committed exercisers, are consistent with the view of many non-exercisers that exercise is an uncomfortable, unpleasant and even painful activity.

An explanation for the acute emotional effects of exercise should be able to account for the opposite effects which exercise can have in different conditions or in different people, given that differences in the level of exertion cannot alone account for differences in emotional experience (Hardy and Rejeski, 1989). Possible explanations can be drawn from two theories. One concerns the effect of expectations or attributions on the 'labelling' of physiological arousal (Schachter and Singer, 1962; Reinsenzein, 1983). In contrived laboratory experiments, it has been shown that the arousal produced by exercise can enhance the experience of, for example, anxiety, anger, romantic attraction, elation or symptoms of illness depending on the salience of appropriate stimuli or the person's expectations (e.g. Zillman *et al.*, 1972; White and Knight, 1984; Meyer *et al.*, 1990). One report in which a similar result was not found used imagery, rather than a more life-like stimulus, to induce an emotional reaction

(Fillingim *et al.*, 1992). It is therefore surprising that very little work has explored the extent to which the emotional effects of 'real-life' exercise depend on exercisers' expectations or the cues present.

Nevertheless, early applications of exercise in the treatment of anxiety are most easily explained by this kind of influence. As part of exposure treatment resembling systematic desensitization, subjects were exposed to phobic stimuli when exhausted by physical exercise (Orwin, 1973; Muller and Armstrong, 1975; Driscoll, 1976). The rationale at the time was that subjects would learn to associate the phobic stimuli with a state of exhaustion, thought to be physiologically incompatible with anxiety. It seems more likely that the exercise caused subjects to reinterpret the physiological arousal produced by phobic stimuli (Muller and Armstrong, 1975). That is, rather than interpreting it in an anxiety-provoking way (for example, as a sign of illness or impending collapse), they might interpret it as a normal physiological arousal response. This would explain why the anxiety improvement which resulted from pairing physical exercise with a phobic stimulus correlated with the subjects' enjoyment of physical activity (Driscoll, 1976). This view is also consistent with current, cognitive accounts of anxiety and its treatment (Clark, 1986), in which individuals' catastrophic interpretations of their physiological arousal are primary.

A second theory which might explain how acute exercise can have opposing emotional effects is opponent-process theory. In his original account of this, Solomon (1980) cited strenuous exercise as an example of a strong aversive stimulus which, according to his theory, recruits an adaptive counter-regulation in the nervous system. Because of its delayed and sluggish time-course this counter-regulatory process was said to result in a 'rebound' improvement in mood after exercise stops. This could obviously explain a tendency for reduced distress by comparison with baseline levels 15 min after strenuous exercise in nonexercisers (Steptoe and Bolton, 1988). With repeated exercise, changes in the response to an individual exercise session were attributed to the growth of the counter-regulatory process. This would lead to tolerance of the unpleasant effects of exercise and, if the counter-regulation eventually outweighed the unpleasant effect, the growth of a positive emotional response. The situation may be even more complex than this, because of evidence that the counter-regulatory reaction to some emotional stimuli, at least, can be classically conditioned; that is, it is elicited not by the original emotional stimulus but by situations which have been associated with it in the past. This has been demonstrated in response to opioid and other drugs (Hinson and Siegel, 1982) although not yet in relation to exercise. Clear and testable predictions about emotional effects of exercise do arise from this theory, however. In particular, the acute emotional effects of exercise should depend on the stimuli which are present, and their previous association with exercise, as well as on the duration of an individual's experience of exercise training.

14.3 EMOTIONAL EFFECTS OF EXERCISE TRAINING

Little information is, however, available about how exercise training affects the emotional response to single sessions. Instead, studies of the long-term effects of exercise training have focused on enduring changes in mental state.

Anecdotally, exercise training is often claimed to make people 'feel better', and higher levels of physical activity correlate with greater wellbeing in community surveys (e.g. Stephens, 1988). In a longitudinal study, Cramer *et al.* (1991) confirmed that a 15-week walking programme improved general well-being in mildly obese women by comparison with a control group which took part in no alternative activity. The subscales which showed the most improvement were ones which measured feelings of energy and freedom from worry over health. However, this review will follow the overwhelming bias of most research in its focus on two specific aspects of unpleasant emotional state: anxiety and depression.

14.3.1 ANXIETY

Apart from the early reports (discussed above) in which acute effects of exercise were used to treat phobias, anxiolytic applications of exercise have been neglected until recently. When exercise training has been compared with a no-treatment control, results have been promising. Blumenthal *et al.* (1982) showed improvements in trait and state anxiety in healthy volunteers after a 10-week walking and jogging programme, and Norris *et al.* (1990) similarly reported that exercise training relieved minor neurotic symptoms. It is generally accepted, however, that psychological treatment trials require controls who receive a 'treatment' which, although designed to be ineffective, is as convincing as the active treatment and which involves the same social aspects. Such trials have suggested that anxiolytic effects are not specific to exercise. For instance, Long (1984) compared a jogging programme with stress inoculation training for volunteers who had responded to an advert seeking subjects who needed 'help in coping with stress'. Both treatments reduced state- and trait-anxiety by comparison with a waiting list control. Moreover, they were equally effective whether symptoms were mainly cognitive or somatic, despite earlier suggestions that exercise would preferentially improve somatic anxiety (Schwartz *et al.*, 1978; Sime, 1984). Later Long and Haney (1988) found that eight weekly group sessions of jogging or relaxation training reduced trait anxiety equally by comparison with baseline levels. Unfortunately, many controlled trials which apparently confirmed that exercise training is anxiolytic used inadequate controls. In some, control activities required less time or involvement than did the exercise (e.g. Goldwater and Collis, 1985); in others, they have been less plausible than exercise as a treatment; for example, Fasting and Gronningsaeter (1986) found a greater decline in trait anxiety in unemployed men after aerobic training than in a control group, but this merely met for discussion of current affairs.

Recently, however, two well-controlled studies have confirmed that a 10-week programme of moderate intensity walking and jogging reduced anxious mood in normal volunteers (Moses *et al.*, 1989) or in volunteers selected for high levels of anxiety (Steptoe *et al.*, 1989) by comparison with control subjects who underwent carefully matched training in flexibility exercises. Two further controlled reports contain some evidence that the people who were initially the most anxious gained the most benefit (Fasting and Gronningsaeter, 1986; Simons and Birkimer, 1988). This points to the need for studies which show whether clinical anxiety – particularly panic anxiety – can be treated by exercise training.

Increased fitness is unlikely to explain these effects. Reductions in anxiety have not correlated with measures of fitness, including estimated VO_2max (Fasting and Gronningsaeter, 1986) or heart rate during low intensity exercise (Simons and Birkimer, 1988). Moreover, Moses *et al.* (1989) found that, whereas aerobic fitness was increased more by a 10-week programme of strenuous walking and jogging than by one of milder exercise, anxious mood was reduced only by the mild exercise.

Perhaps because of the long-standing assumption that enhanced fitness underlies emotional benefits of exercise, other explanations have rarely been considered. Many of the reported effects might be explained by social factors since, as explained above, it is unlikely that typical control activities maximize social interaction as much as does confronting new and strenuous physical challenges. When Wilson *et al.* (1981) used a different physiological activity (eating together) as their control for exercise sessions, they found similar reductions in anxiety in both groups! The well-controlled studies by Steptoe *et al.* (1989) and Moses *et al.* (1989) require a different explanation, however.

Broadly, two types of explanation are possible. One is that the long-term anxiolytic effect represents merely the accumulation of acute effects, such as the relabelling of physiological arousal or the short-term 'rebound' improvement in mood discussed above; this possibility will be considered later. Alternatively, an entirely separate process might develop as exercise training continues. For example, Hollandsworth (1979) speculated that training reduces anxiety by improving the ability to control physiological responses, although he did not describe the mechanism by which this might occur. A suggestion to which we shall return is that exercise might increase people's resistance to anxiety-provoking events in daily life, perhaps by increasing their feelings of competence to deal with such events (Long, 1984). However, attempts to explain specific effects on anxiety may be less helpful than theories formulated in the light of evidence about effects of exercise on other aspects of negative emotional state; in particular, depression.

14.3.2 DEPRESSION

The use of exercise to treat depression has been a long-standing theme in both scientific and popular literature. Recommendations for its use as an antidepressant treatment (Brown *et al.*, 1978) have apparently been unrestrained by the lack of evidence, until recently, as to its effectiveness (Folkins and Sime, 1981; Weinstein and Meyers, 1983; Simons *et al.*, 1985). Although the elaboration of guidelines for treatment has continued (e.g. Eischen and Griest, 1984; Clearing-Sky, 1988), the evidence is still far from complete.

The view that exercise is antidepressive arose in a report by Morgan (1969) that a sample of depressed psychiatric patients were less physically fit than non-depressed patients. Morgan *et al.* (1970) went on to suggest that physical exercise could dispel depression on the grounds that the 'feel-good' response to exercise seemed incompatible with depression. Although it is possible to show more conclusively than did Morgan (1969) that people who habitually exercise are less depressed than sedentary people (e.g. Lobstein *et al.*, 1989), such evidence says nothing about the direction of cause and effect; depression may simply prevent people from exercising.

The most dramatic experimental evidence in support has been from single case studies. In one, Doyne et al. (1983) examined the effects on four severely clinically depressed women of a six-week programme of stationary cycling four days per week. Self-ratings suggested that each woman became happier by comparison with a baseline period during which a spurious 'subliminal' task controlled for attention and expectations of improvement – although not for the skill mastery involved in exercise training.

Unfortunately, many controlled treatment trials in depressed people, although at first sight supporting the view that exercise relieves depression, are compromised by inadequate controls. Exercisers often turn out to have had greater contact with the therapist, or with each other, than did control patients (Griest et al., 1979), or control activities have been less involving than the exercise programme. For example, Martinsen (1987) reported that nine weeks of exercise training improved mood in depressed inpatients, but the 'occupational therapy' received by controls seems hardly comparable to the training in jogging, cycling, skiing and swimming enjoyed by the exercise group! When control procedures have involved acquiring mastery of a physical skill, benefits no longer seem specific to exercise. Klein et al. (1985) found that self-rated depression was reduced equally in clinically depressed subjects who were trained in physical exercise or in meditation and relaxation by comparison with psychotherapy. The skill mastery entailed in exercise training, therefore, provides one possible explanation for many of the reported antidepressive effects. The social aspects of exercising in groups provide a further explanation: in one of the few trials in which exercise has been solitary, no antidepressive effect was found (Hughes et al., 1986).

Such arguments do not invalidate exercise as a clinical tool; it might provide an effective vehicle for increasing feelings of competence or social integration to treat depression. Nevertheless, the available evidence does not yet justify claims of an antidepressive effect *specific* to exercise in *clinical* depression. That there are so few well-controlled studies may reflect the unacceptability of exercise to severely depressed people; it is impractical to imagine that severely depressed patients could readily be motivated to engage in activities such as jogging, cycling or skiing.

Controlled trials in people from the normal population who are not clinically depressed (and are therefore easier to motivate) evade this problem, albeit at the expense of compromising clinical relevance. Difficulties with control groups remain, however. For instance, McCann and Holmes (1984) found that self-rated depression declined more in response to nine weeks of twice-weekly exercise than in controls undergoing relaxation training, but the latter involved merely receiving unsupervised instructions which would be unlikely to control for social and skill-mastery aspects of the exercise. Two well-controlled studies have, however, now confirmed antidepressive effects in non-clinical groups. Roth and Holmes (1987) selected students who reported a high level of recent life events, and subjected them to 11 weeks of training in running and walking, to relaxation training carefully matched with this for social content and skill mastery or to a no-treatment control. Although relaxation subjects felt more positive about their training than did exercised subjects at the fifth week, depression was reduced at this time only in exercisers. Steptoe et al. (1989) used strength and flexibility training to control for the mastery and structured activity inherent in their 10-week exercise training (walking and jogging) in anxious adults drawn from the normal population. Exercise reduced feelings of depression by comparison with controls.

Within some of these studies there are indications that, as with anxiety, subjects who are more disturbed gain the most benefit (Morgan et al., 1970; Roth and Holmes, 1987; Simons and Birkimer, 1988). Therefore, well-controlled studies of clinically depressed patients should be attempted.

In principle, as has been pointed out, both the antidepressant and anxiolytic effects of long-term exercise training could result from no more than the accumulation of short-term mood-improvement; no additional, long-term mechanism need be postulated. Indeed, reports of rapid deterioration of mood when exercise regimes are interrupted suggest that short term effects might be all important. For example, in a single case report, Sime (1987) described a man who used running and jogging to combat depression and in whom depression was apparently alleviated only on the days that he actually exercised. Despite other, anecdotal reports of rapidly worsening mood and subjective physical state in regular exercisers when their programmes of activity have been interrupted for a day or two (Baekeland, 1970; Thaxton, 1982), only one systematic report of relatively prolonged deprivation is available (Morris et al., 1990a). In this we showed that, notwithstanding rapid deterioration in physical symptoms and feelings of being unable to cope, deterioration in depression and anxiety was apparent only after two weeks of deprivation. Because of the evidence for a high level of psychopathology in people who take up regular running, we suggested that this represented a gradual loss of an enduring protection which exercise training had conferred against anxiety and depression.

The explanation for a long-term antidepressive action remains unclear. As with the anxiolytic effect, it is probably unrelated to changes in aerobic fitness. In Roth and Holmes' (1987) study, aerobic fitness is unlikely to have changed by the time that improvement in depression appeared in the exercise group (five weeks after the start of training). Simons and Birkimer (1988) found no correlation between changes in exercise heart rate and changes in mood. Similarly, Steptoe et al. (1989) found no correlation of increased fitness after their 10-week walking and jogging programme with the reduction in depression. Although alternative explanations have been proposed, none has much evidence to support it. Weinstein and Meyers (1983) proposed that exercise combats depression simply by increasing levels of reinforcement. Antidepressive effects of exercise could also be understood in terms of contemporary cognitive explanations of depression, in which a negative view of one's own competence is central. Consistent with this, there are many reports, reviewed by Folkins and Sime (1981), that exercise training improves self-concept and self-esteem. Most of these are, however, uncontrolled or only poorly controlled, and it is possible that improvements in self-concept are not a direct result of physical exercise but are mediated by social approval of the exercise (Heaps, 1978; Hilyer and Mitchell, 1979).

Finally, one process which is known to maintain depression is the vicious circle whereby depressed mood increases the probability that depressive, rather than cheerful, memories will be retrieved (Teasdale et al., 1980). Exercise could interfere with this either by distracting people from negative thoughts, as Morgan (1985, 1987) has suggested, or by promoting the retrieval of positive thoughts, as Clark et al. (1983) have speculated.

None of these theories for the antidepressant effect of exercise has appreciable empirical support. A further limitation of them all is that they have been developed in isolation from attempts to explain the anxiolytic effects of exercise. This may be mistaken: symptoms of anxiety and depression coexist and the syndromes are hard to separate in community samples; some drugs can relieve both and the neurochemical substrates may be similar (Gary, 1982; Dobson, 1985; Stavrakaki and Vargo, 1986). A further link is that both anxiety and depression are often regarded, in part, as responses to stress. Perhaps, therefore, a unified explanation for anxiolytic and antidepressive effects should also accommodate what is known about the effects of exercise training on vulnerability to stress. This was anticipated by Long (1984) in her speculation that anxiolytic effects of exercise reflect an increased resistance to anxiety-provoking events.

14.4 EXERCISE TRAINING AND RESISTANCE TO STRESS

Broadly, there are two ways in which exercise might increase resistance to stress. First, it might provide a way of coping palliatively with individual instances of stress. In animals, at least, there is evidence which is most easily understood in this way. Concurrent access to an exercise wheel reduced the increases in plasma corticosterone levels or blood pressure induced by periodic electric shock or social isolation (Starzec et al., 1983; Mills and Ward, 1986). The effect of exercise on concurrent stress has, however, been neglected in human literature (Gal and Lazarus, 1975). Instead, this has focused on a second way in which exercise training might help people to withstand stress: that is, by conferring an enduring ability to withstand future instances of stress.

14.4.1 REAL-LIFE STRESS

One way in which the ability of exercise training to enhance resistance to stress has been tested has been by relating naturally occurring variability in sensitivity to stress between individuals to their different histories of physical exercise. For example, in business executives selected for experience of a relatively large number of life events, those who habitually exercised most suffered the fewest symptoms of physical and psychiatric illness (Kobasa et al., 1985). Roth and Holmes (1985) went further than this in supporting a 'stress-buffering' role of physical exercise, by showing that the relationship between recent life-event scores and illness was smaller in fit than in unfit undergraduates. A subsequent study failed to reproduce these effects, but used subjects' own estimates of their fitness rather than an objective assessment (Roth et al., 1989). More recently, Brown (1991) reported that a high frequency of negative life events was associated with more episodes of recent illness in relatively unfit undergraduates, but not in fit ones. This followed a similar report in adolescents, using exercise participation rather than fitness as the moderator variable (Brown and Siegel, 1988). An obvious interpretation of this pattern of findings is that fitness, or another variable correlated with it such as amount of exercise, protects against the harmful physical effects of psychological stress. Nevertheless, the direction of cause and effect in correlational evidence of this sort is ambiguous. The pattern of results could be explained by other, less interesting, hypotheses: for example, that people who cope better with life stress have more time to take up exercise.

Experimental tests of the effects of exercise training on resistance to stress are obviously necessary to disentangle cause and effect. However, the study of major life events in this way is obviously problematic because they are too rare and diverse for responsivity to them to be useful as an outcome measure. It is, however, easier to find out how training affects the experience of more routine daily 'hassles'. A short-lived reduction in hassles was reported by Cramer *et al.* (1991) during a 15-week exercise programme, but this did not endure to the end of the programme. An alternative, but rarely used, strategy has been to contrive exposure to intense forms of real-life stress in relatively controlled ways. For instance, Brooke and Long (1987) showed that recovery of subjective anxiety and plasma adrenaline levels after rappelling (a manouevre similar to abseiling) was faster in fit than in unfit novices. Use of this strategy is obviously limited by the difficulty of contriving forms of stress which are of general relevance to the population at large.

14.4.2 STRESS IN THE LABORATORY

Therefore the most common approach has been to study controlled, mild and easily reproducible forms of stress in the laboratory on the assumption that responses to such tasks may indicate sensitivity to real-life stress. It should be noted, however, that the reliability of such a link is controversial (Gannon *et al.*, 1989; Turner *et al.*, 1990; Pollak, 1991; Steptoe and Vogele, 1991).

The usual design has been to compare stress responses between subjects who already differ in fitness. Once again, of course, it is not clear whether the crucial difference between the subjects is in fitness or in amount of exercise or, indeed, in some other physiological, psychological or social variable which is correlated with fitness. In an early study, Zillman *et al.* (1974) measured the intensity of anger elicited by verbal insult (by recording the intensity of electric shock that subjects thought they were administering to their tormentor). When subjects had been moderately aroused by strenuous exercise a few minutes previously, the less fit ones retaliated more intensely. In subsequent studies, cardiovascular responses have provided the most common measures. Concern with the cardiovascular system in this context owes much to the assumption that cardiovascular responses to psychological stress are similar to those to physical exertion which are, of course, reduced by exercise training. There are, however, important differences between the two sorts of response (Van Doornen *et al.*, 1988) and it may be more useful to regard the cardiovascular system as simply providing one kind of 'marker' for a psychological stress response (see Chapter 4). Nevertheless, their significance might be wider than this. Because of the role of abnormal or exaggerated cardiovascular responses to psychological stress in the development of cardiovascular disease (see Chapter 4), such results may also be relevant to the protective effect of exercise against this.

The following discussion is predicated on the widespread assumption that smaller responses to stress are better; i.e. that they reflect adaptation or resistance to stress.

However, this has not always been believed. An early paper suggested that exercise might improve adaptation to stress by promoting *greater* physiological sensitivity (Michael, 1957). There is a growing view that whether smaller or greater responses are 'better' may depend on the nature of the stress (Sothmann *et al.*, 1991). For example, Dienstbier (1991) has argued, on the basis of evidence about relationships of physiological responses to performance, that greater or more rapid catecholamine responses represent 'toughness' or stress-adaptation if stressors are unfamiliar, very challenging and last longer than half an hour or so. By contrast, the more adaptive response to brief and mild stressors would be a small one. Claytor (1991) has also reported some evidence to support his own suggestion that the less familiar a stress task is, the less likely are fit people to show a smaller cardiovascular response than unfit ones. Nevertheless, the relevance of these distinctions to understanding effects of exercise training on laboratory stress tasks remains unclear because these have almost always been mild and brief.

These laboratory studies have used forms of stress such as paced mental arithmetic or simple psychomotor tasks which, although perhaps less realistic, are more readily reproducible than Zillman *et al.*'s provocation by insult. At first, many negative results were reported using blood pressure or heart rate to index cardiovascular responses in fit and unfit people (Zimmerman and Fulton, 1981; Sinyor *et al.*, 1983; Hollander and Seraganian, 1984; Hull *et al.*, 1984; Keller and Seraganian, 1984; Plante and Karpowitz, 1987; Seraganian *et al.*, 1987) although two recent studies which used more specific measures of cardiovascular function did find that smaller responses to psychological stress were associated with greater aerobic fitness (Shulhan *et al.*, 1986; de Geus *et al.*, 1990). Positive results using heart rate and blood pressure have been more reliably obtained when groups differing greatly in fitness have been contrasted. Two such studies have shown smaller heart rate responses to mental stress in very fit than in very unfit undergraduates (Holmes and Roth, 1985; Light *et al.*, 1987). Similarly, Van Doornen and de Geus (1989) contrasted responses to a reaction time task between eight athletes and seven sedentary men; increases in heart rate, diastolic blood pressure and total peripheral resistance were all smaller in the fit groups. Nevertheless, even when such highly selected groups are used, negative results have sometimes been reported (Claytor *et al.*, 1988).

Another way in which positive results may be more reliably found is by studying subjects known to be at risk for showing greater cardiovascular responses to stress. Holmes and Cappo (1987) studied subjects whose family history put them at risk of hypertension. As has been found before, cardiovascular responses to a psychomotor task (the Stroop colour-word conflict) were greater in these subjects than in controls with no such family history. Among a highly fit subgroup, however, there was no difference. Similarly, fitness may be a more significant protector against stress in older than in young subjects; in a report by Hull *et al.* (1984), the only evidence for a fitness effect on stress responses was in a subgroup of subjects aged over 40 years.

Controlled studies of Type A men (i.e. men with a competitive, hard-driving coronary-prone personality) have also tended to yield clearer results. Blumenthal *et al.* (1988) showed that a 12-week walking and jogging programme was more effective than a similarly structured strength and flexibility programme at reducing heart rate, blood pressure and estimated myocardial oxygen consumption during mental arithmetic and recovery from it. In a similar study, Sherwood *et al.* (1989) found that training in jogging and walking reduced the blood pressure response to a competitive reaction time task in a subset of Type A subjects who were also borderline hypertensive, but not in normotensive subjects. Studying 'at-risk' groups has, however, not always delivered significant effects of exercise training. Cleroux *et al.* (1985) found no difference in heart rate responses to a video game in borderline hypertensive subjects before and after a 20-week exercise programme. Seraganian *et al.* (1987) found no effect of a 10-week programme of stationary jogging or cycling on cardiovascular responses of Type A males to a series of mental stress tasks including the Stroop colour-word conflict.

A second concern of this type of study has been with the speed of recovery from stress because of the view that more rapid recovery indicates more effective coping. Despite some evidence that fitness is associated with faster recovery, even where it does not correlate with the size of the response (Sinyor *et al.*, 1983), it would be premature to propose a fundamental dissociation between effects on responses and on recovery from them. The most plausible explanation for the pattern of findings on laboratory stress is that spontaneously occurring variation in fitness is associated with slight differences in cardiovascular function both during and after exposure to stressors. Whether the effects are strong enough to be apparent in any instance may depend on the size of the stress response (Holmes and Roth, 1985) or, as we have seen, on the selection of the sample.

The problem remains as to whether even this small effect of fitness results from exercise training or from some other difference between fit and unfit people. Fully experimental studies, in which stress responses are measured before and after exercise training or a control activity, are necessary to test this. Holmes and McGilley (1987) found that heart rate responses to a 3 min digit-recall task were reduced in initially unfit undergraduates after a 13-week exercise programme by comparison with 'controls' attending psychology classes instead. The groups were, however, not randomly allocated. In studies in which exercise and control groups *were* randomly allocated, Sinyor *et al.* (1986) and Steptoe *et al.* (1990) failed to show any effect of exercise training. In others, the tonically lower heart rate and blood pressure of regular exercisers has made differences in responsivity impossible to detect (e.g. Plante and Karpowitz, 1987; Holmes and Roth, 1988). In view of the results of the cross-sectional studies, it is possible that studies of more 'at risk' populations might be more productive.

Crews and Landers (1987) have reported a meta-analysis of the literature (in which the magnitude of previously reported effects are expressed in terms of the standard error of the outcome measures in the samples so that they can be subjected to a single analysis); according to this, exercise training is, on balance, associated with reduced physiological responsivity to laboratory mental stress. Nevertheless, the effects are clearly so variable as to suggest that the effect of exercise depends on properties of the subject population or stressor that are not yet appreciated. Alternatively, it could be that results are inconsistent because the cardiovascular response system may be an unreliable index of resistance to psychological stress. Other sorts of response have sometimes been recorded: in particular, emotional ratings and plasma catecholamines, although no consistent picture has emerged. One difficulty is in choosing psychological responses which are appropriate to particular stressors. For example, Zillman *et al.*'s (1974) measurement of aggression in response to provocation is clearly appropriate. However, the more familiar self-ratings of anxiety used in relation to pressured mental tasks may not be so relevant. Animal studies of resistance to stress, to which we shall turn presently, have focused on perseverance in the face of adversity. In this context, one cross-sectional comparison of fit and unfit subjects is of particular interest because it found that, immediately after completion of a 12 min Stroop colour-word conflict task, unfit subjects were slower to complete a series of anagrams than were their fit counterparts (Sothmann *et al.*, 1987). Anagram solution has been used as a way of detecting deficits caused by exposure to uncontrollable stressors in human analogues of 'learned helplessness' experiments in animals.

In the present state of the literature, it is therefore safest to regard the reduction of cardiovascular responsivity by exercise training as one, relatively weak, expression of a general effect of training to reduce sensitivity to stress. As suggested above, there is a *prima facie* case for suspecting that the antidepressive and anxiolytic effects of exercise training may be further expressions of the same process. A theoretical account for this follows.

14.5 EXERCISE AS TOUGHENING UP

Despite its psychological and physical benefits, exercise is not a popular activity. From surveys in Western industrialized countries, only 30–50% of the population engage in significant amounts of exercise weekly. Around 50% of participants in exercise programmes are lost within 3–6 months of starting. One way of explaining the unpopularity of exercise has been to blame it on deficits of personality or character. Many such suggestions have been made, such as lack of self-motivation, lack of early experience of physical activity, inappropriate health beliefs or an absence of internal locus of control (Dishman *et al.*, 1980; Enstrom, 1986; Sonstroem, 1988). A simpler explanation is that, as we have seen when considering its short-term emotional effects, exercise is an unpleasant activity for many people. Just as the interpretation of acute effects of exercise has to accommodate both its pleasant and unpleasant nature so, on this reasoning, should an explanation of the long-term effects.

A theoretical framework which can encompass both the unpleasantness of exercise and its putative long-term stress-protecting effects derives from a phenomenon known variously as 'stress-adaptation' or 'toughening up'. It has been studied almost exclusively in animals because it concerns the systematic repetition of stressful experiences. The basis of the theory is the observation that animals habituate to repeated exposure to stress; that is, with repetition, any of a variety of stressors including handling and stroking, immobilization, disappointment or electric shocks come to have a less disruptive effect on an animal's behaviour. The phenomenon is important because repetition of one form of stress can also confer a resistance to other forms of stress. In different contexts, stress-adaptation has been discussed as a model of resistance to depression or anxiety (Gray, 1985). Its relevance to the present discussion is to suggest that the aversive, or stressful, nature of exercise could itself underlie its ability to enhance resistance to stress in general (Salmon *et al.*, 1988).

A few experiments using animals have tried to test whether repeated exercise does increase resistance to other forms of stress. Two studies have found no effect of prior swim-training or wheel-running on physiological responses to restraint or electric tail or foot shock (Cox et al., 1985a; Kant et al., 1985). Using a strain of rat known to develop hypertension in response to stress, however, Cox et al. (1985b) found that daily swimming did prevent stress-induced hypertension. More recently, it has been reported that 40 daily sessions of forced treadmill exercise did reduce blood pressure and heart rate responses to loud noise (Overton et al., 1991). Furthermore, the same responses to intracerebrally administered corticotropin releasing factor (see Chapter 13) were also reduced by training. This points to an action of exercise training on the 'output' side of the stress response system, although it does not preclude an effect on the 'input' also (i.e. the extent to which a stimulus is appraised as stressful).

Studies of behavioural responses have been more consistent. Following an earlier report that experience of subsequent wheel running, motivated by a shock-avoidance schedule, reduced the hypothermic effect of restraint (Bartlett, 1956), Weber and Lee (1968) examined the effects of a 35 day programme of 30 min daily swimming on emotionality in an open field. In this apparatus, the rat is subjected to the stress of placement in a novel, open arena. The conventional measure of its resistance to the stress is the extent to which it moves around, exploring the arena. The swimming group moved more, signifying greater resistance to stress, by comparison either with sedentary controls or a group merely given free access to a running wheel. This experiment suffers from deficiencies in controls just as we have seen with many of the human experiments. In particular, the greater experience of handling in the swimming group could well have lowered their emotionality in the open field (Salmon and Stanford, 1989). Tharp and Carson (1975) tried to control for exposure to handling and other extraneous stimuli by comparing rats required to swim in deep warm water with weights fastened to their backs to rats wading through shallow water. Similarly, rats forced to run in a treadmill were compared with rats rotated at a leisurely pace. When tested in an open field, the runners and swimmers were both more mobile than their controls signifying, once again, less emotionality. Interpretation of even this experiment is difficult because the exercise groups differ from their controls in ways other than amount of exercise. Fear, pain and discomfort resulting from placement in deep water or confinement in a fast-rotating wheel, or the perception of lack of control over events, could also distinguish the exercise groups. One way of eliminating many of these factors would be to induce high or low levels of exercise in different groups by positive reinforcement. Alternatively, high spontaneous levels of exercise can be induced in laboratory animals over a long period merely by making suitable running wheels available (Shyu et al., 1984) or by food deprivation (Rockman and Glavin, 1986).

Some of these 'extraneous' properties of exercise regimes could, of course, turn out to be crucial to their stress-reducing effects. Indeed, we should anticipate this in the light of the hypotheses which have been proposed to explain stress-adaptation. On the whole, these are concerned not with intrinsic properties of the stressors, but with 'extraneous' variables such as their controllability or aversiveness. A number of variables can be identified which may be crucial to 'stress-adaptation'. The first, derived from the literature on learned helplessness, is that a single exposure to stress disrupts behaviour to the extent that the stress is experienced as uncontrollable or unpredictable (see Chapter 9); by contrast, one way in which resistance can be acquired relatively rapidly is by experience of controllable stress. To apply the notion of controllability to people is difficult: there is obviously no direct parallel to the uncontrollable ways in which stress can be experienced in the animal laboratory. It is important to remember, however, that the learned helplessness theory from which the concept is derived is a 'cognitive' theory; that is, the central component is the *perception* that stress is uncontrollable or controllable. It is obvious that, in some people, exercise is carried out from a feeling of coercion; to avoid losing face in front of peers or to keep cardiovascular disease at bay, for example. In others, it is clearly taken up out of a feeling of choice. Whether participants' perception of choice or coercion influences the emotional effects of exercise, although intuitively plausible, has apparently never been investigated.

A second account of stress-adaptation, derived from the work of Amsel (1972) and Gray (1975), is based on counterconditioning; this is the view that stress and stimuli associated with it gradually lose their disruptive effects on behaviour, and even acquire positive motivational properties, by being repeatedly experienced in association with rewarding stimuli. In the animal laboratory, this is arranged by pairing with a rewarding event, such as food, signals which the animal has come to associate with stressful events such as electric shock or nonreward. Again, experience of one form of stress in this way confers resistance to others. Extrapolation of this theory to people is, however, also not straightforward. One obvious difference is that, in people, effects resembling counterconditioning can be achieved merely by the receipt of information: for instance, unpleasant aspects of exercise could be viewed positively because of people's knowledge that they signify progress to improved health or appearance (Lees and Dygdon, 1988). This theory does, however, point to the importance of studying the way in which exercise is linked to other rewarding activities in the exerciser's life. For example, its association with social or other rewards might be important, as might the way in which an exerciser administers 'self-reinforcement' contingent on their performance of exercise. It is obvious that social variables are important in maintaining regular exercise for many people; therefore it is surprising that their importance to the emotional effects of exercise has not yet been studied.

Although rarely made explicit in the literature on stress-adaptation, there are other features which distinguish stress regimes which produce tolerance from those which do not and which are of obvious relevance to exercise. The most obvious is the unpleasantness of the stress. Clearly a minimal level of aversiveness is necessary, although it is not clear how the extent of stress-adaptation relates to severity of stress beyond this level. The simplest way of extrapolating this to exercise would be to find whether beneficial effects of exercise training depend on its initial unpleasantness. A further factor that distinguishes stress regimes which produce adaptation from those which do not is the regularity and predictability with which stress is presented. The regularity of conventional exercise programmes may prove essential to their ability to confer resistance to stress. To demonstrate the applicability to exercise of predictions drawn from the stress-adaptation literature would have importance beyond exercise psychology. It would provide, for the first time, a way of studying a phenomenon in people which, although perhaps fundamental in understanding emotional behaviour, has so far seemed beyond ethical bounds.

14.6 EFFECTS OF EXERCISE ON CENTRAL NORADRENERGIC AND OPIOID SYSTEMS

14.6.1 NORADRENALINE

Interest in effects of exercise on central noradrenaline has been led by attempts to explain the antidepressive effect of exercise in neurochemical terms. However, the effects are relevant also to the view that exercise training may be a case of stress-adaptation, since this process has characteristic correlates in central noradrenergic systems.

Acute exposure to procedures which involve strenuous exercise, such as swimming or wheel-running, increases brain noradrenaline turnover (Chaouloff, 1989); as a result, brain noradrenaline levels are sometimes depleted (Barchas and Freedman, 1963; Gordon *et al.*, 1966). Rough parallels have been drawn between these effects of exercise and those of electro-convulsive therapy and some antidepressant drugs (Ransford, 1982; de Castro and Duncan, 1985). Such analogies are unlikely to be helpful, however. The effect of antidepressant drugs on brain noradrenaline systems is complex, and comparisons based on isolated parameters are unlikely to be as helpful as those based on techniques measuring postsynaptic responses to stimulation.

When repeated exposure to exercise regimes of the kind which have been found to produce stress-adaptation is examined, levels of noradrenaline in the brain recover or are even increased. For example, brain noradrenaline concentration was increased by eight weeks of daily wheel-running sessions (either forced by a treadmill or motivated by shock-avoidance; Brown and van Huss, 1973; Brown *et al.*, 1979) and by a 17-week regime of regular swimming in warm water (Ostman and Nyback, 1976). In light of the view that depression reflects a failure to withstand stress which, in turn, reflects a failure of central noradrenaline levels to meet demands upon them (Anisman and Zacharko, 1982), Morgan (1985) has attributed the antidepressive action of exercise to its ability to maintain or increase central neurotransmitter levels. These results also align exercise with other forms of stress such as electric shock or immobilization: whereas acute stress often reduces brain noradrenaline levels, chronic stress regimes which are assumed to produce stress-adaptation preserve or increase them (Anisman and Zacharko, 1982). Tsuda *et al.* (1983) used a paradigm in which a high level of exercise is induced by a regime of food deprivation. In this case, stress adaptation very clearly does *not* result; on the contrary, gastric ulceration and death are common when this is continued for some weeks. They found that brain noradrenaline levels were rapidly depleted by this regime and that the depletion increased, rather than reversing, over five days.

Once again, however, the experimental designs mean that it is often impossible to disentangle the contribution of exercise from that of other aversive aspects of the procedures used to administer it. Without 'extraneous' features such as shock-avoidance or forced treadmill rotation, exercise might prove to be a less potent neurochemical stimulus. In one experiment which used positive reinforcement to train high levels of running over an eight week period there was no evidence for effects either on noradrenaline levels or on β-adrenoceptor density in the brain by comparison with rats that exercised spontaneously at a low rate (de Castro and Duncan, 1985). By contrast, brain dopamine concentrations were higher, and dopamine receptor densities lower, in the exercised animals. This could reflect the well-established involvement of central dopamine systems in motor behaviour and might illustrate a dissociation at the neurochemical level between exercise as motor activity (with involvement of dopaminergic systems in positively rewarded exercise) and exercise as an emotional stimulus (where an aversive component may be necessary to recruit noradrenergic involvement).

14.6.2 ENDORPHINS

A desire to reduce the complex emotional effects of physical exercise to a single neurochemical mechanism has apparently led some writers to build more on the idea that exercise stimulates central endorphin systems than the evidence can support. This is an area where, for many years, speculation has run ahead of empirical evidence. Ostensibly the simplest type of evidence cited is that exercise acutely increases circulating β-endorphin levels (Harber and Sutton, 1984), and that this effect apparently grows during training (Carr et al., 1981). Most of the β-endorphin found in plasma reflects pituitary co-release with adrenocorticotropic hormone (ACTH), however, and changes in peripheral circulation of endorphins are no guide to changes in the central nervous system. Nevertheless, animal studies have shown increases also in β-endorphin concentrations in cerebrospinal fluid after high levels of spontaneous exercise (Hoffman et al., 1990).

Indirect evidence that exercise activates central opioid mechanisms is that administration of an opioid antagonist, naloxone, precipitated clear signs of opiate withdrawal in mice which had been forced to swim for 3 min at 2 h intervals in warm water over the preceding two days (Christie and Chesher, 1992). When given to nonexercised controls, naloxone had no effect. It is therefore possible that, in running voluntarily at high rates in the absence of any obvious reward for doing so, rats are 'self-administering' endogenous opioids. Once stable levels of spontaneous running have been reached, naloxone reduces the distance run (Boer et al., 1990). Whether this is because it interferes with the 'internal reward' process is a matter of speculation. A potentially important clinical implication follows from the evidence linking exercise to endorphin systems: that exercise might help in treatment of opiate addiction by substituting endogenous for exogenous opiates (Ramsay and Farmer, 1988; Thoren et al., 1990). Despite attempts to use exercise training in the treatment of addiction to a different drug, alcohol (Sinyor et al., 1982), this remains to be tested.

In popular literature, further support for the idea that exercise stimulates endorphin systems has been drawn from the claim that exercise produces intense feelings of wellbeing (e.g. the 'runners' high; Kostrubala, 1976), of the kind popularly associated with the administration of opiate drugs. Despite the enthusiasm of its adherents this is, however, the weakest part of the case for opioid involvement. First, we have seen that the view that exercise is intrinsically pleasant is not supportable. Of course, opiate administration is also not necessarily pleasant but long-term addicts are the least likely to experience pleasant effects (O'Brien *et al.*, 1986) whereas, as was discussed, long-term exercisers appear more likely than others to experience exercise as pleasant. Moreover, even in committed runners, the evidence for the 'runner's high' is largely anecdotal (Sachs, 1984). Nevertheless, where mood does improve after exercise in regular exercisers, opioid mechanisms might be involved. Janal *et al.* (1984) and Allen and Coen (1987) were able to attenuate the mood improvement which running causes in regular runners by injecting naloxone intravenously. An earlier failure to do this (Markoff *et al.*, 1982) involved a lower dose of naloxone given subcutaneously, which is a less effective route.

Another effect of acute exercise in regular exercisers in which opioid mechanisms have been implicated is analgesia. In regular runners, an increase in ischaemic pain threshold can be detected immediately after a run but is blocked by prior administration of naloxone (Janal *et al.*, 1984), as can analgesia to pain caused by pressure on a finger-joint (Haier *et al.*, 1981). Dental and cutaneous stimulation are not consistently affected, however (Olausson *et al.*, 1986; Droste *et al.*, 1991). Animal studies provide further evidence for opioid involvement in exercise-induced analgesia. Christie and Chesher (1982) and Christie *et al.* (1981, 1982) found analgesia which was cross-tolerant with morphine after a 3 min swim in warm water, i.e. repeated exposure to swimming produced tolerance to the analgesic effects of swimming but also to the analgesic effects of a subsequent administration of morphine. Shyu *et al.* (1982) has implicated opioid involvement in analgesic effects of spontaneous exercise, also. Rats given access to a running wheel during the 12 h dark periods of a 24 h cycle run an average of 6 km/night after 20 days. An increase was greatest in the final hour of the dark phase (during which the animals were most active) and declined over the first few hours of light; further, the size of the increase in threshold in any particular dark period correlated with the distance run during that period. The threshold was decreased by naloxone.

Exercise-induced analgesia helps to align exercise with other forms of stress which also are analgesic. Although opioid activation by exercise may be part of a more general opioid activation by stress, it is unlikely that the extraneous stressors which are commonly part of exercise procedures in animals, such as shock or immersion in water, are necessary to provoke the activation. Similarities between opioid and analgesic effects of exercise and of sciatic nerve or gastrocnemius muscle stimulation have led Thoren *et al.* (1990) to suggest that the engagement of large muscle groups, causing continuous activation of A-delta afferent fibres, is a sufficient stimulus.

A further putative property of exercise which is popularly thought to result from endorphin mechanisms is its addictive quality, but this effect is little better supported by evidence than is the runners' high. The occurrence of withdrawal symptoms which remit once running resumes has been a central component in the belief that exercise is addictive (Veale, 1987). Supporting anecdotal reports, deterioration in mood has been shown after periods of deprivation lasting only a day or two (Baekeland, 1970; Thaxton, 1982). In the study of Morris *et al.* (1990a), deterioration in subjective physical state and in feelings of ability to cope was apparent within the first week and may represent a 'withdrawal syndrome', albeit one which does not approach the severity of that after cessation of opiate administration. Increases in anxiety and depression were only apparent after two weeks, possibly reflecting the gradual loss of the stress-reducing effect of exercise.

Despite being more popularly associated with the poorly substantiated phenomena of 'runner's high' and exercise addiction, opioid mechanisms may prove to be the key to understanding the long-term adaptation which occurs when exercise is repeated. This possibility was foreseen when, in applying opponent-process theory to exercise, Solomon (1980) suggested that the recruitment of endorphin systems accounted for the growth of tolerance to the aversive effect of exercise and the development of positive effects on mood in experienced exercisers. Unfortunately, no evidence is available to confirm how opioid involvement in mood effects changes with adaptation to exercise.

More evidence is, however, available about opioid involvement in physiological effects of exercise. Although very sketchy, the hypothesis which emerges is that these may result from the increase of opioid tone by regular exercise. On the basis of their studies of spontaneously hypertensive rats, Thoren et al. (1990) have suggested that the demonstrated reductions of blood pressure and sympathetic nerve activity by exercise training in hypertensive animals and, by extrapolation, people are mediated by opioid systems. In the peripheral nervous system, sympathoadrenal responses to physical challenge and to psychological stress are tonically inhibited by endorphin systems (Grossman and Moretti, 1986; Morris et al., 1990b), but whether exercise training reduces cardiovascular stress responses by increasing opioid tone is not established. That exercise has similar effects on noradrenergic function in the central nervous system is possible in view of the inhibitory role of opioid systems on central noradrenergic responses to stress (Tanaka et al., 1983).

14.7 CONCLUDING REMARKS

It is clear that physical exercise is a potent psychological stimulus and that its influence on sensitivity to stress extends far beyond the specific effects, particularly short-term mood improvement, with which it is commonly associated. It is clear, also, that its effects on physical fitness are not the basis for this influence. Instead, it is possible that exercise might provide a way of studying processes concerned with stress-adaptation in people which, although potentially of fundamental importance to understanding anxiety and depression as well as variability in resistance to stress in the normal population, have previously seemed inaccessible to investigation. One aim for future research is to test the applications of exercise in treating a variety of disorders, from depression to opiate-addiction. Whether or not these clinical predictions are fulfilled, arguably a more important task will be to explore the value of exercise as a paradigm for understanding the relevance of stress-adaptation to human behaviour and to investigate the mechanisms which subserve it.

14.8 REFERENCES

Allen, M.E. and Coen, D. (1987) Naloxone blocking of running-induced mood changes. *Ann. Sports Med.*, **3**: 190–5

Amsel, A. (1972) Behavioral habituation, counterconditioning and a general theory of persistence, in A.H. Black and W.F. Prokasy (eds.), *Classical Conditioning II: Current Research and Theory*, Appleton-Century-Crofts, New York, pp 409–26

Anisman, H. and Zacharko, R.M. (1982) Depression: the predisposing influence of stress. *Behav. Brain Sci.*, **5:** 89–137

Baekeland, F. (1970) Exercise deprivation. *Arch. Gen. Psychiatry*, **22:** 365–9

Barchas, J.D. and Freedman, D.X. (1963) Brain amines: response to physiological stress. *Biochem. Pharmacol.*, **12:** 1232–5

Bartlett, R.G. (1956) Stress adaptation and inhibition of restraint induced (emotional) hyperthermia. *J. Appl. Physiol.*, **8:** 661–3

Berger, B.G. and Owen, D.R. (1983) Mood alteration with swimming – swimmers really do 'feel better'. *Psychosomatic Med.*, **45:** 425–33

Blumenthal, J.A., Williams, R.S., Williams, R.B. and Wallace, A.G. (1980) Effects of exercise on the Type A (coronary prone) behavior pattern. *Psychosomatic Med.*, **42:** 289–96

Blumenthal, J.A., Williams, R.S., Needels, T.L. and Wallace, A.G. (1982) Psychological changes accompanying aerobic exercise in healthy middle-aged adults. *Psychosomatic Med.*, **44:** 529

Blumenthal, J.A., Emery, C.F., Walsh, M.A. *et al.* (1988) Exercise training in healthy Type A middle-aged men: effects on behavioral and cardiovascular responses. *Psychosomatic Med.*, **50:** 418–33

Boer, D.P., Epling, W.F., Pierce, W.D. and Russell, J.C. (1990) Suppression of food deprivation-induced high-rate wheel running in rats. *Physiol. Behav.*, **48:** 339–47

Boutcher, S.H. and Landers, D.M. (1988) The effects of vigorous exercise on anxiety, heart rate, and alpha activity of runners and nonrunners. *Psychophysiol.*, **25:** 696–702

Brooke, S.T. and Long, B.C. (1987) Efficiency of coping with a real-life stressor: a multimodal comparison of aerobic fitness. *Psychophysiol.*, **24:** 173–80

Brown, B.S. and van Huss, W. (1973) Exercise and rat brain catecholamines. *J. Appl. Physiol.*, **30:** 664–9

Brown, B.S., Payne, T., Kim, C. *et al.* (1979) Chronic response of rat brain norepinephrine and serotonin levels to endurance training. *J. Appl. Physiol.*, **46:** 19–23

Brown, J.D. (1991) Staying fit and staying well: physical fitness as a moderator of life stress. *J. Personality Social Psychol.*, **60:** 555–61

Brown, J.D. and Siegel, J.M. (1988) Exercise as a buffer of life stress: a prospective study of adolescent health. *Health Psychol.*, **7:** 341–53

Brown, R.S., Ramirez, D.E. and Taub, J.M. (1978) The prescription of exercise for depression. *Physician Sports Med.*, **6:** 34–45

Carr, D.B., Bullen, B.A., Skrinar, G.S. *et al.* (1981) Physical conditioning facilitates the exercise-induced secretion of beta-endorphin and beta-lipotropin in women. *New England J. Med.*, **305:** 560–3

Chaouloff, F. (1989) Physical exercise and brain monoamines: a review. *Acta Physiol. Scand.*, **137:** 1–13

Christie, M.J. and Chesher, G.B. (1982) Physical dependence on physiologically released endogenous opiates. *Life Sci.*, **30:** 1173–7

Christie, M.J., Chesher, G.B. and Bird, K.D. (1981) The correlation between swim stress induced antinociception and [^{3}H]leu-enkephalin binding to brain homogenates in mice. *Pharmacol. Biochem. Behav.*, **15:** 853–7

Christie, M.J., Trisdikoon, P. and Chesher, G.B. (1982) Tolerance and cross-tolerance with morphine resulting from physiological release of endogenous opiates. *Life Sci.*, **31:** 839–84

Clark, D.M. (1986) A cognitive approach to panic. *Behav. Res. Ther.*, **24:** 461–70

Clark, M.S., Milberg, S. and Ross, J. (1983) Arousal cues arousal-related material in memory: implications for understanding effects of mood on memory. *J. Verbal Learn. Verbal Behav.*, **22:** 633–49

Claytor, R.A. (1991) Stress reactivity: hemodynamic adjustments in trained and untrained humans. *Med. Sci. in Sports and Exercise*, **23:** 873–81

Claytor, R.P., Cox, R.H., Howley, E.T. *et al.* (1988) Aerobic power and cardiovascular response to stress. *J. Appl. Physiol.*, **65:** 1416–23

Clearing-Sky, M. (1988) Exercise: issues for prescribing psychologists. *Psychol. Health*, **2:** 189–207

Cleroux, J., Peronnet, F. and De Champlain, J. (1985) Sympathetic indices during psychological and physical stimuli before and after training. *Physiol. Behav.*, **35:** 271–5

Cox, R.H., Hubbard, J.W., Lawler, J.E. *et al.* (1985a) Cardiovascular and sympathoadrenal responses to stress in swim-trained rats. *J. Appl. Physiol.*, **58:** 1207–14

Cox, R.H., Hubbard, J.W., Lawler, J.E. *et al.* (1985b) Exercise training attenuates stress-induced hypertension in the rat. *Hypertension*, **7:** 747–51

Cramer, S.R., Nieman, D.C. and Lee, J.W. (1991) The effects of moderate exercise training on psychological well-being and mood state in women. *J. Psychosomatic Res.*, **35:** 437–49

Crews, D.J. and Landers, D.M. (1987) A meta-analytic review of aerobic fitness and reactivity to psychosocial stressors. *Med. Sci. in Sports and Exercise*, **19:** S114–20

de Castro, J.M. and Duncan, G. (1985) Operantly conditioned running: effects on brain catecholamine concentrations and receptor densities in the rat. *Pharmacol. Biochem. Behav.*, **23:** 495–500

de Geus, E.J.C., van Doornen, L.J.P., de Visser, D.C. and Orlebeke, J.F. (1990) Existing and training induced differences in aerobic fitness: the relationship to physiological response patterns during different types of stress. *Psychophysiol.*, **27:** 457–78

Dienstbier, R. (1991) Behavioral correlates of sympathoadrenal reactivity: the toughness model. *Med. Sci. in Sports and Exercise*, **23:** 846–52

Dishman, R.K., Ickes, W. and Morgan, W.P. (1980) Self-motivation and adherence to habitual physical activity. *J. Appl. Social Psychol.*, **10:** 115–32

Dobson, K.S. (1985) The relationship between anxiety and depression. *Clin. Psychol. Rev.*, **5:** 307–24

Doyne, E.J., Chambless, D.C. and Beutler, L.E. (1983) Aerobic exercise as a treatment for depression in women. *Behav. Ther.*, **14:** 434–40

Driscoll, R. (1976) Anxiety reduction using physical exertion and positive images. *Psychol. Rec.*, **26:** 87–94

Droste, C., Greenlee, M.W., Schreck, M. and Roskamin H. (1991) Experimental pain thresholds and plasma beta-endorphin levels during exercise. *Med. Sci. in Sports and Exercise*, **23:** 334–42

Eischen, R.R. and Greist, J.H. (1984) Beginning and continuing running: steps to psychological well-being, in M.L. Sachs and G.W. Buffone (eds.), *Running as Therapy: an Integrated Approach*, Nebraska University Press, Lincoln, pp 63–82

Engstrom, L.-M. (1986) The process of socialization into keep-fit activities. *Scand. J. Sports Sci.*, **8:** 89–97

Fasting, K. and Gronningsaeter, H. (1986) Unemployment, trait-anxiety and physical exercise. *Scand. J. Sports Sci.*, **8:** 99–103

Fillingim, R.B., Roth, D.L. and Cook, E.W. (1992) The effects of aerobic exercise on cardiovascular, facial EMG, and self-report responses to emotional imagery. *Psychosomatic Med.*, **54:** 109–20

Folkins, C.H. and Sime, W.E. (1981) Physical fitness training and mental health. *Am. Psychol.*, **36:** 373–89

Gal, R. and Lazarus, R.S. (1975) The role of activity in anticipating and confronting stressful situations. *J. Human Stress*, **1:** 4–20

Gannon, L., Banks, J. and Shelton, D. (1989) The mediating effects of psychophysiological reactivity and recovery on the relationship between environmental stress and illness. *J. Psychosomatic Res.*, **33:** 67–175

Goldwater, B.C., and Collis, M.L. (1985) Psychologic effects of cardiovascular conditioning: a controlled experiment. *Psychosomatic Med.*, **47:** 174–81

Gordon, R., Spector, S., Sjoerdsma, A. and Udenfriend, S. (1966) Increased synthesis of norepinephrine and epinephrine in the intact rat during exercise and exposure to cold. *J. Pharmacol. Exp. Ther.*, **153:** 440–7

Gray, J.A. (1975) *Elements of a Two-process Theory of Learning*, Academic Press, London

Gray, J.A. (1982) *The Neuropsychology of Anxiety*, Clarendon Press, Oxford

Gray, J.A. (1985) Issues in the neuropsychology of anxiety, in A.H. Tuma and J.D. Maser (eds.), *Anxiety and the Anxiety Disorders*, Lawrence Erlbaum, Hillsdale, New Jersey pp 5–25

Griest, J.H., Klein, M.H., Eischens, R.R. *et al.* (1979) Running as a treatment for depression. *Comprehensive Psychiatry*, **20**: 41–54

Grossman A. and Moretti, A. (1986) Opioid peptides and their relationship to hormonal changes during acute exercise, in G. Benzi, L. Packer and N. Siliprandi (eds.), *Biochemical Aspects of Physical Exercise*, Elsevier, Amsterdam, pp 375–86

Haier, R.J., Quaid, K. and Mills, J.S.C. (1981) Naloxone alters pain perception after jogging. *Psychiatry Res.*, **5**: 231–2

Harber, V.J. and Sutton, J.R. (1984) Endorphins and exercise. *Sports Med.*, **1**: 154–71

Hardy, C.J. and Rejeski, W.J. (1989) Not what, but how one feels: the measurement of affect during exercise. *J. Sport Exercise Psychol.*, **11**: 304–17

Heaps, R.A. (1978) Relating physical and psychological fitness: a psychological point of view. *J. Sports Med. Physical Fitness*, **18**: 399–408

Hilyer, J. and Mitchell, W. (1979) Effect of systematic physical fitness training combined with counseling on the self-concept of college students. *J. Counseling Psychol.*, **26**: 427–36

Hinson, R.E. and Siegel, S. (1982) Nonpharmacological bases of drug tolerance and dependence. *J. Psychosomatic Res.*, **26**: 495–503

Hoffman, P., Terenius, L. and Thoren, P. (1990) Cerebrospinal fluid immunoreactive beta-endorphin concentration is increased by long-lasting voluntary exercise in the spontaneously hypertensive rat. *Regulatory Peptides*, **28**: 233–9

Hollander, B.J. and Seraganian, P. (1984) Aerobic fitness and psychophysiological reactivity. *Canad. J. Behav. Sci.*, **16**: 257–61

Hollandsworth, J.G. (1979) Some thoughts on distance running as training in biofeedback. *J. Sport Behav.*, **2**: 71–82

Holmes, D.S. and Cappo, B.M. (1987) Prophylactic effect of aerobic fitness on cardiovascular arousal among individuals with a family history of hypertension. *J. Psychosomatic Res.*, **31**: 601–5

Holmes, D.S. and McGilley, B.M. (1987) Influence of a brief aerobic training program on heart rate and subjective response to a psychologic stressor. *Psychosomatic Med.*, **49**: 366–74

Holmes, D.S. and Roth, D.L. (1985) Association of aerobic fitness with pulse rate and subjective responses to psychological stress. *Psychophysiology*, **22**: 525–9

Holmes, D.S. and Roth, D.L. (1988) Effects of aerobic exercise training and relaxation training on cardiovascular activity during psychosomical stress. *J. Psychosomatic Res.*, **32**: 469–74

Hughes, J.R., Casal, D.C. and Leon, A.S. (1986) Psychological effects of exercise: a randomized cross-over trial. *J. Psychosomatic Res.*, **30**: 355–60

Hull, E.M., Young, S.H. and Ziegler, M.G. (1984) Aerobic fitness affects cardiovascular and catecholamine responses to stressors. *Psychophysiology*, **21**: 353–60

Janal, M.N., Colt, E.W.D., Clark, W.C. and Glusman, M. (1984) Pain sensitivity, mood and plasma endocrine levels in man following long distance running: effects of naloxone. *Pain*, **19**: 13–25

Kant, G.J., Eggleston, T., Landman-Roberts, L. *et al.* (1985) Habituation to repeated stress is stressor specific. *Pharmacol. Biochem. Behav.*, **22**: 631–4

Keller, S. and Seraganian, P. (1984) Physical fitness level and autonomic reactivity to psychological stress. *J. Psychosomatic Res.*, **28**: 279–87

Klein, M.H., Greist, J.H., Gurman, A.S. *et al.* (1985) A comparative outcome study of group psychotherapy vs exercise treatments for depression *Int. J. Mental Health*, **13**: 148–77

Kobasa, S.C.., Maddi, S.R., Puccetti, M.C. and Zola, M.A. (1985) Effectiveness of hardiness, exercise and social support as resources against illness. *J. Psychosomatic Res.*, **29**: 525–33

Kostrubala, T. (1976) *The Joy of Running*, Lippincott, Philadelphia

Lees, L.A. and Dygdon, J.A. (1988) The initiation and maintenance of exercise behaviour: a learning theory conceptualization. *Clin. Psychol. Rev.*, **8**: 345–53

Light, K.C., Obrist, P.A., James, S.A. and Strogatz. (1987) Cardiovascular responses to stress: II. Relationships to aerobic exercise patterns. *Psychophysiology*, **24**: 79–86

Lobstein, D.D., Rasmussen, C.L., Dunphy, G.E. and Dunphy, M.J. (1989) Beta-endorphin and components of depression as powerful discriminators between joggers and sedentary middle-aged men. *J. Psychosomatic Res.*, **33**: 293–305

Long, B.C. (1984) Aerobic conditioning and stress inoculation: a comparison of stress-management interventions. *Cogn. Ther. Res.*, **8**: 517–42

Long, B.C. and Haney, C.J. (1988) Long-term follow-up of stressed working women: a comparison of aerobic exercise and progressive relaxation. *J. Sport Exercise Psychol.*, **10:** 461–70

Markoff, R.A., Ryan, P. and Young T. (1982) Endorphin and mood changes in long-distance running. *Med. Sci. Sports and Exercise*, **14:** 11–15

Martinsen, E.W. (1987) Exercise and medication in the psychiatric patient, in W.P. Morgan and S.E. Goldston (eds.), *Exercise and Mental Health*, Hemisphere, Washington, pp 85–95

McCann, I.L. and Holmes, D.S. (1984) Influence of aerobic exercise on depression. *J. Personality Social Psychol.*, **46:** 1142–7

McIntyre, C.W., Watson, D. and Cunningham, A.C. (1990) The effects of social interaction, exercise, and test stress on positive and negative affect. *Bull. Psychonomic Soc.*, **28:** 141–3

Meyer, R., Kroner-Herwig, B. and Sporkel, H. (1990) The effect of exercise and induced expectations on visceral perception in asthmatic patients. *J. Psychosomatic Res.*, **34:** 455–60

Michael, E.D. (1957) Stress adaptation through exercise. *Res. Quart.*, **28:** 50–4

Mills, D.W. and Ward, R.P. (1986) Attenuation of stress-induced hypertension by exercise independent of training effects: an animal model. *J. Behav. Med.*, **9:** 599–605

Morgan, W.P. (1969) A pilot investigation of physical working capacity in depressed and non depressed psychiatric males. *Res. Quart.*, **40:** 859–61

Morgan, W.P. (1985) Affective beneficence of vigorous physical activity. *Med. Sci. Sports and Exercise*, **17:** 94–100

Morgan, W.P. (1987) Reduction of state anxiety following acute physical activity, in W.P. Morgan and S.E. Goldston (eds.), *Exercise and Mental Health*, Hemisphere, Washington, pp 105–9

Morgan, W.P., Roberts, J.A., Brand, F.R and Feinerman, A.D. (1970) Psychological effect of chronic physical activity. *Med. Sci. Sports*, **2:** 213–17

Morris, M., Salmon, P. and Steinberg, H. (1988) *The 'runner's high': dimensional structure of mood before and after running.* Proceedings of the Symposium on Sport, Health Psychology and Exercise. Bisham Abbey, near Maidenhead, UK, pp 147–52

Morris, M., Salmon, P., Steinbergh, H. *et al.* (1990b) Endogenous opioids modulate the cardiovascular response to mental stress. *Psychoneuroendocrinology*, **15:** 185–92

Morris, M., Steinberg, H., Sykes, E.A. and Salmon, P. (1990a) Effects of temporary withdrawal from regular running. *J. Psychosomatic Res.*, **34:** 493–500

Moses, J., Steptoe, A., Mathews, A. and Edwards, S. (1989) The effects of exercise training on mental well-being in the normal population: a controlled trial. *J. Psychosomatic Res.*, **33:** 47–61

Muller, B. and Armstrong, H.E. (1975) A further note on the 'running treatment' for anxiety. *Psychotherapy: Theory, Res. Pract.*, **12:** 385–7

Norris, R., Carroll, D. and Cochrane, R. (1990) The effects of aerobic and anaerobic training on fitness, blood pressure and psychological stress and well-being. *J. Psychosomatic Res.*, **34:** 367–75

O'Brien, C.P., Ehrman, R.N. and Ternes, J.W. (1986) Classical conditioning in human opioid dependence, in S.R. Goldberg and I.P. Stolerman (eds.), *Behavioral Analysis of Drug Dependence*, Acadamic Press, Orlando

Olausson, B., Eriksson, E., Ellmarker, B. *et al.* (1986) Effects of naloxone on dental pain threshold following muscle exercise and low frequency transcutaneous nerve stimulation: a comparative study in man. *Acta Physiol. Scand.*, **126:** 299–305

Orwin, A. (1973) 'The running treatment': a preliminary communication on a new use for an old therapy (physical activity) in the agoraphobic syndrome. *Br. J. Psychiatry*, **122:** 175–9

Ostman, I. and Nyback, H. (1976) Adaptive changes in central and peripheral noradrenergic neurons in rats, following chronic exercise. *Neuroscience*, **1:** 41–7

Overton, J.M., Kregol, K.C., Davis-Gorman, G. *et al.* (1991) Effects of exercise training on responses to central injection of CRF and noise stress. *Physiol. Behav.*, **49:** 93–8

Paffenbarger, R.S., Jr. and Hyde, R.T. Exercise adherence, coronary heart disease and longevity, in R.K. Dishman (ed.), *Exercise Adherence: Its Impact on Public Health*, Human Kinetics Books, Champaign, IL, pp 41–73

Plante, T.G. and Karpowitz, D. (1987) The influence of aerobic exercise on physiological stress responsivity. *Psychophysiology*, **24:** 670–7

Pollak, M.H. (1991) Heart rate reactivity to laboratory tasks and ambulatory heart rate in daily life. *Psychosomatic Med.*, **53**: 25–35

Ramsay, R. and Farmer, R. (1978) Physical exercise and mental health. *Br. Medical J.*, **296**: 1069–70

Ransford, C.P. (1982) A role for amines in the antidepressant effect of exercise: a review. *Med. Sci. Sports and Exercise*, **14**: 1–10

Reinsenzein, R. (1983) The Schachter theory of emotion: two decades later. *Psychological Bull.*, **94**: 239–64

Rockman, G.E. and Glavin, G.B. (1986) Activity stress effects on voluntary ethanol consumption, mortality and ulcer development in rats. *Pharmacol. Biochem. Behav.*, **24**: 869–73

Roth, D.L. and Holmes, D.S. (1985) Influence of physical fitness in determining the impact of stressful life events on physical and psychological health. *Psychosomatic Med.*, **47**: 164–73

Roth, D.L. and Holmes, D.S. (1987) Influence of aerobic exercise training and relaxation on physical and psychologic health following stressful life events. *Psychosomatic Med.*, **49**: 355–65

Roth, D.L., Wiebe, D.J., Fillingim, R.B. and Shay, K.A. (1989) Life events, fitness, hardiness and health: a simultaneous analysis of proposed stress-resistance effects. *J. Personality Social Psychol.*, **57**: 136–42

Sachs, M.C. (1984) The runner's high, in M.L. Sachs and G.W. Buffone (eds.), *Running as Therapy: an Integrated Approach*, Nebraska Press, Lincoln, pp 273–87

Salmon, P. and Stanford, S.C. (1989) β-Adrenoceptor binding correlates with behaviour of rats in the open field. *Psychophysiology*, **98**: 412–16

Salmon, P., Steinberg, H., Morris, M. and Sykes, E.A. (1988) *Physical exercise as a form of psychological stress*. Annual Conference of the British Psychological Society, December, London

Schachter, S. and Singer, J. (1962) Cognitive, social and physiological determinants of emotional state. *Psychol. Rev.*, **69**: 379–99

Schwartz, G.E., Davidson, R.J. Goleman, D.J. (1978) Patterning of cognitive and somatic processes in the self-regulation of anxiety: effects of meditation versus exercise. *Psychosomatic Med.*, **40**: 321–8

Serganian, P., Roskies, E., Hanley, J.A. et al. (1987) Failure to alter psychophysiological reactivity in Type A men with physical exercise or stress management programs. *Psychol. Health*, **1**: 195–213

Sherwood, A., Light, K.C. and Blumenthal, J.A. (1989) Effects of aerobic exercise training on hemodynamic responses during psychosocial stress in normotensive and borderline hypertensive type A men: a preliminary report. *Psychosomatic Med.*, **51**: 123–36

Shulhan, D., Scher, H. and Furedy, J.F. (1986) Phasic cardiac reactivity to psychological stress as a function of aerobic fitness level. *Psychophysiology*, **23**: 562–6

Shyu, B.C., Andersson, S.A. and Thoren, P. (1982) Endorphin mediated increase in pain threshold induced by long-lasting exercise in rats. *Life Sci.*, **30**: 833–40

Shyu, B.C., Andersson, S.A. and Thoren, P. (1984) Spontaneous running in wheels. A microprocessor assisted method for measuring physiological parameters during exercise in rodents. *Acta Physiol. Scand.*, **121**: 103–9

Sime, W.E. (1984) Psychological benefits of exercise training in the healthy individual, in J.D. Matarazzo, S.M. Weiss, J.A. Herd et al. (eds.), *Behavioral Health: a Handbook of Health Enhancement and Disease Prevention*, Wiley, New York, pp 488–508

Sime, W.E. (1987) Exercise in the prevention and treatment of depression, in W.P. Morgan and S.E. Goldston (eds.), *Exercise and Mental Health*, Hemisphere, Washington, pp 145–52

Simons, C.W. and Birkimer, J.C. (1988) An exploration of factors predicting the effects of aerobic conditioning on mood state. *J. Psychosomatic Res.*, **32**: 63–75

Simons, A.D., McGowan, C.R., Epstein, L.H. and Kupfer, D.J. (1985) Exercise as a treatment for depression: an update. *Clin. Psychol. Rev.*, **5**: 553–68

Sinyor, D., Brown, T., Rostant, L. and Seraganian, P. (1982) The role of a physical fitness program in the treatment of alcoholism. *J. Stud. Alcohol*, **43**: 380–6

Sinyor, D., Schwartz, S.G., Peronnet, F. et al. (1983) Aerobic fitness level and reactivity to psychosocial stress: physiological, biochemical, and subjective measures. *Psychosomatic Med.*, **45**: 205–17

Sinyor, D., Golden, M., Steinert, Y. and Seraganian, P. (1986) Experimental manipulation of aerobic fitness and the response to psychological stress: heart rate and self-report measures. *Psychosomatic Med.*, **48:** 324–37

Solomon, R.L. (1980) The opponent-process theory of acquired motivation: the costs of pleasure and the benefits of pain. *Am. Psychol.*, **35:** 691–712

Sonstroem, R.J. (1988) Psychological models, in R.K. Dishman (ed.), *Exercise Adherence: Its Impact on Public Health*, Human Kinetics Books, Champaign, Il, pp 125–53

Sothmann, M., Hart, B.A. and Horn, T.S. (1991) Plasma catecholamine response to acute psychological stress in humans: relation to aerobic fitness and exercise training. *Med. Sci. Sports and Exercise*, **23:** 860–7

Southmann, M., Horn, T., Hart, B. and Gustafson, A. (1987) Comparison of discrete cardiovascular fitness groups on plasma catecholamine and selected behavioral responses to psychological stress. *Psychophysiology*, **24:** 47–54

Starzec, J.J., Berger, D.F. and Hesse, R. (1983) Effects of stress and exercise on plasma corticosterone, plasma cholesterol and aortic cholesterol levels in rats. *Psychosomatic Med.*, **45:** 219–26

Stavrakaki, C, and Vargo, B. (1986) The relationship of anxiety and depression: a review of the literature. *Br. J. Psychiatry*, **49:** 7–16

Stephens, T. (1988) Physical activity and mental health in the United States and Canada: evidence from four popular surveys. *Preventive Med.*, **17:** 35–47

Steptoe, A. and Bolton, J. (1988) The short-term influence of high and low intensity physical exercise on mood. *Psychol. Health*, **2:** 91–106

Steptoe, A. and Cox, S. (1988). The acute effects of aerobic exercise on mood: a controlled study. *Health Psychol.*, **7:** 329–40

Steptoe, A. and Vogel, C. (1991) Methodology of mental stress testing in cardiovascular research. *Circulation Supplement II*, **83:** 14–24.

Steptoe, A., Edwards, S., Moses, J. and Mathews, A. (1989) The effects of exercise training on mood and perceived coping ability in anxious adults from the general population. *J. Psychosomatic Res.*, **33:** 537–47

Steptoe, A., Moses, J., Mathews, A. and Edwards, S. (1990) Aerobic fitness, physical activity and psychophysiological reactions to mental tasks. *Psychophysiology*, **27:** 264–74

Tanaka, M., Kohno, Y., Tsuda, A. *et al.* (1983) Differential effects of morphine on noradrenaline release in brain regions of stressed and non-stressed rats. *Brain Res.*, **275:** 105–15

Teasdale, J.D., Taylor, R. and Fogarty, S.J. (1980) Effects of induced elation-depression on the accessibility of memories of happy and unhappy experiences. *Behav. Res. Therapy*, **18:** 339–46

Tharp, G.D. and Carson, W.H. (1975) Emotionality changes in rats following chronic exercise. *Med. Sci. in Sports*, **7:** 123–6

Thaxton, L. (1982) Physiological and psychological effects of short-term exercise addiction on habitual runners. *J. Sport Psychol.*, **4:** 73–80

Thayer, R.E. (1987) Energy, tiredness and tension effects of a sugar snack versus moderate exercise. *J. Personality Social Psychol.*, **52:** 119–25

Thoren, P., Floras, J.S., Hoffman, P. and Seals, D.R. (1990) Endorphins and exercise: physiological mechanisms and clinical implications. *Med. Sci. in Sports and Exercise*, **22:** 417–28

Tsuda A., Tanaka, M., Kohno, Y. *et al.* (1983) Daily increase in noradrenaline turnover in brain regions of activity-stressed rats. *Pharmacol. Biochem. Behav.*, **19:** 393–6

Turner, J.R., Girdler, S.S., Sherwood, A. and Light, K.C. (1990) Cardiovascular responses to behavioral stressors: laboratory-field generalization and inter-task consistency. *J. Psychosomatic Res.*, **34:** 581–9

Van Doornen, L.J.P. and de Geus, E.J.C. (1989) Aerobic fitness and the cardiovascular response to stress. *Psychophysiology*, **26:** 17–28

Van Doornen, L.J.P., de Geus, E.J.C. and Orlebeke, J.F. (1988) Aerobic fitness and the physiological stress response: a critical evaluation. *Social Sci. Med.*, **26:** 303–7

Veale, D.M.W. de C. (1987) Exercise dependence. *Br. J. Addiction*, **82:** 735–40

Watson, D. (1988) Intraindividual and interindividual analyses of positive and negative affect: their relation to health complaints, perceived stress, and daily activities. *J. of Personality Social Psychol.*, **54:** 1020–30

Weber, J.C. and Lee, R.A. (1968) Effects of differing prepuberty exercise programs on the emotionality of male albino rats. *Res. Quart.*, **39:** 748–51

Weinstein, W.S. and Meyers, A.W. (1983) Running as a treatment for depression: is it worth it? *J. Sport Psychol.*, **5:** 288–301

White, G.L. and Knight, T.D. (1984) Misattribution of arousal and attraction: effects of salience of explanations for arousal. *J. Exper. Social Psychol.*, **20:** 55–64

Wilson, V.E., Berger, B.G. and Bird, E.J. (1981) Effects of running and of an exercise class on anxiety. *Perceptual and Motor Skills*, **53:** 472–4

Zillman, D., Johnson, R.C. and Day, K.D. (1974) Attribution of apparent arousal and proficiency of recovery from sympathetic activation affecting excitation transfer to aggressive behaviour. *J. Exper. Social Psychol.*, **10:** 503–15

Zillman, D., Katcher, A.H. and Milavsky, B. (1972) Excitation transfer from physical exercise to subsequent aggressive behaviour. *J. Exper. Social Psychol.*, **8:** 247–59

Zimmerman, J.D. and Fulton M. (1981) Aerobic fitness and emotional arousal: a critical attempt at replication. *Psychol. Rep.*, **48:** 911–18

15

Posttraumatic stress disorder

DAVID STURGEON

DEPARTMENT OF PSYCHOLOGICAL MEDICINE
UNIVERSITY COLLEGE HOSPITAL
GOWER STREET
LONDON, UK

15.1 INTRODUCTION

Major disasters and their catastrophic effects have been known throughout the history of mankind. They usually involve large numbers of people, or whole communities; they usually take place suddenly and unexpectedly, leading to loss of life, serious injuries and sometimes to loss of home or property. Some disasters develop more slowly, for example during famines, or affect only a small number of individuals. The common theme of disasters is that they are so catastrophic and overwhelming that they go beyond anything the individuals involved would normally have to cope with. As a result, their psychological capacity to function is stretched beyond the limits of endurance. Kingston and Rosser (1974) have

ISBN 0–12–663370–3

reviewed some of the earlier literature on the mental and physical effects of disasters.

15.2 DISASTERS IN THE TWENTIETH CENTURY

Some disasters are due to natural events, for example fires, earthquakes, floods, volcanic eruptions, drought, famine or disease. Others are manmade: wars are an obvious source of disaster. The destruction of Hiroshima and Nagasaki by the atomic bomb and the effects on the survivors, the Hibakusha, have been vividly described by Lifton (1967). Many Vietnam veterans suffer from the after-effects of their overwhelming experiences during the war.

Other manmade disasters include major accidents, fires and industrial accidents, including leakage of nuclear materials, and can include an element of human error or misjudgment (Weisath, 1989). In 1966, the waste from a giant tip above the Welsh mining village of Aberfan suddenly collapsed and hit the village school killing 116 school children. Almost every family in the village was bereaved, and the effects on all the inhabitants persisted for months or even years (Lacey, 1972). The sudden collapse of the Buffalo creek dam in West Virginia led to torrential floods which carried with them whole communities, houses and their inhabitants. Both these disasters were made worse in the eyes of the survivors because, in each case, the disaster could have been prevented if proper precautions had been taken in time.

The fire in the Coconut Grove night club in Boston in 1942 caused many deaths and severe burns among young people; the victims were taken to the Massachusetts General Hospital. Cobb and Lindemann (1943) and Lindemann (1944) have described the grief reactions and other psychiatric effects among the survivors and relatives whom they saw there and later followed up. The deaths by drowning in March 1987 of 193 people when the channel ferry *Herald of Free Enterprise* suddenly capsized outside Zeebrugge; the fire at a football ground in Bradford, England in 1985; the deaths of 31 people during the fire of the Kings Cross underground station in London in November 1987; the destruction of the PanAm jet above Locherbie in 1988. All these disasters remain very much in people's minds today. Even more recent are the *Marchioness* disaster, in which a pleasure boat sank in the river Thames, and that at the Hillsborough football ground in Sheffield where spectators were crushed to death. Again, a common view is that these disasters could have been prevented had safety procedures been adequate.

In reviewing the effects of disasters, and how to provide the help that is needed, Raphael (1986) described the effects of bush fires in Australia, including the Ash Wednesday fire which threatened and destroyed some of the suburbs of Adelaide. However, her review also shows that disasters are not confined to acute crises, but can be made up of chronic or repeated challenges such as those experienced by the Jews during the Nazi holocaust.

15.3 WHAT IS PTSD?

Posttraumatic Stress Disorder (PTSD) has been recognized, but under different names, for quite some time (Ramsay, 1990). For example, many survivors of the

Vietnam war suffered from 'battle fatigue' and veterans from the First World War suffered from 'shell shock'. The so-called 'concentration camp syndrome' was another manifestation of PTSD.

The definition of PTSD has been changing over recent years and the current definition has been in clinical use for only five years. It is likely to take some time before an accurate natural history of the specific phenomena of PTSD can be recorded. Psychiatrists, particularly those in America, have tended to classify the after-effects of disaster, and have coined the term 'posttraumatic stress disorder' which is now included in the Diagnostic and Statistical Manual-III-R (American Psychiatric Association, 1987), an American classification of psychiatric disorders.

In making a diagnosis, the time course of the syndrome may be important: PTSD may start soon after the event, or have a delayed presentation. It may also be demonstrated that, although most individuals have stable symptoms, some may have fluctuating symptoms which meet the diagnostic criteria at some times but not at others (Curran et al., 1990). In summary, the criteria for a diagnosis of PTSD include the following:

1.   The person has experienced a stressful event that is outside the range of usual human experience, and would be markedly distressing to almost anyone.
2.   The event is re-experienced in vivid dreams, intrusive recollections and flashbacks, usually in response to some triggering stimulus. Illusions and short-lived hallucinatory experiences may also occur.
3.   The person avoids stimuli which would be reminiscent of the disaster, leading to emotional numbness, unresponsiveness and withdrawal.
4.   The person experiences increased arousal which was not present before the event. This includes difficulty in going to sleep, irritability or outbursts of anger, difficulty in concentrating, hypervigilance and an exaggerated startle response.
5.   The condition may be acute, with an early onset but clearing up within six months of the dramatic event, or it may be chronic, in which case the onset may be delayed and the person may never fully recover.

Study of the psychological responses to disasters demonstrates that the primary phenomena can be distinguished from the secondary and associated ones (Rosser and Jackson, 1991). The most characteristic phenomenon is re-experiencing the mental state which existed at the time of the disaster, or during its aftermath. This includes disturbing dreams, abnormal perceptions and obsessive and compulsive phenomena often accompanied by intense emotion. Hallucinations are almost universal in victims of disaster, at least in the immediate aftermath. In the past, these have been grossly underestimated, according to Rosser and Jackson (1991). Many survivors are unwilling to talk about such experiences because they find this intensely distressing and feel that it is a threat to their sanity. Survivors of the Kings Cross fire have described a recurrence of the nauseous smell of burning flesh (Sturgeon et al., 1991). Auditory and tactile perceptual distortions and deceptions are very common. Some survivors experience obsessional thoughts which they try to resist using a great deal of mental effort.

The secondary phenomena include generalized anxiety, phobias, panics, hypervigilance and agitation. Stimuli may be specific as, for example, in the Zeebrugge ferry survivor who had a panic attack when she heard the sound of bath water running. Alternatively, they may be generalized so that not only does a Kings Cross survivor avoid the underground, but also any public transport or place from which rapid escape would be difficult. Behavioural changes may occur so that triggering experiences are avoided. Sometimes there is an enormous increase in risk-taking, such as drinking excessive alcohol, drunk driving, speeding, substance abuse or overspending.

Some frequently encountered symptoms have been dropped from the operational diagnostic criteria of PTSD in the DSM-III-R and ICD 10 (International Classification of Diseases: World Health Organization, 1989). These include survivor guilt, which was vividly described by Lifton (1967). Depersonalization is common and is often a defence against survivor guilt. Tremendous anger is seen in victims of both manmade and natural disasters. The euphoria of the 'Hero Syndrome' almost invariably occurs in at least one victim of such a disaster and is usually followed by a period of severe depression (Sturgeon et al., 1991).

It is important to recognize that the clinical features of PTSD are not the only human reaction to overwhelming trauma: alienation, or feeling different and misunderstood are also common experiences.

15.4 PREVALENCE AND NATURAL HISTORY

The point prevalence of PTSD in the general population is about 1%, although the disorder is much more common in high risk groups. For example, among uninjured war veterans it is 3.5% (Helzer et al., 1987) and among injured combat veterans 20–40% (Pitman et al., 1989); among victims of personal attack it is 3.5% (Helzer et al., 1987); among fire fighters exposed to a major bush fire it is 30% (MacFarlane, 1987). The prevalence of the disorder in prisoners of war is around 65%, varying with the severity and duration of their internment (Kluznik et al., 1986). Many more survivors have subthreshold levels of symptoms which are, nevertheless, debilitating.

15.5 NATURE OF THE STRESSFUL EXPERIENCE

If the disaster starts suddenly and unexpectedly, as is often the case, those involved experience intense fear, but panic is rare. Fear of death and an intense wish to escape usually predominate. Individuals become alert and hypervigilant, searching for a way out, protecting themselves with their hands, crouching and holding onto others. Panic is more likely to ensue when people are unsure about what is happening and become confused or isolated. Often they hold onto family members, for example parents to their children, in a desperate attempt to protect them. Some try heroically to save others, although this applies more often to rescuers when they reach the scene, than to the victims.

The flight itself can be most distressing. Some of the survivors of the Kings Cross fire in London, who were familiar with the underground station and tried to find the exits through dense smoke and fire, knew that they were scrambling over bodies of injured or dying people to escape. Some victims become helpless, stunned and apathetic as a result of the severe shock and confusion, so that they wander around dazed and unable to help themselves: the so-called 'disaster syndrome'. This may persist for hours or even days.

Those who have escaped death, including some who may be seriously injured, are likely to remain preoccupied for some time with the thought of how near to dying they have been. They keep going over the events, wondering how they managed to survive, but perhaps feeling guilty and deeply distressed that others close to them, such as their parents, children, husband or wife have died. In the days following the disaster, survivors may re-experience the trauma by having flashbacks which are intrusive, uncontrollable and disturbing. This may happen when they are awake or during sleep in dreams or nightmares. These flashbacks, sometimes accompanied by panic, may be triggered by stimuli that are reminiscent of the disaster, for example a sudden noise or an explosion, a screeching of brakes, a smell of burning, or the sound of running water. As a result, some survivors attempt to avoid exposure to anything that might remind them of the disaster and consequently become withdrawn or isolated.

Others may experience a sense of vulnerability and remain frightened of a similar event happening again, or they may anticipate new danger. General irritability and anger at the senselessness of the disaster are common, and may be directed at people's lack of understanding, or at the inefficiency and failure of a system that went tragically wrong.

Not only the victims, but also the rescuers may be affected and, of course, relatives of some of the victims who died. For example, having to identify the mutilated body of a relative can be extremely stressful; so may the waiting period when relatives do not yet know whether someone close to them has died in the disaster or, perhaps, the fact that the body has never been found.

15.6 WHO DEVELOPS PTSD?

A useful indicator of vulnerability to PTSD is how the person has coped with previous loss or trauma in their life. People who have become intensely anxious or depressed following major life events may be particularly vulnerable. The debate continues between those who argue that the amount (dose) of trauma is the critical element (Shore et al., 1986) and those who see personality vulnerability factors as more important (MacFarlane, 1987). It is likely that both views have some validity. Those persons who are physically damaged or scarred by the trauma have a constant bodily reminder and reinforcer of the disaster to which they succumbed. In the presence of borderline trauma it is those individuals who have other vulnerability factors who may be the ones to succumb to PTSD. By contrast, faced with intense and massive trauma, it is likely that anyone may develop PTSD.

What is undoubtedly clear is that behaviour during the disaster carries no predictive value for vulnerability to developing PTSD. The heroic survivor of the Zeebrugge ferry sinking who allowed his body to be used as a human bridge, so that others could cross to safety, subsequently developed massive PTSD.

15.7 QUANTIFYING THE TRAUMA

Research into PTSD is still at an early stage, but increasingly research workers are agreeing on the usefulness of specific measures. The questionnaires most frequently used are the Impact of Events Scale, the General Health Questionnaire and the Symptoms Check List 90 R.

The Impact of Events Scale (IES; Horowitz, 1979) was specifically designed to measure the effects of traumatic events. The items in the scale are drawn from inter-views with survivors and encompass customary thoughts and memories of the event. The scale, which has 15 items in total, is divided into two parts. The first sub-scale, which is labelled 'intrusion', measures the extent to which unwanted memories of the traumatic event continue to impinge upon a person's mind and the extent to which other things remind the person about the event. The basic theme of this scale is to evaluate how much memories of the disaster intrude upon the person's consciousness. The other scale is called 'Avoidance' and measures people's attempts to put the bad memories out of their mind. It contains items about deliberately trying to avoid getting upset, pushing the memory out of one's mind, and avoidance of anything which serves as a reminder of the event. These two scales, when added together, give a total Impact of Events score. This score is a useful indicator of the extent to which the disaster is reverberating in people's minds.

The General Health Questionnaire – 28 item version (GHQ; Goldberg and Hillies, 1979) – has the merit of being designed for use with the general population. It acts as a screening instrument, giving a probability estimate that an individual is a psychiatric case. There is now a large body of work on the results which have been obtained on these questionnaires with different populations. The original standardization work also linked scores on this screening approach with psychiatric diagnoses carried out by independent assessors. This gives us an understanding of the link between self report and an external observer's assessment. The questionnaire provides sub-totals on four scales and a final total. The four scales were derived from the statistical analysis of responses (factor analysis). The first scale contains items about people's feelings of health and fatigue. It provides a measure of all those bodily sensations which often accompany emotional upsets. The second set of questions contains items which relate to anxiety and sleeplessness. The third scale measures the extent to which the respondent is able to cope with the demands of their work and life. It picks up people's feelings about whether they are able to deal with the usual challenges of life. The final scale contains items which are to do with severe depression and suicide. The final total is achieved by summing the four scales.

Another screening questionnaire which has been used is the Symptom Check List-90-R (SCL-90-R; Derogatis, 1977). The SCL-90-R is a multidimensional self-report symptom inventory designed to measure symptomatic psychological distress. It provides three global indices of distress. The General Severity Index combines information on numbers of symptoms and intensity of distress, while the Positive Symptom Total reflects only numbers of symptoms; the Positive Symptom Distress Index is a pure intensity measure, adjusted for numbers of symptoms present.

Collecting such data can yield interesting findings. The Impact of Events Scale (IES; Weiss *et al.*, 1984) and the General Health Questionnaire were completed by four of the most severely damaged Kings Cross fire survivors. Four months after the disaster, mean IES was much lower for the survivors than for passengers and bystanders, as was the GHQ total score. Numbers are too small to allow statistical analyses, but it is tempting to speculate that the psychological help these victims had received whilst in hospital had some protective value.

15.8 LEGAL SIGNIFICANCE OF PTSD

Increasingly, PTSD is being recognized as having both clinical and legal significance. Until quite recently, the only term to denote psychological distress which was acceptable in legal proceedings was 'nervous shock'. Following the Second World War many individuals were not compensated for psychological damage, partly because of the (incorrect) assumption that neurotic symptoms inevitably reflected a primary intrinsic weakness in the victim, rather than a genuine response of a normal individual to overwhelming trauma (Turner, 1991).

PTSD has now been accepted as a basis for compensation in recent agreements following major incidents, although the amount of money allocated is often small. There have even been attempts to secure compensation for bereaved relatives who witnessed a fatal disaster on television. It is likely that there will be further developments in case law in this area.

15.9 RESCUERS' PSYCHOLOGICAL RESPONSES TO DISASTERS

Many people other than the primary victims may be affected psychologically after a disaster (Raphael *et al.*, 1991). There are many rescue workers who fall victim to stresses created by the work which they have to do. Although, in most cases, the symptoms of posttraumatic stress will settle, in some the distress does not quieten and severe morbidity develops. The stress of rescue operations can often be severe and workers may confront scenes which bring physical revulsion; transient physical, emotional and behavioural reactions are common. Sometimes the trauma can overwhelm them and rescuers may then feel helpless and retreat from, or misinterpret, what they find. Alternatively, the excitement of involvement may generate a 'high' which can extend to over-involvement and a sense of omnipotence.

The most sensitive indicators of continuing impairment are cognitive impairment and disturbed interpersonal relationships, as well as increased arousal, irritability and loss of interest or withdrawal. Alcohol may be taken in excess in an effort to forget or dampen distress. Families may also need support, particularly if the rescuer remains locked into his or her experience.

The emotional impact of the disaster appears to have a direct relationship to the degree of stress experienced by the rescue worker. For example, gruesome disasters, particularly when there are multiple deaths, mutilated bodies or the deaths of children, are stressful for most workers; up to 20% still show signs of stress two years after the event. In addition, the experience of disaster is likely to aggravate many ordinary work-related stresses (James, 1988) and may provide an additional psychological burden. Hodgkinson and Stewart (1991) found primary stressors for rescue workers to be personal loss or injury, encounters with death and 'mission failure'. This includes frustration about lives that cannot be saved, failure of equipment, delays and overwhelming demands, all of which contribute to psychological distress. Indeed, symptoms experienced by rescue workers may reflect this conflict: guilt, anxiety, general irritability, focused resentment and loss of interest in work (Duckworth, 1986).

Powerful antidotes to those stressful effects appear to be good management and careful organization. Emergency workers who helped to retrieve bodies after the Piper Alpha oil disaster were provided with detailed induction to their tasks. This explained the importance of what they were doing, their possible reactions and the need to attend to their own welfare. They worked in pairs, one of whom was an older, more experienced officer. Their shifts were limited, they were debriefed each day and psychiatric support was available. The workers showed no long-term effects from their stress. Even when stress is experienced at the level of clinical 'caseness' by disaster workers, as in the fire at Bradford football stadium, brief counselling sessions can facilitate recovery (Duckworth, 1986).

Certain factors have emerged as being protective for rescue workers faced with stress. Simply being older and more experienced is, in itself, protective. Hodgkinson and Stewart (1991) highlight seemingly helpful coping styles that emphasize sharing problems, constructive use of humour and the use of social support. Conversely, those who are drawn to action, but deny their vulnerability, will find it difficult to admit stress or to seek help. Fear that workmates will think them inadequate, or that their career prospects will be damaged, are the most common reasons for distressed workers not taking advantage of stress counselling.

Like the police and ambulance workers, medical and hospital staff are often thought to be immune to stress because of their training. However, we know that that they are also affected, but are less likely to have access to support programmes. Studies of debriefing programmes after the Hillsborough disaster, where 95 people were crushed to death, showed that hospital staff could benefit from such sessions: two-thirds of those attending found them helpful. Those who remained distressed six to nine months later had experienced high levels of exposure, showed more distress symptoms on systemic measures and were concerned about personal and organizational performance. Nevertheless, an appreciable minority found the experience positive with a renewed appraisal of their value of life (Shapiro and Kunkler, 1990).

Increased interest in the reactions of rescue workers has been accompanied by the development of programmes such as critical incident (or stress) debriefing. This is usually provided to groups by mental health professionals and peer support workers in the first 24–72 h after the disaster (Mitchell, 1983). This technique aims at early and comprehensive stress reduction but should only be part of a range of organizational, educational and support responses.

15.10 CO-MORBIDITY

Alcohol use following major trauma is a particular risk. Survivors often use the tranquillizing effects of alcohol as self-medication, but this carries the further hazard of inducing a secondary alcohol dependence syndrome. Drinking patterns among survivors should be carefully monitored where possible and steps taken to reduce habitual drinking.

Dependence on opiates and benzodiazepines has been reported in disaster victims, usually those who have sustained severe injuries which have needed hospital treatment. Disaster victims have often been discharged from hospital on drugs which induce dependence and with no provision for supervised withdrawal. Compensation for such action has recently been the case of litigation.

Major depression can occur in up to 25% of PTSD sufferers. The onset is usually several months after the event, often when the symptoms of PTSD have been severe. Biological symptoms predominate, commonly: sleep disturbance, diurnal variation in mood, anorexia and weight loss. Withdrawal and social isolation are frequently seen after traumatic stress, but are better regarded as symptoms of a major depression, rather than symptoms of PTSD *per se*.

Anxiety symptoms, although commonly occurring with depressive features, may predominate. They may take the form of panic disorder, generalized anxiety, hyperventilation disorder or other somatoform disorders. Treatment for anxiety is predominantly psychological; it is aimed at anxiety management and, if the anxiety is situational, desensitization. Thought should also be given to ascertaining morbidity in other family members who may have experienced the trauma in other, less direct ways.

15.11 TREATMENT

Although there is currently insufficient research evidence on which to base an informed approach to treatment, there are various themes on which different workers in this area are agreed.

15.11.1 MUTUAL SUPPORT

After group trauma, there is often a banding together of survivors, e.g. in survivors' self-help groups, relatives' groups. These talk about and share experiences and sometimes produce newsletters for those further afield (Dyregrov, 1989). It is important to remember that disaster victims are normal individuals who have been adversely affected by circumstance. They already possess their own strengths and the encouragement to use them by groups can be a big step towards recovery.

15.11.2 COUNSELLING

The central element of most treatment approaches is for the victims to go over their experience and for the therapist to receive it, but with a view to effecting changes. This may be formally acknowledged in a cognitive-behavioural treatment (Rosser *et al.*, 1990), or this may be informally acknowledged in other psychodynamic approaches. Relating the history of the trauma often re-awakens associated emotions; the job of the therapist is to help the survivor to tolerate experience of the disaster in such a way that they do not have to retreat into avoidance and despair.

This means that interviews must take place in a calm and relaxed setting, free from interruptions, with the possibility of prolonging sessions for up to several hours. By contrast, brief contacts which do no more than raise the emotional temperature, thereby causing further distress, may be positively harmful. Some media contact with distressed disaster victims can fall into this latter category.

Often the emotional impact of disaster will stir up previous anxieties of a person's life. For example, a woman who lost her best friend in the Kings Cross fire found that, after several months, she was also grieving the loss of her own mother who had died from breast cancer four years earlier. She discovered that many of the feelings and memories about her friend, that she had not had an opportunity of talking through, also applied to her mother with whom she had had a difficult relationship. Much of the therapeutic work had to concentrate on this earlier bereavement which, in turn, helped her to cope with the loss of her friend.

15.11.3 REHEARSAL

This treatment encourages exposure through verbal recall of the events. It aims to achieve processing or habituation (Richards *et al.*, 1990). Exposure *in vivo* has also been used (e.g. after natural disasters), and may be used in preventative work as part of critical incident debriefing. This is particulary valuable following localized events, and has been used after armed robberies and terrorist incidents. A meeting is held within one or two days of the event with individuals who were involved. Events are explored in group discussion and basic ground rules concerning tolerance of silence and confidentiality are agreed. By exploring the emotions and distress, which are still very near the surface at this early stage, and by doing this with other individuals who are also affected, the distress reaction can be brought more easily within normal range and the traumatic impact of the event reduced.

15.11.4 MEDICATION

Research and clinical evidence increasingly point to the efficacy of antidepressant drugs in the management of PTSD (Frank *et al.*, 1988). Drugs such as the tricyclic antidepressant, dothiepin, taken in a single dose at night-time, appear both to facilitate sleep and to reduce intrusive thoughts. Some of the newer antidepressants, such as paroxetine and fluoxetine, which selectively inhibit reuptake of 5-hydroxytryptamine, also appear to be helpful in PTSD, but lack the sedative effects of some tricyclic antidepressants. This often means that hypnotics may also have to be given at night if sleep disturbance is a problem.

15.12 CONCLUDING REMARKS

PTSD is increasingly being recognized as a common reaction to major trauma. It occurs in normal individuals who have been exposed to overwhelming stress, and it represents the overloading of a normal psychological mechanism. Intervention can be successful, given adequate time and appropriate discussion of the event and its psychological impact. Successful management requires prior planning and training. It is a matter of serious concern that so few hospital disaster plans consider the psychosocial responses needed after major incidents. Too much of the planning takes place after the event, and this usually serves to inhibit the early use of the available resources.

Emergency organizations need policies that identify stressful circumstances and teach their staff to cope with them. They should also provide an effective safety-net of debriefing and counselling when disasters occur. Support for rescue workers should be based on the expectation that they will master their own stress given appropriate understanding and help from their organization. Such policies can provide a positive environment for the smaller disasters which confront workers every day.

This chapter has not suggested links to the different levels of analysis of stress which precede it in this book. Such links would allow us to apply insights from other sources of work, ranging from life events research or neuropharmacology, to posttraumatic stress. However, to draw on such work at the present time would be merely speculative. For this reason, the present chapter has focused on the clinical picture; it is this that must ultimately be explained or informed by an integration of the different approaches to understanding stress.

15.13 REFERENCES

American Psychiatric Association (1987) *Diagnostic and Statistical Manual of Mental Disorders. 3rd rev. edn.* Washington, DC

Cobb, S. and Lindemann, E. (1943) Management of the Coconut Grove burns: neuropsychiatric observations. *Ann. Surg.*, **117**: 814–24

Curran, P.S., Bell, P., Murray, A. *et al.* (1990) Psychological consequences of the Enniskillen bombing. *Br. J. Psychiat.* **156**: 479–82

Derogatis, L.R. (1977) *SCL-90-R: Administration Scoring and Procedure Manual I*, Clinical Psychometrics Research, Baltimore

Duckworth, P. (1986) Psychological problems arising from disaster work. *Stress Med.*, **2**: 315–23

Dyregrov, A. (1989) Caring for helpers in disaster situations: psychological debriefing. *Disaster Management*, **2**(1)

Frank, J.B., Koster, T.R., Gilles, E.L. and Dan, E. (1988) A randomized clinical trial of phenelzine and imipramine for post-traumatic stress disorder. *Am. J. Psychiat.*, **145**: 1289–91

Goldberg, D.P. and Hillies, V.F. (1979) A scaled version of the General Health Questionnaire. *Psychol. Med.*, **9**: 139–45

Helzer, J.E., Robins, L.N. and McEvoy, L. (1987) PTSD in the general population: findings of the Epidemiological Catchment Area Study. *New England J. Med.*, **317**: 1630–4

Hodgkinson, P.E. and Stewart, M. (1991) *Coping with Catastrophes: A Handbook of Disaster Management*, Routledge, London

Horowitz, M., Wilner, N. and Alvarez, W. (1979) Impact of Events Scale: a measure of subjective stress. *Psychosomatic Med.*, **41**(3): 209–18

James, A. (1988) Perceptions of stress in British ambulance personnel. *Work and Stress*, **2**: 319–26

Kingston, W. and Rosser, R. (1974) Disaster: effects on physical and mental state. *J. Psychosomatic Res.*, **18**: 437–56

Kluznik, J., Speed, N. and Van Valkenburg, C. (1986) Forty year follow-up of US prisoners of war. *Am. J. Psychiat.* **143**: 1443–6

Lacey, G.N. (1972) Observations on Aberfan. *J. Psychosomatic Res.*, **144**: 35–7

Lifton, R.J. (1967) *Death in Life: Survivors of Hiroshima*, Ransom Howe, New York.

Lindemann, E. (1944) Symptomatology and management of acute grief. *Am. J. Psychiat.*, **101:** 141–8

MacFarlane, A.C. (1987) Life events and psychiatric disorder: the role of a natural disaster. *Br. J. Psychiat.* **151:** 362–7

Mitchell, J.T. (1983) When disaster strikes: the critical incident stress debriefing process. *J. Emer. Med. Serv.*, **8:** 36–9

Pitman, R., Altman, B. and Macklin, M. (1989) The prevalence of post-traumatic stress disorders in wounded Vietnam veterans. *Am. J. Psychiat.* **146:** 667–9

Ramsay, R. (1990) Post-traumatic stress disorder: a new clinical entity. *J. Psychosomatic Res.*, **34:** 355–65

Raphael, B. (1986) *When Disaster Strikes*, Hutchinson, London

Raphael, B., Meldrum, L. and O'Toole, B. (1991) Rescuers' psychological responses to disasters. *Br. Med. J.*, **303:** 1346–7

Richards, D., Rose, J. and Lovell, K. (1990) *In-vivo and imaginal exposure in the treatment of PTSD*. Proceedings of the Second European Conference on Traumatic Stress, Leeuwenhurst, Holland

Rosser, R. and Jackson, G. (1991) The disaster response cycle, in Punukolly, (ed.), *Recent Advances in Crisis Intervention: Vol. 1*, International Institute of Crisis Intervention & Community Psychiatry Publications

Rosser, R.M., Thompson, J.M. and Jackson, G. (1990) *The Kings Cross fire*. Proceedings of the Second European Conference on Traumatic Stress, Leeuwenhurst, Holland

Shapiro, D. and Kunkler, J. (1990) Psychological support for hospital staff initiated by clinical psychologists in the aftermath of the Hillsborough disaster. Sheffield: *Report for Clinical Psychology*, Northern General Hospital

Shore, J.H., Tatum, E.L. and Vollmer, W.M. (1986) Psychiatric reactions to disaster: the Mount St Helens experience. *Am. J. Psychiat.* **143:** 590–5

Sturgeon, D., Rosser, R. and Shoenberg, P. (1991) The Kings Cross fire: the psychological injuries. *Burns*, **17**(1): 10–13

Turner, S. (1991) Post-traumatic stress disorder. *Hospital Up-date*, **17:** 644–9

Weisath, L. (1989) The stressors and the post-traumatic stress syndrome after an industrial disaster. *Acta Psychiatr. Scand.*, **335** (supl): 25–37

Weiss, D., Horowitz, M.J. and Wilner, N. (1984) The stress response rating scale: a clinician's measure for rating the response to various life events. *Br. J. Clin. Psychol.*, **23:** 205–15

World Health Organization (1989) *International Classification of Diseases*, World Health Organisation Press, Geneva

Index